Springer Complexity

Springer Complexity is an interdisciplinary program publishing the best research and academic-level teaching on both fundamental and applied aspects of complex systems – cutting across all traditional disciplines of the natural and life sciences, engineering, economics, medicine, neuroscience, social and computer science.

Complex Systems are systems that comprise many interacting parts with the ability to generate a new quality of macroscopic collective behavior the manifestations of which are the spontaneous formation of distinctive temporal, spatial or functional structures. Models of such systems can be successfully mapped onto quite diverse "real-life" situations like the climate, the coherent emission of light from lasers, chemical reaction-diffusion systems, biological cellular networks, the dynamics of stock markets and of the internet, earthquake statistics and prediction, freeway traffic, the human brain, or the formation of opinions in social systems, to name just some of the popular applications.

Although their scope and methodologies overlap somewhat, one can distinguish the following main concepts and tools: self-organization, nonlinear dynamics, synergetics, turbulence, dynamical systems, catastrophes, instabilities, stochastic processes, chaos, graphs and networks, cellular automata, adaptive systems, genetic algorithms and computational intelligence.

The two major book publication platforms of the Springer Complexity program are the monograph series "Understanding Complex Systems" focusing on the various applications of complexity, and the "Springer Series in Synergetics", which is devoted to the quantitative theoretical and methodological foundations. In addition to the books in these two core series, the program also incorporates individual titles ranging from textbooks to major reference works.

Springer Series in Synergetics

Founding Editor: H. Haken

The Springer Series in Synergetics was founded by Herman Haken in 1977. Since then, the series has evolved into a substantial reference library for the quantitative, theoretical and methodological foundations of the science of complex systems.

Through many enduring classic texts, such as Haken's *Synergetics* and *Information and Self-Organization,* Gardiner's *Handbook of Stochastic Methods,* Risken's *Planck-Equation* or Haake's *Quantum Signatures of Chaos,* the series has made, and continues to make, important contributions to shaping the foundations of the field.

The series publishes monographs and graduate-level textbooks of broad and general interest, with a pronounced emphasis on the physico-mathematical approach.

Crispin Gardiner

Stochastic Methods

A Handbook for the Natural and Social Sciences

Fourth Edition

Springer

Prof. Crispin Gardiner
University of Otago
Physics Department
Dunedin
New Zealand

ISBN: 978-3-642-08962-6 e-ISBN: 978-3-540-70713-4

Softcover reprint of the hardcover 4th edition 2009

Cover design: WMXDesign GmbH, Heidelberg

Printed on acid-free paper

9 8 7 6 5 4 3 2 1

springer.com

Preface to the Fourth Edition

This fourth edition of *Stochastic Methods* is thoroughly revised and augmented, and has been completely reset. While keeping to the spirit of the book I wrote originally, I have reorganised the chapters of Fokker-Planck equations and those on approximation methods, and introduced new material on the white noise limit of driven stochastic systems, and on applications and validity of simulation methods based on the Poisson representation. Further, in response to the revolution in financial markets following from the discovery by Fischer Black and Myron Scholes of a reliable option pricing formula, I have written a chapter on the application of stochastic methods to financial markets. In doing this, I have not restricted myself to the geometric Brownian motion model, but have also attempted to give some flavour of the kinds of methods used to take account of the realities of financial markets. This means that I have also given a treatment of Lévy processes and their applications to finance, since these are central to most current thinking.

Since this book was written the rigorous mathematical formulation of stochastic processes has developed considerably, most particularly towards greater precision and generality, and this has been reflected in the way the subject is presented in modern applications, particularly in finance. Nevertheless, I have decided to adhere to my original decision, to use relatively simple language without excessive rigour; indeed I am not convinced that the increase in rigour and precision has been of significant help to those who want to use stochastic methods as a practical tool.

The new organisation of the material in the book is as in the figure on the next page. Instead of the original ten chapters, there are now fifteen. Some of the increase is a result of my decision to divide up some of the larger chapters into tighter and more logically structured smaller chapters, but Chapters 8 and 10 are completely new. The basic structure of the book is much the same, building on the basis of Ito stochastic differential equations, and then extending into Fokker-Planck equations and jump processes. I have put all of the work on the Poisson representation into a single chapter, and augmented this chapter with new material.

Stochastic Methods, although originally conceived as a book for physicists, chemists and similar scientists, has developed a readership with far more varied tastes, and this new edition is designed to cater better for the wider readership, as well as to those I originally had in mind. At the same time, I have tried hard to maintain "look and feel" of the original, and the same degree of accessibility.

University of Otago, New Zealand — *C.W. Gardiner*
July, 2008

From the Preface to the First Edition

My intention in writing this book was to put down in relatively simple language and in a reasonably deductive form, all those formulae and methods which have been

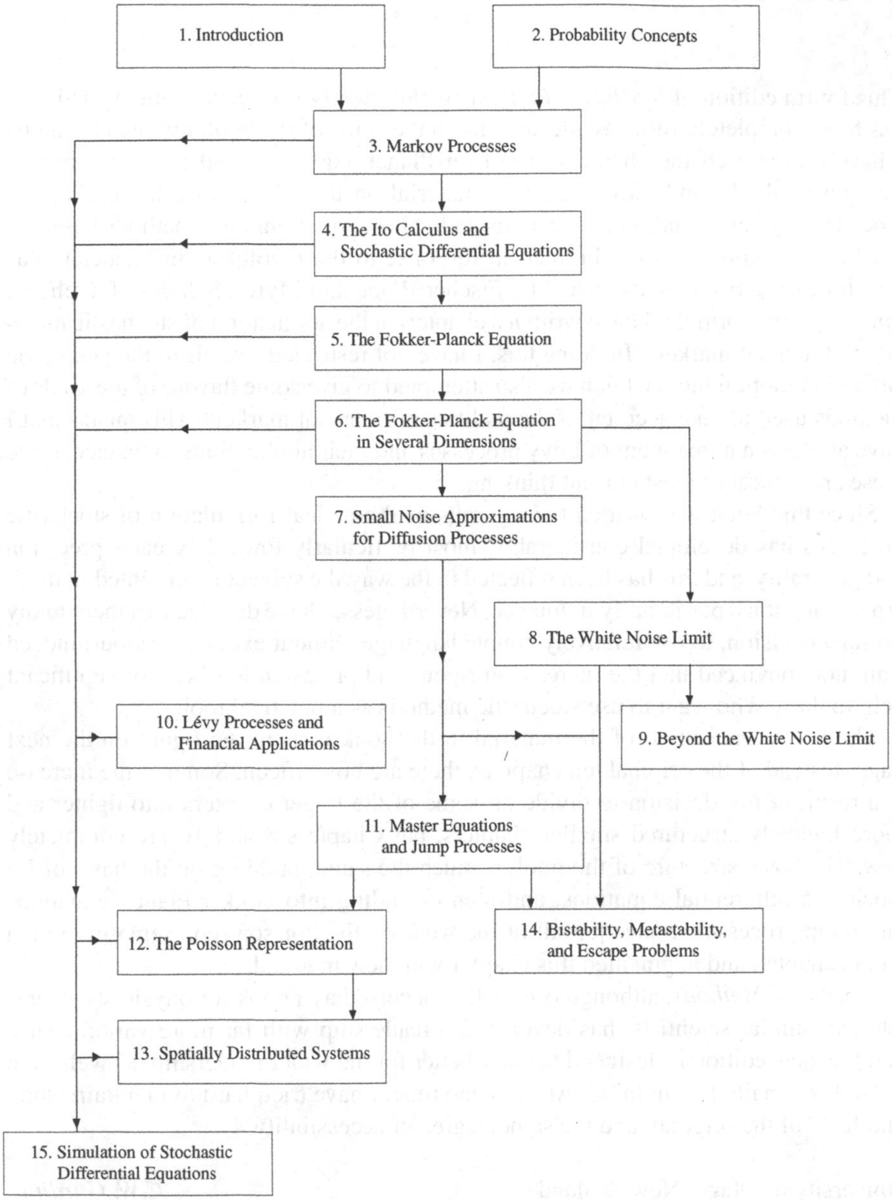
1. Introduction
2. Probability Concepts
3. Markov Processes
4. The Ito Calculus and Stochastic Differential Equations
5. The Fokker-Planck Equation
6. The Fokker-Planck Equation in Several Dimensions
7. Small Noise Approximations for Diffusion Processes
8. The White Noise Limit
9. Beyond the White Noise Limit
10. Lévy Processes and Financial Applications
11. Master Equations and Jump Processes
12. The Poisson Representation
13. Spatially Distributed Systems
14. Bistability, Metastability, and Escape Problems
15. Simulation of Stochastic Differential Equations

scattered throughout the scientific literature on stochastic methods throughout the eighty years that they have been in use. This might seem an unnecessary aim since there are scores of books entitled "Stochastic Processes", and similar titles, but careful perusal of these soon shows that their aim does not coincide with mine. There are purely theoretical and highly mathematical books, there are books related to electrical engineering or communication theory, and there are books for biologists—many of them very good, but none of them covering the kind of applications that appear nowadays so frequently in Statistical Physics, Physical Chemistry, Quantum Optics and Electronics, and a host of other theoretical subjects.

The main new point of view here is the amount of space which deals with methods of approximating problems, or transforming them for the purpose of approximating them. I am fully aware that many workers will not see their methods here. But my criterion here has been whether an approximation is *systematic*. Many approximations are based on unjustifiable or uncontrollable assumptions, and are justified *a posteriori*. Such approximations are not the subject of a systematic book—at least, not until they are properly formulated, and their range of validity controlled. In some cases I have been able to put certain approximations on a systematic basis, and they appear here—in other cases I have not.

A word on the background assumed. The reader must have a good knowledge of practical calculus including contour integration, matrix algebra, differential equations, both ordinary and partial, at the level expected of a first degree in applied mathematics, physics or theoretical chemistry.

I expect the readership to consist mainly of theoretical physicists and chemists, and thus the general standard is that of these people. This is not a rigorous book in the mathematical sense, but it contains results, all of which I am confident are provable rigorously, and whose proofs can be developed out of the demonstrations given. The organisation of the book is as in the following table, and might raise some eyebrows. For, after introducing the general properties of Markov processes, I have chosen to base the treatment on the conceptually difficult but intuitively appealing concept of the stochastic differential equation. I do this because of my own experience of the simplicity of stochastic differential equation methods, once one has become familiar with the Ito calculus, which I have presented in Chap. 4 in a rather straightforward manner, such as I have not seen in any previous text.

For the sake of compactness and simplicity I have normally presented only one way of formulating certain methods. For example, there are several different ways of formulating the adiabatic elimination results, though few have been used in this context. To have given a survey of all formulations would have required an enormous and almost unreadable book. However, where appropriate I have included specific references, and further relevant matter can be found in the general bibliography.

Hamilton, New Zealand
January, 1983

C.W. Gardiner

Acknowledgements

My warmest appreciation must go to Professor Hermann Haken for inviting me to write this book for the Springer Series in Synergetics, and for helping support a sabbatical leave in Stuttgart in 1979–1980 where I did most of the initial exploration of the subject and commenced writing the book.

The physical production of the manuscript would not have been possible without the thoroughness of Christine Coates, whose ability to produce a beautiful typescript, in spite of my handwriting and changes of mind, has never ceased to arouse my admiration. The thorough assistance of Moira Steyn-Ross in checking formulae and the consistency of the manuscript has been a service whose essential nature can only be appreciated by an author.

Many of the diagrams, and some computations, were prepared with the assistance of Craig Savage, for whose assistance I am very grateful.

To my former colleagues and students at the University of Waikato must go a considerable amount of credit for much of the work in this book; in particular to the late Bruce Liley, whose encouragement and provision of departmental support I appreciated so much. I want to express my appreciation to the late Dan Walls who first introduced me to this field, and with whom I enjoyed a fruitful collaboration for many years; to Howard Carmichael, Peter Drummond, Ken McNeil, Gerard Milburn, Moira Steyn-Ross, and above all, to Subhash Chaturvedi, whose insights into and knowledge of this field have been of particular value.

Since I first became interested in stochastic phenomena, I have benefited greatly from contact with a large number of people, and in particular I wish to thank Ludwig Arnold, Robert Graham, Siegfried Grossman, Fritz Haake, Pierre Hohenberg, Werner Horsthemke, Nicco van Kampen, the late Rolf Landauer, René Lefever, Mohammed Malek-Mansour, Gregoire Nicolis, Abraham Nitzan, Peter Ortoleva, John Ross, Friedrich Schlögl, Urbaan Titulaer and Peter Zoller

In preparing Chap. 15 of the this edition I was greatly helped by discussions with Peter Drummond, whose expertise on numerical simulation of stochastic differential equations has been invaluable, and by Ashton Bradley, who carefully checked the both content and the proofs of this edition.

The extract from the paper by Albert Einstein which appears in Sect. 1.2.1 is reprinted with the permission of the Hebrew University, Jerusalem, Israel, who hold the copyright.

The diagram which appears as Fig. 1.3 b) is reprinted with permission of Princeton University Press.

Contents

1. A Historical Introduction

1.1 Motivation

Theoretical science up to the end of the nineteenth century can be viewed as the study of solutions of differential equations and the modelling of natural phenomena by deterministic solutions of these differential equations. It was at that time commonly thought that if all initial data could only be collected, one would be able to predict the future with certainty.

We now know this is not so, in at least two ways. Firstly, the advent of quantum mechanics within a quarter of a century gave rise to a new physics, and hence a new theoretical basis for all science, which had as an essential basis a purely statistical element. Secondly, more recently, the concept of chaos has arisen, in which even quite simple differential equation systems have the rather alarming property of giving rise to essentially unpredictable behaviour. To be sure, one can predict the future of such a system given its initial conditions, but any error in the initial conditions is so rapidly magnified that no practical predictability is left. In fact, the existence of chaos is really not surprising, since it agrees with more of our everyday experience than does pure predictability—but it is surprising perhaps that it has taken so long for the point to be made.

Chaos and quantum mechanics are not the subject of this chapter. Here I wish to give a semihistorical outline of how a phenomenological theory of fluctuating phenomena arose and what its essential points are. The very usefulness of predictable models indicates that life is not entirely chaos. But there is a limit to predictability, and what we shall be most concerned with in this book are models of limited predictability. The experience of careful measurements in science normally gives us data like that of Fig. 1.1, representing the growth of the number of molecules of a

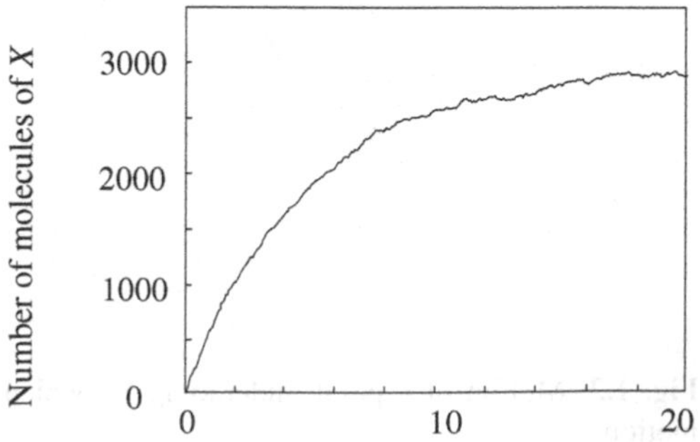

Fig. 1.1. Stochastic simulation of an isomerisation reaction $X \rightleftharpoons A$

substance X formed by a chemical reaction of the form $X \rightleftharpoons A$. A quite well defined deterministic motion is evident, and this is reproducible, unlike the fluctuations around this motion, which are not.

1.2 Some Historical Examples

1.2.1 Brownian Motion

The observation that, when suspended in water, small pollen grains are found to be in a very animated and irregular state of motion, was first systematically investigated by Robert Brown in 1827 [1.1], and the observed phenomenon took the name *Brownian Motion* because of his fundamental pioneering work. Brown was botanist—indeed a very famous botanist—and he was examining pollen grains in order to elucidate the mechanism which by which the grains moved towards the ova when fertilising flowers. At first he thought this motion was a manifestation of life he was seeking, but when he found that this motion was present in apparently dead pollen, some over a century old, some even extracted from fossils, and then even in any suspension of fine particles—glass, minerals and even a fragment of the sphinx—he ruled out any specifically organic origin of this motion. The motion is illustrated in Fig. 1.2.

The riddle of Brownian motion was not quickly solved, and a satisfactory explanation did not come until 1905, when *Einstein* published an explanation under the rather modest title "Über die von der molekular-kinetischen Theorie der Wärme geforderte Bewegung von in ruhenden Flüssigkeiten suspendierten Teilchen" (concerning the motion, as required by the molecular-kinetic theory of heat, of particles suspended in liquids at rest) [1.2]. The same explanation was independently developed by *Smoluchowski* [1.3], who was responsible for much of the later systematic development and for much of the experimental verification of Brownian motion theory.

There were two major points in Einstein's solution to the problem of Brownian motion.

Fig. 1.2. Motion of a point undergoing Brownian motion

i) The motion is caused by the exceedingly frequent impacts on the pollen grain of the incessantly moving molecules of liquid in which it is suspended.

ii) The motion of these molecules is so complicated that its effect on the pollen grain can only be described probabilistically in terms of exceedingly frequent statistically independent impacts.

The existence of fluctuations like these ones calls out for a statistical explanation of this kind of phenomenon. Statistics had already been used by Maxwell and Boltzmann in their famous gas theories, but only as a description of possible states and the likelihood of their achievement and not as an intrinsic part of the time evolution of the system. *Rayleigh* [1.4] was in fact the first to consider a statistical description in this context, but for one reason or another, very little arose out of his work. For practical purposes, Einstein's explanation of the nature of Brownian motion must be regarded as the beginning of stochastic modelling of natural phenomena.

Einstein's reasoning is very clear and elegant. It contains all the basic concepts which will make up the subject matter of this book. Rather than paraphrase a classic piece of work, I shall simply give an extended excerpt from Einstein's paper (author's translation):

> "It must clearly be assumed that each individual particle executes a motion which is independent of the motions of all other particles; it will also be considered that the movements of one and the same particle in different time intervals are independent processes, as long as these time intervals are not chosen too small.
>
> "We introduce a time interval τ into consideration, which is very small compared to the observable time intervals, but nevertheless so large that in two successive time intervals τ, the motions executed by the particle can be thought of as events which are independent of each other.
>
> "Now let there be a total of n particles suspended in a liquid. In a time interval τ, the X-coordinates of the individual particles will increase by an amount Δ, where for each particle Δ has a different (positive or negative) value. There will be a certain *frequency law* for Δ; the number dn of the particles which experience a shift which is between Δ and $\Delta + d\Delta$ will be expressible by an equation of the form
>
> $$dn = n\phi(\Delta)d\Delta\,, \tag{1.2.1}$$
>
> where
>
> $$\int_{-\infty}^{\infty} \phi(\Delta)d\Delta = 1\,, \tag{1.2.2}$$
>
> and ϕ is only different from zero for very small values of Δ, and satisfies the condition
>
> $$\phi(\Delta) = \phi(-\Delta)\,. \tag{1.2.3}$$
>
> "We now investigate how the diffusion coefficient depends on ϕ. We shall once more restrict ourselves to the case where the number ν of particles per unit volume depends only on x and t.

"Let $\nu = f(x,t)$ be the number of particles per unit volume. We compute the distribution of particles at the time $t+\tau$ from the distribution at time t. From the definition of the function $\phi(\Delta)$, it is easy to find the number of particles which at time $t+\tau$ are found between two planes perpendicular to the x-axis and passing through points x and $x+dx$. One obtains

$$f(x,t+\tau)dx = dx \int_{-\infty}^{\infty} f(x+\Delta,t)\phi(\Delta)d\Delta \,. \tag{1.2.4}$$

But since τ is very small, we can set

$$f(x,t+\tau) = f(x,t) + \tau\frac{\partial f}{\partial t} \,. \tag{1.2.5}$$

Furthermore, we develop $f(x+\Delta,t)$ in powers of Δ:

$$f(x+\Delta,t) = f(x,t) + \Delta\frac{\partial f(x,t)}{\partial x} + \frac{\Delta^2}{2!}\frac{\partial^2 f(x,t)}{\partial x^2} + \cdots \,. \tag{1.2.6}$$

We can use this series under the integral, because only small values of Δ contribute to this equation. We obtain

$$f + \frac{\partial f}{\partial \tau}\tau = f\int_{-\infty}^{\infty}\phi(\Delta)d\Delta + \frac{\partial f}{\partial x}\int_{-\infty}^{\infty}\Delta\phi(\Delta)d\Delta + \frac{\partial^2 f}{\partial x^2}\int_{-\infty}^{\infty}\frac{\Delta^2}{2}\phi(\Delta)d\Delta \,. \tag{1.2.7}$$

Because $\phi(x) = \phi(-x)$, the second, fourth, etc., terms on the right-hand side vanish, while out of the 1st, 3rd, 5th, etc., terms, each one is very small compared with the previous. We obtain from this equation, by taking into consideration

$$\int_{-\infty}^{\infty}\phi(\Delta)d\Delta = 1 \,, \tag{1.2.8}$$

and setting

$$\frac{1}{\tau}\int_{-\infty}^{\infty}\frac{\Delta^2}{2}\phi(\Delta)d\Delta = D \,, \tag{1.2.9}$$

and keeping only the 1st and third terms of the right-hand side,

$$\frac{\partial f}{\partial t} = D\frac{\partial^2 f}{\partial x^2} \cdots \,. \tag{1.2.10}$$

This is already known as the differential equation of diffusion and it can be seen that D is the diffusion coefficient. ...

"The problem, which corresponds to the problem of diffusion from a single point (neglecting the interaction between the diffusing particles), is now completely determined mathematically: its solution is

$$f(x,t) = \frac{n}{\sqrt{4\pi D}} = \frac{e^{-x^2/4Dt}}{\sqrt{t}} \cdots . \tag{1.2.11}$$

"We now calculate, with the help of this equation, the displacement λ_x in the direction of the X-axis that a particle experiences on the average or, more exactly, the square root of the arithmetic mean of the square of the displacement in the direction of the X-axis; it is

$$\lambda_x = \sqrt{\bar{x}^2} = \sqrt{2Dt}\,.\text{''} \tag{1.2.12}$$

Einstein's derivation is really based on a discrete time assumption, that impacts happen only at times $0, \tau, 2\tau, 3\tau \ldots$, and his resulting equation (1.2.10) for the distribution function $f(x,t)$ and its solution (1.2.11) are to be regarded as approximations, in which τ is considered so small that t may be considered as being continuous. Nevertheless, his description contains very many of the major concepts which have been developed more and more generally and rigorously since then, and which will be central to this book. For example:

i) *The Chapman-Kolmogorov Equation* occurs as Einstein's equation (1.2.4). It states that the probability of the particle being at point x at time $t + \tau$ is given by the sum of the probability of all possible "pushes" Δ from positions $x + \Delta$, multiplied by the probability of being at $x + \Delta$ at time t. This assumption is based on the independence of the push Δ of any previous history of the motion: it is only necessary to know the initial position of the particle at time t—not at any previous time. This is the *Markov postulate* and the Chapman Kolmogorov equation, of which (1.2.4) is a special form, is the central dynamical equation to all Markov processes. These will be studied in detail in Chap. 3.

ii) *The Fokker-Planck Equation*: Eq. (1.2.10) is the diffusion equation, a special case of the Fokker-Planck equation (also known as *Kolmogorov's equation*) which describes a large class of very interesting stochastic processes in which the system has a continuous sample path. In this case, that means that the pollen grain's position, if thought of as obeying a probabilistic law given by solving the diffusion equation (1.2.10), in which time t is continuous (not discrete, as assumed by Einstein), can be written $x(t)$, where $x(t)$ is a *continuous function of time*—but a random function. This leads us to consider the possibility of describing the dynamics of the system in some direct probabilistic way, so that we would have a *random* or *stochastic differential equation for the path*. This procedure was initiated by Langevin with the famous equation that to this day bears his name. We will discuss this in Sect. 1.2.2, and in detail in Chap. 4.

iii) *The Kramers-Moyal* and similar expansions are essentially the same as that used by Einstein to go from (1.2.4) (the Chapman-Kolmogorov equation) to the diffusion equation (1.2.10). The use of this type of approximation, which effectively replaces a process whose sample paths need not be continuous with one whose paths are continuous, is very common and convenient. Its use and validity will be discussed in Chap. 11.

1.2.2 Langevin's Equation

Some time after Einstein's original derivation, *Langevin* [1.5] presented a new method which was quite different from Einstein's and, according to him, "infinitely more simple." His reasoning was as follows.

From statistical mechanics, it was known that the mean kinetic energy of the Brownian particle should, in equilibrium, reach a value

$$\left\langle \tfrac{1}{2}mv^2 \right\rangle = \tfrac{1}{2}kT\,, \tag{1.2.13}$$

(T; absolute temperature, k; Boltzmann's constant). (Both Einstein and Smoluchowski had used this fact). Acting on the particle, of mass m there should be two forces:

i) A viscous drag: assuming this is given by the same formula as in macroscopic hydrodynamics, this is $-6\pi\eta a\, dx/dt$ where η is the viscosity and a the diameter of the particle, assumed spherical.

ii) Another *fluctuating force* X which represents the incessant impacts of the molecules of the liquid on the Brownian particle. All that is known about it is that fact, and that it should be positive and negative with equal probability. Thus, the equation of motion for the position of the particle is given by Newton's law as

$$m\frac{d^2x}{dt^2} = -6\pi\eta a\frac{dx}{dt} + X\,, \tag{1.2.14}$$

and multiplying by x, this can be written

$$\frac{m}{2}\frac{d^2}{dt^2}(x^2) - mv^2 = -3\pi\eta a\frac{d(x^2)}{dt} + Xx\,, \tag{1.2.15}$$

where $v = dx/dt$. We now average over a large number of different particles and use (1.2.13) to obtain an equation for $\langle x^2 \rangle$:

$$\frac{m}{2}\frac{d^2\langle x^2\rangle}{dt^2} + 3\pi\eta a\frac{d\langle x^2\rangle}{dt} = kT\,, \tag{1.2.16}$$

where the term $\langle xX\rangle$ has been set equal to zero because (to quote Langevin) "of the irregularity of the quantity X". One then finds the general solution

$$\frac{d\langle x^2\rangle}{dt} = kT/(3\pi\eta a) + C\exp(-6\pi\eta a t/m)\,, \tag{1.2.17}$$

where C is an arbitrary constant. Langevin estimated that the decaying exponential approaches zero with a time constant of the order of 10^{-8} s, which for any practical observation at that time, was essentially immediately. Thus, for practical purposes, we can neglect this term and integrate once more to get

$$\langle x^2\rangle - \langle x_0^2\rangle = [kT/(3\pi\eta a)]t\,. \tag{1.2.18}$$

This corresponds to (1.2.12) as deduced by Einstein, provided we identify

$$D = kT/(6\pi\eta a)\,, \tag{1.2.19}$$

a result which Einstein derived in the same paper but by independent means.

Langevin's equation was the first example of the *stochastic differential equation*—a differential equation with a random term X and hence whose solution is, in some sense, a random function. Each solution of Langevin's equation represents a different random trajectory and, using only rather simple properties of X (his fluctuating force), measurable results can be derived.

One question arises: Einstein explicitly required that (on a sufficiently large time scale) the change Δ be completely independent of the preceding value of Δ. Langevin did not mention such a concept explicitly, but it is there, implicitly, when one sets $\langle Xx \rangle$ equal to zero. The concept that X is extremely irregular *and* (which is not mentioned by Langevin, but is implicit) that X and x are *independent* of each other—that the irregularities in x as a function of time, do not somehow conspire to be always in the same direction as those of X, so that it would not be valid to set the average of the product equal to zero. These are really equivalent to Einstein's independence assumption. The method of Langevin equations is clearly very much more direct, at least at first glance, and gives a very natural way of generalising a dynamical equation to a probabilistic equation. An adequate mathematical grounding for the approach of Langevin, however, was not available until more than 40 years later, when Ito formulated his concepts of stochastic differential equations. And in this formulation, a precise statement of the independence of X and x led to the calculus of stochastic differentials, which now bears his name and which will be fully developed in Chap. 4.

As a physical subject, Brownian motion had its heyday in the first two decades of last century, when Smoluchowski in particular, and many others carried out extensive theoretical and experimental investigations, which showed complete agreement with the original formulation of the subject as initiated by himself and *Einstein*, see [1.6]. More recently, with the development of laser light scattering spectroscopy, Brownian motion has become very much more quantitatively measurable. The technique is to shine intense, coherent laser light into a small volume of liquid containing Brownian particles, and to study the fluctuations in the intensity of the scattered light, which are directly related to the motions of the Brownian particles. By these means it is possible to observe Brownian motion of much smaller particles than the traditional pollen, and to derive useful data about the sizes of viruses and macromolecules. With the preparation of more concentrated suspensions, interactions between the particles appear, generating interesting and quite complex problems related to macromolecular suspensions and colloids [1.7].

The general concept of fluctuations describable by such equations has developed very extensively in a very wide range of situations. The advantages of a continuous description turn out to be very significant, since only a very few parameters are required, i.e., essentially the coefficients of the derivatives in (1.2.7):

$$\int_{-\infty}^{\infty} \Delta\phi(\Delta)d\Delta, \quad \text{and} \quad \int_{-\infty}^{\infty} \Delta^2\phi(\Delta)d\Delta\,. \tag{1.2.20}$$

It is rare to find a problem which cannot be specified, in at least some degree of approximation, by such a system, and for qualitative simple analysis of problems it is normally quite sufficient to consider an appropriate Fokker-Planck equation, of a form obtained by allowing both coefficients (1.2.20) to depend on x, and in a space of an appropriate number of dimensions.

1.3 The Stock Market

The equations of Brownian motion were in fact first derived by *Bachelier* in his doctoral thesis [1.8], in which he applied the ideas of probability to the pricing of shares and options in the stock market. He introduced the idea of the *relative value* $x = X - X_0$ of a share, that is the difference between its absolute value X and the *most probable value* X_0. He then considered the probability distribution $p_{x,t}$ of relative share prices x at time t, and then deduced the "law of composition" of these probabilities

$$p_{x,t_1+t_2} = \int p_{x,t_1} p_{z-x,t_2}\, dz\,. \tag{1.3.1}$$

This is the Chapman-Kolmogorov equation, that is, it is essentially Einstein's equation, (1.2.4), and the reasoning used to deduce it is basically the same as that of Einstein. Bachelier then sought a solution of the form

$$p = A e^{-B^2 x^2}, \tag{1.3.2}$$

and showed that A and B would be functions of time, concluding:

"The definitive expression for the probability is thus

$$p = \frac{1}{2\pi k\sqrt{t}}\, e^{-\frac{x^2}{4\pi k^2 t}}\,. \tag{1.3.3}$$

The mathematical expectation

$$\int_0^\infty p\, x\, dx = k\sqrt{t}\,. \tag{1.3.4}$$

is proportional to the square root of the time."

Bachelier gave another derivation rather more similar to Einstein's, in which he divided time into discrete intervals, and considered discrete jumps in the share prices, arriving finally at the *heat equation*, (1.2.10) as the differential equation for the probability distribution. The thesis then considers applications of this probability law to a range of the kind of financial transactions current on the Paris stock exchange of the early 1900s. The value of the work lies in the ideas, rather than the actual results, since Bachelier's use of the Gaussian form for the distribution $p_{x,t}$ clearly has the defect that there is a finite probability that the stock price can become negative, a possibility that he considers, but prefers to treat as negligible.

1.3.1 Statistics of Returns

That the price changes x can have a Gaussian distribution is a reasonable result only if these changes are small compared with the mean price—but this must clearly break down with increasing time if $\langle x^2 \rangle \sim t$. Bachelier's work did not generate much interest in finance circles until the 1960s, when *Samuelson* [1.9] decided to develop the approach further. Samuelson rather unfairly criticised Bachelier for "forgetting" that negative prices of shares were not permissible, and suggested a solution to this problem by proposing that changes in prices are most reasonably described as percentages. Explicitly, he proposes the correct quantity to consider is what has become known as the *return* on the share price, given by

$$r = \frac{x}{X}, \tag{1.3.5}$$

that is the fractional gain or loss in the share price. This leads to a formulation in which

$$p = \log X, \tag{1.3.6}$$

is regarded as the quantity that undergoes Brownian motion. This has the obvious advantage that $p \to -\infty$ means $X \to 0$, so the natural range $(0, \infty)$ of prices is recovered.

There is also a certain human logic in the description. Prices move as a result of judgments by buyers and sellers, to whom the natural measure of a price change is not the absolute size of the change, but the fractional change. The improvement over Bachelier's result is so significant, and the resulting description in terms of the logarithm of the price and the fractional price change so simple, that this is the preferred model to this day. Samuelson termed the process *geometric Brownian motion* or alternatively *economic Brownian motion.*

1.3.2 Financial Derivatives

In order to smooth the running of business, it is often helpful to fix in advance the price of a commodity which will be needed in the future—for example, the price of wheat which has not yet been grown and harvested is moderately uncertain. A baker could choose to pay a fixed sum now for the future delivery of wheat. Rather than deal with an individual grower, the baker can buy the ungrown wheat from a dealer in *wheat futures*, who charges a premium and arranges appropriate contracts with growers. However, the contract to deliver wheat at a certain price on a future date can itself become a tradable item. Having purchased such a contract, the baker can sell it to another baker, or indeed, to anyone else, who may buy it with the view to selling it at a future date, without ever having had anything to do with any wheat at all.

Such a contract is known as a *derivative* security. The wheat future exists only because there is a market for real wheat, but nevertheless can develop an existence of its own. Another kind of derivative is an *option*, in which one buys the *right* to purchase something at a future date at a definite price. If the market price on the date at which the option is exercised is larger than the option price, one exercises the option. If the market price turns out to be below the option price, one discards the option and pays the market price. Purchasing the option limits exposure to price rises, transferring the risk to the seller of the option, who charges appropriately, and specializes in balancing risks. Options to purchase other securities, such as shares and stocks, are very common, and indeed there are options markets which trade under standardized conditions.

1.3.3 The Black-Scholes Formula

Although a description of market processes in terms of stochastic processes was well-known by the 1970s, it was not clear how it could be used as a tool for making investment decisions. The breakthrough came with the realization that a *portfolio* containing an appropriate mix of cash, stocks and options could be devised in which the short term fluctuations in the various values could be cancelled, and that this gave a relatively simple formula for valuing options—the *Black-Scholes Formula*—which would be of very significant value in making investment decisions. This formula has truly revolutionized the practice of finance; to quote Samuelson [1.10]

> "A great economist of an earlier generation said that, useful though economic theory is for understanding the world, no one would go to an economic theorist for advice on how to run a brewery or produce a mousetrap. Today that sage would have to change his tune: economic principles really do apply and woe the accountant or marketer who runs counter to economic law. Paradoxically, one of our most elegant and complex sectors of economic analysis—the modern theory of finance—is confirmed daily by millions of statistical observations. When today's associate professor of security analysis is asked 'Young man, if you're so smart why ain't you rich?', he replies by laughing all the way to the bank or his appointment as a high-paid consultant to Wall Street."

The derivation was given first in the paper of *Black* and *Scholes* [1.11], and a different derivation was given by *Merton* [1.12]. The formula depends critically on description of the *returns* on securities as a Brownian motion process, which is of limited accuracy. Nevertheless, the formula is sufficiently realistic to make investing in stocks and options a logical and rational process, justifying Samuelson's perhaps over-dramatised view of modern financial theory.

1.3.4 Heavy Tailed Distributions

There is, however, no doubt that the geometric Brownian motion model of financial markets is not exact, and even misses out very important features. One need only study the empirical values of the returns in stock market records (as well as other kinds of markets) and check what kinds of distributions are in practice observed. The results are not really in agreement with a Gaussian distribution of returns—rather, the observed distribution of returns is usually approximately Gaussian for small values of r, but the probability of large values of r is always observed to be significantly larger than the Gaussian prediction—the observed distributions are said to have *heavy tails*.

The field of *Continuous Time Finance* [1.10] is an impressive theoretical edifice built on this flawed foundation of Brownian motion, but so far it appears to be the most practical method of modelling financial markets. With modern electronic banking and transfer of funds, it is possible to trade over very short time intervals, during which perhaps, in spite of the overall increase of trading activity which results, a Brownian description is valid.

It is certainly sufficiently valued for its practitioners to be highly valued, as Samuelson notes. However, every so often one of these practitioners makes a spectacular loss, threatening financial institutions. While there is public alarm about billion dollar losses, those who acknowledge the significance of heavy tails are unsurprised.

1.4 Birth-Death Processes

A wide variety of phenomena can be modelled by a particular class of process called a birth-death process. The name obviously stems from the modelling of human or animal populations in which individuals are born, or die. One of the most entertaining models is that of the prey-predator system consisting of two kinds of animal, one of which preys on the other, which is itself supplied with an inexhaustible food supply. Thus letting X symbolise the prey, Y the predator, and A the food of the prey, the process under consideration might be

$$X + A \to 2X\,, \tag{1.4.1a}$$

$$X + Y \to 2Y\,, \tag{1.4.1b}$$

$$Y \to B\,, \tag{1.4.1c}$$

which have the following naive, but charming interpretation. The first equation symbolises the prey eating one unit of food, and reproducing immediately. The second equation symbolises a predator consuming a prey (who thereby dies—this is the only death mechanism considered for the prey) and immediately reproducing. The final equation symbolises the death of the predator by natural causes. It is easy to guess model differential equations for x and y, the numbers of X and Y. One might assume that the first reaction symbolises a rate of production of X proportional to the product of x and the amount of food; the second equation a production of Y (and an equal rate of consumption of X) proportional to xy, and the last equation a death rate of Y, in which the rate of death of Y is simply proportional to y; thus we might write

$$\frac{dx}{dt} = k_1 a x - k_2 x y\,, \tag{1.4.2a}$$

$$\frac{dy}{dt} = k_2 x y - k_3 y\,. \tag{1.4.2b}$$

The solutions of these equations, which were independently developed by *Lotka* [1.13] and *Volterra* [1.14] have very interesting oscillating solutions, as presented in Fig. 1.3a. These oscillations are qualitatively easily explicable. In the absence of significant numbers of predators, the prey population grows rapidly until the presence of so much prey for the predators to eat stimulates their rapid reproduction, at

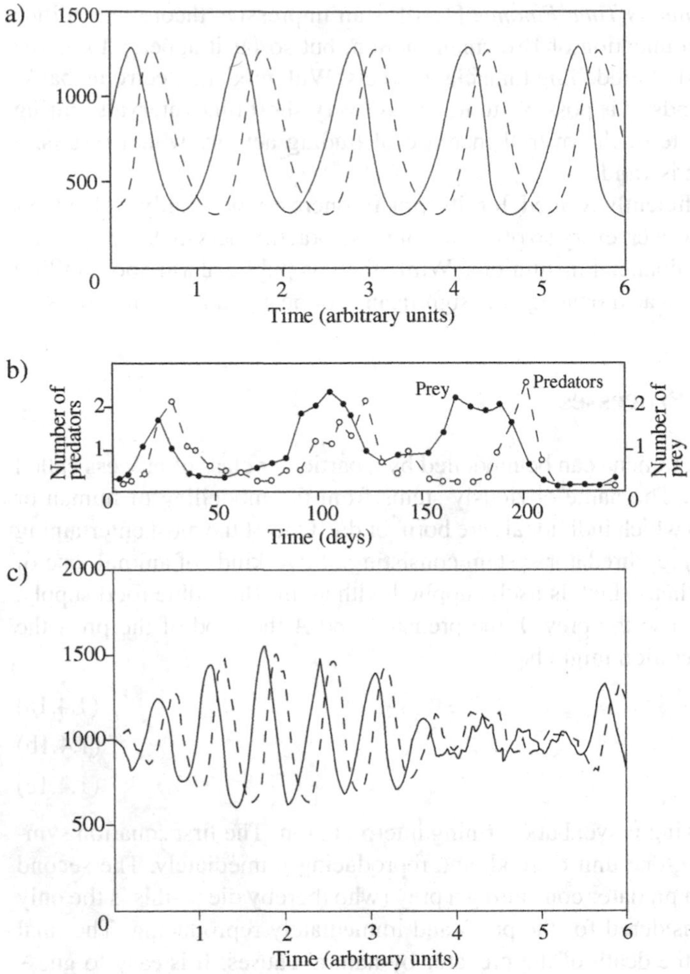

Fig. 1.3a–c. Time development in prey-predator systems. (**a**) Plot of solutions of the deterministic equations (1.4.2a, 1.4.2b) (x = solid line, y = dashed line). (**b**) Data for a real prey-predator system. Here the predator is a mite (Eotetranychus sexmaculatus—dashed line) which feeds on oranges, and the prey is another mite (Typhlodromus occidentalis). Data from [1.15, 1.16]. (**c**) Simulation of stochastic equations (1.4.3a–1.4.3d).

the same time reducing the number of prey which get eaten. Because a large number of prey have been eaten, there are no longer enough to maintain the population of predators, which then die out, returning us to our initial situation. The cycles repeat indefinitely and are indeed, at least qualitatively, a feature of many real prey-predator systems. An example is given in Fig. 1.3b.

Of course, the realistic systems do not follow the solutions of differential equations exactly—they fluctuate about such curves. One must include these fluctuations and the simplest way to do this is by means of a *birth-death master equation*. We assume a probability distribution, $P(x, y, t)$, for the number of individuals at a given time and ask for a probabilistic law corresponding to (1.4.2a, 1.4.2b). This is done by assuming that in an infinitesimal time Δt, the following *transition probability* laws holds.

$$\text{Prob}\,(x \to x+1; y \to y) = k_1 ax\Delta t\,, \tag{1.4.3a}$$

$$\text{Prob}\,(x \to x-1; y \to y+1) = k_2 xy\Delta t\,, \tag{1.4.3b}$$

$$\text{Prob}\,(x \to x; y \to y-1) = k_3 y\Delta t\,, \tag{1.4.3c}$$

$$\text{Prob}\,(x \to x; y \to y) = 1 - (k_1 ax + k_2 xy + k_3 y)\Delta t\,. \tag{1.4.3d}$$

Thus, we simply, for example, replace the simple rate laws by probability laws. We then employ what amounts to the same equation as Einstein and others used, i.e., the Chapman-Kolmogorov equation, namely, we write the probability at $t + \Delta t$ as a sum of terms, each of which represents the probability of a previous state multiplied by the probability of a transition to the state (x, y). Thus, we find by letting $\Delta t \to 0$:

$$\begin{aligned}\frac{P(x,y,t+\Delta t) - P(x,y,t)}{\Delta t} &\to \frac{\partial P(x,y,t)}{\partial t}\\ &= k_1 a(x-1)P(x-1,y,t) + k_2(x+1)(y-1)\\ &\quad\times P(x+1,y-1,t) + k_3(y+1)P(x,y+1,t)\\ &\quad- (k_1 ax + k_2 xy + k_3 y)P(x,y,t).\end{aligned} \tag{1.4.4}$$

In writing the assumed probability laws (1.4.3a–1.4.3d), we are assuming that the probability of each of the events occurring can be determined simply from the knowledge of x and y. This is again the Markov postulate which we mentioned in Sect. 1.2.1. In the case of Brownian motion, very convincing arguments can be made in favour of this Markov assumption. Here it is by no means clear. The concept of heredity, i.e., that the behaviour of progeny is related to that of parents, clearly contradicts this assumption. How to *include* heredity is another matter; by no means does a unique prescription exist.

The assumption of the Markov postulate in this context is valid to the extent that different individuals of the same species are similar; it is invalid to the extent that, nevertheless, perceptible inheritable differences do exist.

This type of model has a wide application—in fact to any system to which a population of individuals may be attributed, for example systems of molecules of various chemical compounds, of electrons, of photons and similar physical particles as well as biological systems. The particular choice of transition probabilities is made on various grounds determined by the degree to which details of the births and deaths involved are known. The simple multiplicative laws, as illustrated in (1.4.3a–1.4.3d), are the most elementary choice, ignoring, as they do, almost all details of the processes involved. In some of the physical processes we can derive the transition probabilities in much greater detail and with greater precision.

Equation (1.4.4) has no simple solution, but one major property differentiates equations like it from an equation of Langevin's type, in which the fluctuation term is simply added to the differential equation. Solutions of (1.4.4) determine both the gross deterministic motion and the fluctuations; the fluctuations are typically of the same order of magnitude as the square roots of the *numbers* of individuals involved. It is not difficult to simulate a sample time development of the process as in Fig. 1.3c. The figure does show the correct general features, but the model is so obviously simplified that exact agreement can never be expected. Thus, in contrast to the situation in Brownian motion, we are not dealing here so much with a theory of a phenomenon, as with a class of mathematical models, which are simple enough to have a very wide range of approximate validity. We will see in Chap. 11 that a theory can be developed which can deal with a wide range of models in this category, and that there is indeed a close connection between this kind of theory and that of stochastic differential equations.

1.5 Noise in Electronic Systems

The early days of radio with low transmission powers and primitive receivers, made it evident to every ear that there were a great number of highly irregular electrical signals which occurred either in the atmosphere, the receiver, or the radio transmitter, and which were given the collective name of "noise", since this is certainly what they sounded like on a radio. Two principal sources of noise are shot noise and Johnson noise.

1.5.1 Shot Noise

In a vacuum tube (and in solid-state devices) we get a nonsteady electrical current, since it is generated by individual electrons, which are accelerated across a distance and deposit their charge one at a time on the anode. The electric current arising from such a process can be written

$$I(t) = \sum_{t_k} F(t - t_k), \tag{1.5.1}$$

where $F(t - t_k)$ represents the contribution to the current of an electron which arrives at time t_k. Each electron is therefore assumed to give rise to the same shaped pulse, but with an appropriate delay, as in Fig. 1.4.

A statistical aspect arises immediately we consider what kind of choice must be made for t_k. The simplest choice is that each electron arrives independently of the previous one—that is, the times t_k are randomly distributed with a certain average number per unit time in the range $(-\infty, \infty)$, or whatever time is under consideration.

The analysis of such noise was developed during the 1920's and 1930's and was summarised and largely completed by *Rice* [1.17]. It was first considered as early as 1918 by *Schottky* [1.18].

We shall find that there is a close connection between shot noise and processes described by birth-death master equations. For, if we consider n, the number of electrons which have arrived up to a time t, to be a statistical quantity described by a probability $P(n,t)$, then the assumption that the electrons arrive independently is clearly the Markov assumption. Then, assuming the probability that an electron will

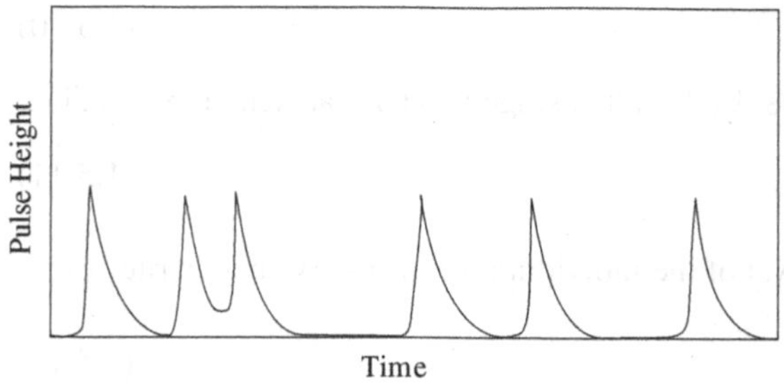

Fig. 1.4. Illustration of shot noise: identical electric pulses arrive at random times

arrive in the time interval between t and $t + \Delta t$ is completely independent of t and n, its only dependence can be on Δt. By choosing an appropriate constant λ, we may write

$$\text{Prob}\,(n \to n+1, \text{in time } \Delta t) = \lambda\,\Delta t\,, \tag{1.5.2}$$

so that

$$P(n, t+\Delta t) = P(n,t)(1-\lambda\,\Delta t) + P(n-1,t)\lambda\,\Delta t\,, \tag{1.5.3}$$

and taking the limit $\Delta t \to 0$

$$\frac{\partial P(n,t)}{\partial t} = \lambda[P(n-1,t) - P(n,t)]\,, \tag{1.5.4}$$

which is a pure birth process. By writing

$$G(s,t) = \sum s^n P(n,t)\,, \tag{1.5.5}$$

[here, $G(s,t)$ is known as the generating function for $P(n,t)$, and the particular technique of solving (1.5.4) is very widely used], we find

$$\frac{\partial G(s,t)}{\partial t} = \lambda(s-1)G(s,t), \tag{1.5.6}$$

so that

$$G(s,t) = \exp[\lambda(s-1)t]G(s,0)\,. \tag{1.5.7}$$

By requiring at time $t = 0$ that no electrons had arrived, it is clear that $P(0,0)$ is 1 and $P(n,0)$ is zero for all $n \geqslant 1$, so that $G(s,0) = 1$. Expanding the solution (1.5.7) in powers of s, we find

$$P(n,t) = \exp(-\lambda t)(\lambda t)^n/n!\,, \tag{1.5.8}$$

which is known as a *Poisson distribution* (Sect. 2.8.3). Let us introduce the variable $N(t)$, which is to be considered as the number of electrons which have arrived up to time t, and is a random quantity. Then,

$$P(n,t) = \text{Prob}\,\{N(t) = n\}\,, \tag{1.5.9}$$

and $N(t)$ can be called a *Poisson process variable*. Then clearly, the quantity $\mu(t)$, formally defined by

$$\mu(t) = dN(t)/dt\,, \tag{1.5.10}$$

is zero, except when $N(t)$ increases by 1; at that stage it is a Dirac delta function, i.e.,

$$\mu(t) = \sum_k \delta(t - t_k)\,, \tag{1.5.11}$$

where the t_k are the times of arrival of the individual electrons. We may write

$$I(t) = \int_{-\infty}^{\infty} dt' F(t - t')\mu(t')\,. \tag{1.5.12}$$

A very reasonable restriction on $F(t - t')$ is that it vanishes if $t < t'$, and that for $t \to \infty$, it also vanishes. This simply means that no current arises from an electron before it arrives, and that the effect of its arrival eventually dies out. We assume then, for simplicity, the very commonly encountered form

$$F(t) = \begin{cases} q\,\mathrm{e}^{-\alpha t}\,, & (t > 0)\,, \\ 0\,, & (t < 0)\,, \end{cases} \tag{1.5.13}$$

so that (1.5.12) can be rewritten as

$$I(t) = \int_{-\infty}^{t} dt'\, q\,\mathrm{e}^{-\alpha(t-t')}\frac{dN(t')}{dt'}\,. \tag{1.5.14}$$

We can derive a simple differential equation. We differentiate $I(t)$ to obtain

$$\frac{dI(t)}{dt} = -\int_{-\infty}^{t} dt'\,\alpha q\,\mathrm{e}^{-\alpha(t-t')}\frac{dN(t')}{dt'} + \left[q\,\mathrm{e}^{-\alpha(t-t')}\frac{dN(t')}{dt}\right]_{t'=t}\,, \tag{1.5.15}$$

so that

$$\frac{dI(t)}{dt} = -\alpha I(t) + q\mu(t)\,. \tag{1.5.16}$$

This is a kind of stochastic differential equation, similar to Langevin's equation, in which, however, the fluctuating force is given by $q\mu(t)$, where $\mu(t)$ is the derivative of the Poisson process, as given by (1.5.11). However, the mean of $\mu(t)$ is nonzero, in fact, from (1.5.10)

$$\langle \mu(t)dt\rangle = \langle dN(t)\rangle = \lambda\, dt\,, \tag{1.5.17}$$

$$\langle [dN(t) - \lambda dt]^2\rangle = \lambda\, dt\,, \tag{1.5.18}$$

from the properties of the Poisson distribution, for which the variance equals the mean. Defining, then, the fluctuation as the difference between the mean value and $dN(t)$, we write

$$d\eta(t) = dN(t) - \lambda dt\,, \tag{1.5.19}$$

so that the stochastic differential equation (1.5.16) takes the form

$$dI(t) = [\lambda q - \alpha I(t)]dt + q\,d\eta(t)\,. \tag{1.5.20}$$

Now how does one solve such an equation? In this case, we have an academic problem anyway since the solution is known, but one would like to have a technique. Suppose we try to follow the method used by Langevin—what will we get as an answer? The short reply to this question is: nonsense. For example, using ordinary calculus and assuming $\langle I(t)d\eta(t)\rangle = 0$, we can derive

$$\frac{d\langle I(t)\rangle}{dt} = \lambda q - \alpha\langle I(t)\rangle\,, \tag{1.5.21}$$

$$\frac{1}{2}\frac{d\langle I^2(t)\rangle}{dt} = \lambda q\langle I(t)\rangle - \alpha\langle I^2(t)\rangle\,. \tag{1.5.22}$$

Solving in the limit $t \to \infty$, where the mean values would reasonably be expected to be constant one finds

$$\langle I(\infty)\rangle = \lambda q/\alpha, \tag{1.5.23}$$

$$\langle I^2(\infty)\rangle = (\lambda q/\alpha)^2. \tag{1.5.24}$$

The first answer is reasonable—it merely gives the average current through the system in a reasonable equation, but the second implies that the mean square current is the same as the square of the mean, i.e., the current at $t \to \infty$ does not fluctuate! This is rather unreasonable, and the solution to the problem will show that stochastic differential equations are rather more subtle than we have so far presented.

Firstly, the notation in terms of differentials used in (1.5.17–1.5.20) has been chosen deliberately. In deriving (1.5.22), one uses ordinary calculus, i.e., one writes

$$d(I^2) \equiv (I + dI)^2 - I^2 = 2I\,dI + (dI)^2\,, \tag{1.5.25}$$

and then one drops the $(dI)^2$ as being of second order in dI. But now look at (1.5.18): this is equivalent to

$$\langle d\eta(t)^2\rangle = \lambda\,dt\,, \tag{1.5.26}$$

so that a *quantity of second order in* $d\eta$ *is actually of first order in* dt. The reason is not difficult to find. Clearly,

$$d\eta(t) = dN(t) - \lambda\,dt\,, \tag{1.5.27}$$

but the curve of $N(t)$ is a step function, discontinuous, and certainly not differentiable, at the times of arrival of the individual electrons. In the ordinary sense, none of these calculus manipulations is permissible. But we can make sense out of them as follows. Let us simply calculate $\langle d(I^2)\rangle$ using (1.5.20, 1.5.25, 1.5.26):

$$\langle d(I)^2\rangle = 2\langle I\{[\lambda q - \alpha I]dt + q\,d\eta(t)\}\rangle + \langle\{[\lambda q - \alpha I]dt + q\,d\eta(t)\}^2\rangle\,. \tag{1.5.28}$$

We now assume again that $\langle I(t)d\eta(t)\rangle = 0$ and expand, after taking averages using the fact that $\langle d\eta(t)^2\rangle = \lambda\,dt$, to 1*st order in dt*. We obtain

$$\tfrac{1}{2}d\langle I^2\rangle = \left[\lambda q\langle I\rangle - \alpha\langle I^2\rangle + \tfrac{1}{2}q^2\lambda\right] dt\,, \tag{1.5.29}$$

and this gives

$$\langle I^2(\infty)\rangle - \langle I(\infty)\rangle^2 = \frac{q^2\lambda}{2\alpha}\,. \tag{1.5.30}$$

Thus, there are fluctuations from this point of view, as $t \to \infty$. The extra term in (1.5.29) as compared to (1.5.22) arises directly out of the statistical considerations implicit in $N(t)$ being a discontinuous random function.

Thus we have discovered a somewhat deeper way of looking at Langevin's kind of equation—the treatment of which, from this point of view, now seems extremely naive. In Langevin's method the fluctuating force X is not specified, but it will become clear in this book that problems such as we have just considered are very widespread in this subject. The moral is that random functions cannot normally be differentiated according to the usual laws of calculus; special rules have to be developed, and a precise specification of what one means by differentiation becomes important. We will specify these problems and their solutions in Chap. 4 which will concern itself with situations in which the fluctuations are Gaussian.

1.5.2 Autocorrelation Functions and Spectra

The measurements which one can carry out on fluctuating systems such as electric circuits are, in practice, not of unlimited variety. So far, we have considered the distribution functions, which tell us, at any time, what the probability distribution of the values of a stochastic quantity are. If we are considering a measurable quantity $x(t)$ which fluctuates with time, in practice we can sometimes determine the distribution of the values of x, though more usually, what is available at one time are the mean $\bar{x}(t)$ and the variance $\mathrm{var}[x(t)]$.

The mean and the variance do not tell a great deal about the underlying dynamics of what is happening. What would be of interest is some quantity which is a measure of the influence of a value of x at time t on the value at time $t+\tau$. Such a quantity is the *autocorrelation function*, which was apparently first introduced by *Taylor* [1.19] as

$$G(\tau) = \lim_{T\to\infty} \frac{1}{T} \int_0^T dt\, x(t)x(t+\tau)\,. \tag{1.5.31}$$

This is the time average of a two-time product over an arbitrary large time T, which is then allowed to become infinite. Using modern computerized data collection technology it is straightforward to construct an autocorrelation from any stream of data, either in real time or from recorded data.

A closely connected approach is to compute the *spectrum* of the quantity $x(t)$. This is defined in two stages. First, define

$$y(\omega) = \int_0^T dt\, e^{-i\omega t} x(t)\,, \tag{1.5.32}$$

then the spectrum is defined by

$$S(\omega) = \lim_{T\to\infty} \frac{1}{2\pi T} |y(\omega)|^2. \tag{1.5.33}$$

The autocorrelation function and the spectrum are closely connected. By a little manipulation one finds

$$S(\omega) = \lim_{T\to\infty}\left[\frac{1}{\pi}\int_0^T \cos(\omega\tau)d\tau\frac{1}{T}\int_0^{T-\tau} x(t)x(t+\tau)dt\right], \tag{1.5.34}$$

and taking the limit $T \to \infty$ (under suitable assumptions to ensure the validity of certain interchanges of order), one finds

$$S(\omega) = \frac{1}{\pi}\int_0^\infty \cos(\omega\tau)G(\tau)d\tau. \tag{1.5.35}$$

This is a fundamental result which relates the Fourier transform of the autocorrelation function to the spectrum. The result may be put in a slightly different form when one notices that

$$G(-\tau) = \lim_{T\to\infty}\frac{1}{T}\int_{-\tau}^{T-\tau} dt\, x(t+\tau)x(t) = G(\tau), \tag{1.5.36}$$

so we obtain

$$S(\omega) = \frac{1}{2\pi}\int_{-\infty}^{\infty} e^{-i\omega\tau}G(\tau)\,d\tau, \tag{1.5.37}$$

with the corresponding inverse

$$G(\tau) = \int_{-\infty}^{\infty} e^{i\omega\tau}S(\omega)\,d\omega. \tag{1.5.38}$$

This result is known as the *Wiener-Khinchin theorem* [1.20, 1.21] and has widespread application.

It means that one may either directly measure the autocorrelation function of a signal, or the spectrum, and convert back and forth, which by means of the fast Fourier transform and computer is relatively straightforward.

1.5.3 Fourier Analysis of Fluctuating Functions: Stationary Systems

The autocorrelation function has been defined so far as a time average of a signal, but we may also consider the *ensemble average*, in which we repeat the same measurement many times, and compute averages, denoted by $\langle\;\rangle$. It will be shown that for very many systems, the time average is equal to the ensemble average; such systems are termed *ergodic*—see Sect. 3.7.1.

If we have such a fluctuating quantity $x(t)$, then we can consider the average of the product of two time-values of x

$$\langle x(t)x(t+\tau)\rangle = G(\tau). \tag{1.5.39}$$

The fact that the result is independent of the absolute time t is a consequence of our ergodic assumption.

Now it is very natural to write a Fourier transform for the stochastic quantity $x(t)$

$$x(t) = \int d\omega\, c(\omega) e^{i\omega t}\,, \tag{1.5.40}$$

and consequently,

$$c(\omega) = \frac{1}{2\pi} \int dt\, x(t) e^{-i\omega t}\,. \tag{1.5.41}$$

Note that $x(t)$ real implies

$$c(\omega) = c^*(-\omega)\,. \tag{1.5.42}$$

If the system is ergodic, we must have a constant $\langle x(t) \rangle$, since the time average is clearly constant. The process is then *stationary* by which we mean that all time-dependent averages are functions only of time differences, i.e., averages of functions $x(t_1)$, $x(t_2)$, $\dots x(t_n)$ are equal to those of $x(t_1 + \Delta)$, $x(t_2 + \Delta)$, $\dots x(t_n + \Delta)$.

For convenience, in what follows we assume $\langle x \rangle = 0$. Hence,

$$\langle c(\omega) \rangle = \frac{1}{2\pi} \int dt\, \langle x \rangle e^{-i\omega t} = 0\,, \tag{1.5.43}$$

$$\begin{aligned}
\langle c(\omega) c^*(\omega') \rangle &= \frac{1}{(2\pi)^2} \int\int dt\, dt' e^{-i\omega t + i\omega' t'} \langle x(t) x(t') \rangle\,, \\
&= \frac{1}{(2\pi)} \delta(\omega - \omega') \int d\tau\, e^{i\omega\tau} G(\tau)\,, \\
&= \delta(\omega - \omega') S(\omega)\,.
\end{aligned} \tag{1.5.44}$$

Here we find not only a relationship between the mean square $\langle |c(\omega)|^2 \rangle$ and the spectrum, but also the result that stationarity alone implies that $c(\omega)$ and $c^*(\omega')$ are uncorrelated, since the term $\delta(\omega - \omega')$ arises because $\langle x(t) x(t') \rangle$ is a function only of $t - t'$.

1.5.4 Johnson Noise and Nyquist's Theorem

Two brief and elegant papers appeared in 1928 in which *Johnson* [1.22] demonstrated experimentally that an electric resistor automatically generated fluctuations of electric voltage, and *Nyquist* [1.23] demonstrated its theoretical derivation, in complete accordance with Johnson's experiment. The principle involved was already known by *Schottky* [1.18] and is the same as that used by Einstein and Langevin. This principle is that of thermal equilibrium. If a resistor R produces electric fluctuations, these will produce a current which will generate heat. The heat produced in the resistor must exactly balance the energy taken out of the fluctuations. The detailed working out of this principle is not the subject of this section, but we will find that such results are common throughout the physics and chemistry of stochastic processes, where the principles of statistical mechanics, whose basis is not essentially stochastic, are brought in to complement those of stochastic processes—such results are known as *fluctuation-dissipation theorems*.

Nyquist's experimental result was the following. We have an electric resistor of resistance R at absolute temperature T. Suppose by means of a suitable filter we measure $E(\omega)d\omega$, the voltage across the resistor with angular frequency in the range $(\omega, \omega + d\omega)$. Then, if k is Boltzmann's constant,

$$\langle E^2(\omega)\rangle = RkT/\pi\,. \tag{1.5.45}$$

This result is known nowadays as *Nyquist's theorem*. Johnson remarked. "The effect is one of the causes of what is called 'tube noise' in vacuum tube amplifiers. Indeed, it is often by far the larger part of the 'noise' of a good amplifier."

Johnson noise is easily described by the formalism of the previous subsection. The mean noise voltage is zero across a resistor, and the system is arranged so that it is in a steady state and is expected to be well represented by a stationary process. Johnson's quantity is, in practice, a limit of the kind (1.5.33) and may be summarised by saying that the voltage spectrum $S(\omega)$ is given by

$$S(\omega) = RkT/\pi, \tag{1.5.46}$$

that is, the spectrum is flat, i.e., a constant function of ω. In the case of light, the frequencies correspond to different colours of light. If we perceive light to be white, it is found that in practice all colours are present in equal proportions—the optical spectrum of white light is thus flat—at least within the visible range. In analogy, the term *white noise* is applied to a noise voltage (or any other fluctuating quantity) whose spectrum is flat.

White noise cannot actually exist. The simplest demonstration is to note that the mean power dissipated in the resistor in the frequency range (ω_1, ω_2) is given by

$$\int_{\omega_1}^{\omega_2} d\omega\, S(\omega)/R = kT(\omega_1 - \omega_2)/\pi\,, \tag{1.5.47}$$

so that the total power dissipated in all frequencies is infinite! Nyquist realised this, and noted that, in practice, there would be quantum corrections which would, at room temperature, make the spectrum flat only up to 7×10^{13} Hz, which is not detectable in practice, in a radio situation. The actual power dissipated in the resistor would be somewhat less than infinite—10^{-10} W in fact! And in practice there are other limiting factors such as the inductance of the system, which would limit the spectrum to even lower frequencies.

From the definition of the spectrum in terms of the autocorrelation function given in Sect. 1.5, we have

$$\langle E(t+\tau)E(t)\rangle = G(\tau)\,, \tag{1.5.48}$$

$$= \frac{1}{2\pi}\int_{-\infty}^{\infty} d\omega\, \mathrm{e}^{-\mathrm{i}\omega\tau} 2RkT\,, \tag{1.5.49}$$

$$= 2RkT\delta(\tau)\,, \tag{1.5.50}$$

which implies that no matter how small the time difference τ, $E(t+\tau)$ and $E(\tau)$ are not correlated. This is, of course, a direct result of the flatness of the spectrum. A typical model of $S(\omega)$ that is almost flat is

$$S(\omega) = RkT/[\pi(\omega^2\tau_C^2 + 1)]\,. \tag{1.5.51}$$

This is flat provided $\omega \ll \tau_C^{-1}$. The Fourier transform can be explicitly evaluated in this case to give

$$\langle E(t+\tau)E(t)\rangle = (R\,kT/\tau_C)\exp(-\tau/\tau_C)\,, \tag{1.5.52}$$

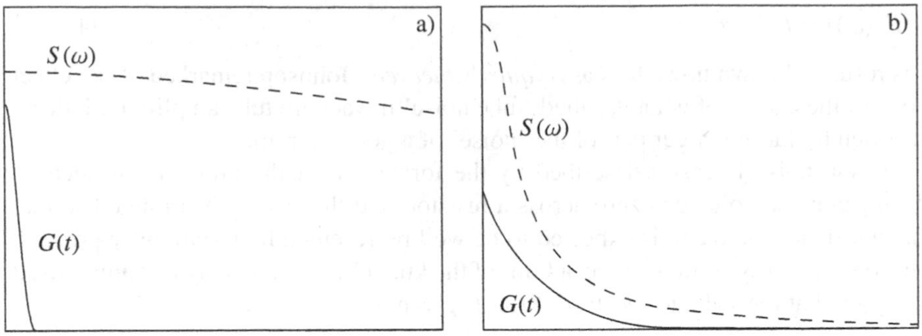

Fig. 1.5. Correlation Functions (———) and corresponding spectra (- - - - -) for (**a**) short correlation time corresponding to an almost flat spectrum; (**b**) long correlation time, giving a quite rapidly decreasing spectrum

so that the autocorrelation function vanishes only for $\tau \gg \tau_C$, which is called the *correlation time* of the fluctuating voltage. Thus, the delta function correlation function appears as an idealisation, only valid on a sufficiently long time scale.

This is very reminiscent of Einstein's assumption regarding Brownian motion and of the behaviour of Langevin's fluctuating force. The idealised *white noise* will play a highly important role in this book but, in just the same way as the fluctuation term that arises in a stochastic differential equation is not the same as an ordinary differential, we will find that differential equations which include white noise as a driving term have to be handled with great care. Such equations arise very naturally in any fluctuating system and it is possible to arrange by means of *Stratonovich's rules* for ordinary calculus rules to apply, but at the cost of imprecise mathematical definition and some difficulties in stochastic manipulation. It turns out to be far better to abandon ordinary calculus and use the *Ito calculus*, which is not very different (it is, in fact, very similar to the calculus presented for shot noise) and to preserve tractable statistical properties. All these matters will be discussed thoroughly in Chap. 4.

White noise, as we have noted above, does not exist as a physically realisable process and the rather singular behaviour it exhibits does not arise in any realisable context. It is, however, fundamental in a mathematical, and indeed in a physical sense, in that it is an idealisation of very many processes that do occur. The slightly strange rules which we will develop for the calculus of white noise are not really very difficult and are very much easier to handle than any method which always deals with a real noise. Furthermore, situations in which white noise is not a good approximation can very often be indirectly expressed quite simply in terms of white noise. In this sense, white noise is the starting point from which a wide range of stochastic descriptions can be derived, and is therefore fundamental to the subject of this book.

2. Probability Concepts

In the preceding chapter, we introduced probability notions without any definitions. In order to formulate essential concepts more precisely, it is necessary to have some more precise expression of these concepts. The intention of this chapter is to provide some background, and to present a number of essential results. It is not a thorough outline of mathematical probability, for which the reader is referred to standard mathematical texts such as those by *Feller* [2.1] and *Papoulis* [2.2].

2.1 Events, and Sets of Events

It is convenient to use a notation which is as general as possible in order to describe those occurrences to which we might wish to assign probabilities. For example, we may wish to talk about a situation in which there are 6.4×10^{14} molecules in a certain region of space; or a situation in which a Brownian particle is at a certain point $\boldsymbol{x}$ in space; or possibly there are 10 mice and 3 owls in a certain region of a forest.

These occurrences are all examples of practical realisations of *events*. More abstractly, an event is simply a member of a certain space, which in the cases most practically occurring can be characterised by a vector of integers

$$\boldsymbol{n} = (n_1, n_2, n_3 \ldots), \tag{2.1.1}$$

or a vector of real numbers

$$\boldsymbol{x} = (x_1, x_2, x_3 \ldots). \tag{2.1.2}$$

The dimension of the vector is arbitrary.

It is convenient to use the language of set theory, introduce the concept of a *set of events*, and use the notation

$$\omega \in A, \tag{2.1.3}$$

to indicate that the event ω is one of events contained in A. For example, one may consider the set $A(25)$ of events in the ecological population in which there are no more than 25 animals present; clearly the event $\bar{\omega}$ that there are 3 mice, a tiger, and no other animals present satisfies

$$\bar{\omega} \in A(25). \tag{2.1.4}$$

More significantly, suppose we define the set of events $A(\boldsymbol{r}, \Delta V)$ that a molecule is within a volume element ΔV centred on a point $\boldsymbol{r}$. In this case, the practical significance of working in terms of sets of events becomes clear, because we should normally be able to determine whether or not a molecule is within a neighbourhood ΔV of $\boldsymbol{r}$, but to determine whether the particle is exactly at $\boldsymbol{r}$ is impossible. Thus, if we define the event $\omega(\boldsymbol{y})$ that the molecule is at point $\boldsymbol{y}$, it makes sense to ask whether

$$\omega(\boldsymbol{y}) \in A(\boldsymbol{r}, \Delta V)\,, \tag{2.1.5}$$

and to assign a certain probability to the *set* $A(\boldsymbol{r}, \Delta V)$, which is to be interpreted as the probability of the occurrence of (2.1.5).

2.2 Probabilities

Most people have an intuitive conception of a probability, based on their own experience. However, a precise formulation of intuitive concepts is fraught with difficulties, and it has been found most convenient to axiomatise probability theory as an essentially abstract science, in which a probability measure $P(A)$ is *assigned* to every set A, in the space of events, including

$$\text{The set of all events} : \Omega\,, \tag{2.2.1}$$

$$\text{The set of no events} : \varnothing\,, \tag{2.2.2}$$

in order to define probability, we need our sets of events to form a closed system (known by mathematicians as a *σ-algebra*) under the set theoretic operations of union and intersection.

2.2.1 Probability Axioms

We introduce the probability of A, $P(A)$, as a function of A satisfying the following *probability axioms*:

i) $P(A) \geqslant 0 \quad \text{for all } A\,,$ (2.2.3)

ii) $P(\Omega) = 1\,,$ (2.2.4)

iii) If $A_i \quad (i = 1, 2, 3, \ldots)$ is a countable (but possibly infinite) collection of nonoverlapping sets, i.e., such that

$$A_i \cap A_j = \varnothing \quad \text{for all} \quad i \neq j\,, \tag{2.2.5}$$

then

$$P(\bigcup_i A) = \sum_i P(A_i)\,. \tag{2.2.6}$$

These are all the axioms needed. Consequentially, however, we have:

iv) if $\bar{A}$ is the complement of A, i.e., the set of all events not contained in A, then

$$P(\bar{A}) = 1 - P(A)\,, \tag{2.2.7}$$

v) $P(\varnothing) = 0\,.$ (2.2.8)

2.2.2 The Meaning of $P(A)$

There is no way of making probability theory correspond to reality without requiring a certain degree of intuition. The probability $P(A)$, as axiomatised above, is the intuitive probability that an "*arbitrary*" event ω, i.e., an event ω "*chosen at random*", will satisfy $\omega \in A$. Or more explicitly, if we choose an event "*at random*" from Ω N times, the relative frequency that the particular event chosen will satisfy $\omega \in A$ approaches $P(A)$ as the number of times, N, we choose the event, approaches infinity. The number of choices N can be visualised as being done one after the other ("*independent*" tosses of one die) or at the same time (N dice are thrown at the same time "*independently*"). All definitions of this kind must be intuitive, as we can see by the way undefined terms ("*arbitrary*", "*at random*", "*independent*") keep turning up. By eliminating what we now think of as intuitive ideas and axiomatising probability, *Kolmogorov* [2.3] cleared the road for a rigorous development of mathematical probability. But the circular definition problems posed by wanting an intuitive understanding remain. The simplest way of looking at axiomatic probability is as a formal method of manipulating probabilities using the axioms. In order to apply the theory, the probability space must be defined *and* the probability measure P assigned. These are *a priori probabilities*, which are simply assumed. Examples of such a priori probabilities abound in applied disciplines. For example, in equilibrium statistical mechanics one assigns equal probabilities to equal volumes of phase space. Einstein's reasoning in Brownian motion assigned a probability $\phi(\Delta)$ to the probability of a "push" Δ from a position x at time t.

The task of applying probability is

i) To assume some set of *a priori* probabilities which seem reasonable and to deduce results from this and from the structure of the probability space,

ii) To measure experimental results with some apparatus which is constructed to measure quantities in accordance with these a priori probabilities.

The structure of the probability space is very important, especially when the space of events is compounded by the additional concept of time. This extension makes the effective probability space infinite-dimensional, since we can construct events such as "the particle was at points $\boldsymbol{x}_n$ at times t_n for $n = 0, 1, 2, \ldots, \infty$".

2.2.3 The Meaning of the Axioms

Any intuitive concept of probability gives rise to nonnegative probabilities, and the probability that an arbitrary event is contained in the set of all events must be 1 no matter what our definition of the word arbitrary. Hence, axioms i) and ii) are understandable. The heart of the matter lies in axiom iii). Suppose we are dealing with only 2 sets A and B, and $A \cap B = \varnothing$. This means there are *no* events contained in both A and B. Therefore, the probability that $\omega \in A \cup B$ is the probability that *either* $\omega \in A$ or $\omega \in B$. Intuitive considerations tell us this probability is the sum of the individual probabilities, i.e.,

$$P(A \cup B) \equiv P\{(\omega \in A) \text{ or } (\omega \in B)\} = P(A) + P(B)\,. \tag{2.2.9}$$

Notice this is not a proof—merely an explanation.

The extension now to any finite number of nonoverlapping sets is obvious, but the extension only to any *countable* number of nonoverlapping sets requires some comment.

This extension must be made restrictive because of the existence of sets labelled by a continuous index, for example, $\boldsymbol{x}$, the position in space. The probability of a molecule being in the set whose only element in $\boldsymbol{x}$ is zero; but the probability of being in a region R of finite volume is nonzero. The region R is a union of sets of the form $\{\boldsymbol{x}\}$—but not a *countable* union. Thus axiom iii) is not applicable and the probability of being in R is *not* equal to the sum of the probabilities of being in $\{\boldsymbol{x}\}$.

2.2.4 Random Variables

The concept of a random variable is a notational convenience which is central to this book. Suppose we have an abstract probability space whose events can be written $\boldsymbol{x}$. Then we can introduce the random variable $F(\boldsymbol{x})$ which is a function of $\boldsymbol{x}$, which takes on certain values for each $\boldsymbol{x}$. In particular, the identity function of $\boldsymbol{x}$, written $\boldsymbol{X}(\boldsymbol{x})$ is of interest; it is given by

$$\boldsymbol{X}(\boldsymbol{x}) = \boldsymbol{x}\,. \tag{2.2.10}$$

We shall normally use capitals in this book to denote random variables and small letters $\boldsymbol{x}$ to denote their values whenever it is necessary to make a distinction.

Very often, we have some quite different underlying probability space Ω with values ω, and talk about $\boldsymbol{X}(\omega)$ which is some function of ω, and then omit explicit mention of ω. This can be for either of two reasons:

i) we specify the events by the values of $\boldsymbol{x}$ anyway, i.e., we identify $\boldsymbol{x}$ and ω;
ii) the underlying events ω are too complicated to describe, or sometimes, even to know.

For example, in the case of the position of a molecule in a liquid, we really should interpret each ω as being capable of specifying all the positions, momenta, and orientations of each molecule in that volume of liquid; but this is simply too difficult to write down, and often unnecessary.

One great advantage of introducing the concept of a random variable is the simplicity with which one may handle functions of random variables, e.g., $\boldsymbol{X}^2$, $\sin(\boldsymbol{a}\cdot\boldsymbol{X})$, etc., and compute means and distributions of these. Further, by defining stochastic differential equations, one can also quite simply talk about time development of random variables in a way which is quite analogous to the classical description by means of differential equations of non-probabilistic systems.

2.3 Joint and Conditional Probabilities: Independence

2.3.1 Joint Probabilities

We explained in Sect. 2.2.3 how the occurrence of mutually exclusive events is related to the concept of nonintersecting sets. We now consider the concept $P(A \cap B)$, where $A \cap B$ is nonempty. An event ω which satisfies $\omega \in A$ will only satisfy $\omega \in A \cap B$ if $\omega \in B$ as well.

$$\text{Thus, } P(A \cap B) = P\{(\omega \in A) \text{ and } (\omega \in B)\}, \tag{2.3.1}$$

and $P(A \cap B)$ is called the *joint probability* that the event ω is contained in both classes, or, alternatively, that both the events $\omega \in A$ and $\omega \in B$ occur. Joint probabilities occur naturally in the context of this book in two ways:

i) *When the event is specified by a vector*, e.g., m mice and n tigers. The probability of this event is the joint probability of [m mice (and any number of tigers)] and [n tigers (and any number of mice)]. All vector specifications are implicitly joint probabilities in this sense.
ii) *When more than one time is considered*: what is the probability that (at time t_1 there are m_1 tigers and n_1 mice) and (at time t_2 there are m_2 tigers and n_2 mice). To consider such a probability, we have effectively created out of the events at time t_1 and events at time t_2, *joint events* involving one event at each time. In essence, there is no difference between these two cases except for the fundamental dynamical role of time.

2.3.2 Conditional Probabilities

We may specify conditions on the events we are interested in and consider only these, e.g., the probability of 21 buffaloes given that we know there are 100 lions. What does this mean? Clearly, we will be interested only in those events contained in the set B = {all events where exactly 100 lions occur}. This means that we to define conditional probabilities, which are defined only on the collection of all sets contained in B. we define the conditional probability as

$$P(A \mid B) = P(A \cap B)/P(B), \tag{2.3.2}$$

and this satisfies our intuitive conception that the conditional probability that $\omega \in A$ (given that we know $\omega \in B$), is given by dividing the probability of joint occurrence by the probability ($\omega \in B$).

We can define in both directions, i.e., we have

$$P(A \cap B) = P(A \mid B)P(B) = P(B \mid A)P(A). \tag{2.3.3}$$

There is no particular conceptual difference between, say, the probability of {(21 buffaloes) given (100 lions)} and the reversed concept. However, when two times are involved, we do see a difference. For example, the probability that a particle is at position $\boldsymbol{x}_1$ at time t_1, given that it was at $\boldsymbol{x}_2$ at the *previous* time t_2, is a very natural thing to consider; indeed, it will turn out to be a central concept in this book.

The converse looks to the past rather than the future; given that a particle is at x_1 at time t_1, what is the probability that that at the previous time t_2 it was at position x_2. The first concept—the *forward* probability—looks at where the particle will go, the second—the *backward* probability—at where it came from.

The forward probability has already occurred in this book, for example, the $\phi(\Delta)d\Delta$ of Einstein (Sect. 1.2.1) is the probability that a particle at x at time t will be in the range $[x+\Delta, x+\Delta+d\Delta]$ at time $t+\tau$, and similarly in the other examples. Our intuition tells us as it told Einstein (as can be seen by reading the extract from his paper) that this kind of conditional probability is directly related to the time development of a probabilistic system.

2.3.3 Relationship Between Joint Probabilities of Different Orders

Suppose we have a collection of sets B_i such that

$$B_i \cap B_j = \varnothing , \tag{2.3.4}$$

$$\bigcup_i B_i = \Omega , \tag{2.3.5}$$

so that the sets divide up the space Ω into nonoverlapping subsets.
Then

$$\bigcup_i (A \cap B_i) = A \cap \Big(\bigcup_i B_i\Big) = A \cap \Omega = A . \tag{2.3.6}$$

Using now the probability axiom iii), we see that $A \cap B_i$ satisfy the conditions on the A_i used there, so that

$$\sum_i P(A \cap B_i) = P\big(\bigcup_i (A \cap B_i)\big) , \tag{2.3.7}$$

$$= P(A) , \tag{2.3.8}$$

and thus

$$\sum_i P(A \,|\, B_i)P(B_i) = P(A) . \tag{2.3.9}$$

Thus, summing over all mutually exclusive possibilities of B in the joint probability eliminates that variable.

Hence, in general,

$$\sum_i P(A_i \cap B_j \cap C_k \ldots) = P(B_j \cap C_k \cap \ldots) . \tag{2.3.10}$$

The result (2.3.9) has very significant consequences in the development of the theory of stochastic processes, which depends heavily on joint probabilities.

2.3.4 Independence

We need a probabilistic way of specifying what we mean by independent events. Two sets of events A and B should represent independent sets of events if the specification that a particular event is contained in B has no influence on the probability of that event belonging to A. Thus, the conditional probability $P(A \,|\, B)$ should be independent of B, and hence

$$P(A \cap B) = P(A)P(B) . \qquad (2.3.11)$$

In the case of several events, we need a somewhat stronger specification. The events $(\omega \in A_i)(i = 1, 2, \ldots, n)$ will be considered to be independent if for any subset $(i_1, i_2, \ldots, i_k)$ of the set $(1, 2, \ldots, n)$,

$$P(A_{i_1} \cap A_{i_2} \ldots A_{i_k}) = P(A_{i_1})P(A_{i_2}) \ldots P(A_{i_k}) . \qquad (2.3.12)$$

It is important to require factorisation for all possible combinations, as in (2.3.12). For example, for three sets A_i, it is quite conceivable that

$$P(A_i \cap A_j) = P(A_i)P(A_j) , \qquad (2.3.13)$$

for all different i and j, but also that

$$A_1 \cap A_2 = A_2 \cap A_3 = A_3 \cap A_1 . \qquad \text{(see Fig. 2.1)} \qquad (2.3.14)$$

This requires

$$\begin{aligned} P(A_1 \cap A_2 \cap A_3) &= P(A_2 \cap A_3 \cap A_3) = P(A_2 \cap A_3) \\ &= P(A_2)P(A_3) \neq P(A_1)P(A_2)P(A_3) . \end{aligned} \qquad (2.3.15)$$

We can see that the occurrence of $\omega \in A_2$ and $\omega \in A_3$ necessarily implies the occurrence of $\omega \in A_1$. In this sense the events are obviously not independent.

Random variables $X_1, X_2, X_3, \ldots$, will be said to be independent random variables, if for all sets of the form $A_i = x$ such that $a_i \leqslant x \leqslant b_i)$ the events $X_1 \in A_1, X_2 \in A_2, X_3 \in A_3, \ldots$ are independent events. This will mean that all values of the X_i are assumed independently of those of the remaining X_i.

2.4 Mean Values and Probability Density

The mean value (or *expectation*) of a random variable $R(\omega)$ in which the basic events ω are countably specifiable is given by

$$\langle R \rangle = \sum_{\omega} P(\omega)R(\omega) , \qquad (2.4.1)$$

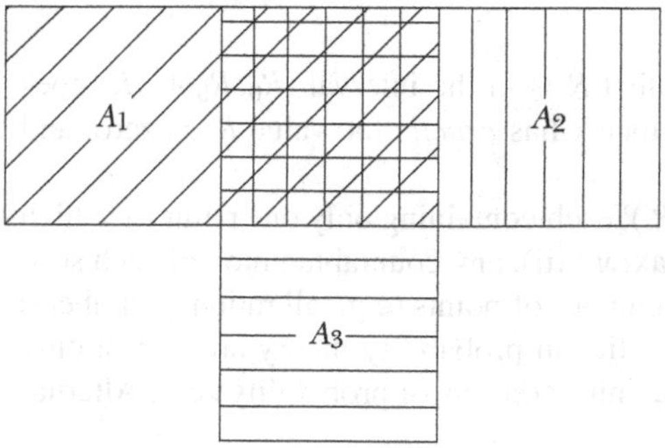

Fig. 2.1. Illustration of statistical independence in pairs, but not in threes. In the three sets $A_j \cap A_i$ is, in all cases, the central region. By appropriate choice of probabilities, it can be arranged that $P(A_i \cap A_j) = P(A_i)P(A_j)$.

where $P(\omega)$ means the probability of the set containing only the single event ω. In the case of a continuous variable, the probability axioms above enable us to define a probability density function $p(\omega)$ such that if $A(\omega_0, d\omega_0)$ is the set

$$(\omega_0 \leqslant \omega < \omega_0 + d\omega_0), \tag{2.4.2}$$

then

$$p(\omega_0)d\omega_0 = P[A(\omega_0, d\omega_0)] \equiv p(\omega_0, d\omega_0). \tag{2.4.3}$$

The last is a notation often used by mathematicians. Details of how this is done have been nicely explained by *Feller* [2.1]. In this case,

$$\langle R \rangle = \int_{\omega \in \Omega} d\omega\, R(\omega) p(\omega). \tag{2.4.4}$$

One can often (as mentioned in Sect. 2.2.4) use R itself to specify the event, so we will often write

$$\langle R \rangle = \int dR\, R\, p(R). \tag{2.4.5}$$

Obviously, $p(R)$ is not the same function of R as $p(\omega)$ is of ω—more precisely

$$p(R_0)\, dR_0 = P\,(R_0 < R < R_0 + dR_0). \tag{2.4.6}$$

2.4.1 Determination of Probability Density by Means of Arbitrary Functions

Suppose for every function $f(R)$ we know

$$\langle f(R) \rangle = \int dR\, f(R) p(R), \tag{2.4.7}$$

then we know $p(R)$, which is known as a *probability density*. The proof follows by choosing

$$f(R) = \begin{cases} 1 & R_0 \leqslant R < R_0 + dR_0, \\ 0 & \text{otherwise}. \end{cases} \tag{2.4.8}$$

Because the expectation of an arbitrary function is sometimes a little easier to work with than a density, this relation will be used occasionally in this book.

Notation: The notation $\langle A \rangle$ for the expectation used in this book is a physicist's notation. The most common mathematical notation is $E(A)$, which is in my opinion a little less intuitive.

2.4.2 Sets of Probability Zero

If a density $p(R)$ exists, the probability that R is in the interval $(R_0, R_0 + dR)$ goes to zero with dR. Hence, the probability that R has *exactly* the value R_0 is zero; and similarly for any other value.

Thus, in such a case, there are sets $S(R_i)$, each containing only one point R_i, which have zero probability. From probability axiom iii), any countable union of such sets, i.e., any set containing only a countable number of points (e.g., all rational numbers) has probability zero. In general, all equalities in probability theory are at best only "*almost certainly true*", i.e., they may be untrue on sets of probability zero. Alternatively, one says, for example,

$$X = Y \text{ with probability } 1\,, \tag{2.4.9}$$

which is by no means the same as saying that

$$X(R) = Y(R) \text{ for all } R\,. \tag{2.4.10}$$

Of course, if the theory is to have any connection with reality, events with probability zero do not occur.

In particular, notice that our previous result if inspected carefully, only implies that we know $p(R)$ only with probability 1, given that we know $\langle f(R)\rangle$ for all $f(R)$.

2.5 The Interpretation of Mean Values

The question of what to measure in a probabilistic system is nontrivial. In practice, one measures either a set of individual values of a random variable (the number of animals of a certain kind in a certain region at certain points in time; the electric current passing through a given circuit element in each of a large number of replicas of that circuit, etc.) or alternatively, the measuring procedure may implicitly construct an average of some kind. For example, to measure an electric current, we may measure the electric charge transferred and divide by the time taken—this gives a measure of the average number of electrons transferred per unit time. It is important to note the essential difference in this case, that it will not normally be possible to measure anything other than a few selected averages and thus, higher moments (for example) will be unavailable.

In contrast, when we measure individual events (as in counting animals), we can then construct averages of the observables by the obvious method

$$\bar{X}_N = \frac{1}{N}\sum_{n=1}^{N} X(n)\,. \tag{2.5.1}$$

The quantities $X(n)$ are the individual observed values of the quantity X. We expect that as the number of samples N becomes very large, the quantity $\bar{X}_N$ approaches the mean $\langle X\rangle$ and that, in fact,

$$\lim_{N\to\infty}\frac{1}{N}\sum_{n=1}^{N} f[X(n)] = \lim_{N\to\infty}\overline{f(X)}_N = \langle f(X)\rangle \tag{2.5.2}$$

and such a procedure will determine the probability density function $p(x)$ of X if we carry out this procedure for all functions f. The validity of this procedure depends on the degree of independence of the successive measurements and is dealt with in Sect. 2.5.2.

In the case where only averages themselves are directly determined by the measuring method, it will not normally be possible to measure $X(n)$ and therefore, it will not, in general, be possible to determine $\overline{f(X)}_N$. All that will be available will be $f(\bar{X}_N)$—quite a different thing unless f is linear. We can often find situations in which measurable quantities are related (by means of some theory) to mean values of certain functions, but to hope to measure, for example, the mean value of an arbitrary function of the number of electrons in a conductor is quite hopeless. The mean number—yes, and indeed even the mean square number, but the measuring methods available are not direct. We do *not* enumerate the individual numbers of electrons at different times and hence arbitrary functions are not attainable.

2.5.1 Moments, Correlations, and Covariances

Quantities of interest are given by the *moments* $\langle X^n \rangle$ since these are often easily calculated. However, probability densities must always vanish as $x \to \pm\infty$, so we see that higher moments tell us only about the properties of unlikely large values of X. In practice we find that the most important quantities are related to the first and second moments. In particular, for a single variable X, *the variance* defined by

$$\mathrm{var}[X] \equiv \{\sigma[X]\}^2 \equiv \langle [X - \langle X \rangle]^2 \rangle \,, \tag{2.5.3}$$

and as is well known, the *variance* var$[X]$ or its square root the *standard deviation* $\sigma[X]$, is a measure of the degree to which the values of X deviate from the mean value $\langle X \rangle$.

In the case of several variables, we define the *covariance matrix* as

$$\langle X_i, X_j \rangle \equiv \langle (X_i - \langle X_i \rangle)(X_j - \langle X_j \rangle) \rangle \equiv \langle X_i X_j \rangle - \langle X_i \rangle \langle X_j \rangle \,. \tag{2.5.4}$$

Obviously,

$$\langle X_i, X_i \rangle = \mathrm{var}[X_i]. \tag{2.5.5}$$

If the variables are independent *in pairs*, the covariance matrix is diagonal.

2.5.2 The Law of Large Numbers

As an application of the previous concepts, let us investigate the following model of measurement. We assume that we measure the same quantity N times, obtaining sample values of the random variable $X(n)$; $(n = 1, 2, \ldots, N)$. Since these are all measurements of the same quantity at successive times, we assume that for every $n, X(n)$ has the same probability distribution but we do not assume the $X(n)$ to be independent. However, provided the covariance matrix $\langle X(n), X(m) \rangle$ vanishes sufficiently rapidly as $|n - m| \to \infty$, then defining

$$\bar{X}_N = \frac{1}{N} \sum_{n=1}^{N} X(n) \,, \tag{2.5.6}$$

we shall show

$$\lim_{N \to \infty} \bar{X}_N = \langle X \rangle \,. \tag{2.5.7}$$

It is clear that

$$\langle \bar{X}_N \rangle = \langle X \rangle \,. \tag{2.5.8}$$

We now calculate the variance of $\bar{X}_N$ and show that as $N \to \infty$ it vanishes under certain conditions:

$$\langle \bar{X}_N \bar{X}_N \rangle - \langle \bar{X}_N \rangle^2 = \frac{1}{N^2} \sum_{n,m=1}^{N} \langle X_n, X_m \rangle \,. \tag{2.5.9}$$

Provided $\langle X_n, X_m \rangle$ falls off sufficiently rapidly as $|n - m| \to \infty$, we find

$$\lim_{N\to\infty} (\mathrm{var}[\bar{X}_N]) = 0 \,, \tag{2.5.10}$$

so that $\lim_{N\to\infty} \bar{X}_N$ is a deterministic variable equal to $\langle X \rangle$.

Two models of $\langle X_n, X_m \rangle$ can be chosen.

a) $\langle X_n, X_m \rangle \sim K\lambda^{|m-n|}, \qquad (\lambda < 1)\,,$ (2.5.11)

for which one finds

$$\mathrm{var}[\bar{X}_N] = \frac{2K}{N^2}\left(\frac{\lambda^{N+2} - N(\lambda - 1) - \lambda}{(\lambda - 1)^2}\right) - \frac{K}{N} \to 0\,. \tag{2.5.12}$$

b) $\langle X_n, X_m \rangle \sim |n - m|^{-1}, \qquad (n \neq m)\,,$ (2.5.13)

and one finds approximately

$$\mathrm{var}[\bar{X}_N] \sim \frac{2}{N}\log N - \frac{1}{N} \to 0\,. \tag{2.5.14}$$

In both these cases, $\mathrm{var}[X_N] \to 0$, but the rate of convergence is very different. Interpreting n, m as the times at which the measurement is carried out, one sees than even very slowly decaying correlations are permissible. The law of large numbers comes in many forms, which are nicely summarised by *Papoulis* [2.2]. The central limit theorem is an even more precise result in which the limiting distribution function of $\bar{X}_N - \langle X \rangle$ is determined (see Sect. 2.8.2).

2.6 Characteristic Function

One would like a condition where the variables are independent, not just in pairs. To this end (and others) we define the characteristic function.

If $\boldsymbol{s}$ is the vector $(s_1, s_2, \ldots, s_n)$, and $\boldsymbol{X} = (X_1, X_2, \ldots, X_n)$ is a vector of random variables, then the characteristic function (or moment generating function) is defined by

$$\phi(\boldsymbol{s}) = \langle \exp(\mathrm{i}\boldsymbol{s}\cdot\boldsymbol{X}) \rangle = \int d\boldsymbol{x}\; p(\boldsymbol{x}) \exp(\mathrm{i}\boldsymbol{s}\cdot\boldsymbol{x})\,. \tag{2.6.1}$$

The characteristic function has the following properties ([2.1], Chap. XV)

i) $\phi(\boldsymbol{0}) = 1$.

ii) $|\phi(\boldsymbol{s})| \leqslant 1$.

iii) $\phi(\boldsymbol{s})$ is a uniformly continuous function of its arguments for all finite real $\boldsymbol{s}$ [2.4].

iv) If the *moments* $\langle \prod_i X_i^{m_i} \rangle$ exist, then

$$\left\langle \prod_i X_i^{m_i} \right\rangle = \left[\prod_i \left(-\mathrm{i}\frac{\partial}{\partial s_i}\right)^{m_i} \phi(\boldsymbol{s})\right]_{\boldsymbol{s}=\boldsymbol{0}} . \tag{2.6.2}$$

v) A sequence of probability densities converges to limiting probability density if and only if the corresponding characteristic functions converge to the corresponding characteristic function of the limiting probability density.

vi) Fourier inversion formula

$$p(\boldsymbol{x}) = (2\pi)^{-n} \int d\boldsymbol{s}\, \phi(\boldsymbol{s}) \exp(-\mathrm{i}\boldsymbol{x} \cdot \boldsymbol{s})\,. \tag{2.6.3}$$

Because of this inversion formula, $\phi(\boldsymbol{s})$ determines $p(\boldsymbol{x})$ with probability 1. Hence, the characteristic function does truly *characterise* the probability density.

vii) Independent random variables: from the definition of independent random variables in Sect. 2.3.4, it follows that the variables $X_1, X_2 \dots$ are independent if and only if

$$p(x_1, x_2, \dots, x_n) = p_1(x_1)p_2(x_2)\dots p_n(x_n)\,, \tag{2.6.4}$$

in which case,

$$\phi(s_1, s_2, \dots s_n) = \phi_1(s_1)\phi_2(s_2)\dots\phi_n(s_n). \tag{2.6.5}$$

viii) Sum of independent random variables: if $X_1, X_2, \dots,$ are independent random variables and if

$$Y = \sum_{i=1}^{n} X_i\,, \tag{2.6.6}$$

and the characteristic function of Y is

$$\phi_y(s) = \langle \exp(\mathrm{i}sY) \rangle\,, \tag{2.6.7}$$

then

$$\phi_y(s) = \prod_{i=1}^{n} \phi_i(s)\,. \tag{2.6.8}$$

The characteristic function plays an important role in this book which arises from the convergence property (v), which allows us to perform limiting processes on the characteristic function rather than the probability distribution itself, and often makes proofs easier. Further, the fact that the characteristic function is truly characteristic, i.e., the inversion formula (vi), shows that different characteristic functions arise from different distributions. As well as this, the straightforward derivation of the moments by (2.6.2) makes any determination of the characteristic function directly relevant to measurable quantities.

2.7 Cumulant Generating Function: Correlation Functions and Cumulants

A further important property of the characteristic function arises by considering its logarithm

$$\Phi(\boldsymbol{s}) = \log \phi(\boldsymbol{s})\,, \tag{2.7.1}$$

which is called the *cumulant generating function*. Let us assume that all moments exist so that $\phi(s)$ and hence, $\Phi(s)$, is expandable in a power series which can be written as

$$\Phi(s) = \sum_{r=1}^{\infty} \mathrm{i}^r \sum_{\{m\}} \langle\langle X_1^{m_1} X_2^{m_2} \dots X_n^{m_n} \rangle\rangle \frac{s_1^{m_1} s_2^{m_2} \dots s_n^{m_n}}{m_1! m_2! \dots m_n!} \, \delta\left(r, \sum_{i=1}^{n} m_i\right), \tag{2.7.2}$$

where the quantities $\langle\langle X_1^{m_1} X_2^{m2} \dots X_n^{m_n} \rangle\rangle$ are called the *cumulants* of the variables $\boldsymbol{X}$. The notation chosen should not be taken to mean that the cumulants are functions of the particular product of powers of the $\boldsymbol{X}$; it rather indicates the moment of highest order which occurs in their expression in terms of moments. *Stratonovich* [2.5] also uses the term *correlation functions*, a term which we shall reserve for cumulants which involve more than one X_i. For, if the $\boldsymbol{X}$ are all independent, the factorisation property (2.6.6) implies that $\Phi(s)$ (the cumulant generating function) is a sum of n terms, each of which is a function of only one s_i and hence the coefficient of mixed terms, i.e., the *correlation functions* (in our terminology) are all zero and the converse is also true. Thus, the magnitude of the correlation functions is a measure of the degree of correlation.

The cumulants and correlation functions can be evaluated in terms of moments by expanding the characteristic function as a power series:

$$\phi(s) = \sum_{r=1}^{\infty} \frac{\mathrm{i}^r}{r!} \sum_{\{m\}} \langle X_1^{m_1} X_2^{m_2} \dots X_n^{m_n} \rangle \frac{r!}{m_1! m_2! \dots m_n!} \, \delta\left(r, \sum_{i=1}^{n} m_i\right) s_1^{m_1} s_2^{m_2} \dots s_n^{m_n} \, . \tag{2.7.3}$$

Expanding the logarithm in a power series, and comparing it with (2.7.2) for $\Phi(s)$, the relationship between the cumulants and the moments can be deduced. No *simple* formula can be given, but the first few cumulants can be exhibited: we find

$$\langle\langle X_i \rangle\rangle = \langle X_i \rangle \, , \tag{2.7.4}$$

$$\langle\langle X_i X_j \rangle\rangle = \langle X_i X_j \rangle - \langle X_i \rangle \langle X_j \rangle \, , \tag{2.7.5}$$

$$\langle\langle X_i X_j X_k \rangle\rangle = \langle X_i X_j X_k \rangle - \langle X_i X_j \rangle \langle X_k \rangle - \langle X_i \rangle \langle X_j X_k \rangle - \langle X_i X_k \rangle \langle X_j \rangle + 2 \langle X_i \rangle \langle X_j \rangle \langle X_k \rangle \, . \tag{2.7.6}$$

Here, all formulae are also valid for any number of equal i, j, k, l. An explicit general formula can be given as follows. Suppose we wish to calculate the cumulant $\langle\langle X_1 X_2 X_3 \dots X_n \rangle\rangle$. The procedure is the following:

i) Write a sequence of n dots ...... ;
ii) Divide into $p + 1$ subsets by inserting angle brackets

$$\langle \dots \rangle \langle . . \rangle \langle \dots \dots \rangle . . \langle . . \rangle \, ; \tag{2.7.7}$$

iii) Distribute the symbols $X_1 \dots X_n$ in place of the dots in such a way that all *different* expressions of this kind occur, e.g.,

$$\langle X_1 \rangle \langle X_2 X_3 \rangle = \langle X_1 \rangle \langle X_3 X_2 \rangle \neq \langle X_3 \rangle \langle X_1 X_2 \rangle \, ; \tag{2.7.8}$$

iv) Take the sum of all such terms for a given p. Call this $C_p(X_1, X_2, \dots, X_n)$;

v) $$\langle\langle X_1X_2\ldots X_n\rangle\rangle = \sum_{p=0}^{n-1}(-1)^p p! C_p(X_1, X_2, \ldots, X_n)\,. \tag{2.7.9}$$

A derivation of this formula was given by *Meeron* [2.6]. The particular procedure is due to *van Kampen* [2.7].

vi) *Cumulants in which there is one or more repeated element*:
For example $\langle\langle X_1^2 X_3 X_2\rangle\rangle$—simply evaluate $\langle\langle X_1X_2X_3X_4\rangle\rangle$ and set $X_4 = X_1$ in the resulting expression.

2.7.1 Example: Cumulant of Order 4: $\langle\langle X_1X_2X_3X_4\rangle\rangle$

a) $p = 0$

Only term is $\langle X_1X_2X_3X_4\rangle = C_0(X_1X_2X_3X_4)$.

b) $p = 1$

Partition $\langle.\rangle\langle\ldots\rangle$
Term $\{\langle X_1\rangle\langle X_2X_3X_4\rangle + \langle X_2\rangle\langle X_3X_4X_1\rangle + \langle X_3\rangle\langle X_4X_1X_2\rangle$
$+\langle X_4\rangle\langle X_1X_2X_3\rangle\} \equiv D_1$

partition $\langle..\rangle\langle..\rangle$
Term $\langle X_1X_2\rangle\langle X_3X_4\rangle + \langle X_1X_3\rangle\langle X_2X_4\rangle + \langle X_1X_4\rangle\langle X_2X_3\rangle \equiv D_2\,.$

Hence,

$$D_1 + D_2 = C_1(X_1X_2X_3X_4)\,. \tag{2.7.10}$$

c) $p = 2$

Partition $\langle.\rangle\langle.\rangle\langle..\rangle$
Term $\langle X_1\rangle\langle X_2\rangle\langle X_3X_4\rangle + \langle X_1\rangle\langle X_3\rangle\langle X_2X_4\rangle + \langle X_1\rangle\langle X_4\rangle\langle X_2X_3\rangle$
$+\langle X_2\rangle\langle X_3\rangle\langle X_1X_4\rangle + \langle X_2\rangle\langle X_4\rangle\langle X_1X_3\rangle + \langle X_3\rangle\langle X_4\rangle\langle X_1X_2\rangle$
$= C_2(X_1X_2X_3X_4)$.

d) $p = 3$

Partition $\langle.\rangle\langle.\rangle\langle.\rangle\langle.\rangle$
Term $\langle X_1\rangle\langle X_2\rangle\langle X_3\rangle\langle X_4\rangle = C_3(X_1X_2X_3X_4)\,.$

Hence,

$$\langle\langle X_1X_2X_3X_4\rangle\rangle = C_0 - C_1 + 2C_2 - 6C_3\,. \tag{2.7.11}$$

2.7.2 Significance of Cumulants

From (2.7.4, 2.7.5) we see that the first two cumulants are the means $\langle X_i\rangle$ and covariances $\langle X_i, X_j\rangle$. Higher-order cumulants contain information of decreasing significance, unlike higher-order moments. We cannot set all *moments* higher than a certain order equal to zero since $\langle X^{2n}\rangle \geqslant \langle X^n\rangle^2$ and thus, all moments contain information about lower moments.

For cumulants, however, we can consistently set

$$\langle\langle X \rangle\rangle = a\,,$$
$$\langle\langle X^2 \rangle\rangle = \sigma^2\,,$$
$$\langle\langle X^n \rangle\rangle = 0\,, \qquad (n > 2)\,,$$

and we can easily deduce by using the inversion formula for the characteristic function that

$$p(x) = \frac{1}{\sigma\sqrt{2\pi}} \exp\left(-\frac{(x-a)^2}{2\sigma^2}\right), \tag{2.7.12}$$

that is, a Gaussian probability distribution. It does not, however, seem possible to give more than this intuitive justification. Indeed, the theorem of *Marcinkiewicz* [2.8, 2.9] shows that the cumulant generating function cannot be a polynomial of degree greater than 2, that is, either all but the first 2 cumulants vanish or there are an infinite number of nonvanishing cumulants. The greatest significance of cumulants lies in the definition of the correlation functions of different variables in terms of them; this leads further to important approximation methods.

2.8 Gaussian and Poissonian Probability Distributions

2.8.1 The Gaussian Distribution

By far the most important probability distribution is the Gaussian, or normal distribution. Here we collect together the most important facts about it.

If $\boldsymbol{X}$ is a vector of n Gaussian random variables, the corresponding multivariate probability density function can be written

$$p(\boldsymbol{x}) = \frac{1}{\sqrt{(2\pi)^n \det(\sigma)}} \exp\left[-\tfrac{1}{2}(\boldsymbol{x}-\bar{\boldsymbol{x}})^{\mathrm{T}}\sigma^{-1}(\boldsymbol{x}-\bar{\boldsymbol{x}})\right], \tag{2.8.1}$$

so that

$$\langle \boldsymbol{X} \rangle = \int d\boldsymbol{x}\, \boldsymbol{x}\, p(\boldsymbol{x}) = \bar{\boldsymbol{x}}, \tag{2.8.2}$$
$$\langle \boldsymbol{X}\boldsymbol{X}^{\mathrm{T}} \rangle = \int d\boldsymbol{x}\, \boldsymbol{x}\boldsymbol{x}^{\mathrm{T}} p(\boldsymbol{x}) = \bar{\boldsymbol{x}}\bar{\boldsymbol{x}}^{\mathrm{T}} + \sigma, \tag{2.8.3}$$

and the characteristic function is given by

$$\phi(\boldsymbol{s}) = \langle \exp(\mathrm{i}\boldsymbol{s}^{\mathrm{T}}\boldsymbol{X}) \rangle = \exp\left(\mathrm{i}\boldsymbol{s}^{\mathrm{T}}\bar{\boldsymbol{x}} - \tfrac{1}{2}\boldsymbol{s}^{\mathrm{T}}\sigma\boldsymbol{s}\right). \tag{2.8.4}$$

This particularly simple characteristic function implies that all cumulants of higher order than 2 vanish, and hence means that all moments of order higher than 2 are expressible in terms of those of order 1 and 2. The relationship (2.8.3) means that σ is the covariance matrix (as defined in Sect. 2.5.1), i.e., the matrix whose elements are the second-order correlation functions. Of course, σ is symmetric.

The precise relationship between the higher moments and the covariance matrix σ can be written down straightforwardly by using the relationship between the moments and the characteristic function [Sect. 2.6 iv)]. The formula is only simple if $\bar{\boldsymbol{x}} = 0$, in which case the odd moments vanish and the even moments satisfy

$$\langle X_i X_j X_k \ldots \rangle = \frac{(2N)!}{N!2^N} \{\sigma_{ij}\sigma_{kl}\sigma_{mn} \ldots\}_{\text{sym}} , \tag{2.8.5}$$

where the subscript "sym" means the symmetrised form of the product of σ's, and $2N$ is the order of the moment. For example,

$$\begin{aligned} \langle X_1 X_2 X_3 X_4 \rangle &= \frac{4!}{4.2!} \left\{ \frac{1}{3} [\sigma_{12}\sigma_{34} + \sigma_{41}\sigma_{23} + \sigma_{13}\sigma_{24}] \right\} , \\ &= \sigma_{12}\sigma_{34} + \sigma_{41}\sigma_{23} + \sigma_{13}\sigma_{24} , \end{aligned} \tag{2.8.6}$$

$$\langle X_1^4 \rangle = \frac{4!}{4.2!} \left\{ \sigma_{11}^2 \right\} = 3\sigma_{11}^2 . \tag{2.8.7}$$

2.8.2 Central Limit Theorem

The Gaussian distribution is important for a variety of reasons. Many variables are, in practice, empirically well approximated by Gaussians and the reason for this arises from the *central limit theorem*, which, roughly speaking, asserts that a random variable composed of the sum of many parts, each independent but arbitrarily distributed, is Gaussian. More precisely, let $X_1, X_2, X_3, \ldots, X_n$ be independent random variables such that

$$\langle X_i \rangle = 0, \qquad \text{var}[X_i] = b_i^2, \tag{2.8.8}$$

and let the distribution function of X_i be $p_i(x_i)$.

Define

$$S_n = \sum_{i=1}^{n} X_i , \tag{2.8.9}$$

and

$$\sigma_n^2 = \text{var}[S_n] = \sum_{i=1}^{n} b_i^2 . \tag{2.8.10}$$

We require further the fulfilment of the *Lindeberg condition*:

$$\lim_{n\to\infty} \left[\frac{1}{\sigma_n^2} \sum_{i=1}^{n} \int_{|x|>t\sigma_n} dx\, x^2 \, p_i(x) \right] = 0, \tag{2.8.11}$$

for any fixed $t > 0$. Then, under these conditions, the distribution of the normalised sums S_n/σ_n tends to the Gaussian with zero mean and unit variance.

The proof of the theorem can be found in [2.1]. It is worthwhile commenting on the hypotheses, however. We first note that the summands X_i are required to be independent. This condition is not absolutely necessary; for example, choose

$$X_i = \sum_{r=i}^{i+j} Y_r, \tag{2.8.12}$$

where the Y_j are independent. Since the sum of the X's can be rewritten as a sum of Y's (with certain finite coefficients), the theorem is still true.

Roughly speaking, as long as the correlation between X_i and X_j goes to zero sufficiently rapidly as $|i-j| \to \infty$, a central limit theorem will be expected. The Lindeberg condition (2.8.11) is not an obviously understandable condition but is the weakest condition which expresses the requirement that the probability for $|X_i|$ to be large is very small. For example, if all the b_i are infinite or greater than some constant C, it is clear that σ_n^2 diverges as $n \to \infty$. The sum of integrals in (2.8.11) is the sum of contributions to variances for all $|X_i| > t\sigma_n$, and it is clear that as $n \to \infty$, each contribution goes to zero. The Lindeberg condition requires the sum of all the contributions not to diverge as fast as σ_n^2. In practice, it is a rather weak requirement; satisfied if $|X_i| < C$ for all X_i, or if $p_i(x)$ go to zero sufficiently rapidly as $x \to \pm\infty$. An exception is

$$p_i(x) = \frac{a_i}{\pi(x^2 + a_i^2)}, \tag{2.8.13}$$

the *Cauchy*, or *Lorentzian* distribution. The variance of this distribution is infinite and, in fact, the sum of all the X_i has a distribution of the same form as (2.8.13) with a_i replaced by $\sum_{i=1}^{n} a_i$. Obviously, the Lindeberg condition is not satisfied.

A related condition, also called the Lindeberg condition, will arise in Sect. 3.3.1, where we discuss the replacement of a discrete process by one with continuous steps.

2.8.3 The Poisson Distribution

A distribution which plays a central role in the study of random variables which take on positive integer values is the Poisson distribution. If X is the relevant variable the Poisson distribution is defined by

$$P(X = x) \equiv P(x) = \frac{e^{-\alpha}\alpha^x}{x!}, \tag{2.8.14}$$

and clearly, the *factorial moments*, defined by

$$\langle X^r \rangle_f = \langle x(x-1)\dots(x-r+1)\rangle, \tag{2.8.15}$$

are given by

$$\langle X^r \rangle_f = \alpha^r. \tag{2.8.16}$$

For variables whose range is nonnegative integral, we can very naturally define the *generating function*

$$G(s) = \sum_{x=0}^{\infty} s^x P(x) = \langle s^x \rangle, \tag{2.8.17}$$

which is related to the characteristic function by

$$G(s) = \phi(-\mathrm{i}\log s). \tag{2.8.18}$$

The generating function has the useful property that

$$\langle X^r \rangle_f = \left[\left(\frac{\partial}{\partial s}\right)^r G(s)\right]_{s=1}. \tag{2.8.19}$$

For the Poisson distribution we have

$$G(s) = \sum_{x=0}^{\infty} \frac{e^{-\alpha}(s\alpha)^x}{x!} = \exp[\alpha(s-1)] \,. \tag{2.8.20}$$

We may also define the factorial cumulant generating function $g(s)$ by

$$g(s) = \log G(s) \tag{2.8.21}$$

and the *factorial cumulants* $\langle\langle X^r \rangle\rangle_f$ by

$$g(s) = \sum_{x=1}^{\infty} \langle\langle X^r \rangle\rangle_f \frac{(s-1)^r}{r!} \,. \tag{2.8.22}$$

We see that the Poisson distribution has all but the first factorial cumulant zero.

The Poisson distribution arises naturally in very many contexts, for example, we have already met it in Sect. 1.5.1 as the solution of a simple master equation. It plays a similar central role in the study of random variables which take on integer values to that occupied by the Gaussian distribution in the study of variables with a continuous range. However, the only simple multivariate generalisation of the Poisson is simply a product of Poissons, i.e., of the form

$$P(x_1, x_2, x_3, \dots) = \prod_{i=1}^{n} \frac{e^{-\alpha_i}(\alpha_i)^{x_i}}{x_i!} \,. \tag{2.8.23}$$

There is no logical concept of a correlated multipoissonian distribution, similar to that of a correlated multivariate Gaussian distribution.

2.9 Limits of Sequences of Random Variables

Much of computational work consists of determining *approximations* to random variables, in which the concept of a *limit of a sequence of random variables* naturally arises. However, there is no unique way of defining such a limit.

For, suppose we have a probability space Ω, and a sequence of random variables X_n defined on Ω. Then by the limit of the sequence as $n \to \infty$

$$X = \lim_{n\to\infty} X_n \,, \tag{2.9.1}$$

we mean a random variable X which, in some sense, is approached by the sequence of random variables X_n. The various possibilities arise when one considers that the probability space Ω has elements ω which have a probability density $p(\omega)$. Then we can choose the following definitions.

2.9.1 Almost Certain Limit

X_n converges *almost certainly* to X if, for all ω except a set of probability zero

$$\lim_{n\to\infty} X_n(\omega) = X(\omega) \,. \tag{2.9.2}$$

Thus each realisation of X_n converges to X and we write

$$\text{ac-}\lim_{n\to\infty} X_n = X. \tag{2.9.3}$$

2.9.2 Mean Square Limit (Limit in the Mean)

Another possibility is to regard the $X_n(\omega)$ as functions of ω, and look for the mean square deviation of $X_n(\omega)$ from $X(\omega)$. Thus, we say that X_n converges to X in the *mean square* if

$$\lim_{n\to\infty} \int d\omega\, p(\omega)[X_n(\omega) - X(\omega)]^2 \equiv \lim_{n\to\infty} \langle (X_n - X)^2 \rangle = 0. \tag{2.9.4}$$

This is the kind of limit which is well known in Hilbert space theory. We write

$$\operatorname*{ms-lim}_{n\to\infty} X_n = X\,. \tag{2.9.5}$$

2.9.3 Stochastic Limit, or Limit in Probability

We can consider the possibility that $X_n(\omega)$ approaches X because the probability of deviation from X approaches zero: precisely, this means that if for any $\varepsilon > 0$

$$\lim_{n\to\infty} P(|X_n - X| > \varepsilon) = 0\,, \tag{2.9.6}$$

then the *stochastic limit* of X_n is X.

In this case, we write

$$\operatorname*{st-lim}_{n\to\infty} X_n = X\,. \tag{2.9.7}$$

2.9.4 Limit in Distribution

An even weaker form of convergence occurs if, for any continuous bounded function $f(x)$

$$\lim_{n\to\infty} \langle f(X_n) \rangle = \langle f(X) \rangle\,. \tag{2.9.8}$$

In this case the convergence of the limit is said to be *in distribution*. In particular, using $\exp(ixs)$ for $f(x)$, we find that the characteristic functions approach each other, and hence the probability density of X_n approaches that of X.

2.9.5 Relationship Between Limits

The following relations can be shown.

1) Almost certain convergence $\Longrightarrow$ stochastic convergence.
2) Convergence in mean square $\Longrightarrow$ stochastic convergence.
3) Stochastic convergence $\Longrightarrow$ convergence in distribution.

All of these limits have uses in applications.

3. Markov Processes

3.1 Stochastic Processes

All of the examples given in Chap. 1 can be mathematically described as *stochastic processes* by which we mean, in a loose sense, systems which evolve probabilistically in time or more precisely, systems in which a certain time-dependent random variable $X(t)$ exists. We can measure values $x_1, x_2, x_3, \ldots$, etc., of $X(t)$ at times $t_1, t_2, t_3, \ldots$ and we assume that a set of joint probability densities exists

$$p(x_1, t_1; x_2, t_2; x_3, t_3; \ldots), \tag{3.1.1}$$

which describe the system completely.

In terms of these joint probability density functions, one can also define conditional probability densities:

$$p(x_1, t_1, ; x_2, t_2; \ldots | y_1, \tau_1; y_2, \tau_2; \ldots) \equiv \frac{p(x_1, t_1; x_2, t_2; \ldots; y_1, \tau_1; y_2, \tau_2; \ldots)}{p(y_1, \tau_1; y_2, \tau_2; \ldots)}. \tag{3.1.2}$$

These definitions are valid independently of the ordering of the times, although it is usual to consider only times which increase from right to left i.e.,

$$t_1 \geqslant t_2 \geqslant t_3 \geqslant \cdots \geqslant \tau_1 \geqslant \tau_2 \geqslant \ldots. \tag{3.1.3}$$

The concept of an evolution equation leads us to consider the conditional probabilities as predictions of the future values of $X(t)$ (i.e., $x_1, x_2, \ldots$ at times $t_1, t_2, \ldots,$) given the knowledge of the past (values $y_1, y_2, \ldots,$ at times $\tau_1, \tau_2, \ldots.$)

3.1.1 Kinds of Stochastic Process

The concept of a general stochastic process is very loose. To define the process we need to know at least all possible joint probabilities of the kind in (3.1.1). If such knowledge does define the process, it is known as a *separable stochastic process*. All the processes considered in this book will be assumed to be separable.

a) Complete Independence: This is the most simple kind of stochastic process; it satisfies the property

$$p(x_1, t_1; x_2, t_2; x_3, t_3; \ldots) = \prod_i p(x_i, t_i), \tag{3.1.4}$$

which means that the value of X at time t is completely independent of its values in the past (or future).

b) Bernoulli Trials: An even more special case occurs when the $p(\boldsymbol{x}_i, t_i)$ are independent of t_i, so that the same probability law governs the process at all times. We then have the *Bernoulli trials*, in which a probabilistic process is repeated at successive times.

c) Martingales: The conditional mean value of $X(t)$ given that $X(t_0) = x_0$ is defined as

$$\langle X(t) \| [x_0, t_0] \rangle \equiv \int dx\, x\, p(x, t \,|\, x_0, t_0) \,. \tag{3.1.5}$$

In a martingale this has the simple property

$$\langle X(t) \| [x_0, t_0] \rangle = x_0 \,. \tag{3.1.6}$$

The martingale property is rather strong property, and is associate with many similar and related processes, such as local martingales, sub-martingales, super-martingales etc., which have come to be extensively studied and used in the past 25 years. *Protter* [3.1] has written the definitive book on their use in stochastic processes.

d) Markov Processes: The next most simple idea is that of the *Markov process* in which knowledge of only the present determines the future, and most of this book is built around this concept.

3.2 Markov Process

The *Markov assumption* is formulated in terms of the conditional probabilities. We require that if the times satisfy the ordering (3.1.3), the conditional probability is determined entirely by the knowledge of the most recent condition, i.e.,

$$p(\boldsymbol{x}_1, t_1; \boldsymbol{x}_2, t_2; \dots \,|\, \boldsymbol{y}_1, \tau_1; \boldsymbol{y}_2, \tau_2; \dots) = p(\boldsymbol{x}_1, t_1; \boldsymbol{x}_2, t_2; \dots \,|\, \boldsymbol{y}_1, \tau_1) \,. \tag{3.2.1}$$

This is simply a more precise statement of the assumptions made by Einstein, Smoluchowski and others. It is, even by itself, extremely powerful. For it means that we can define everything in terms of the simple conditional probabilities $p(\boldsymbol{x}_1, t_1 \,|\, \boldsymbol{y}_1, \tau_1)$. For example, by definition of the conditional probability density $p(\boldsymbol{x}_1, t_1; \boldsymbol{x}_2, t_2 \,|\, \boldsymbol{y}_1, \tau_1) = p(\boldsymbol{x}_1, t_1 \,|\, \boldsymbol{x}_2, t_2;\, \boldsymbol{y}_1, \tau_1) p(\boldsymbol{x}_2, t_2 \,|\, \boldsymbol{y}_1, \tau_1)$ and using the Markov assumption (3.2.1), we find

$$p(\boldsymbol{x}_1, t_1; \boldsymbol{x}_2, t_2; \boldsymbol{y}_1, \tau_1) = p(\boldsymbol{x}_1, t_1 \,|\, \boldsymbol{x}_2, t_2) p(\boldsymbol{x}_2, t_2 \,|\, \boldsymbol{y}_1, \tau_1) \,, \tag{3.2.2}$$

and it is not difficult to see that an arbitrary joint probability can be expressed simply as

$$\begin{aligned} &p(\boldsymbol{x}_1, t_1; \boldsymbol{x}_2, t_2; \boldsymbol{x}_3, \tau_3; \dots \boldsymbol{x}_n, t_n) \\ &\quad = p(\boldsymbol{x}_1, t_1, \,|\, \boldsymbol{x}_2, t_2) p(\boldsymbol{x}_2, t_2 \,|\, \boldsymbol{x}_3, t_3) p(\boldsymbol{x}_3, t_3 \,|\, \boldsymbol{x}_4, t_4) \dots \\ &\quad\quad \dots p(\boldsymbol{x}_{n-1}, t_{n-1} \,|\, \boldsymbol{x}_n, t_n) p(\boldsymbol{x}_n, t_n) \,, \end{aligned} \tag{3.2.3}$$

provided

$$t_1 \geqslant t_2 \geqslant t_3 \geqslant \dots \geqslant t_{n-1} \geqslant t_n \,. \tag{3.2.4}$$

3.2.1 Consistency—the Chapman-Kolmogorov Equation

From Sect. 2.3.3 we require that summing over all mutually exclusive events of one kind in a joint probability eliminates that variable, i.e.,

$$\sum_B P(A \cap B \cap C \ldots) = P(A \cap C \ldots), \tag{3.2.5}$$

and when this is applied to stochastic processes, we get two deceptively similar equations:

$$p(\boldsymbol{x}_1, t_1) = \int d\boldsymbol{x}_2\, p(\boldsymbol{x}_1, t_1; \boldsymbol{x}_2, t_2) = \int d\boldsymbol{x}_2\, p(\boldsymbol{x}_1, t_1 \,|\, \boldsymbol{x}_2, t_2) p(\boldsymbol{x}_2, t_2). \tag{3.2.6}$$

This equation is an identity valid for all stochastic processes and is the first in a hierarchy of equations, the second of which is

$$\begin{aligned} p(\boldsymbol{x}_1, t_1 \,|\, \boldsymbol{x}_3, t_3) &= \int d\boldsymbol{x}_2\, p(\boldsymbol{x}_1, t_1; \boldsymbol{x}_2, t_2 \,|\, \boldsymbol{x}_3, t_3) \\ &= \int d\boldsymbol{x}_2\, p(\boldsymbol{x}_1, t_1 \,|\, \boldsymbol{x}_2, t_2; \boldsymbol{x}_3, t_3) p(\boldsymbol{x}_2, t_2 \,|\, \boldsymbol{x}_3, t_3). \end{aligned} \tag{3.2.7}$$

This equation is also always valid. We now introduce the Markov assumption. If $t_1 \geqslant t_2 \geqslant t_3$, we can drop the t_3 dependence in the doubly conditioned probability and write

$$p(\boldsymbol{x}_1, t_1 \,|\, \boldsymbol{x}_3, t_3) = \int d\boldsymbol{x}_2\, p(\boldsymbol{x}_1, t_1 \,|\, \boldsymbol{x}_2, t_2) p(\boldsymbol{x}_2, t_2 \,|\, \boldsymbol{x}_3, t_3), \tag{3.2.8}$$

which is the *Chapman-Kolmogorov equation.*

What is the essential difference between (3.2.8) and (3.2.6)? The obvious answer is that (3.2.6) is for unconditioned probabilities, whereas (3.2.7) is for conditional probabilities. Equation (3.2.8) is a rather complex nonlinear functional equation relating all conditional probabilities $p(\boldsymbol{x}_i, t_i \,|\, \boldsymbol{x}_j, t_j)$ to each other, whereas (3.2.6) simply constructs the one time probabilities in the future t_1 of t_2, given the conditional probability $p(\boldsymbol{x}_1, t_1 \,|\, \boldsymbol{x}_2, t_2)$.

The Chapman-Kolmogorov equation has many solutions. These are best understood by deriving the differential form which is done in Sect. 3.4.1 under certain rather mild conditions.

3.2.2 Discrete State Spaces

In the case where we have a discrete variable, we will use the symbol $\boldsymbol{N} = (N_1, N_2, N_3 \ldots)$, where the N_i are random variables which take on integral values. Clearly, we now replace

$$\int d\boldsymbol{x} \longleftrightarrow \sum_{\boldsymbol{n}}, \tag{3.2.9}$$

and we can now write the Chapman-Kolmogorov equation for such a process as

$$P(\boldsymbol{n}_1, t_1 \,|\, \boldsymbol{n}_3, t_3) = \sum_{\boldsymbol{n}_2} P(\boldsymbol{n}_1, t_1 \,|\, \boldsymbol{n}_2, t_2) P(\boldsymbol{n}_2, t_2 \,|\, \boldsymbol{n}_3, t_3). \tag{3.2.10}$$

This is now a matrix multiplication, with possibly infinite matrices.

3.2.3 More General Measures

A more general formulation would assume a measure $d\mu(\boldsymbol{x})$ instead of $d\boldsymbol{x}$ where a variety of choices can be made. For example, if $\mu(\boldsymbol{x})$ is a step function with steps at integral values of $\boldsymbol{x}$, we recover the discrete state space form. Most mathematical works attempt to be as general as possible. For applications, such generality can lead to lack of clarity so, where possible, we will favour a more specific notation.

3.3 Continuity in Stochastic Processes

Whether or not the random variable $\boldsymbol{X}(t)$ has a continuous range of possible values is a completely different question from whether the sample path of $\boldsymbol{X}(t)$ is a continuous function of t. For example, in a gas composed of molecules with velocities $\boldsymbol{V}(t)$, it is clear that all possible values of $\boldsymbol{V}(t)$ are in principle realisable, so that the *range* of $\boldsymbol{V}(t)$ is continuous. However, a model of collisions in a gas of hard spheres as occurring instantaneously is often considered, and in such a model the velocity before the collision, $\boldsymbol{v}_i$, will change instantaneously at the time of impact to another value $\boldsymbol{v}_f$, so the sample path of $\boldsymbol{V}(t)$ is not continuous. Nevertheless, in such a model, the *position* of a gas molecule $\boldsymbol{X}(t)$ would be expected to change continuously.

A major question now arises. Do *Markov* processes with *continuous sample paths* actually exist in reality? Notice the combination of *Markov* and *continuous*. It is almost certainly the case that in a classical picture (i.e., not quantum mechanical), all variables with a continuous range have continuous sample paths. Even the hard sphere gas mentioned above is an idealisation and more realistically, one should allow some potential to act which would continuously deflect the molecules during a collision. But it would also be the case that, if we observe on such a fine time scale, the process will probably not be Markovian. The immediate history of the whole system will almost certainly be required to predict even the probabilistic future. This is certainly born out in all attempts to derive Markovian probabilistic equations from mechanics. Equations which are derived are rarely truly Markovian—rather there is a certain characteristic memory time during which the previous history is important [3.2, 3.3].

This means that in the real world there is really no such thing as a Markov process; rather, there may be systems whose memory time is so small that, on the time scale on which we carry out observations, it is fair to regard them as being well approximated by a Markov process. But in this case, the question of whether the sample paths are actually continuous is not relevant. The sample paths of the approximating Markov process certainly need not be continuous. Even if collisions of molecules are not accurately modelled by hard spheres, during the time taken for a collision, a finite change of velocity takes place and this will appear in the approximating Markov process as a discrete step. On this time scale, even the position may change discontinuously, thus giving the picture of Brownian motion as modelled by Einstein.

In chemical reactions, for example, the time taken for an individual reaction to proceed to completion—roughly of the same order of magnitude as the collision time

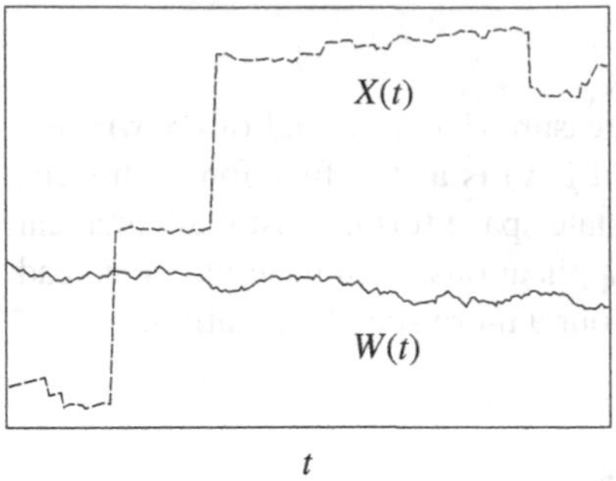

Fig. 3.1. Illustration of sample paths of the Cauchy process $X(t)$ (dashed line) and Brownian motion $W(t)$ (solid line.)

for molecules—provides yet another minimum time, since during this time, states which cannot be described in terms of individual molecules exist. Here, therefore, the very description of the state in terms of individual molecules requires a certain minimum time scale to be considered.

However, Markov processes with continuous sample paths do exist *mathematically* and are useful in describing reality. The model of the gas mentioned above provides a useful example. The position of the molecule is indeed probably best modelled as changing discontinuously by discrete jumps. Compared to the distances travelled, however, these jumps are infinitesimal and a continuous curve provides a good approximation to the sample path. On the other hand, the velocities can change by amounts which are of the same order of magnitude as typical values attained in practice. The average velocity of a molecule in a gas is about 1000 m/s and during a collision can easily reverse its sign. The velocities simply cannot reach (with any significant probability) values for which the changes of velocity can be regarded as very small. Hence, there is no sense in a continuous path description of velocities in a gas.

3.3.1 Mathematical Definition of a Continuous Markov Process

For a *Markov process*, it can be shown [3.4] that with probability one, the sample paths are continuous functions of t, if for any $\varepsilon > 0$ we have

$$\lim_{\Delta t \to 0} \frac{1}{\Delta t} \int_{|x-z|>\varepsilon} dx\, p(x, t+\Delta t \,|\, z, t) = 0\,, \tag{3.3.1}$$

uniformly in z, t and Δt.

This means that the probability for the final position x to be finitely different from z goes to zero *faster* that Δt, as Δt goes to zero. Equation (3.3.1) is sometimes called the *Lindeberg condition.*

Examples

i) *Einstein's solution* for his $f(x,t)$ (Sect. 1.2.1) is really the conditional probability $p(x,t\,|\,0,0)$. Following his method we would find

$$p(x, t + \Delta t \,|\, z, t) = \frac{1}{\sqrt{4\pi D \Delta t)}} \exp\left\{ -\frac{(x-z)^2}{4D\Delta t} \right\}, \tag{3.3.2}$$

and it is easy to check that (3.3.1) is satisfied in this case. Thus, Brownian motion in Einstein's formulation has continuous sample paths.

ii) *Cauchy Process* : Suppose

$$p(x, t + \Delta t \,|\, z, t) = \frac{\Delta t}{\pi[(x-z)^2 + \Delta t^2]} . \tag{3.3.3}$$

Then this does not satisfy (3.3.1) so the sample paths are discontinuous.

However, in both cases, we have as required for consistency

$$\lim_{\Delta t \to 0} p(x, t + \Delta t \,|\, z, t) = \delta(x - z), \tag{3.3.4}$$

and it is easy to show that in both cases, the Chapman-Kolomogorov equation is satisfied.

The difference between the two processes just described is illustrated in Fig. 3.1 in which simulations of both processes are given. The difference between the two is striking. Notice, however, that even the Brownian motion curve is extremely irregular, even though continuous—in fact it is nowhere differentiable. The Cauchy process curve, however, is only piecewise continuous.

3.4 Differential Chapman-Kolmogorov Equation

Under appropriate assumptions, the Chapman-Kolmogorov equation can be reduced to a differential equation. The assumptions made are closely connected with the continuity properties of the process under consideration. Because of the form of the continuity condition (3.3.1), one is led to consider a method of dividing the differentiability conditions into parts, one corresponding to continuous motion of a representative point and the other to discontinuous motion.

We require the following conditions *for all* $\varepsilon > 0$:

i) $$\lim_{\Delta t \to 0} p(\boldsymbol{x}, t + \Delta t | \boldsymbol{z}, t)/\Delta t = W(\boldsymbol{x} \,|\, \boldsymbol{z}, t), \tag{3.4.1}$$

uniformly in $\boldsymbol{x}$, $\boldsymbol{z}$, and t for $|\boldsymbol{x} - \boldsymbol{z}| \geqslant \varepsilon$;

ii) $$\lim_{\Delta t \to 0} \frac{1}{\Delta t} \int_{|x-z|<\varepsilon} d\boldsymbol{x}(x_i - z_i) p(\boldsymbol{x}, t + \Delta t \,|\, \boldsymbol{z}, t) = A_i(\boldsymbol{z}, t) + O(\varepsilon), \tag{3.4.2}$$

iii) $$\lim_{\Delta t \to 0} \frac{1}{\Delta t} \int_{|x-z|<\varepsilon} d\boldsymbol{x}(x_i - z_i)(x_j - z_j) p(\boldsymbol{x}, t + \Delta t \,|\, \boldsymbol{z}, t) = B_{ij}(\boldsymbol{z}, t) + O(\varepsilon), \tag{3.4.3}$$

the last two being uniform in $\boldsymbol{z}$, ε, and t.

Notice that all higher-order coefficients of the form in (3.4.2, 3.4.3) must vanish. For example, consider the third-order quantity defined by

$$\begin{aligned} &\lim_{\Delta t \to 0} \frac{1}{\Delta t} \int_{|x-z|<\varepsilon} d\boldsymbol{x}(x_i - z_i)(x_j - z_j)(x_k - z_k) p(\boldsymbol{x}, t + \Delta t \,|\, \boldsymbol{z}, t), \\ &\quad \equiv C_{ijk}(\boldsymbol{z}, t) + O(\varepsilon) . \end{aligned} \tag{3.4.4}$$

Since C_{ijk} is symmetric in i, j, k, consider

$$\sum_{i,j,k} \alpha_i\alpha_j\alpha_k C_{ijk}(z,t) \equiv \bar{C}(\alpha, z, t), \tag{3.4.5}$$

so that

$$C_{ijk}(z,t) = \frac{1}{3!}\frac{\partial^3}{\partial\alpha_i\partial\alpha_j\partial\alpha_k}\bar{C}(\alpha, z, t). \tag{3.4.6}$$

Then,

$$\begin{aligned}
|\bar{C}(\alpha, z, t)| &\leqslant \lim_{\Delta t\to 0}\frac{1}{\Delta t}\int_{|x-z|<\varepsilon} |\alpha\cdot(x-z)|\,[\alpha\cdot(x-z)]^2 p(x, t+\Delta t\,|\,z, t)\,dx + O(\varepsilon),\\
&\leqslant |\alpha|\,\varepsilon \lim_{\Delta t\to 0}\int[\alpha\cdot(x-z)]^2 p(x, t+\Delta t\,|\,z, t)dx + O(\varepsilon),\\
&= \varepsilon\,|\alpha|\,[\alpha_i\alpha_j B_{ij}(z,t) + O(\varepsilon)] + O(\varepsilon),\\
&= O(\varepsilon).
\end{aligned} \tag{3.4.7}$$

so that C is zero. Similarly, we can show that all corresponding higher-order quantities also vanish.

According to the condition for continuity (3.3.1), the process can only have continuous paths if $W(x\,|\,z, t)$ vanishes for all $x \neq z$. Thus, this function must in some way describe discontinuous motion, while the quantities A_i and B_{ij} must be connected with continuous motion.

3.4.1 Derivation of the Differential Chapman-Kolmogorov Equation

We consider the time evolution of the expectation of a function $f(z)$ which is twice continuously differentiable.

Thus,

$$\begin{aligned}
&\partial_t \int dx\, f(x)p(x, t\,|\,y, t')\\
&= \lim_{\Delta t\to 0}\frac{1}{\Delta t}\left\{\int dx\, f(x)\big[p(x, t+\Delta t\,|\,y, t') - p(x, t\,|\,y, t')\big]\right\},
\end{aligned} \tag{3.4.8}$$

$$= \lim_{\Delta t\to 0}\frac{1}{\Delta t}\left\{\int dx\int dz\, f(x)\, p(x, t+\Delta t\,|\,z, t)\, p(z, t\,|\,y, t') - \int dz f(z)\, p(z, t\,|\,y, t')\right\}, \tag{3.4.9}$$

where we have used the Chapman-Kolmogorov equation in the positive term of (3.4.8) to produce the corresponding term in (3.4.9).

We now divide the integral over x into two regions $|x-z| \geqslant \varepsilon$ and $|x-z| < \varepsilon$. When $|x-z| < \varepsilon$, since $f(z)$ is, by assumption, twice continuously differentiable, we may write

$$f(x) = f(z) + \sum_i \frac{\partial f(z)}{\partial z_i}(x_i - z_i) + \tfrac{1}{2}\sum_{i,j}\frac{\partial^2 f(z)}{\partial z_i\partial z_j}(x_i - z_i)(x_j - z_j) + |x-z|^2 R(x, z), \tag{3.4.10}$$

where we have (again by the twice continuous differentiability)

$$|R(x,z)| \to 0 \quad \text{as} \quad |x-z| \to 0\,. \tag{3.4.11}$$

Now substitute in (3.4.9):

$$\begin{aligned}
(3.4.9) = \lim_{\Delta t\to 0}\frac{1}{\Delta t}\Bigg\{ & \iint\limits_{|x-z|<\varepsilon} dx\,dz\left[\sum_i (x_i - z_i)\frac{\partial f}{\partial z_i} + \tfrac{1}{2}\sum_{i,j}(x_i - z_i)(x_j - z_j)\frac{\partial^2 f}{\partial z_i \partial z_j}\right] \\
&\qquad \times p(x, t+\Delta t\,|\,z, t)p(z, t\,|\,y, t') \\
&+ \iint\limits_{|x-z|<\varepsilon} dx\,dz\,|x-z|^2 R(x,z)\,p(x, t+\Delta t\,|\,z, t)\,p(z, t\,|\,y, t') \\
&+ \iint\limits_{|x-z|\geqslant\varepsilon} dx\,dz\,f(x)\,p(x, t+\Delta t\,|\,z, t)\,p(z, t\,|\,y, t') \\
&+ \iint\limits_{|x-z|<\varepsilon} dx\,dz\,f(z)\,p(x, t+\Delta t\,|\,z, t)\,p(z, t\,|\,y, t') \\
&- \iint dx\,dz\,f(z)\,p(x, t+\Delta t\,|\,z, t)\,p(z, t\,|\,y, t')\Bigg\}.
\end{aligned} \tag{3.4.12}$$

Note that since $p(x, t+\Delta t\,|\,z, t)$ is a probability, the integral over x in the last term gives 1—this is simply the last term in (3.4.9).

We now consider these line by line.

Lines 1 and 2: By the assumed uniform convergence, we take the limit inside the integral to obtain [using conditions (ii) and (iii) of Sect. 3.4]

$$\int dz\left[\sum_i A_i(z)\frac{\partial f}{\partial z_i} + \tfrac{1}{2}\sum_{i,j}B_{ij}(z)\frac{\partial^2 f}{\partial z_i \partial z_j}\right]p(z, t\,|\,y, t') + O(\varepsilon)\,. \tag{3.4.13}$$

Line 3: This is a remainder term and vanishes as $\varepsilon \to 0$. For

$$\begin{aligned}
&\left|\frac{1}{\Delta t}\int\limits_{|x-z|<\varepsilon} dx\,|x-z|^2 R(x,z)p(x, t+\Delta t\,|\,z, t)\right| \\
&\leqslant \left[\frac{1}{\Delta t}\int\limits_{|x-z|<\varepsilon} dx|x-z|^2 p(x, t+\Delta t|z, t)\right]\max_{|x-z|<\varepsilon}|R(x,z)| \\
&\to \left[\sum_i B_{ii}(z,t) + O(\varepsilon)\right]\left\{\max_{|x-z|<\varepsilon}|R(x,z)|\right\}.
\end{aligned} \tag{3.4.14}$$

From (3.4.11) we can see that as $\varepsilon \to 0$, the factor in curly brackets vanishes.

Lines 4–6: We can put these all together to obtain

$$\iint\limits_{|x-z|\geqslant\varepsilon} dxdz\,f(z)[W(z\,|\,x, t)p(x, t\,|\,y, t') - W(x\,|\,z, t)p(z, t\,|\,y, t')]\,. \tag{3.4.15}$$

The whole right-hand side of (3.4.12) is independent of ε. Hence, taking the limit $\varepsilon \to 0$, we find

$$\begin{aligned}
\partial_t \int dz f(z)p(z, t\,|\,y, t') = {} & \int dz\left[\sum_i A_i(z,t)\frac{\partial f(z)}{\partial z_i} + \tfrac{1}{2}\sum_{i,j}\boldsymbol{B}_{ij}(z)\frac{\partial^2 f(z)}{\partial z_i \partial z_j}\right]p(z, t\,|\,y, t') \\
&+ \int dz f(z)\left\{ ⨍ dx[W(z\,|\,x, t)p(x, t\,|\,y, t') - W(x\,|\,z, t)p(z, t\,|\,y, t')]\right\}.
\end{aligned} \tag{3.4.16}$$

Notice that we use the definition

$$\lim_{\varepsilon \to 0} \int_{|x-z|>\varepsilon} dx\, F(x,z) \equiv ⨍ dx\, F(x,z), \tag{3.4.17}$$

for a principal value integral of a function $F(x,z)$. For (3.4.16) to have any meaning, this integral should exist. Equation (3.4.1) defines $W(x\,|\,z,t)$ only for $x \neq z$ and hence leaves open the possibility that it is infinite at $x = z$, as is indeed the case for the Cauchy process, discussed in Sect. 3.3.1, for which

$$W(x\,|\,z,t) = 1/[\pi(x-z)^2]\,. \tag{3.4.18}$$

However, if $p(x,t\,|\,y,t')$ is continuous and once differentiable, then the principal value integral exists. In the remainder of the book we shall not write this integral explicitly as a principal value integral since one rarely considers the singular cases for which it is necessary.

The final step now is to integrate by parts. We find

$$\begin{aligned}
&\int dz f(z)\partial_t p(z,t\,|\,y,t') \\
&= \int dz f(z)\Bigg\{ -\sum_i \frac{\partial}{\partial z_i} A_i(z,t)p(z,t\,|\,y,t') + \tfrac{1}{2}\sum_{i,j}\frac{\partial^2}{\partial z_i \partial z_j} B_{ij}(z,t)p(z,t\,|\,y,t') \\
&\qquad + \int dx \Big[W(z\,|\,x,t)p(x,t\,|\,y,t') - W(x\,|\,z,t)p(z,t\,|\,y,t')\Big]\Bigg\} \\
&\qquad + \text{surface terms}\,.
\end{aligned} \tag{3.4.19}$$

We have not specified the range of the integrals. Suppose the process is confined to a region R with surface S. Then clearly,

$$p(x,t\,|\,z,t') = 0 \text{ unless both } x \text{ and } z \in R\,. \tag{3.4.20}$$

It is clear that by definition we have

$$W(x\,|\,z,t) = 0 \text{ unless both } x \text{ and } y \in R\,. \tag{3.4.21}$$

But the conditions on $A_i(z,t)$ and $B_{ij}(z,t)$ can result in discontinuities in these functions as defined by (3.4.2) and (3.4.3) since $p(x,t+\Delta t\,|\,z,t')$, the conditional probability, can very reasonably change discontinuously as z crosses the boundary of R, reflecting the fact that no transitions are allowed from outside R to inside R.

In integrating by parts, we are forced to differentiate both A_i and B_{ij} and by our reasoning above, one cannot assume that this is possible on the boundary of the region. Hence, let us choose $f(z)$ to be arbitrary but nonvanishing only in an arbitrary region R' entirely contained in R. We can then deduce that for all z in *the interior* of R,

$$\begin{aligned}
\frac{\partial p(z,t\,|\,y,t')}{\partial t} = &-\sum_i \frac{\partial}{\partial z_i}[A_i(z,t)p(z,t\,|\,y,t')] + \tfrac{1}{2}\sum_{i,j}\frac{\partial^2}{\partial z_i \partial z_j}[B_{ij}(z,t)p(z,t\,|\,y,t')] \\
&+ \int dx[W(z\,|\,x,t)p(x,t\,|\,y,t') - W(x\,|\,z,t)p(z,t\,|\,y,t')]\,.
\end{aligned} \tag{3.4.22}$$

Surface terms do not arise, since they necessarily vanish.

This equation does not seem to have any agreed name in the literature. Since it is purely a differential form of the Chapman-Kolmogorov equation, I propose to call it the *differential Chapman-Kolmogorov* equation.

3.4.2 Status of the Differential Chapman-Kolmogorov Equation

From our derivation it is not clear to what extent solutions of the differential Chapman-Kolmogorov equation are solutions of the Chapman-Kolmogorov equation itself or indeed, to what extent solutions exist. It is certainly true, however, that a set of conditional probabilities which obey the Chapman-Kolmogorov equation does generate a Markov process, in the sense that the joint probabilities so generated satisfy all probability axioms.

It can be shown [3.5] that, under certain conditions, if we specify $\boldsymbol{A}(\boldsymbol{x},t)$, $\mathsf{B}(\boldsymbol{x},t)$ (which must be positive semi-definite), and $W(\boldsymbol{x}\,|\,\boldsymbol{y},t)$ (which must be non-negative), that a non-negative solution to the differential Chapman-Kolmogorov equation exists, and this solution also satisfies the Chapman-Kolmogorov equation. The conditions to be satisfied are the *initial condition*,

$$p(\boldsymbol{z},t\,|\,\boldsymbol{y},t) = \delta(\boldsymbol{y}-\boldsymbol{z})\,, \tag{3.4.23}$$

which follows from the definition of the conditional probability density, and any appropriate boundary conditions. These are very difficult to specify in the full equation, but in the case of the *Fokker-Planck equation* (Sect. 3.5.2) are given in Chap. 5 and Chap. 6.

3.5 Interpretation of Conditions and Results

Each of the conditions (i), (ii), (iii) of Sect. 3.4 can now be seen to give rise to a distinctive part of the equation, whose interpretation is rather straightforward. We can identify three processes taking place, which are known as jumps, drift and diffusion.

3.5.1 Jump Processes: The Master Equation

We consider a case in which

$$A_i(\boldsymbol{z},t) = B_{ij}(\boldsymbol{z},t) = 0, \tag{3.5.1}$$

so that we now have the *master equation:*

$$\partial_t p(\boldsymbol{z},t\,|\,\boldsymbol{y},t') = \int d\boldsymbol{x}[W(\boldsymbol{z}\,|\,\boldsymbol{x},t)p(\boldsymbol{x},t\,|\,\boldsymbol{y},t') - W(\boldsymbol{x}\,|\,\boldsymbol{z},t)p(\boldsymbol{z},t\,|\,\boldsymbol{y},t')]\,. \tag{3.5.2}$$

To first order in Δt we solve approximately, as follows. Notice that

$$p(\boldsymbol{z},t\,|\,\boldsymbol{y},t) = \delta(\boldsymbol{y}-\boldsymbol{z})\,. \tag{3.5.3}$$

Hence,

$$p(z, t + \Delta t \,|\, \boldsymbol{y}, t) = \delta(\boldsymbol{y} - z)\,[1 - \int d\boldsymbol{x}\, W(\boldsymbol{x} \,|\, \boldsymbol{y}, t)\Delta t] + W(z \,|\, \boldsymbol{y}, t)\Delta t\,. \tag{3.5.4}$$

We see that for any Δt there is a finite probability, given by the coefficient of the $\delta(\boldsymbol{y} - z)$ in (3.5.4), for the particle to stay at the original position $\boldsymbol{y}$. The distribution of those particles which do not remain at $\boldsymbol{y}$ is given by $W(z \,|\, \boldsymbol{y}, t)$ after appropriate normalisation. Thus, a typical path $\boldsymbol{X}(t)$ will consist of sections of straight lines $\boldsymbol{X}(t)$ = constant, interspersed with discontinuous jumps whose distribution is given by $W(z \,|\, \boldsymbol{y}, t)$. For this reason, the process is known as a jump process. The paths are discontinuous at discrete points.

a) Time Between Jumps: The probability $Q(\boldsymbol{y}, t, t_0)$ that, given that we start from point $\boldsymbol{y}$ at time t_0, we are still at point $\boldsymbol{y}$ at time t is given for infinitesimal Δt by

$$Q(\boldsymbol{y}, t_0 + \Delta t, t_0)) = 1 - \int d\boldsymbol{x}\, W(\boldsymbol{x} \,|\, \boldsymbol{y}, t_0)\Delta t\,. \tag{3.5.5}$$

Clearly this means that

$$\frac{\partial Q(\boldsymbol{y}, t, t_0)}{\partial t} = -\int d\boldsymbol{x}\, W(\boldsymbol{x} \,|\, \boldsymbol{y}, t)\, Q(\boldsymbol{y}, t, t_0)\,. \tag{3.5.6}$$

If the jump probability is independent of t, then this has the simple solution

$$Q(\boldsymbol{y}, t, t_0) = \exp(-\lambda t)\,, \tag{3.5.7}$$

$$\text{with } \lambda = \int d\boldsymbol{x}\, W(\boldsymbol{x} \,|\, \boldsymbol{y})\,. \tag{3.5.8}$$

Thus, the jump times are exponentially distributed, and can be simulated very simply. To simulate, one first chooses a jump time according to the probability law (3.5.7), and then chooses the value of $\boldsymbol{x}$ to which the jump was made according to a probability law

$$w(\boldsymbol{x} \,|\, \boldsymbol{y}) = W(\boldsymbol{x} \,|\, \boldsymbol{y})/\lambda\,. \tag{3.5.9}$$

If the jump probability is not independent of t, the same procedure is in principle possible, but instead of using the exponential form (3.5.7), one must solve the differential equation (3.5.6).

b) Integer State Space: In the case where the state space consists of integers only, the master equation takes the form

$$\partial_t P(\boldsymbol{n}, t \,|\, \boldsymbol{n}', t') = \sum_{m} [W(\boldsymbol{n} \,|\, \boldsymbol{m}, t) P(\boldsymbol{m}, t \,|\, \boldsymbol{n}', t') - W(\boldsymbol{m} \,|\, \boldsymbol{n}, t) P(\boldsymbol{n}, t \,|\, \boldsymbol{n}', t')]\,. \tag{3.5.10}$$

There is no longer any question that only jumps can occur, since only discrete values of the state variable $\boldsymbol{N}(t)$ are allowed. It is most important, however, to be aware that a pure jump process can occur even though the variable $\boldsymbol{X}(t)$ can take on a continuous range of variables.

3.5.2 Diffusion Processes—the Fokker-Planck Equation

When the quantities $W(z \,|\, \boldsymbol{x}, t)$ are zero, the differential Chapman-Kolmogorov equation reduces to the *Fokker-Planck equation:*

$$\frac{\partial p(\boldsymbol{z},t\,|\,\boldsymbol{y},t')}{\partial t} = -\sum_i \frac{\partial}{\partial z_i}[A_i(\boldsymbol{z},t)p(\boldsymbol{z},t\,|\,\boldsymbol{y},t')] + \frac{1}{2}\sum_{i,j}\frac{\partial^2}{\partial z_i \partial z_j}[B_{ij}(\boldsymbol{z},t)p(\boldsymbol{z},t\,|\,\boldsymbol{y},t')]\,. \tag{3.5.11}$$

and the corresponding process is known mathematically as a *diffusion process*. The vector $\boldsymbol{A}(\boldsymbol{z},t)$ is known as the *drift vector* and the matrix $\mathsf{B}(\boldsymbol{z},t)$ as the *diffusion matrix*. The diffusion matrix is positive semidefinite and symmetric as a result of its definition in (3.4.3). It is easy to see from (3.4.1), the definition of $W(\boldsymbol{x}\,|\,\boldsymbol{z},t)$, that the requirement (3.3.1) for continuity of the sample paths is satisfied if $W(\boldsymbol{x}\,|\,\boldsymbol{z},t)$ is zero. Hence, the Fokker-Planck equation describes a process in which $\boldsymbol{X}(t)$ has continuous sample paths.

In fact, we can heuristically give a much more definite description of the process. Let us consider computing $p(\boldsymbol{z},t+\Delta t\,|\,\boldsymbol{y},t)$, given that

$$p(\boldsymbol{z},t\,|\,\boldsymbol{y},t) = \delta(\boldsymbol{z}-\boldsymbol{y})\,. \tag{3.5.12}$$

For a small Δt, the solution of the Fokker-Planck equation will still be on the whole sharply peaked, and hence derivatives of $A_i(\boldsymbol{z},t)$ and $B_{ij}(\boldsymbol{z},t)$ will be negligible compared to those of p. We are thus reduced to solving, approximately

$$\frac{\partial p(\boldsymbol{z},t\,|\,\boldsymbol{y},t')}{\partial t} = -\sum_i A_i(\boldsymbol{y},t)\frac{\partial p(\boldsymbol{z},t\,|\,\boldsymbol{y},t')}{\partial z_i} + \sum_{i,j}\tfrac{1}{2}B_{ij}(\boldsymbol{y},t)\frac{\partial^2 p(\boldsymbol{z},t\,|\,\boldsymbol{y},t')}{\partial z_i \partial z_j}\,, \tag{3.5.13}$$

where we have also neglected the time dependence of A_i and B_{ij} for small $t-t'$. Equation (3.5.13) can now be solved, subject to the initial condition (3.5.12), and we get

$$p(\boldsymbol{z},t+\Delta t\,|\,\boldsymbol{y},t) = \left\{(2\pi)^N \det[\mathsf{B}(\boldsymbol{y},t)\Delta t]\right\}^{-1/2}$$

$$\times\exp\left\{-\frac{1}{2}\frac{[\boldsymbol{z}-\boldsymbol{y}-\boldsymbol{A}(\boldsymbol{y},t)\Delta t]^{\mathrm{T}}[\mathsf{B}(\boldsymbol{y},t)]^{-1}[\boldsymbol{z}-\boldsymbol{y}-\boldsymbol{A}(\boldsymbol{y},t)\Delta t]}{\Delta t}\right\}, \tag{3.5.14}$$

that is, a Gaussian distribution with variance matrix $\mathsf{B}(\boldsymbol{y},t)$ and mean $\boldsymbol{y}+\boldsymbol{A}(\boldsymbol{y},t)\Delta t$. We get the picture of the system moving with a systematic drift, whose velocity is $\boldsymbol{A}(\boldsymbol{y},t)$, on which is superimposed a Gaussian fluctuation with covariance matrix $\mathsf{B}(\boldsymbol{y},t)\Delta t$, that is, we can write

$$\boldsymbol{y}(t+\Delta t) = \boldsymbol{y}(t) + \boldsymbol{A}(\boldsymbol{y}(t),t)\Delta t + \boldsymbol{\eta}(t)\Delta t^{1/2}\,, \tag{3.5.15}$$

where

$$\langle\boldsymbol{\eta}(t)\rangle = 0, \tag{3.5.16}$$

$$\langle\boldsymbol{\eta}(t)\boldsymbol{\eta}(t)^{\mathrm{T}}\rangle = \mathsf{B}(\boldsymbol{y},t)\,. \tag{3.5.17}$$

It is easy to see that this picture gives

i) Sample paths which are always continuous, since it is clear that $\boldsymbol{y}(t+\Delta t)\to\boldsymbol{y}(t)$ as $\Delta t\to 0$;

ii) Sample paths which are nowhere differentiable, because of the $\Delta t^{1/2}$ occurring in (3.5.15).

We shall see later, in Chap. 4 that the heuristic picture of (3.5.15) can be made much more precise and leads to the concept of the *stochastic differential equation.*

3.5.3 Deterministic Processes—Liouville's Equation

It is possible that in the differential Chapman-Kolmogorov equation (3.4.22) only the first term is nonzero, so we are led to the special case of a *Liouville equation:*

$$\frac{\partial p(z,t\,|\,\boldsymbol{y},t')}{\partial t} = -\sum_i \frac{\partial}{\partial z_i}[A_i(z,t)\,p(z,t\,|\,\boldsymbol{y},t')]\,, \tag{3.5.18}$$

which occurs in classical mechanics. This equation describes a completely deterministic motion, i.e., if $\boldsymbol{x}(\boldsymbol{y},t)$ is the solution of the ordinary differential equation

$$\frac{d\boldsymbol{x}(t)}{dt} = \boldsymbol{A}[\boldsymbol{x}(t),t]\,, \tag{3.5.19}$$

with

$$\boldsymbol{x}(\boldsymbol{y},t') = \boldsymbol{y}\,, \tag{3.5.20}$$

then the solution to (3.5.18) with initial condition

$$p(z,t'\,|\,\boldsymbol{y},t') = \delta(z-\boldsymbol{y}), \tag{3.5.21}$$

is

$$p(z,t\,|\,\boldsymbol{y},t') = \delta[z-\boldsymbol{x}(\boldsymbol{y},t)]\,. \tag{3.5.22}$$

The proof of this assertion is best obtained by direct substitution. For

$$\sum_i \frac{\partial}{\partial z_i}\Big\{A_i(z,t)\delta[z-\boldsymbol{x}(\boldsymbol{y},t)]\Big\} = \sum_i \frac{\partial}{\partial z_i}\Big\{A_i[\boldsymbol{x}(\boldsymbol{y},t),t]\delta[z-\boldsymbol{x}(\boldsymbol{y},t)]\Big\}, \tag{3.5.23}$$

$$= \sum_i \left\{A_i[\boldsymbol{x}(\boldsymbol{y},t),t]\frac{\partial}{\partial z_i}\delta[z-\boldsymbol{x}(\boldsymbol{y},t)]\right\}, \tag{3.5.24}$$

and

$$\frac{\partial}{\partial t}\delta[z-\boldsymbol{x}(\boldsymbol{y},t)] = -\sum_i \frac{\partial}{\partial z_i}\delta[z-\boldsymbol{x}(\boldsymbol{y},t)]\frac{dx_i(\boldsymbol{y},t)}{dt}, \tag{3.5.25}$$

and by use of (3.5.19), we see that (3.5.24) and (3.5.25) are equal. Thus, if the particle is in a well-defined initial position $\boldsymbol{y}$ at time t', it stays on the trajectory obtained by solving the ordinary differential equation (3.5.19).

Hence, deterministic motion, as defined by a first-order differential equation of the form (3.5.19), is an elementary form of Markov process. The solution (3.5.22) is, of course, merely a special case of the kind of process approximated by equations like (3.5.14) in which the Gaussian part is zero.

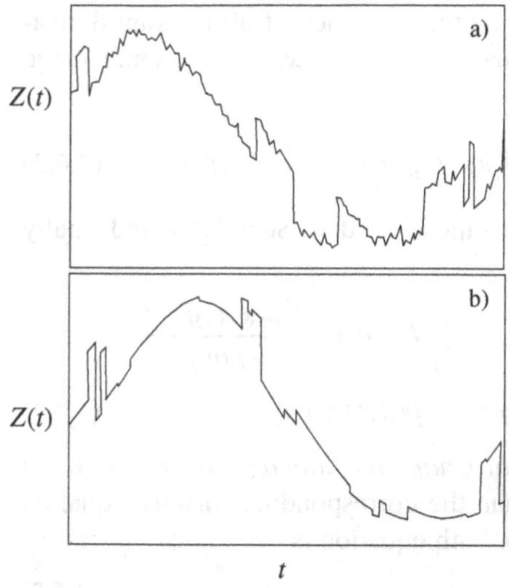

Fig. 3.2. a) Illustration of a sample path of a general Markov process, in which drift, diffusion and jumps exist; **b)** Sample path of a Markov process with only drift and jumps.

3.5.4 General Processes

In general, none of the quantities in $\boldsymbol{A}(z,t)$, $\mathsf{B}(z,t)$ and $W(\boldsymbol{x}\,|\,z,t)$ need vanish, and in this case we obtain a process whose sample paths are as illustrated in Fig. 3.2a, i.e., a piecewise continuous path made up of pieces which correspond to a diffusion process with a nonzero drift, onto which is superimposed a fluctuating part.

It is also possible that $\boldsymbol{A}(z,t)$ is nonzero, but $\mathsf{B}(z,t)$ is zero and here the sample paths are, as in Fig. 3.2b composed of pieces of smooth curve [solutions of (3.5.19)] with discontinuities superimposed. This is very like the picture one would expect in a dilute gas where the particles move freely between collisions which cause an instantaneous change in momentum, though not position.

3.6 Equations for Time Development in Initial Time—Backward Equations

We can derive much more simply than in Sect. 3.4, some equations which give the time development with respect to the initial variables $\boldsymbol{y}, t'$ of $p(\boldsymbol{x},t\,|\,\boldsymbol{y},t')$.

We consider

$$\lim_{\Delta t'\to 0}\frac{1}{\Delta t'}\big\{p(\boldsymbol{x},t\,|\,\boldsymbol{y},t'+\Delta t') - p(\boldsymbol{x},t\,|\,\boldsymbol{y},t')\big\}, \tag{3.6.1}$$

$$= \lim_{\Delta t'\to 0}\frac{1}{\Delta t'}\int dz\; p(z,t'+\Delta t'\,|\,\boldsymbol{y},t')\big\{p(\boldsymbol{x},t\,|\,\boldsymbol{y},t'+\Delta t') - p(\boldsymbol{x},t\,|\,z,t'+\Delta t')\big\}. \tag{3.6.2}$$

The second line follows by use of the Chapman-Kolmogorov equation in the second term and by noting that the first term gives $1 \times p(\boldsymbol{x},t\,|\,\boldsymbol{y},t'+\Delta t')$.

The assumptions that are necessary are now the existence of all relevant derivatives, and that $p(\boldsymbol{x},t\,|\,\boldsymbol{y},t')$ is continuous and bounded in $\boldsymbol{x},t,t'$ for some range $t-t'>\delta>0$. We may then write

$$(3.6.2) = \lim_{\Delta t'\to 0}\frac{1}{\Delta t'}\int dz\, p(z,t'+\Delta t'\,|\,\boldsymbol{y},t')\big\{p(\boldsymbol{x},t\,|\,\boldsymbol{y},t') - p(\boldsymbol{x},t\,|\,z,t')\big\}. \qquad (3.6.3)$$

We now proceed using similar techniques to those used in Sect. 3.4.1 and finally derive

$$\frac{\partial p(\boldsymbol{x},t\,|\,\boldsymbol{y},t')}{\partial t'} = -\sum_i A_i(\boldsymbol{y},t')\frac{\partial p(\boldsymbol{x},t\,|\,\boldsymbol{y},t')}{\partial y_i} - \tfrac{1}{2}\sum_{ij}\boldsymbol{B}_{ij}(\boldsymbol{y},t')\frac{\partial^2 p(\boldsymbol{x},t\,|\,\boldsymbol{y},t')}{\partial y_i \partial y_j}$$
$$+\int dz\, W(z\,|\,\boldsymbol{y},t')\big\{p(\boldsymbol{x},t\,|\,\boldsymbol{y},t') - p(\boldsymbol{x},t\,|\,z,t')\big\}. \qquad (3.6.4)$$

This will be called the *backward differential Chapman-Kolmogorov equation.* In a mathematical sense, it is better defined than the corresponding forward equation (3.4.22). The appropriate initial condition for both equation is

$$p(\boldsymbol{x},t\,|\,\boldsymbol{y},t) = \delta(\boldsymbol{x}-\boldsymbol{y}) \text{ for all } t\,, \qquad (3.6.5)$$

representing the obvious fact that if the particle is at $\boldsymbol{y}$ at time t, the probability density for finding it at $\boldsymbol{x}$ at the same time is $\delta(\boldsymbol{x}-\boldsymbol{y})$.

The forward and the backward equations are equivalent to each other. For, solutions of the forward equation, subject to the initial condition (3.6.5) and any appropriate boundary conditions, yield solutions of the Chapman-Kolmogorov equation, as noted in Sect. 3.4.2. But these have just been shown to yield the backward equation. (The relation between appropriate boundary conditions for the Fokker-Planck equations is dealt with in Sects. 5.1,5.1.2). The basic difference is which set of variables is held fixed. In the case of the forward equation, we hold $\boldsymbol{y}$ and t' fixed, and solutions exist for $t \geqslant t'$, so that (3.6.5) is an *initial condition* for the forward equation. For the backward equation, solutions exist for $t' \leqslant t$, so that since the backward equation expresses development in t', (3.6.5) is better termed *final condition* in this case.

Since they are equivalent, the forward and backward equations are both useful. The forward equation gives more directly the values of measurable quantities as a function of the observed time, t, and tends to be used more commonly in applications. The backward equation finds most application in the study of *first passage time or exit problems*, in which we find the probability that a particle leaves a region in a given time.

3.7 Stationary and Homogeneous Markov Processes

In Sect. 1.5.3 we met the concept of a stationary process, which represents the stochastic motion of a system which has settled down to a steady state, and whose stochastic properties are independent of when they are measured. Stationarity can be defined in various degrees, but we shall reserve the term "*stationary process*" for a strict definition, namely, a stochastic process $\boldsymbol{X}(t)$ is stationary if $\boldsymbol{X}(t)$ and the process

$X(t+\varepsilon)$ have the same statistics for any ε. This is equivalent to saying that all joint probability densities satisfy time translation invariance, i.e.,

$$p(\boldsymbol{x}_1,t_1;\boldsymbol{x}_2,t_2;\boldsymbol{x}_3,t_3;\ldots;\boldsymbol{x}_n,t_n)$$
$$= p(\boldsymbol{x}_1,t_1+\varepsilon;\boldsymbol{x}_2,t_2+\varepsilon;\boldsymbol{x}_3,t_3+\varepsilon;\ldots;\boldsymbol{x}_n,t_n+\varepsilon)\,, \qquad (3.7.1)$$

and hence such probabilities are only functions of the time differences, $t_i - t_j$. In particular, the one-time probability is independent of time and can be simply written as

$$p_s(\boldsymbol{x}), \qquad (3.7.2)$$

and the two-time joint probability as

$$p_s(\boldsymbol{x}_1,t_1-t_2;\boldsymbol{x}_2,0)\,. \qquad (3.7.3)$$

Finally, the conditional probability can also be written as

$$p_s(\boldsymbol{x}_1,t_1-t_2\,|\,\boldsymbol{x}_2,0)\,. \qquad (3.7.4)$$

For a *Markov process*, since all joint probabilities can be written as products of the two-time conditional probability and the one-time probability, a necessary and sufficient condition for stationarity is the ability to write the one and two-time probabilities in the forms given in (3.7.1–3.7.3).

3.7.1 Ergodic Properties

If we have a stationary process, it is reasonable to expect that average measurements could be constructed by taking values of the variable x at successive times, and averaging various functions of these. This is effectively a belief that the law of large numbers (as explained in Sect. 2.5.2) applies to the variables defined by successive measurements in a stochastic process.

a) Ergodic Property of the Mean: Let us define the variable $\bar{X}(T)$ by

$$\bar{X}(T) = \frac{1}{2T}\int_{-T}^{T} dt\; x(t), \qquad (3.7.5)$$

where $x(t)$ is a stationary process, and consider the limit $T\to\infty$. This represents a possible model of measurement of the mean by averaging over all times. Clearly

$$\langle \bar{X}(t)\rangle = \langle x\rangle_s\,. \qquad (3.7.6)$$

We now calculate the variance of $\bar{X}(T)$. Thus,

$$\langle \bar{X}(T)^2\rangle = \frac{1}{4T^2}\int_{-T}^{T}\int_{-T}^{T} dt_1 dt_2 \langle x(t_1)x(t_2)\rangle, \qquad (3.7.7)$$

and if the process is stationary,

$$\langle x(t_1)x(t_2)\rangle \equiv R(t_1-t_2) + \langle x\rangle^2, \qquad (3.7.8)$$

where R is the two-time correlation function. Hence,

$$\langle \bar{X}(T)^2 \rangle - \langle x \rangle^2 = \frac{1}{4T^2} \int_{-2T}^{2T} d\tau\, R(\tau)(2T - |\tau|), \tag{3.7.9}$$

where the last factor follows by changing variables to

$$\tau = t_1 - t_2, \quad t = t_1 \,, \tag{3.7.10}$$

and integrating t.

The left-hand side is now the variance of $\bar{X}(T)$ and we will show that under certain conditions, this vanishes as $T \to \infty$. Most straightforwardly, all we require is that

$$\lim_{T\to\infty} \frac{1}{T} \int_{-2T}^{2T} d\tau \left(1 - \frac{|\tau|}{2T}\right) R(\tau) = 0, \tag{3.7.11}$$

which is a little obscure. However, it is clear that a sufficient condition for this limit to be zero is for

$$\int_0^\infty d\tau\, |R(\tau)| < \infty \,, \tag{3.7.12}$$

in which case, we simply require that the correlation function $\langle x(t_1), x(t_2) \rangle$ should tend to zero sufficiently rapidly as $|t_1 - t_2| \to \infty$. In cases of interest it is frequently found that the asymptotic behavior of $R(\tau)$ is

$$R(\tau) \sim \operatorname{Re}\{A \exp(-\tau/\tau_c)\} \,, \tag{3.7.13}$$

where τ_c is a (possibly complex) parameter known as the *correlation time*. Clearly the criterion of (3.7.12) is satisfied, and we find in this case that the variance in $\bar{X}(T)$ approaches zero, so that using (3.7.6) and (2.9.4), we may write

$$\underset{T\to\infty}{\text{ms-lim}}\, \bar{X}(T) = \langle x \rangle_s \,. \tag{3.7.14}$$

This means that the averaging procedure (3.7.5) is indeed valid. It is not difficult to extend the result to an average of an infinite set of measurements at discrete times $t_n = t_0 + n\Delta t$.

Other ergodic hypotheses can easily be stated, and the two quantities that are of most interest are the autocorrelation function and the distribution function.

b) Ergodic Property of the Autocorrelation Function: As already mentioned in Sect. 1.5.2, the most natural way of measuring an autocorrelation function is through the definition

$$G(\tau, T) = \frac{1}{T} \int_0^T dt\, x(t)x(t+\tau), \tag{3.7.15}$$

and we can rather easily carry through similar reasoning to show that

$$\underset{T\to\infty}{\text{ms-lim}}\, G(\tau, T) = \langle x(t)x(t+\tau) \rangle_s \,, \tag{3.7.16}$$

provided the following condition is satisfied. Namely, define $\rho(\tau, \lambda)$ by

$$\langle x(t+\lambda+\tau)x(t+\lambda)x(t+\tau)x(t) \rangle_s = \rho(\tau, \lambda) + \langle x(t+\tau)x(t) \rangle_s^2 \,. \tag{3.7.17}$$

Then we require

$$\lim_{T\to\infty} \frac{1}{2T} \int_{-2T}^{2T} \left(1 - \frac{|\lambda|}{2T}\right) \rho(\tau, \lambda) d\lambda = 0 \,. \tag{3.7.18}$$

We can see that this means that for sufficiently large λ, the four-time average (3.7.17) factorises into a product of two-time averages, and that the "error term" $\rho(\tau, \lambda)$ must vanish sufficiently rapidly for $\lambda \to \infty$. Exponential behaviour, such as given in (3.7.13) is sufficient, and usually found.

b) Ergodic Property of the Spectrum: We similarly find that the spectrum, given by the Fourier transform

$$S(\omega) = \frac{1}{2\pi} \int_{-\infty}^{\infty} \mathrm{e}^{-\mathrm{i}\omega\tau} G(\tau)\, d\tau, \tag{3.7.19}$$

as in Sect. 1.5.2, is also given by the procedure

$$S(\omega) = \lim_{T\to\infty} \frac{1}{2\pi T} \left| \int_0^T dt\, \mathrm{e}^{-\mathrm{i}\omega t} x(t) \right|^2 . \tag{3.7.20}$$

d) Ergodic Property of the Distribution Function: Finally, the practical method of measuring the distribution function is to consider an interval (x_1, x_2) and measure $x(t)$ repeatedly to determine whether it is in this range or not. This gives a measure of $\int_{x_1}^{x_2} dx\, p_s(x)$. Essentially, we are then measuring the time average value of the function $\chi(x)$ defined by

$$\chi(x) = \begin{cases} 1, & x_1 < x < x_2\,, \\ 0, & \text{otherwise.} \end{cases} \tag{3.7.21}$$

and we adapt the method of proving the ergodicity of $\langle x \rangle$ to find that the distribution is ergodic provided

$$\lim_{T\to\infty} \frac{1}{2T} \int_{-2T}^{2T} d\tau \left(1 - \frac{|\tau|}{2T}\right) \int_{x_1}^{x_2} dx'\, p_s(x') \left\{ \int_{x_1}^{x_2} dx [p(x, \tau | x', 0) - p_s(x)] \right\} = 0\,. \tag{3.7.22}$$

The most obvious sufficient condition here is that

$$\lim_{T\to\infty} p(x, \tau \,|\, x', 0) = p_s(x), \tag{3.7.23}$$

and that this limit is approached sufficiently rapidly. In practice, an exponential approach is frequently found and this is, as in the case of the mean, quite sufficiently rapid.

This condition is, in fact, sufficient for ergodicity of the mean and autocorrelation function for a Markov process, since all means can be expressed in terms of conditional probabilities and the sufficiently rapid achievement of the limit (3.7.23) can be readily seen to be sufficient to guarantee both (3.7.18) and (3.7.11). We will call a Markov process simply *ergodic* if this rather strong condition is satisfied.

3.7.2 Homogeneous Processes

If the condition (3.7.23) is satisfied for a stationary Markov process, then we clearly have a way of constructing from the stationary Markov process a nonstationary process whose limit as time becomes large is the stationary process. We simply define the process for $t, t' > t_0$ by

$$p(\boldsymbol{x},t) = p_s(\boldsymbol{x},t\,|\,\boldsymbol{x}_0,t_0)\,, \tag{3.7.24}$$

$$p(\boldsymbol{x},t\,|\,\boldsymbol{x}',t') = p_s(\boldsymbol{x}',t\,|\,\boldsymbol{x}',t')\,, \tag{3.7.25}$$

and all other joint probabilities are obtained from these in the usual manner for a Markov process. Clearly, if (3.7.23) is satisfied, we find that as $t \to \infty$ or as $t_0 \to -\infty$,

$$p(\boldsymbol{x},t) \to p_s(\boldsymbol{x}) \tag{3.7.26}$$

and all other probabilities become stationary because the conditional probability is stationary. Such a process is known as a *homogeneous* process.

The physical interpretation is rather obvious. We have a stochastic system whose variable $\boldsymbol{x}$ is by some external agency fixed to have a value $\boldsymbol{x}_0$ at time t_0. It then evolves back to a stationary system with the passage of time. This is how many stationary systems are created in practice.

From the point of view of the differential Chapman-Kolmogorov equation, we will find that the stationary distribution function $p_s(x)$ is a solution of the stationary differential Chapman-Kolmogorov equation, which takes the form

$$\begin{aligned} 0 = &-\sum_i \frac{\partial}{\partial z_i}[A_i(z)p(z,t\,|\,\boldsymbol{y},t')] + \tfrac{1}{2}\sum_{i,j}\frac{\partial^2}{\partial z_i \partial z_j}[B_{ij}(z)p(z,t\,|\,\boldsymbol{y},t')] \\ &+ \int dx[W(z\,|\,x)p(x,t\,|\,\boldsymbol{y},t') - W(x\,|\,z)p(z,t\,|\,\boldsymbol{y},t')]\,, \end{aligned} \tag{3.7.27}$$

where we have used the fact that the process is homogeneous to note that $\boldsymbol{A}$, B and W, as defined in (3.4.1–3.4.3), are independent of t. This is an alternative definition of a homogeneous process.

3.7.3 Approach to a Stationary Process

A converse problem also exists. Suppose $\boldsymbol{A}$, B and W are independent of time and $p_s(z)$ satisfies (3.7.27). Under what conditions does a solution of the differential Chapman-Kolmogorov equation approach the stationary solution $p_s(z)$?

There does not appear to be a complete answer to this problem. However, we can give a reasonably good picture as follows. We define a *Lyapunov functional* K of any two solutions p_1 and p_2 of the differential Chapman-Kolgomorov equation by

$$K = \int dx\, p_1(\boldsymbol{x},t\log[p_1(\boldsymbol{x},t)/p_2(\boldsymbol{x},t)]\,, \tag{3.7.28}$$

and assume for the moment that neither p_1 nor p_2 are zero anywhere. We will now show that K is always positive and dK/dt is always negative.

Firstly, noting that both $p_2(\boldsymbol{x},t)$ and $p_1(\boldsymbol{x},t)$ are normalised to one, we write

$$K[p_1, p_2, t] = \int dx\, p_1(x,t)\Big\{\log[p_1(x,t)/p_2(x,t)] + p_2(x,t)/p_1(x,t) - 1\Big\} \tag{3.7.29}$$

and use the inequality valid for all $z > 0$,

$$-\log z + z - 1 \geqslant 0\,, \tag{3.7.30}$$

to show that $K \geqslant 0$.

Let us now show that $dK/dt \leqslant 0$. We can write (using an abbreviated notation)

$$\frac{dK}{dT} = \int dx \left\{ \frac{\partial p_1}{\partial t}[\log p_1 + 1 - \log p_2] - \frac{\partial p_2}{\partial t}[p_1/p_2] \right\}. \tag{3.7.31}$$

We now calculate one by one the contributions to dK/dt from drift, diffusion, and jump terms in the differential Chapman-Kolmogorov equation:

$$\left(\frac{dK}{dt}\right)_{\text{drift}} = \sum_i \int dx \left\{ -[\log(p_1/p_2) + 1]\frac{\partial}{\partial x_i}(A_i p_1) + (p_1/p_2)\frac{\partial}{\partial x_i}(A_i p_2) \right\} \tag{3.7.32}$$

which can be rearranged to give

$$\left(\frac{dK}{dt}\right)_{\text{drift}} = \sum_i \int dx \frac{\partial}{\partial x_i}[-A_i p_1 \log(p_1/p_2)]\,. \tag{3.7.33}$$

Similarly, we may calculate

$$\left(\frac{dK}{dt}\right)_{\text{diff}} = -\tfrac{1}{2}\sum_{i,j}\int dx \left\{ [\log(p_1/p_2) + 1]\frac{\partial^2}{\partial x_i \partial x_j}(B_{ij}p_1) - (p_1/p_2)\frac{\partial^2}{\partial x_i \partial x_j}(B_{ij}p_2) \right\} \tag{3.7.34}$$

and after some rearranging we may write

$$\left(\frac{dK}{dt}\right)_{\text{diff}} = -\tfrac{1}{2}\sum_{i,j}\int dx\, p_1 B_{ij} \left\{ \frac{\partial}{\partial x_i}[\log(p_1/p_2)] \right\} \left\{ \frac{\partial}{\partial x_i}[\log(p_1/p_2)] \right\}$$
$$+\tfrac{1}{2}\sum_{i,j}\int dx \frac{\partial^2}{\partial x_i \partial x_j}\Big[p_1 B_{ij}\log(p_1/p_2)\Big]\,. \tag{3.7.35}$$

Finally, we may calculate the jump contribution similarly:

$$\left(\frac{dK}{dt}\right)_{\text{jump}} = \int dx\, dx' \Big\{ [W(x\,|\,x')p_1(x't) - W(x'\,|\,x)p_1(x,t)]$$
$$\times \{\log[p_1(x,t)/p_2(x,t)] + 1\}$$
$$- \Big[W(x\,|\,x')p_2(x',t) - W(x'\,|\,x)p_2(x,t)\Big]p_1(x,t)/p_2(x,t)\Big\}, \tag{3.7.36}$$

and after some rearrangement,

$$\left(\frac{dK}{dt}\right)_{\text{jump}} = \int dx\, dx'\, W(x\,|\,x')\{p_2(x',t)[\phi'\log[\phi/\phi'] - \phi + \phi']\}, \tag{3.7.37}$$

where

$$\phi = p_1(x,t)/p_2(x,t), \tag{3.7.38}$$

and ϕ' is similarly defined in terms of $\boldsymbol{x}'$.

We now consider the simplest case. Suppose a stationary solution $p_s(\boldsymbol{x})$ exists which is nonzero everywhere, except at infinity, where it and its first derivative vanish. Then we may choose $p_2(\boldsymbol{x}, t) = p_s(\boldsymbol{x})$. The contributions to dK/dt from (3.7.33) and the second term in (3.7.35) can be integrated to give surface terms which vanish at infinity so that we find

$$\left(\frac{dK}{dt}\right)_{\text{drift}} = 0, \tag{3.7.39a}$$

$$\left(\frac{dK}{dt}\right)_{\text{diff}} \leqslant 0, \tag{3.7.39b}$$

$$\left(\frac{dK}{dt}\right)_{\text{jump}} \leqslant 0, \tag{3.7.39c}$$

where the last inequality comes by setting $z = \phi'/\phi$ in (3.7.30).

We must now consider under what situations the *equalities* in (3.7.39c) are actually achieved. Inspection of (3.7.37) shows that this term will be zero if and only if $\phi = \phi'$ for almost all $\boldsymbol{x}$ and $\boldsymbol{x}'$ which are such that $W(\boldsymbol{x}\,|\,\boldsymbol{x}') \neq 0$. Thus, if $W(\boldsymbol{x}\,|\,\boldsymbol{x}')$ is never zero, i.e., if transitions can take place in *both directions* between any pair of states, the vanishing of the jump contribution implies that $\phi(x) = \phi(x')$ for all x and x', that is $\phi(x)$ is independent of x, so that

$$p_1(\boldsymbol{x}, t)/p_s(\boldsymbol{x}) = \text{constant}. \tag{3.7.40}$$

The constant must equal one since $p_1(\boldsymbol{x}, t)$ and $p_s(\boldsymbol{x})$ are both normalised.

The term arising from diffusion will be strictly negative if B_{ij} is almost everywhere positive definite. Hence, we have now shown that under rather strong conditions, namely,

$$\left.\begin{array}{lll} p_s(\boldsymbol{x}) & \neq 0 & \text{with probability 1,} \\ W(\boldsymbol{x}\,|\,\boldsymbol{x}') & \neq 0 & \text{with probability 1,} \\ B_{ij}(\boldsymbol{x}) & \text{positive definite} & \text{with probability 1,} \end{array}\right\} \tag{3.7.41}$$

that any solution of the differential Chapman-Kolmogorov equation approaches the stationary solution $p_s(\boldsymbol{x})$ at $t \to \infty$.

The result fails in two basic kinds of systems.

a) Disconnected State Space: The result is best illustrated when A_i and B_{ij} vanish, so we have a pure jump system. Suppose the space divides into two regions R_1 and R_2 such that transitions from R_1 to R_2 and back are impossible; hence, $W(\boldsymbol{x}\,|\,\boldsymbol{x}') = 0$ if $\boldsymbol{x}$ and $\boldsymbol{x}'$ are not both in R_1 or R_2. Then it is possible to have $dK/dt = 0$ if

$$p_1(\boldsymbol{x}, t) = \begin{cases} \lambda_1 p_s(\boldsymbol{x}), & x \in R_1, \\ \lambda_2 p_s(\boldsymbol{x}), & x \in R_2, \end{cases} \tag{3.7.42}$$

so that there is no unique stationary distribution. The two regions are disconnected and separate stochastic processes take place in each, and in each of these, there is a unique stationary solution. The relative probability of being R_1 or R_2 is not changed by the process.

A similar result holds, in general, if as well we have B_{ij} and A_i vanishing on the boundary between R_1 and R_2.

b) $p_s(x)$ Vanishes in Some Definite Region: If we have

$$p_s(\boldsymbol{x}) \begin{cases} = 0, & \boldsymbol{x} \in R_1, \\ \neq 0, & \boldsymbol{x} \in R_2, \end{cases} \tag{3.7.43}$$

and again A_i and B_{ij} vanish, then it follows that, since $p_s(\boldsymbol{x})$ satisfies the stationary equation (3.7.27),

$$W(\boldsymbol{x} \,|\, \boldsymbol{y}) = 0, \quad \boldsymbol{x} \in R_1, \boldsymbol{y} \in R_2\,. \tag{3.7.44}$$

In other words, no transitions are possible from the region R_2 where the stationary distribution is positive to R_1, where the stationary distribution vanishes.

3.7.4 Autocorrelation Function for Markov Processes

For any Markov process, we can write a very elegant formula for the autocorrelation function. We define

$$\langle \boldsymbol{X}(t) \,|\, [\boldsymbol{x}_0, t_0] \rangle = \int d\boldsymbol{x}\, \boldsymbol{x}\, p(\boldsymbol{x}, t \,|\, \boldsymbol{x}_0, t_0)\,, \tag{3.7.45}$$

then the autocorrelation matrix

$$\langle \boldsymbol{X}(t)\boldsymbol{X}(t_0)^T \rangle = \int d\boldsymbol{x} \int d\boldsymbol{x}_0\, \boldsymbol{x}\boldsymbol{x}_0^T p(\boldsymbol{x}, t; \boldsymbol{x}_0, t_0)\,, \tag{3.7.46}$$

$$= \int d\boldsymbol{x}_0\, \langle \boldsymbol{X}(t) \,|\, [\boldsymbol{x}_0, t_0] \rangle \boldsymbol{x}_0^T p(\boldsymbol{x}_0, t_0)\,. \tag{3.7.47}$$

Thus we see that (3.7.45) defines the mean of $\boldsymbol{X}(t)$ under the condition that $\boldsymbol{X}$ had the value $\boldsymbol{x}_0$ at time t_0, and (3.7.47) tells us that the autocorrelation matrix is obtained by averaging this conditional average (multiplied by $\boldsymbol{x}_0^T$) at time t_0. These results are true by definition for any stochastic process.

In a Markov process we have, however, a unique conditional probability which determines the whole process. Thus, for a Markov process, $\langle \boldsymbol{X}(t) \,|\, [\boldsymbol{x}_0, t_0] \rangle$ is a uniquely defined quantity, since the knowledge of $\boldsymbol{x}_0$ at time t_0 completely determines the future of the process.

a) Stationary Autocorrelation Function: The most notable use of this property is in the computation of the stationary autocorrelation function. To illustrate how this uniqueness is important, let us consider a non-Markov stationary process with joint probabilities

$$p_s(\boldsymbol{x}_1, t_1; \boldsymbol{x}_2, t_2; \ldots \boldsymbol{x}_n, t_n)\,, \tag{3.7.48}$$

which, of course, depend only on time differences. Let us now create a corresponding nonstationary process by selecting only sample paths which pass through the point $x = a$ at time $t = 0$. Thus, we define

$$p_a(\boldsymbol{x}_1, t_1; \boldsymbol{x}_2, t_2; \ldots \boldsymbol{x}_n, t_n) = p_s(\boldsymbol{x}_1, t_1, \boldsymbol{x}_2, t_2; \ldots \boldsymbol{x}_n, t_n \,|\, \boldsymbol{a}, 0)\,. \tag{3.7.49}$$

Then for this process we note that

$$\langle \boldsymbol{X}(t) \,|\, [\boldsymbol{x}_0, t_0]\rangle_{\boldsymbol{a}} = \int d\boldsymbol{x}\, \boldsymbol{x}\, p_s(\boldsymbol{x}, t \,|\, \boldsymbol{x}_0, t_0; \boldsymbol{a}, 0)\,, \tag{3.7.50}$$

which contains a dependence on $\boldsymbol{a}$ symbolised by the subscript $\boldsymbol{a}$ on the average bracket. If the original stationary process possesses appropriate ergodic properties, then

$$\lim_{\tau\to\infty} p_s(\boldsymbol{x}, t+\tau \,|\, \boldsymbol{x}_0, t_0+\tau; \boldsymbol{a}, 0) = p_s(\boldsymbol{x}, t-t_0 \,|\, \boldsymbol{x}_0, 0)\,, \tag{3.7.51}$$

so that we will also have a stationary conditional average of $\boldsymbol{x}$

$$\langle \boldsymbol{X}(t) \,|\, [\boldsymbol{x}_0, t_0]\rangle_s = \lim_{\tau\to\infty} \langle \boldsymbol{X}(t+\tau) \,|\, [\boldsymbol{x}_0, t_0+\tau]\rangle_{\boldsymbol{a}}\,, \tag{3.7.52}$$

and the stationary autocorrelation matrix is given by

$$\begin{aligned}\langle \boldsymbol{X}(t)\boldsymbol{X}(t_0)^{\mathrm{T}}\rangle_s &= \int d\boldsymbol{x}_0\, \boldsymbol{x}_0^{\mathrm{T}} \langle \boldsymbol{X}(t) \,|\, [\boldsymbol{x}_0, t_0]\rangle_s p_s(\boldsymbol{x}_0)\,, \qquad (3.7.53)\\ &= \lim_{\tau\to\infty} \langle \boldsymbol{X}(t+\tau)\boldsymbol{X}(t_0+\tau)^{\mathrm{T}}\rangle_{\boldsymbol{a}}\,,\\ &= \lim_{\tau\to\infty} \int d\boldsymbol{x}_0\, \boldsymbol{x}_0^{\mathrm{T}} \langle \boldsymbol{x}(t+\tau) \,|\, [\boldsymbol{x}_0, t_0+\tau]\rangle_{\boldsymbol{a}} p_{\boldsymbol{a}}(\boldsymbol{x}_0, t_0+\tau)\,. \qquad (3.7.54)\end{aligned}$$

However, when the process is Markovian, this cumbersome limiting procedure is not necessary since

$$\begin{aligned}\text{Markov} \Rightarrow \langle \boldsymbol{X}(t) \,|\, [\boldsymbol{x}_0, t_0]\rangle_s &= \langle \boldsymbol{X}(t) \,|\, [\boldsymbol{x}_0, t_0]\rangle_{\boldsymbol{a}}\,,\\ &= \langle \boldsymbol{X}(t) \,|\, [\boldsymbol{x}_0, t_0]\rangle\,. \qquad (3.7.55)\end{aligned}$$

b) Regression Theorem: Equation (3.7.47) is a *regression theorem* when applied to a Markov process and is the basis of a more powerful regression theorem for *linear systems*. By this we mean systems such that a linear equation of motion exists for the means, i.e.,

$$d\langle \boldsymbol{X}(t) \,|\, [\boldsymbol{x}_0, t_0]\rangle / dt = -\boldsymbol{A}\langle \boldsymbol{X}(t) \,|\, [\boldsymbol{x}_0, t_0]\rangle\,, \tag{3.7.56}$$

which is very often the case in systems of practical interest, either as an exact result or as an approximation. The initial conditions for (3.7.56) are clearly

$$\langle \boldsymbol{X}(t_0) \,|\, [\boldsymbol{x}_0, t_0]\rangle = \boldsymbol{x}_0\,. \tag{3.7.57}$$

Then from (3.7.55,3.7.56)

$$\frac{d}{dt}\langle \boldsymbol{X}(t)\boldsymbol{X}(t_0)^{\mathrm{T}}\rangle = -\boldsymbol{A}\langle \boldsymbol{X}(t)\boldsymbol{X}(t_0)^{\mathrm{T}}\rangle\,, \tag{3.7.58}$$

with initial conditions $\langle \boldsymbol{X}(t_0)\boldsymbol{X}(t_0)^{\mathrm{T}}\rangle$. The time correlation matrix

$$\langle \boldsymbol{X}(t)\boldsymbol{X}(t_0)^{\mathrm{T}}\rangle - \langle \boldsymbol{X}(t)\rangle\langle \boldsymbol{X}(t_0)^{\mathrm{T}}\rangle = \langle \boldsymbol{X}(t), \boldsymbol{X}(t_0)^{\mathrm{T}}\rangle\,, \tag{3.7.59}$$

obviously obeys the same equation, with the initial condition given by the covariance matrix at time t_0. In a stationary system, we have the result that if $\boldsymbol{G}(t)$ is the stationary time correlation function and $\boldsymbol{\sigma}$ the stationary covariance matrix, then

$$\frac{dG(t)}{dt} = -A\,G(t)\,, \tag{3.7.60}$$

$$G(0) = \sigma\,, \tag{3.7.61}$$

so that

$$G(t) = \exp[-At]\sigma, \tag{3.7.62}$$

which is the *regression theorem* in its simplest form. We again stress that it is valid for the *Markov processes* in which the mean values obey *linear* evolution equations like (3.7.56).

For non-Markov processes there is no simple procedure. We must carry out the complicated procedure implicit in (3.7.54).

3.8 Examples of Markov Processes

We present here for reference some fundamental solutions of certain cases of the differential Chapman-Kolmogorov equation. These will have a wide application throughout the remainder of this book.

3.8.1 The Wiener Process

This takes its name from *Norbert Wiener* who studied it extensively. From the point of view of this chapter, it is the solution of the Fokker-Planck equation as discussed in Sect. 3.5.2, in which there is only one variable $W(t)$, the drift coefficient is zero and the diffusion coefficient is 1. Thus, the Fokker-Planck equation for this case is

$$\frac{\partial}{\partial t}p(w,t\,|\,w_0,t_0) = \frac{1}{2}\frac{\partial^2}{\partial w^2}p(w,t\,|\,w_0,t_0)\,. \tag{3.8.1}$$

Utilising the initial condition

$$p(w,t_0\,|\,w_0,t_0) = \delta(w-w_0), \tag{3.8.2}$$

on the conditional probability, we solve (3.8.1) by use of the characteristic function

$$\phi(s,t) = \int dw\, p(w,t\,|\,w_0,t_0)\exp(isw), \tag{3.8.3}$$

which satisfies

$$\frac{\partial\phi}{\partial t} = -\tfrac{1}{2}s^2\phi, \tag{3.8.4}$$

so that

$$\phi(s,t) = \exp\left[-\tfrac{1}{2}s^2(t-t_0)\right]\phi(s,t_0)\,. \tag{3.8.5}$$

From (3.8.2), the initial condition is

$$\phi(s,t_0) = \exp(isw_0), \tag{3.8.6}$$

so that

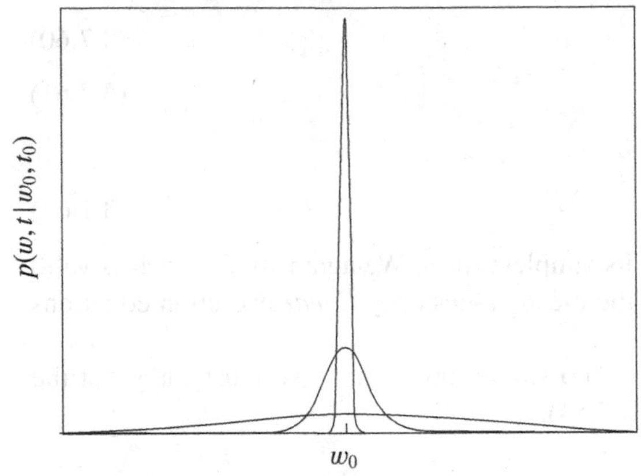

Fig. 3.3. Wiener process: spreading of an initially sharp distribution $p(w, t\,|\,w_0, t_0)$ with increasing time $t - t_0$

$$\phi(s, t) = \exp\left[isw_0 - \tfrac{1}{2}s^2(t - t_0)\right]. \tag{3.8.7}$$

Performing the Fourier inversion, we have the solution to (3.8.1):

$$p(w, t\,|\,w_0, t_0) = [2\pi(t - t_0)]^{-1/2} \exp\left(-\frac{(w - w_0)^2}{2(t - t_0)}\right). \tag{3.8.8}$$

This represents a Gaussian, with

$$\langle W(t) \rangle = w_0, \tag{3.8.9}$$

$$\langle [W(t) - w_0]^2 \rangle = t - t_0, \tag{3.8.10}$$

so that an initially sharp distribution spreads in time, as graphed in Fig. 3.3.

A multivariate Wiener process can be defined as

$$\boldsymbol{W}(t) = [W_1(t), W_2(t), \dots, W_n(t)], \tag{3.8.11}$$

which satisfies the multivariable Fokker-Planck equation

$$\frac{\partial}{\partial t} p(\boldsymbol{w}, t\,|\,\boldsymbol{w}_0, t_0) = \frac{1}{2} \sum_i \frac{\partial^2}{\partial w_i^2} p(\boldsymbol{w}, t\,|\,\boldsymbol{w}_0, t_0) \tag{3.8.12}$$

whose solution is

$$p(\boldsymbol{w}, t\,|\,\boldsymbol{w}_0, t_0) = [2\pi(t - t_0)]^{-n/2} \exp\left(-\frac{(\boldsymbol{w} - \boldsymbol{w}_0)^2}{2(t - t_0)}\right), \tag{3.8.13}$$

a multivariate Gaussian with

$$\langle \boldsymbol{W}(t) \rangle = \boldsymbol{w}_0, \tag{3.8.14}$$

and

$$\langle [W_i(t) - w_{0i}][W_j(t) - w_{0j}] \rangle = (t - t_0)\delta_{ij}. \tag{3.8.15}$$

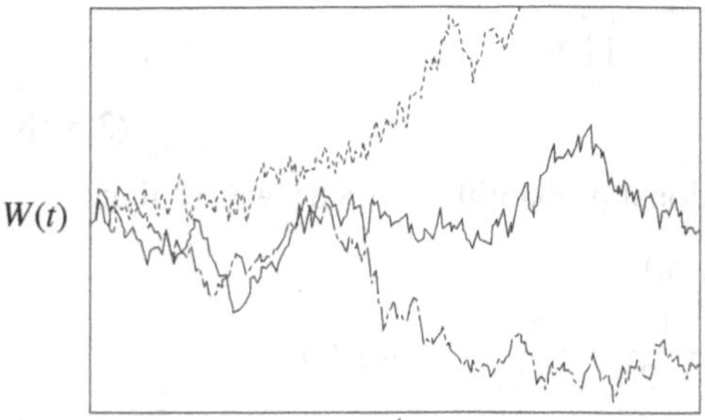

Fig. 3.4. Three simulated sample paths of the Wiener process, illustrating their great variability

The one-variable Wiener process is often simply called Brownian motion, since the Wiener process equation (3.8.1) is exactly the same as the differential equation of diffusion, shown by Einstein to be obeyed by Brownian motion, as we noted in Sect. 1.2. The terminology is, however, not universal.

Points of note concerning the Wiener process are:

a) Irregularity of Sample Paths: Although the mean value of $W(t)$ is zero, the mean square becomes infinite as $t \to \infty$. This means that the sample paths of $W(t)$ are *very* variable, indeed surprisingly so. In Fig. 3.4, we have given a few different sample paths with the same initial point to illustrate the extreme non-reproducibility of the paths.

b) Non-differentiability of Sample Paths: The Wiener process is a diffusion process and hence the sample paths of $W(t)$ are continuous. However, they are *not* differentiable. Consider

$$\text{Prob}\left\{\left|\frac{[W(t+h) - W(t)]}{h}\right| > k\right\}. \tag{3.8.16}$$

From the solution for the conditional probability, this probability is

$$2\int_{kh}^{\infty} dw(2\pi h)^{-1/2}\exp(-w^2/2h), \tag{3.8.17}$$

and in the limit as $h \to 0$ this is one. This means that no matter what value of k choose, $|[W(t+h) - W(t)]/h|$ is almost certain to be greater than this, i.e., the derivative at any point is almost certainly infinite. This is in agreement with the similar intuitive picture presented in Sect. 3.5.2 and the simulated paths given in Fig. 3.4 illustrate in point dramatically. This corresponds, of course, to the well-known experimental fact that the Brownian particles have an exceedingly irregular motion. However, this clearly an idealisation, since if $W(t)$ represents the position of the Brownian particle, this means that its speed is almost certainly infinite. The *Ornstein-Uhlenbeck* process is a more realistic model of Brownian motion (Sect. 3.8.4).

c) Independence of Increment: The Wiener process is fundamental to the study of diffusion processes, and by means of stochastic differential equations, we can express any diffusion process in terms of the Wiener process. Of particular importance is the statistical independence of the increments of $W(t)$. More precisely, since the Wiener process is a Markov process, the joint probability density can be written

$$p(w_n, t_n; w_{n-1}, t_{n-1}; w_{n-2}, t_{n-2}; \dots; w_0, t_0) = \prod_{i-0}^{n-1} p(w_{i+1}, t_{i+1} \,|\, w_i, t_i) p(w_0, t_0) , \tag{3.8.18}$$

and using the explicit form of the conditional probabilities (3.8.8), we see that

$$p(w_n, t_n; w_{n-1}, t_{n-1}; w_{n-2}, t_{n-2}; \dots; w_0, t_0) = \prod_{i=0}^{n} \left\{ \frac{1}{\sqrt{2\pi(t_{i+1} - t_i)}} \exp\left(-\frac{(w_{i+1} - w_t)^2}{2(t_{i+1} - t_i)} \right) \right\} p(w_0, t_0) . \tag{3.8.19}$$

If we define the variables

$$\Delta W_i \equiv W(t_i) - W(t_{i-1}) , \tag{3.8.20}$$

$$\Delta t_i \equiv t_i - t_{i-1} , \tag{3.8.21}$$

then the joint probability density for the ΔW_i is

$$p(\Delta w_n; \Delta w_{n-1}; \Delta w_{n-2}; \dots \Delta w_1; w_0) = \prod_{i=1}^{n} \left\{ (2\pi \Delta t_i)^{-1/2} \exp(-\Delta w_i^2 / 2\Delta t_i) \right\} p(w_0, t_0) , \tag{3.8.22}$$

which shows, from the definition of statistical independence given in Sect. 2.3.4, that the variables ΔW_i are independent of each other and of $W(t_0)$.

The aspect of having independent increments ΔW_i is very important in the definition of stochastic integration which is carried out in Sect. 4.2.

d) Autocorrelation Functions: A quantity of great interest is the autocorrelation function, already discussed in Sects. 1.5.2 and 3.7.4. The formal definition is

$$\langle W(t) W(s) \,|\, [w_0, t_0] \rangle = \int dw_1 \, dw_2 \, w_1 \, w_2 \, p(w_1, t; w_2, s \,|\, w_0, t_0) , \tag{3.8.23}$$

which is the mean product of $W(t)$ and $W(s)$ on the condition that the initial value is $W(t_0) = w_0$, and we can see, assuming $t > s$, that

$$\langle W(t) W(s) \,|\, [W_0, t_0] \rangle = \langle [W(t) - W(s)] W(s) \rangle + \langle [W(s)]^2 \rangle . \tag{3.8.24}$$

Using the independence of increments, the first average is zero and the second is given by (3.8.10) so that we have, in general,

$$\langle W(t) W(s) \,|\, [W_0, t_0] \rangle = \min(t - t_0, s - t_0) + w_0^2 , \tag{3.8.25}$$

which is correct for both $t > s$ and $t < s$.

3.8.2 The Random Walk in One Dimension

A man moves along a line, taking, at random, steps to the left or the right with equal probability. The steps are of length l so that his position can take on only the value nl, where n is integral. We want to know the probability that he reaches a given point a distance nl from the origin after a given elapsed time.

The problem can be defined in two ways. The first, which is more traditional, is to allow the walker to take steps at times $N\tau$ (N integral) at which times he *must* step either left or right, with equal probability. The second is to allow the walker to take steps left or right with a probability per unit time d which means that the walker waits at each point for a *variable* time. The second method is describable by a master equation.

a) Continuous Time Random Walk: To do a master equation treatment of the problem, we consider that the transition probability per unit time is given by the form

$$W(n+1\,|\,n,t) = W(n-1\,|\,n,t) = d\,. \tag{3.8.26}$$

otherwise, $W(n\,|\,m,t) = 0$ so that, according to Sect. 3.5.1, the master equation for the man to be at the position nl, given that he started at $n'l$, is

$$\partial_t P(n,t\,|\,n',t') = d[P(n+1,t\,|\,n',t') + P(n-1,t\,|\,n',t')\,. \tag{3.8.27}$$

b) Discrete Time Random Walk: The more classical form of the random walk does not assume that the man makes his jump to the left or right according to a master equation, but that he jumps left or right with equal probability at times $N\tau$, so that time is a discrete variable. In this case, we can write

$$P(n,(N+1)\tau\,|\,n',N'\tau) = \tfrac{1}{2}\left\{P(n+1,N\tau\,|\,n',N'\tau) + P(n-1,N\tau\,|\,n',N'\tau)\right\}. \tag{3.8.28}$$

If τ is small, we can view (3.8.27) and (3.8.28) as approximations to each other by writing

$$P(n,(N+1)\tau\,|\,n',N',\tau) \simeq P(n,N\tau\,|\,n',N'\tau) + \tau\partial_t P(n,t\,|\,n',t')\,, \tag{3.8.29}$$

with $t = N\tau$, $t' = N'\tau$ and $d = \frac{1}{2}\tau^{-1}$, so that the transition probability per unit time in the master equation model corresponds to half of the inverse waiting time τ in the discrete time model.

c) Solutions Using the Characteristic Function: Both systems can be easily solved by introducing the characteristic function

$$G(s,t) = \langle e^{ins}\rangle = \sum_n P(n,t\,|\,n',t')e^{ins}\,, \tag{3.8.30}$$

in which case the master equation gives

$$\partial_t G(s,t) = d(e^{is} + e^{-is} - 2)G(s,t)\,, \tag{3.8.31}$$

and the discrete time equation becomes

$$G(s,(N+1)\tau) = \tfrac{1}{2}(e^{is} + e^{-is})G(s,N\tau)\,. \tag{3.8.32}$$

Assuming the man starts at the origin $n' = 0$ at time $t' = 0$, we find

$$G(s,0) = 1\,, \tag{3.8.33}$$

in both cases, so that the solution to (3.8.31) is

$$G_1(s,t) = \exp[(e^{is} + e^{-is} - 2)td]\,, \tag{3.8.34}$$

and to (3.8.32)

$$G_2(s, N\tau) = \left[\tfrac{1}{2}(e^{is} + e^{-is})\right]^N . \tag{3.8.35}$$

The appropriate probability distributions can be obtained by expanding $G_1(s, N\tau)$ and $G_2(s, t)$ in powers of $\exp(is)$; we find

$$P_1(n, t\,|\,0, 0) \quad = e^{-2td} I_n(4td) , \tag{3.8.36}$$

$$P_2(n, N\tau\,|\,0, 0) = \left(\frac{1}{2}\right)^N N! \left[\left(\frac{N-n}{2}\right)! \left(\frac{N+n}{2}\right)!\right]^{-1} . \tag{3.8.37}$$

The discrete time distribution is also known as the Bernoulli distribution; it gives the probability of a total of n heads in tossing an unbiased coin N times.

d) Continuous Space Limit: For both kinds of random walk, the limit of continuous space—that is, very many steps of very small size—gives the Wiener process. If we set the distance travelled as

$$x = nl , \tag{3.8.38}$$

so that the characteristic function of the distribution of x is

$$\phi_1(s, t) = \langle e^{isx} \rangle = G_1(ls, t) = \exp[(e^{ils} + e^{-ils} - 2)td] . \tag{3.8.39}$$

Then the limit of infinitesimally small steps $l \to 0$ is

$$\phi_1(s, t) \to \exp(-s^2 tD) , \tag{3.8.40}$$

where

$$D = \lim_{l \to 0}(l^2 d) . \tag{3.8.41}$$

This is the characteristic function of a Gaussian (Sect. 2.8.1) of the form

$$P(x, t\,|\,0, 0) = (4\pi Dt)^{-1/2} \exp(-x^2/4Dt) , \tag{3.8.42}$$

and is of course the distribution for the Wiener process (Sect. 3.8.1) or Brownian motion, as mentioned in Sect. 1.2. Thus, the Wiener process can be regarded as the limit of a continuous time random walk in the limit of infinitesimally small step size.

The limit

$$l \to 0, \tau \to 0, \quad \text{with} \quad D = \lim_{l \to 0}(l^2/\tau), \tag{3.8.43}$$

of the discrete time random walk gives the same result. From this form, we see clearly the expression of D as the mean square distance travelled per unit time.

We can also see more directly that expanding the right-hand side of (3.8.27) as a function of x up to second order in l gives

$$\partial_t p(x, t\,|\,0, 0) = l^2 d\, \partial_x^2 P(x, t\,|\,0, 0) . \tag{3.8.44}$$

The three processes are thus intimately connected with each other at two levels, namely, under the limits considered, the stochastic equations approach each other and under those same limits, the solutions to these equations approach each other. These limits are exactly those used by Einstein. Comparison with Sect. 1.2 shows that he modelled Brownian motion by a discrete time and space random walk, but nevertheless, derived the Wiener process model by expanding the equations for time development of the distribution function.

The limit results of this section are a slightly more rigorous version of Einstein's method. There are generalisations of these results to less specialised situations and it is a fair statement to make that almost any jump process has some kind of limit which is a diffusion process. However, the precise limits are not always so simple, and there are limits in which certain jump processes become deterministic and are governed by Liouville's equation (Sect. 3.5.3) rather than the full Fokker-Planck equation. These results are presented in Sect. 11.2.

3.8.3 Poisson Process

We have already noted the Poisson process in Sect. 1.5.1. The process in which electrons arrive at an anode or customers arrive at a shop with a probability per unit time λ of arriving, is governed by the master equation for which

$$W(n+1\,|\,n,t) = \lambda\,, \tag{3.8.45}$$

otherwise,

$$W(n\,|\,m,t) = 0\,. \tag{3.8.46}$$

This master equation becomes

$$\partial_t P(n,t\,|\,n',t') = \lambda[P(n-1,t\,|\,n',t',) - P(n,t\,|\,n',t')]\,, \tag{3.8.47}$$

and by comparison with (3.8.27) also represents a "one-sided" random walk, in which the walker steps to the right only with probability per unit time equal to λ.

The characteristic function equation is similar to (3.8.31):

$$\partial_t G(s,t) = \lambda[\exp(is) - 1]G(s,t)\,, \tag{3.8.48}$$

with the solution

$$G(s,t) = \exp\{\lambda t\,[\exp(is) - 1]\}\,, \tag{3.8.49}$$

for the initial condition that there are initially no customers (or electrons) at time $t = 0$, yielding

$$P(n,t\,|\,0,0) = \frac{\exp(-\lambda t)(\lambda t)^n}{n!}\,, \tag{3.8.50}$$

a Poisson distribution with mean given by

$$\langle N(t)\rangle = \lambda t\,. \tag{3.8.51}$$

a) Continuous Space Limit: In contrast to the random walk, the only limit that exists is $l \to 0$, with

$$\lambda l \equiv v\,, \tag{3.8.52}$$

held fixed, and the limiting characteristic function is

$$\lim_{l\to 0}\{\exp[\lambda t(e^{ils}-1)]\} = \exp(itvs), \tag{3.8.53}$$

with the solution

$$p(x,t\,|\,0,0) = \delta(x-vt). \tag{3.8.54}$$

We also see that in this limit the master equation (3.8.47) would become Liouville's equation, whose solution would be the deterministic motion we have derived.

b) Asymptotic Approximation: We can do a slightly more refined analysis. We expand the characteristic function up to second order in s in the exponent and find

$$\phi(s,t) = G(ls,t) \simeq \exp[t(ivs - s^2D/2)], \tag{3.8.55}$$

where, as in the previous section,

$$D = l^2\lambda. \tag{3.8.56}$$

This is the characteristic function of a Gaussian with variance Dt and mean vt, so that we now have

$$p(x,t\,|\,0,0) \simeq \frac{1}{\sqrt{2\pi Dt}}\exp\left(-\frac{(x-vt)^2}{2Dt}\right). \tag{3.8.57}$$

It is also clear that this solution is the solution of

$$\partial_t p(x,t\,|\,0,0) = -v\,\partial_x p(x,t\,|\,0,0) + \frac{1}{2}D\,\partial_x^2 p(x,t\,|\,0,0), \tag{3.8.58}$$

which is obtained by expanding the master equation (3.8.47) to order l^2, by writing

$$\begin{aligned} P(n-1,t\,|\,0,0) &= \lambda p(x-l,t\,|\,0,0), \\ &\simeq \lambda\, p(x,t\,|\,0,0) - l\lambda\,\partial_x p(x,t\,|\,0,0) + \tfrac{1}{2}l^2\lambda\,\partial_x^2 p(x,t\,|\,0,0). \end{aligned} \tag{3.8.59}$$

However, this is an *approximation* or an *expansion* and not a limit. The limit $l \to 0$ gives Liouville's equation with the purely deterministic solution (3.8.54). Effectively, the limit $l \to 0$ with well-defined v corresponds to $D = 0$. The kind of approximation just mentioned is a special case of van Kampen's system size expansion which we treat fully in Sect. 11.2.3.

3.8.4 The Ornstein-Uhlenbeck Process

All the examples so far have had no stationary distribution, that is, as $t \to \infty$, the distribution at any finite point approaches zero and we see that, with probability one, the point moves to infinity.

If we add a linear drift term to the Wiener process, we have a Fokker-Planck equation of the form

$$\partial_t p = \partial_x(kxp) + \tfrac{1}{2}D\partial_x^2 p, \tag{3.8.60}$$

where by p we mean $p(x,t\,|\,x_0,0)$. This is the *Ornstein-Uhlenbeck process* [3.6].

a) Characteristic Function Solution: The equation for the characteristic function

$$\phi(s) = \int_{-\infty}^{\infty} e^{isx} p(x,t\,|\,x_0,0)\,dx\,, \tag{3.8.61}$$

is

$$\partial_t \phi + ks\partial_s \phi = -\frac{1}{2}Ds^2\phi\,. \tag{3.8.62}$$

The method of characteristics can be used to solve this equation, namely, if

$$u(s,t,\phi) = a\,, \quad \text{and} \quad v(s,t,\phi) = b\,, \tag{3.8.63}$$

are two integrals of the subsidiary equation (with a and b arbitrary constants)

$$\frac{dt}{1} = \frac{ds}{ks} = -\frac{d\phi}{\frac{1}{2}Ds^2\phi}\,, \tag{3.8.64}$$

then a general solution of (3.8.62) is given by

$$f(u,v) = 0\,. \tag{3.8.65}$$

The particular integrals are readily found by integrating the equation involving dt and ds and that involving ds and $d\phi$; they are

$$u(s,t,\phi) = s\exp(-kt)\,, \tag{3.8.66}$$

$$v(s,t,\phi) = \phi\exp(Ds^2/4k)\,. \tag{3.8.67}$$

The general solution can clearly be put in the form $v = g(u)$ with $g(u)$ an arbitrary function of u. Thus, the general solution is

$$\phi(s,t) = \exp(-Ds^2/4k)g[s\exp(-kt)]\,. \tag{3.8.68}$$

The boundary condition

$$p(x,0\,|\,x_0,0) = \delta(x-x_0)\,, \tag{3.8.69}$$

clearly requires

$$\phi(s,0) = \exp(ix_0 s)\,, \tag{3.8.70}$$

and gives

$$g(s) = \exp(Ds^2/4k + ix_0 s)\,. \tag{3.8.71}$$

Hence

$$\phi(s,t) = \exp\left[\frac{-Ds^2}{4k}(1-e^{-2kt}) + isx_0e^{-kt}\right]. \tag{3.8.72}$$

From Sect. 2.8.1 this corresponds to a Gaussian with

$$\langle X(t)\rangle = x_0\exp(-kt)\,, \tag{3.8.73}$$

$$\operatorname{var}\{X(t)\} = \frac{D}{2k}[1-\exp(-2kt)]\,. \tag{3.8.74}$$

b) Stationary Solution: Clearly, as $t \to \infty$, the mean and variance approach limits 0 and $D/2k$, respectively, which gives a limiting stationary solution. This solution can also be obtained directly by requiring $\partial_t p = 0$, so that p satisfies the stationary Fokker-Planck equation

$$\partial_x \left[kxp + \frac{1}{2} D \partial_x p \right] = 0 , \tag{3.8.75}$$

and integrating once, we find

$$\left[kxp + \frac{1}{2} D \partial_x p \right]_{-\infty}^{x} = 0 . \tag{3.8.76}$$

The requirement that p vanish at $-\infty$ together with its derivative, is necessary for normalisation. Hence, we have

$$\frac{1}{p} \partial_x p = -\frac{2kx}{D} , \tag{3.8.77}$$

so that

$$p_s(x) = \sqrt{\frac{k}{\pi D}} \exp\left(-\frac{kx^2}{D} \right) . \tag{3.8.78}$$

This is a Gaussian with mean 0 and variance $D/2k$, as predicted from the time-dependent solution.

It is clear that a stationary solution can always be obtained for a one variable system by this integration method if such a stationary solution exists. If a stationary solution does not exist, this method gives an unnormalisable solution.

c) Time Correlation Functions: The time correlation function analogous to that mentioned in connection with the Wiener process can be calculated and is a measurable piece of data in most stochastic systems. However, we have no easy way of computing it other than by definition

$$\langle X(t)X(s) \,|\, [x_0, t_0] \rangle = \int\int dx_1 dx_2 \, x_1 x_2 p(x_1, t; x_2, s \,|\, x_0, t_0) , \tag{3.8.79}$$

and using the Markov property

$$= \int\int dx_1 dx_2 x_1 x_2 p(x_1, t \,|\, x_2, s) p(x_2, s \,|\, x_0, t_0) , \tag{3.8.80}$$

on the assumption that

$$t \geqslant s \geqslant t_0 . \tag{3.8.81}$$

The correlation function with a definite initial condition is not normally of as much interest as the *stationary correlation function*, which is obtained by allowing the system to approach the stationary distribution. It is achieved by putting the initial condition in the remote past, as pointed out in Sect. 3.7.2. Letting $t_0 \to -\infty$, we find

$$\lim_{t_0 \to -\infty} p(x_2, s \,|\, x_0, t_0) = p_s(x_2) = (\pi D/k)^{-1/2} \exp(-k x_2^2 / D) . \tag{3.8.82}$$

and by straightforward substitution and integration and noting that the stationary mean is zero, we get

$$\langle X(t)X(s)\rangle_s = \langle X(t), X(s)\rangle_s = \frac{D}{2k}\exp(-k|t-s|)\,. \tag{3.8.83}$$

This result demonstrates the general property of stationary processes: that the correlation functions depend only on time differences. It is also a general result [3.7] that the process we have described in this section is the only stationary Gaussian Markov process in one real variable.

The results of this subsection are very easily obtained by the stochastic differential equation methods which will be developed in Chap. 4.

The Ornstein-Uhlenbeck process is a simple, explicitly representable process, which has a stationary solution. In its stationary state, it is often used to model a realistic noise signal, in which $X(t)$ and $X(s)$ are only significantly correlated if

$$|t-s| \sim 1/k \equiv \tau\,. \tag{3.8.84}$$

More precisely, τ, known as the *correlation time* can be defined for arbitrary processes $X(s)$ by

$$\tau = \int_0^\infty dt \langle X(t), X(0)\rangle_s / \mathrm{var}\{X\}_s\,, \tag{3.8.85}$$

which is independent of the precise functional form of the correlation function.

3.8.5 Random Telegraph Process

We consider a signal $X(t)$ which can have either of two values a and b and switches from one to the other with certain probabilities per unit time. Thus, we have a master equation

$$\left.\begin{aligned} \partial_t P(a,t\,|\,x,t_0) &= -\lambda P(a,t\,|\,x,t_0) + \mu P(b,t\,|\,x,t_0)\,,\\ \partial_t P(b,t\,|\,x,t_0) &= \lambda P(a,t\,|\,x,t_0) - \mu P(b,t\,|\,x,t_0)\,. \end{aligned}\right\} \tag{3.8.86}$$

a) Time-Dependent Solutions: These can simply be found by noting that

$$P(a,t\,|\,x,t_0) + P(b,t\,|\,x,t_0) = 1\,, \tag{3.8.87}$$

and using the initial condition

$$P(x',t_0\,|\,x,t_0) = \delta_{x,x'}\,. \tag{3.8.88}$$

A simple equation can then be derived for $\lambda P(a,t\,|\,x,t_0) - \mu P(b,t\,|\,x,t_0)$, whose solution is

$$\lambda P(a,t\,|\,x,t_0) - \mu P(b,t\,|\,x,t_0) = \exp\left[-(\lambda+\mu)(t-t_0)\right](\lambda\delta_{a,x} - \mu\delta_{bx})\,. \tag{3.8.89}$$

The solution for the probabilities then takes the form

$$\left.\begin{aligned} P(a,t\,|\,x,t_0) &= \frac{\mu}{\lambda+\mu} + \mathrm{e}^{-(\lambda+\mu)(t-t_0)}\left(\frac{\lambda}{\lambda+\mu}\delta_{a,x} - \frac{\mu}{\lambda+\mu}\delta_{b,x}\right),\\ P(b,t\,|\,x,t_0) &= \frac{\lambda}{\lambda+\mu} - \mathrm{e}^{-(\lambda+\mu)(t-t_0)}\left(\frac{\lambda}{\lambda+\mu}\delta_{a,x} - \frac{\mu}{\lambda+\mu}\delta_{b,x}\right). \end{aligned}\right\} \tag{3.8.90}$$

The mean of $X(t)$ is straightforwardly computed:

$$\langle X(t) | [x_0, t_0] \rangle = \sum x P(x, t | x_0, t_0),$$
$$= \frac{a\mu + b\lambda}{\mu + \lambda} + e^{-(\lambda+\mu)(t-t_0)} \left(x_0 - \frac{a\mu + b\lambda}{\mu + \lambda} \right). \tag{3.8.91}$$

The variance can also be computed but is a very messy expression.

b) Stationary Solutions: This process has the stationary solution obtained by letting $t_0 \to -\infty$:

$$P_s(a) = \frac{\mu}{\lambda + \mu}, \qquad P_s(b) = \frac{\lambda}{\lambda + \mu}, \tag{3.8.92}$$

which is obvious from the master equation.

The stationary mean and variance are

$$\langle X \rangle_s = \frac{a\mu + b\lambda}{\mu + \lambda}, \tag{3.8.93}$$

$$\text{var}\,\{X\}_s = \frac{(a-b)^2 \mu\lambda}{(\lambda + \mu)^2}. \tag{3.8.94}$$

c) Stationary Correlation Functions: To compute the stationary time correlation function, let $t \geqslant s$, and write

$$\langle X(t)X(s) \rangle_s = \sum_{xx'} xx' P(x, t | x', s) P_s(x'), \tag{3.8.95}$$
$$= \sum_{x'} x' \langle X(t) | [x', s] \rangle P_s(x'). \tag{3.8.96}$$

Now use (3.8.91–3.8.94) to obtain

$$\langle X(t)X(s) \rangle_s = \langle X \rangle_s^2 + \exp[-(\lambda + \mu)(t - s)](\langle X^2 \rangle_s - \langle X \rangle_s^2), \tag{3.8.97}$$
$$= \left(\frac{a\mu + b\lambda}{\mu + \lambda} \right)^2 + \exp[-(\lambda + \mu)(t - s)] \frac{(a-b)^2 \mu\lambda}{(\lambda + \mu)^2}. \tag{3.8.98}$$

Hence,

$$\langle X(t), X(s) \rangle_s = \langle X(t)X(s) \rangle_s - \langle X \rangle_s^2 = \frac{(a-b)^2 \mu\lambda}{(\lambda + \mu)^2} e^{(\lambda+\mu)|t-s|}. \tag{3.8.99}$$

Notice that this time correlation function is of exactly the same form as that of the Ornstein-Uhlenbeck process. Higher-order correlation functions are not the same of course, but because of this simple correlation function and the simplicity of the two state process, the random telegraph signal also finds wide application in model building.

4. The Ito Calculus and Stochastic Differential Equations

4.1 Motivation

In Sect. 1.2.2 we met for the first time the equation which is the prototype of what is now known as a Langevin equation, which can be described heuristically as an ordinary differential equation in which a rapidly and irregularly fluctuating random function of time [the term $X(t)$ in Langevin's original equation] occurs. The simplicity of Langevin's derivation of Einstein's results is in itself sufficient motivation to attempt to put the concept of such an equation on a reasonably precise footing.

The simple-minded Langevin equation that turns up most often can be written in the form

$$\frac{dx}{dt} = a(x,t) + b(x,t)\xi(t)\,, \tag{4.1.1}$$

where x is the variable of interest, $a(x,t)$ and $b(x,t)$ are certain known functions and $\xi(t)$ is the rapidly fluctuating random term. An idealised mathematical formulation of the concept of a "rapidly varying, highly irregular function" is that for $t \neq t'$, $\xi(t)$ and $\xi(t')$ are statistically independent. We also require $\langle \xi(t) \rangle = 0$, since any nonzero mean can be absorbed into the definition of $a(x,t)$, and thus require that

$$\langle \xi(t)\xi(t') \rangle = \delta(t-t')\,, \tag{4.1.2}$$

which satisfies the requirement of no correlation at different times and furthermore, has the rather pathological result that $\xi(t)$ has infinite variance. From a realistic point of view, we know that no quantity can have such an infinite variance, but the concept of *white noise* as an *idealisation* of a realistic fluctuating signal does have some meaning, and has already been mentioned in Sect. 1.5.2 in connection with Johnson noise in electrical circuits. We have already met two sources which might be considered realistic versions of almost uncorrelated noise, namely, the Ornstein-Uhlenbeck process and the random telegraph signal. For both of these the second-order correlation function can, up to a constant factor, be put in the form

$$\langle X(t), X(t') \rangle = \frac{\gamma}{2}\,\mathrm{e}^{-\gamma|t-t'|}\,. \tag{4.1.3}$$

Now the essential difference between these two is that the sample paths of the random telegraph signal are discontinuous, while those of the Ornstein-Uhlenbeck process are not. If (4.1.1) is to be regarded as a real differential equation, in which $\xi(t)$ is not white noise with a delta function correlation, but rather a noise with a finite correlation time, then the choice of a continuous function for $\xi(t)$ seems essential to make this equation realistic: we do not expect dx/dt to change discontinuously. The limit as $\gamma \to \infty$ of the correlation function (4.1.3) is clearly the Dirac delta function since

$$\int_{-\infty}^{\infty} \frac{\gamma}{2} e^{-\gamma|t-t'|} dt' = 1 , \tag{4.1.4}$$

and for $t \neq t'$,

$$\lim_{\gamma\to\infty} \frac{\gamma}{2} e^{-\gamma|t-t'|} = 0 . \tag{4.1.5}$$

This means that a possible model of the $\xi(t)$ could be obtained by taking some kind of limit as $\gamma \to \infty$ of the Ornstein-Uhlenbeck process. This would correspond, in the notation of Sect. 3.8.4, to the limit $k \to \infty$ with $D = k^2$.

This limit simply does not exist. Any such limit must clearly be taken after calculating measurable quantities. Such a procedure is possible but too cumbersome to use as a calculational tool.

An alternative approach is called for. Since we write the differential equation (4.1.1), we must expect it to be integrable and hence must expect that

$$u(t) = \int_0^t dt' \xi(t') , \tag{4.1.6}$$

exists.

Suppose we now demand the ordinary property of an integral, that $u(t)$ is a continuous function of t. This implies that $u(t)$ is a Markov process since we can write

$$u(t') = \int_0^t ds\, \xi(s) + \int_t^{t'} ds\, \xi(s) , \tag{4.1.7}$$

$$= \lim_{\varepsilon\to 0}\Big[\int_0^{t-\varepsilon} ds\, \xi(s)\Big] + \int_t^{t'} ds\, \xi(s) , \tag{4.1.8}$$

and for any $\varepsilon > 0$, the $\xi(s)$ in the first integral are independent of the $\xi(s)$ in the second integral. Hence, by continuity, $u(t)$ and $u(t') - u(t)$ are statistically independent and further, $u(t') - u(t)$ is independent of $u(t'')$ for all $t'' < t$. This means that $u(t')$ is fully determined (probabilistically) from the knowledge of the value of $u(t)$ and not by any past values. Hence, $u(t)$ is a Markov process.

Since the sample functions of $u(t)$ are continuous, we must be able to describe $u(t)$ by a Fokker-Planck equation. We can compute the drift and diffusion coefficients for this process by using the formulae of Sect. 3.5.2. We can write

$$\langle u(t+\Delta t) - u_0 \,|\, [u_0, t]\rangle = \Big\langle \int_t^{t+\Delta t} \xi(s) ds \Big\rangle = 0 , \tag{4.1.9}$$

and

$$\langle [u(t+\Delta t) - u_0]^2 \,|\, [u_0, t]\rangle = \int_t^{t+\Delta t} ds \int_t^{t+\Delta t} ds' \langle \xi(s)\xi(s')\rangle , \tag{4.1.10}$$

$$= \int_t^{t+\Delta t} ds \int_t^{t+\Delta t} ds' \delta(s-s') = \Delta t . \tag{4.1.11}$$

This means that the drift and diffusion coefficients are

$$A(u_0, t) = \lim_{\Delta t \to 0} \frac{\langle u(t + \Delta t) - u_0 \,|\, [u_0, t] \rangle}{\Delta t} = 0\,, \tag{4.1.12}$$

$$B(u_0, t) = \lim_{\Delta t \to 0} \frac{\langle [u(t + \Delta t) - u_0]^2 \,|\, [u_0, t] \rangle}{\Delta t} = 1\,. \tag{4.1.13}$$

The corresponding Fokker-Planck equation is that of the Wiener process and we can write

$$\int_0^t \xi(t')dt' = u(t) = W(t)\,. \tag{4.1.14}$$

Thus, we have the paradox that the integral of $\xi(t)$ is $W(t)$, which is itself not differentiable, as shown in Sect. 3.8.1. This means that mathematically speaking, the Langevin equation (4.1.1) does not exist. However, the corresponding *integral equation*

$$x(t) - x(0) = \int_0^t a[x(s), s]\, ds + \int_0^t b[x(s), s]\xi(s)\, ds\,, \tag{4.1.15}$$

can be interpreted consistently.

We make the replacement, which follows directly from the interpretation of the integral of $\xi(t)$ as the Wiener process $W(t)$, that

$$dW(t) \equiv W(t + dt) - W(t) = \xi(t)dt \tag{4.1.16}$$

and thus write the second integral as

$$\int_0^t b[x(s), s]dW(s)\,, \tag{4.1.17}$$

which is a kind of stochastic Stieltjes integral with respect to a sample function $W(t)$. Such an integral can be defined and we will carry this out in the next section.

Before doing so, it should be noted that the requirement that $u(t)$ be continuous, while very natural, can be relaxed to yield a way of defining jump processes as stochastic differential equations. This has already been hinted at in the treatment of shot noise in Sect. 1.5.1. However, it does not seem to be nearly so useful and will not be treated in this book. The interested reader is referred to [4.1].

As a final point, we should note that one normally *assumes* that $\xi(t)$ is Gaussian, and satisfies the conditions (4.1.2) as well. The above did not require this: the Gaussian nature follows in fact from the *assumed* continuity of $u(t)$. Which of these assumptions is made is, in a strict sense, a matter of taste. However, the continuity of $u(t)$ seems a much more natural assumption to make than the Gaussian nature of $\xi(t)$, which involves in principle the determination of moments of arbitrarily high order.

4.2 Stochastic Integration

4.2.1 Definition of the Stochastic Integral

Suppose $G(t)$ is an arbitrary function of time and $W(t)$ is the Wiener process. We define the stochastic integral $\int_{t_0}^{t} G(t')dW(t')$ as a kind of Riemann-Stieltjes integral.

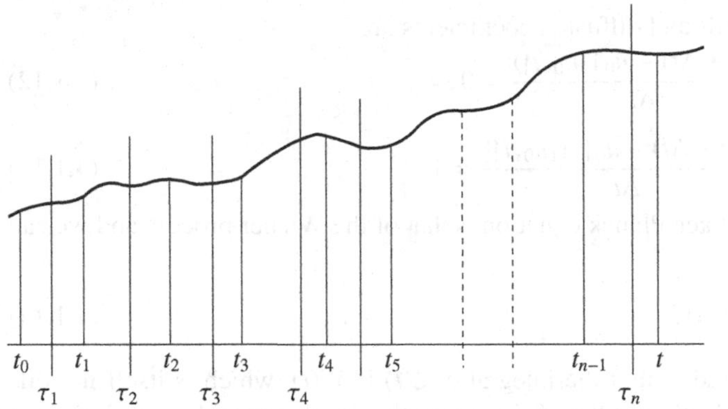

Fig. 4.1. Partitioning of the time interval used in the definition of stochastic integration

Namely, we divide the interval $[t_0, t]$ into n subintervals by means of partitioning points (as in Fig. 4.1)

$$t_0 \leqslant t_1 \leqslant t_2 \leqslant \cdots \leqslant t_{n-1} \leqslant t \,, \tag{4.2.1}$$

and define intermediate points τ_i such that

$$t_{i-1} \leqslant \tau_i \leqslant t_i \,. \tag{4.2.2}$$

The stochastic integral $\int_{t_0}^{t} G(t')dW(t')$ is defined as a limit of the partial sums

$$S_n = \sum_{i=1}^{n} G(\tau_i)[W(t_i) - W(t_{i-1})] \,. \tag{4.2.3}$$

It is heuristically quite easy to see that, in general, the integral defined as the limit of S_n *depends on the particular choice of intermediate point* τ_i. For example, if we take the choice of $G(\tau_i) = W(\tau_i)$,

$$\langle S_n \rangle = \left\langle \sum_{i=1}^{n} W(\tau_i)[W(t_i) - W(t_{i-1})] \right\rangle, \tag{4.2.4}$$

$$= \sum_{i=1}^{n} [\min(\tau_i, t_i) - \min(\tau_i, t_{i-1})] \,, \tag{4.2.5}$$

$$= \sum_{i=1}^{n} (\tau_i - t_{i-1}) \,. \tag{4.2.6}$$

If, for example, we choose for all i

$$\tau_i = \alpha t_i + (1 - \alpha)t_{i-1} \quad (0 < \alpha < 1) \,, \tag{4.2.7}$$

$$\text{then } \langle S_n \rangle = \sum_{i=1}^{n} (t_i - t_{i-1})\,\alpha = (t - t_0)\alpha \,, \tag{4.2.8}$$

So that the mean value of the integral can be anything between zero and $(t - t_0)$, depending on the choice of intermediate points.

4.2.2 Ito Stochastic Integral

The choice of intermediate points characterised by $\alpha = 0$, that is the choice

$$\tau_i = t_{i-1} \tag{4.2.9}$$

defines the *Ito stochastic integral* of the function $G(t)$ by

$$\int_{t_0}^{t} G(t')dW(t') = \text{ms-}\lim_{n\to\infty}\Big\{\sum_{i=1}^{n} G(t_{i-1})[W(t_i) - W(t_{i-i})]\Big\}. \tag{4.2.10}$$

By ms-lim we mean the *mean square limit*, as defined in Sect. 2.9.2.

4.2.3 Example $\int_{t_0}^{t} W(t')dW(t')$

An exact calculation is possible. We write [writing W_i for $W(t_i)$]

$$S_n = \sum_{i=1}^{n} W_{i-1}(W_t - W_{i-1}) \equiv \sum_{i=1}^{n} W_{i-1}\Delta W_i\,, \tag{4.2.12}$$

$$= \tfrac{1}{2}\sum_{i=1}^{n}[(W_{i-1} + \Delta W_i)^2 - (W_{i-1})^2 - (\Delta W_i)^2]\,, \tag{4.2.13}$$

$$= \tfrac{1}{2}[W(t)^2 - W(t_0)^2] - \tfrac{1}{2}\sum_{i=1}^{n}(\Delta W_i)^2\,. \tag{4.2.14}$$

We can calculate the mean square limit of the last term. Notice that

$$\Big\langle \sum \Delta W_i^2 \Big\rangle = \sum_i \langle (W_i - W_{i-1})^2\rangle = \sum_i (t_i - t_{i-1}) = t - t_0\,. \tag{4.2.15}$$

Because of this,

$$\Big\langle \Big[\sum_i (W_i - W_{i-1})^2 - (t-t_0)^2\Big]^2\Big\rangle = \Big\langle \sum_i (W_i - W_{i-1})^4$$
$$+2\sum_{i<j}(W_i - W_{i-1})^2(W_j - W_{j-1})^2 - 2(t-t_0)\sum_i (W_i - W_{i-1})^2 + (t-t_0)^2\Big\rangle. \tag{4.2.16}$$

Notice the $W_i - W_{i-1}$ is a Gaussian variable and is independent of $W_j - W_{j-1}$. Hence, we can factorise. Thus,

$$\langle (W_i - W_{i-1})^2(W_j - W_{j-1})^2\rangle = (t_i - t_{i-1})(t_j - t_{j-1})\,, \tag{4.2.17}$$

and also, using the formula (2.8.7) for the fourth moment of a Gaussian variable

$$\langle (W_i - W_{i-1})^4\rangle = 3\,\langle (W_i - W_{i-1})^2\rangle^2 = 3(t_i - t_{i-1})^2\,, \tag{4.2.18}$$

which combined with (4.2.17) gives

$$\Big\langle \Big[\sum_i (W_i - W_{i-1}) - (t-t_0)\Big]^2\Big\rangle$$
$$= 2\sum_i (t_i - t_{i-i})^2 + \sum_{i,j}[(t_i - t_{i-1}) - (t-t_0)][(t_j - t_{j-1}) - (t-t_0)]\,,$$
$$= 2\sum_i (t_i - t_{i-1})^2 \to 0 \text{ as } n\to\infty\,. \tag{4.2.19}$$

Thus,

$$\text{ms-lim}_{n\to\infty} \sum_i (W_i - W_{i-1})^2 = t - t_0 \,, \tag{4.2.20}$$

by definition of the mean square limit, so

$$\int_{t_0}^{t} W(t')\, dW(t') = \tfrac{1}{2}[W(t)^2 - W(t_0)^2 - (t - t_0)] \,. \tag{4.2.21}$$

Comments

i) $$\left\langle \int_{t_0}^{t} W(t)\, dW(t) \right\rangle = \tfrac{1}{2}[\langle W(t)^2\rangle - \langle W(t_0)^2\rangle - (t - t_0)] = 0 \,. \tag{4.2.22}$$

This is also obvious by definition, since the individual terms are $\langle W_{i-1}\Delta W_i\rangle$, which vanishes because ΔW_i is statistically independent of W_{i-1}, as was demonstrated in Sect. 3.8.1.

ii) The result for the integral is no longer the same as for the ordinary Riemann-Stieltjes integral in which the term $(t - t_0)$ would be absent. The reason for this is that $|W(t + \Delta t) - W(t)|$ is almost always of the order $\sqrt{t}$, so that in contrast to ordinary integration, terms of second order in $\Delta W(t)$ do not vanish on taking the limit.

4.2.4 The Stratonovich Integral

An alternative definition was introduced by *Stratonovich* [4.2] as a stochastic integral in which the anomalous term $(t-t_0)$ does not occur. We define this fully in Sect. 4.4—in the cases considered so far, it amounts to evaluating the integrand as a function of $W(t)$ at the value $\frac{1}{2}[W(t_i) + W(t_{i-1})]$. It is straightforward to show that

$$(S) \int_{t_0}^{t} W(t')\, dW(t') = \text{ms-lim}_{n\to\infty} \sum_{i=1}^{n} \frac{W(t_i) + W(t_{i-1})}{2}[W(t_i) - W(t_{i-1})] \,, \tag{4.2.23}$$

$$= \tfrac{1}{2}[W(t)^2 - W(t_0)^2] \,. \tag{4.2.24}$$

However, the integral as defined by Stratonovich [which we will always designate by a prefixed (S) as in (4.2.23)] has no general relationship with that defined by Ito. That is, for *arbitrary* functions $G(t)$, there is no connection between the two integrals. [In the case, however, where we can specify that $G(t)$ is related to some stochastic differential equation, a formula can be given relating one to the other, see Sect. 4.4].

4.2.5 Nonanticipating Functions

The concept of a nonanticipating function can be easily made quite obscure by complex notation, but is really quite simple. We have in mind a situation in which all functions can be expressed as functions or functionals of a certain Wiener process $W(t)$ through the mechanism of a stochastic differential (or integral) equation of the form

$$x(t) - x(t_0) = \int_{t_0}^{t} a[x(t'), t']dt' + \int_{t_0}^{t} b[x(t'), t']dW(t') . \tag{4.2.25}$$

A function $G(t)$ is called a *nonanticipating function of t* if $G(t)$ is statistically independent of $W(s) - W(t)$ for all s and t such that $t < s$. This means that $G(t)$ is independent of the behaviour of the Wiener process in the future of t. This is clearly a rather reasonable requirement for a physical function which could be a solution of an equation like (4.2.25) in which it seems heuristically obvious that $x(t)$ involves $W(t')$ only for $t' \leqslant t$.

For example, specific nonanticipating functions of t are:

i) $W(t)$,

ii) $\int_{t_0}^{t} F[W(t')]\, dt'$,

iii) $\int_{t_0}^{t} F[W(t')]\, dW(t')$,

iv) $\int_{t_0}^{t} G(t')\, dt'$,

v) $\int_{t_0}^{t} G(t')\, dW(t')$,

(iv), (v): when $G(t)$ is itself a nonanticipating function. (4.2.26)

Results (iii) and (v) depend on the fact that the Ito stochastic integral, as defined in (4.2.10), is a limit of the sequence in which only $G(t')$ for $t' < t$ and $W(t')$ for $t' \leqslant t$ are involved.

The reasons for considering nonanticipating functions specifically are:

i) Many results can be derived, which are only true for such functions;
ii) They occur naturally in situations, such as in the study of differential equations involving time, in which some kind of *causality* is expected in the sense that the unknown future cannot affect the present;
iii) The definition of stochastic differential equations requires such functions.

4.2.6 Proof that $dW(t)^2 = dt$ and $dW(t)^{2+N} = 0$

The formulae in the heading are the key to the use of the Ito calculus as an ordinary computational tool. However, as written they are not very precise and what is really meant is that for an arbitrary *nonanticipating function* $G(t)$

$$\int_{t_0}^{t} [dW(t')]^{2+N} G(t') \equiv \underset{n\to\infty}{\text{ms-lim}} \sum_i G_{i-1}\Delta W_i^{2+N} ,$$

$$= \begin{cases} \int_{t_0}^{t} dt' G(t') , & \text{for } N = 0 , \\ 0 , & \text{for } N > 0 . \end{cases} \tag{4.2.27}$$

The proof is quite straightforward. For $N = 0$, let us define

$$I = \lim_{n\to\infty} \left\langle \left[\sum_i G_{i-1}\left(\Delta W_i^2 - \Delta t_i\right)\right]^2 \right\rangle \tag{4.2.28}$$

$$= \lim_{n\to\infty} \left\{ \left\langle \sum_i \underbrace{(G_{i-1})^2}\, \underbrace{(\Delta W_i^2 - \Delta t_i)^2} + \sum_{i>j} \underbrace{2G_{i-1}G_{j-1}(\Delta W_j^2 - \Delta t_j)}\, \underbrace{(\Delta W_i^2 - \Delta t_i)} \right\rangle \right\}. \tag{4.2.29}$$

The horizontal braces indicate factors which are statistically independent of each other because of the properties of the Wiener process, and because the G_i are values of a nonanticipating function which are independent of all ΔW_j for $j > i$.

Using this independence, we can factorise the means, and also using

i) $\langle \Delta W_t^2 \rangle = \Delta t_i$,

ii) $\langle (\Delta W_t^2 - \Delta t_i)^2 \rangle = 2\Delta t_i^2$ (from Gaussian nature of ΔW_i),

we find

$$I = 2 \lim_{n\to\infty} \left[\sum_i \Delta t_i^2 \langle (G_{i-1})^2 \rangle \right]. \tag{4.2.30}$$

Under reasonably mild conditions on $G(t)$ (e.g., boundedness), this means that

$$\operatorname*{ms-lim}_{n\to\infty} \left(\sum_i G_{i-1}\Delta W_i^2 - \sum_i G_{i-1}\Delta t_i \right) = 0\,, \tag{4.2.31}$$

and since

$$\operatorname*{ms-lim}_{n\to\infty} \sum_i G_{i-1}\Delta t_i = \int_{t_0}^{t} dt' G(t')\,, \tag{4.2.32}$$

we have

$$\boxed{\int_{t_0}^{t} [dW(t')]^2 G(t') = \int_{t_0}^{t} dt' G(t')\,.} \tag{4.2.33}$$

Comments

i) The proof $\int_{t_0}^t G(t)[dW(t)]^{2+N} = 0$ for $N > 0$ is similar and uses the explicit expressions for the higher moments of a Gaussian given in Sect. 2.8.1.

ii) $dW(t)$ only occurs in integrals so that when we restrict ourselves to nonanticipating functions, we can simply write

$$dW(t)^2 \equiv dt\,, \tag{4.2.34}$$

$$dW(t)^{2+N} \equiv 0\,, \quad (N > 0)\,. \tag{4.2.35}$$

iii) The results are only valid for the Ito integral, since we have used the fact that ΔW_i is independent of G_{i-1}. In the Stratonovich integral,

$$\Delta W_i = W(t_i) - W(t_{i-1})\,, \tag{4.2.36}$$

$$G_{i-1} = G\left(\tfrac{1}{2}(t_i + t_{i-1})\right), \tag{4.2.37}$$

and although $G(t)$ is nonanticipating, this is not sufficient to guarantee the independence of ΔW_i, and G_{i-1} as thus defined.

iv) By similar methods one can prove that

$$\int_{t_0}^{t} G(t')\,dt'\,dW(t') \equiv \underset{n\to\infty}{\text{ms-lim}} \sum G_{i-1}\Delta W_i \Delta t_i = 0\,, \tag{4.2.38}$$

and similarly for higher powers. The simplest way of characterising these results is to say that $dW(t)$ is an infinitesimal order of $\frac{1}{2}$ and that in calculating differentials, infinitesimals of order higher than 1 are discarded.

4.2.7 Properties of the Ito Stochastic Integral

a) Existence: One can show that the Ito stochastic integral $\int_{t_0}^{t} G(t')dW(t')$ exists whenever the function $G(t')$ is *continuous* and *nonanticipating* on the closed interval $[t_0, t]$ [4.3].

b) Integration of Polynomials: We can formally use the result of Sect. 4.2.6:

$$d[W(t)]^n = [W(t) + dW(t)]^n - W(t)^n = \sum_{r=1}^{n} \binom{n}{r} W(t)^{n-r}\,dW(t)^r\,, \tag{4.2.39}$$

and using the fact that $dW(t)^r \to 0$ for all $r > 2$,

$$= nW(t)^{n-1}\,dW(t) + \frac{n(n-1)}{2} W(t)^{n-2}\,dt\,, \tag{4.2.40}$$

so that

$$\int_{t_0}^{t} W(t')^n\,dW(t') = \frac{1}{n+1}[W(t)^{n+1} - W(t_0)^{n+1}] - \frac{n}{2}\int_{t_0}^{t} W(t')^{n-1}\,dt'\,. \tag{4.2.41}$$

c) Two Kinds of Integral: We note that for each $G(t)$ there are two kinds of integrals, namely,

$$\int_{t_0}^{t} G(t')dt' \quad \text{and} \quad \int_{t_0}^{t} G(t')dW(t')\,, \tag{4.2.42}$$

both of which occur in the previous equation. There is, in general, no connection between these two kinds of integral.

d) General Differentiation Rules: In forming differentials, as in (b) above, one must keep all terms up to second order in $dW(t)$. This means that, for example,

$$d\{\exp[W(t)]\} = \exp[W(t) + dW(t)] - \exp[W(t)]\,, \tag{4.2.43}$$

$$= \exp[W(t)]\left[dW(t) + \tfrac{1}{2}dW(t)^2\right], \tag{4.2.44}$$

$$= \exp[W(t)]\left[dW(t) + \tfrac{1}{2}dt\right]. \tag{4.2.45}$$

For an arbitrary function

$$df[W(t),t] = \frac{\partial f}{\partial t}dt + \frac{1}{2}\frac{\partial^2 f}{\partial t^2}(dt)^2 + \frac{\partial f}{\partial W}dW(t) + \frac{1}{2}\frac{\partial^2 f}{\partial W^2}[dW(t)]^2 + \frac{\partial^2 f}{\partial W \partial t}dt\, dW(t) + \dots \tag{4.2.46}$$

and we use

$$[dW(t)]^2 \to dt\,, \tag{4.2.47}$$

$$dt\, dW(t) \to 0\,, \quad \text{[Sect. 4.2.6, comment (iv)]} \tag{4.2.48}$$

$$(dt)^2 \to 0\,, \tag{4.2.49}$$

and all higher powers vanish, to arrive at

$$df[W(t),t] = \left(\frac{\partial f}{\partial t} + \frac{1}{2}\frac{\partial^2 f}{\partial W^2}\right)dt + \frac{\partial f}{\partial W}dW(t)\,. \tag{4.2.50}$$

e) Mean Value Formula: For nonanticipating $G(t)$,

$$\left\langle \int_{t_0}^{t} G(t')\,dW(t') \right\rangle = 0\,. \tag{4.2.51}$$

Proof: Since $G(t)$ is nonanticipating, in the definition of the stochastic integral,

$$\left\langle \sum_i G_{i-1}\Delta W_i \right\rangle = \sum_i \langle G_{i-1}\rangle\langle \Delta W_i\rangle = 0\,. \tag{4.2.52}$$

We know from Sect. 2.9.5 that operations of ms-lim and $\langle\ \rangle$ may be interchanged. Hence, taking the limit of (4.2.52), we have the result.

This result is *not true* for Stratonovich's integral, since the value of G_{i-1} is chosen in the middle of the interval, and may be correlated with ΔW_i.

f) Correlation Formula: If $G(t)$ and $H(t)$ are arbitrary continuous nonanticipating functions,

$$\left\langle \int_{t_0}^{t} G(t')\,dW(t') \int_{t_0}^{t} H(t')\,dW(t') \right\rangle = \int_{t_0}^{t} \langle G(t')H(t')\rangle\,dt'\,. \tag{4.2.53}$$

Proof: Notice that

$$\left\langle \sum_i G_{i-1}\Delta W_i \sum_j H_{j-1}\Delta W_j \right\rangle = \left\langle \sum_i G_{i-1}H_{i-1}(\Delta W_i)^2 \right\rangle + \left\langle \sum_{i>j} (G_{i-1}H_{j-1} + G_{j-1}H_{i-1})\Delta W_j \Delta W_i \right\rangle. \tag{4.2.54}$$

In the second term, ΔW_i is independent of all other terms since $j < i$, and G and H are nonanticipating. Hence, we may factorise out the term $\langle \Delta W_i \rangle = 0$ so that this term vanishes. Using

$$\langle \Delta W_i^2 \rangle = \Delta t_i \,, \tag{4.2.55}$$

and interchanging mean and limit operations, the result follows.

g) Relation to Delta-Correlated White Noise: Formally, this is equivalent to the idea that Langevin terms $\xi(t)$ are delta correlated and uncorrelated with $F(t)$ and $G(t)$. For, rewriting

$$dW(t) \to \xi(t)dt \,, \tag{4.2.56}$$

it is clear that if $F(t)$ and $G(t)$ are nonanticipating, $\xi(t)$ is independent of them, and we get

$$\begin{aligned}\int_{t_0}^{t} dt' \int_{t_0}^{t} ds' \langle G(t')H(s')\xi(t')\xi(s')\rangle &= \int_{t_0}^{t}\int_{t_0}^{t} dt'ds' \langle G(t')H(s')\rangle\langle \xi(t')\xi(s')\rangle \,, \\ &= \int_{t_0}^{t} dt' \langle G(t')H(t')\rangle \,,\end{aligned} \tag{4.2.57}$$

which implies

$$\langle \xi(t)\xi(s)\rangle = \delta(t-s) \,. \tag{4.2.58}$$

An important point of definition arises here, however. In integrals involving delta functions, it frequently occurs in the study of stochastic differential equations that the argument of the delta function is equal to either the upper or the lower limit of the integral, that is, we find integrals like

$$I_1 = \int_{t_1}^{t_2} dt f(t)\delta(t-t_1) \,, \tag{4.2.59}$$

or

$$I_2 = \int_{t_1}^{t_2} dt f(t)\delta(t-t_2) . \tag{4.2.60}$$

Various conventions can be made concerning the value of such integrals. We will show that in the present context, we must always make the interpretation

$$I_1 = f(t_1) \,, \tag{4.2.61}$$

$$I_2 = 0 \,, \tag{4.2.62}$$

corresponding to counting all the weight of a delta function at the lower limit of an integral, and none of the weight at the upper limit. To demonstrate this, note that

$$\left\langle \int_{t_0}^{t} G(t')dW(t') \left[\int_{t_0}^{t'} H(s')dW(s') \right] \right\rangle = 0 \,. \tag{4.2.63}$$

This follows, since the function defined by the integral inside the square bracket is, by Sect. 4.2.5 comment (v), a nonanticipating function and hence the complete integrand, [obtained by multiplying by $G(t')$ which is also nonanticipating] is itself nonanticipating. Hence the average vanishes by the result of Sect. 4.2.7e.

Now using the formulation in terms of the Langevin source $\xi(t)$, we can rewrite (4.2.63) as

$$\int_{t_0}^{t} dt' \int_{t_0}^{t'} ds' \langle G(t')H(s')\rangle \delta(t'-s') = 0\,, \tag{4.2.64}$$

which corresponds to not counting the weight of the delta function at the upper limit. Consequently, the full weight must be counted at the lower limit.

This property is a direct consequence of the definition of the Ito integral as in (4.2.10), in which the increment points "towards the future". That is, we can interpret

$$dW(t) = W(t+dt) - W(t)\,. \tag{4.2.65}$$

In the case of the Stratonovich integral, we get quite a different formula, which is by no means as simple to prove as in the Ito case, but which amounts to choosing

$$\left.\begin{aligned} I_1 &= \tfrac{1}{2} f(t_1)\,, \\ I_2 &= \tfrac{1}{2} f(t_2). \end{aligned}\right\} \quad \text{(Stratonovich)} \tag{4.2.66}$$

This means that in both cases, the delta function occurring at the limit of an integral has half its weight counted. This formula, although intuitively more satisfying than the Ito form, is more complicated to use, especially in the perturbation theory of stochastic differential equations, where the Ito method makes very many terms vanish.

4.3 Stochastic Differential Equations (SDE)

We concluded in Sect. 4.1, that the most satisfactory interpretation of the Langevin equation

$$\frac{dx}{dt} = a(x,t) + b(x,t)\xi(t)\,, \tag{4.3.1}$$

is a stochastic integral equation

$$x(t) - x(0) = \int_0^t dt' a[x(t'),t'] + \int_0^t dW(t') b[x(t'),t']\,. \tag{4.3.2}$$

Unfortunately, the kind of stochastic integral to be used is not given by the reasoning of Sect. 4.1. The Ito integral is mathematically and technically the most satisfactory, but it is not always the most natural choice physically. The Stratonovich integral is the natural choice for an interpretation which assumes $\xi(t)$ is a real noise (not a white noise) with finite correlation time, which is then allowed to become infinitesimally small after calculating measurable quantities. Furthermore, a Stratonovich interpretation enables us to use ordinary calculus, which is not possible for an Ito interpretation.

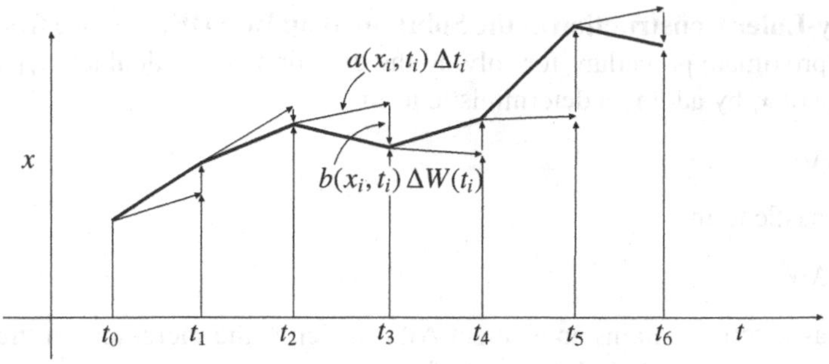

Fig. 4.2. Illustration of the Cauchy-Euler procedure for constructing an approximate solution of the stochastic differential equation $dx(t) = a[x(t), t]dt + b[x(t), t]dW(t)$

From a mathematical point of view, the choice is made clear by the near impossibility of carrying out proofs using the Stratonovich integral. We will therefore define the Ito SDE, develop its equivalence with the Stratonovich SDE, and use either form depending on circumstances. The relationship between white noise stochastic differential equations and the real noise systems is explained in Sect. 8.1.

4.3.1 Ito Stochastic Differential Equation: Definition

A stochastic quantity $x(t)$ obeys an Ito SDE written as

$$dx(t) = a[x(t), t]\,dt + b[x(t), t]\,dW(t)\,, \tag{4.3.3}$$

if for all t and t_0,

$$x(t) = x(t_0) + \int_{t_0}^{t} a[x(t'), t']\,dt' + \int_{t_0}^{t} b[x(t'), t']\,dW(t')\,. \tag{4.3.4}$$

Before considering what conditions must be satisfied by the coefficients in (4.3.4), it is wise to consider what one means by a solution of such an equation and what uniqueness of solution would mean in this context. For this purpose, we can consider a discretised version of the SDE obtained by taking a mesh of points t_i (as illustrated in Fig. 4.2) such that

$$t_0 < t_1 < t_2 < \cdots < t_{n-1} < t_n = t\,, \tag{4.3.5}$$

and writing the equation as

$$x_{i+1} = x_i + a(x_i, t_i)\Delta t_i + b(x_i, t_i)\Delta W_i\,. \tag{4.3.6}$$

Here,

$$\left.\begin{aligned} x_i &= x(t_i)\,, \\ \Delta t_i &= t_{i+1} - t_i\,, \\ \Delta W_i &= W(t_{i+1}) - W(t_i)\,. \end{aligned}\right\} \tag{4.3.7}$$

a) Cauchy-Euler Construction of the Solution of an Ito SDE: We see from (4.3.6) that an approximate procedure for solving the equation is to calculate x_{i+1} from the knowledge of x_i by adding a deterministic term

$$a(x_i, t_i)\Delta t_i\,, \tag{4.3.8}$$

and a stochastic term

$$b(x_i, t_i)\Delta W_i\,. \tag{4.3.9}$$

The stochastic term contains an element ΔW_i, which is the increment of the Wiener process, but is statistically independent of x_i if

i) x_0 is itself independent of all $W(t) - W(t_0)$ for $t > t_0$ (thus, the initial conditions if considered random, must be nonanticipating), and

ii) $a(x, t)$ is a nonanticipating function of t for any fixed x.

Constructing an approximate solution iteratively by use of (4.3.6), we see that x_i is always independent of ΔW_j for $j \geqslant i$.

The solution is then formally constructed by letting the mesh size go to zero. To say that the solution is unique means that for a given sample function $\tilde{W}(t)$ of the random Wiener process $W(t)$, the particular solution of the equation which arises is unique. To say that the solution exists means that with probability one, a solution exists for any choice of sample function $\tilde{W}(t)$ of the Wiener process $W(t)$.

This method of constructing a solution is called the *Cauchy-Euler* method, and can be used to generate simulations. However, there are significantly better algorithms, as is explained in Chap. 10.

b) Existence and Uniqueness of Solutions of an Ito SDE: Existence and uniqueness will not be proved here. The interested reader will find proofs in [4.3]. The conditions which are required for existence and uniqueness in a time interval $[t_0, T]$ are:

i) *Lipschitz condition:* a K exists such that

$$|a(x,t) - a(y,t)| + |b(x,t) - b(y,t)| \leqslant K|x - y|\,, \tag{4.3.10}$$

for all x and y, and all t in the range $[t_0, T]$.

ii) *Growth condition:* a K exists such that for all t in the range $[t_0, T]$,

$$|a(x,t)|^2 + |b(x,t)|^2 \leqslant K^2(1 + |x|^2)\,. \tag{4.3.11}$$

Under these conditions there will be a unique nonanticipating solution $x(t)$ in the range $[t_0, T]$.

Almost every stochastic differential equation encountered in practice satisfies the Lipschitz condition since it is essentially a smoothness condition. However, the growth condition is often violated. This does not mean that no solution exists; rather, it means the solution may "explode" to infinity, that is, the value of x can become infinite in a finite time; in practice, a finite random time. This phenomenon occurs in ordinary differential equations, for example,

$$\frac{dx}{dt} = \tfrac{1}{2} a x^3 \tag{4.3.12}$$

has the general solution with an initial condition $x = x_0$ at $t = 0$,

$$x(t) = (-at + 1/x_0^2)^{-1/2} . \tag{4.3.13}$$

If a is positive, this becomes infinite when $x_0 = (at)^{-1/2}$ but if a is negative, the solution never explodes. Failing to satisfy the Lipschitz condition does not guarantee the solution will explode. More precise stability results are required for one to be certain of that [4.3].

4.3.2 Dependence on Initial Conditions and Parameters

In exactly the same way as in the case of deterministic differential equations, if the functions which occur in a stochastic differential equation depend continuously on parameters, then the solution normally depends continuously on that parameter. Similarly, the solution depends continuously on the initial conditions. Let us formulate this more precisely. Consider a one-variable equation

$$dx = a(\lambda, x, t)\, dt + b(\lambda, x, t)\, dW(t) , \tag{4.3.14}$$

with initial condition

$$x(t_0) = c(\lambda) , \tag{4.3.15}$$

where λ is a parameter. Let the solution of (4.3.14) be $x(\lambda, t)$. Suppose

i) $\displaystyle \operatorname*{st-lim}_{\lambda \to \lambda_0} c(\lambda) = c(\lambda_0) ,$ (4.3.16)

ii) For every $N > 0$

$$\lim_{\lambda \to \lambda_0} \left\{ \sup_{t \in [t_0, T],\ |x| < N} \left[|a(\lambda, x, t) - a(\lambda_0, x, t)| + |b(\lambda, x, t) - b(\lambda_0, x, t)| \right] \right\} = 0 , \tag{4.3.17}$$

iii) There exists a K independent of λ such that

$$|a(\lambda, x, t)|^2 + |b(\lambda, x, t)|^2 \leqslant K^2(1 + |x^2|) . \tag{4.3.18}$$

Then,

$$\operatorname*{st-lim}_{\lambda \to \lambda_0} \left\{ \sup_{t \in [t_0, T]} |x(\lambda, t) - x(\lambda_0, t)| \right\} = 0 . \tag{4.3.19}$$

For a proof see [4.1]. Check ii)

Comments

i) Recalling the definition of stochastic limit, the interpretation of the limit (4.3.19) is that as $\lambda \to \lambda_0$, the probability that the maximum deviation over any finite interval $[t_0, T]$ between $x(\lambda, t)$ and $x(\lambda_0, t)$ is nonzero, goes to zero.

ii) Dependence on the initial condition is achieved by letting a and b be independent of λ.

iii) The result will be very useful in justifying perturbation expansions.

iv) Condition (ii) is written in the most natural form for the case that the functions $a(x,t)$ and $b(x,t)$ are not themselves stochastic. It often arises that $a(x,t)$ and $b(x,t)$ are themselves stochastic (nonanticipating) functions. In this case, condition (ii) must be replaced by a probabilistic statement. It is, in fact, sufficient to replace $\lim\limits_{\lambda\to\lambda_0}$ by $\operatorname{st-lim}\limits_{\lambda\to\lambda_0}$.

4.3.3 Markov Property of the Solution of an Ito SDE

We now show that $x(t)$, the solution to the stochastic differential equation (4.3.4), is a Markov Process. Heuristically, the result is obvious, since with a given initial condition $x(t_0)$, the future time development is uniquely (stochastically) determined, that is, $x(t)$ for $t > t_0$ is determined only by

i) The particular sample path of $W(t)$ for $t > t_0$;

ii) The value of $x(t_0)$.

Since $x(t)$ is a nonanticipating function of t, $W(t)$ for $t > t_0$ is independent of $x(t)$ for $t < t_0$. Thus, the time development of $x(t)$ for $t > t_0$ is independent of $x(t)$ for $t < t_0$ provided $x(t_0)$ is known. Hence, $x(t)$ is a Markov process. For a precise proof see [4.3].

4.3.4 Change of Variables: Ito's Formula

Consider an arbitrary function of $x(t) : f[x(t)]$. What stochastic differential equation does it obey? We use the results of Sect. 4.2.6 to expand $df[x(t)]$ to second order in $dW(t)$:

$$df[x(t)] = f[x(t) + dx(t)] - f[x(t)]\,, \tag{4.3.20}$$

$$= f'[x(t)]dx(t) + \tfrac{1}{2}f''[x(t)]dx(t)^2 + \dots\,, \tag{4.3.21}$$

$$= f'[x(t)]\big\{a[x(t),t]dt + b[x(t),t]dW(t)\big\} + \tfrac{1}{2}f''[x(t)]b[x(t),t]^2\,dW(t)^2. \tag{4.3.22}$$

where all other terms have been discarded since they are of higher order. Now use $dW(t)^2 = dt$ to obtain

$$df[x(t)] = \big\{a[x(t),t]f'[x(t)] + \tfrac{1}{2}b[x(t),t]^2 f''[x(t)]\big\}\,dt + b[x(t),t]f'[x(t)]\,dW(t)\,. \tag{4.3.23}$$

This formula is known as Ito's *formula* and shows that changing variables is not given by ordinary calculus unless $f[x(t)]$ is merely linear in $x(t)$.

Many Variables: In practice, Ito's formula becomes very complicated and the easiest method is to simply use the multivariate form of the rule that $dW(t)$ is an infinitesimal of order $\frac{1}{2}$. By similar methods to those used in Sect. 4.2.6, we can show that for an n dimensional Wiener process $\boldsymbol{W}(t)$,

$$\begin{aligned}
dW_i(t)\,dW_j(t) &= \delta_{ij}\,dt\,, && && (4.3.24a)\\
dW_i(t)^{N+2} &= 0\,, && (N>0)\,, && (4.3.24b)\\
dW_i(t)\,dt &= 0\,, && && (4.3.24c)\\
dt^{1+N} &= 0\,, && (N>0)\,. && (4.3.24d)
\end{aligned}$$

which imply that $dW_i(t)$ is an infinitesimal of order $\frac{1}{2}$. Note, however, that (4.3.24a) is a consequence of the independence of $dW_i(t)$ and $dW_j(t)$. To develop Ito's formula for functions of an n dimensional vector $\boldsymbol{x}(t)$ satisfying the stochastic differential equation

$$d\boldsymbol{x} = \boldsymbol{A}(\boldsymbol{x},t)dt + \mathsf{B}(\boldsymbol{x},t)d\boldsymbol{W}(t)\,, \tag{4.3.25}$$

we simply follow this procedure. The result is

$$\begin{aligned}
df(\boldsymbol{x}) = &\Big\{\sum_i A_i(\boldsymbol{x},t)\partial_i f(\boldsymbol{x}) + \tfrac{1}{2}\sum_{i,j}[\mathsf{B}(\boldsymbol{x},t)\mathsf{B}^{\mathrm{T}}(\boldsymbol{x},t)]_{ij}\partial_i\partial_j f(\boldsymbol{x})\Big\}dt\\
&+ \sum_{i,j} B_{ij}(\boldsymbol{x},t)\partial_i f(\boldsymbol{x})\,dW_j(t)\,.
\end{aligned} \tag{4.3.26}$$

4.3.5 Connection Between Fokker-Planck Equation and Stochastic Differential Equation

a) Forward Fokker-Planck Equation: We now consider the time development of an arbitrary $f(x(t))$. Using Ito's formula

$$\begin{aligned}
\frac{\langle df[x(t)]\rangle}{dt} &= \left\langle\frac{df[x(t)]}{dt}\right\rangle = \frac{d}{dt}\langle f[x(t)]\rangle\,,\\
&= \left\langle a[x(t),t]\,\partial_x f + \tfrac{1}{2}b[x(t),t]^2\partial_x^2 f\right\rangle.
\end{aligned} \tag{4.3.27}$$

However, $x(t)$ has a conditional probability density $p(x,t\,|\,x_0,t_0)$ and

$$\begin{aligned}
\frac{d}{dt}\langle f[x(t)]\rangle &= \int dx\, f(x)\partial_t p(x,t\,|\,x_0,t_0)\,,\\
&= \int dx\left[a(x,t)\partial_x f + \tfrac{1}{2}b(x,t)^2\partial_x^2 f\right]p(x,t\,|\,x_0,t_0)\,.
\end{aligned} \tag{4.3.28}$$

This is now of the same form as (3.4.16) Sect. 3.4.1. Under the same conditions as there, we integrate by parts and discard surface terms to obtain

$$\int dx f(x)\partial_t p = \int dx f(x)\Big\{-\partial_x[a(x,t)p] + \tfrac{1}{2}\partial_x^2[b(x,t)^2 p]\Big\}\,, \tag{4.3.29}$$

and hence, since $f(x)$ is arbitrary,

$$\partial_t p(x,t\,|\,x_0,t_0) = -\partial_x[a(x,t)p(x,t\,|\,x_0,t_0)] + \tfrac{1}{2}\partial_x^2[b(x,t)^2 p(x,t\,|\,x_0,t_0)]\,. \quad (4.3.30)$$

We have thus a complete equivalence to a diffusion process defined by a drift coefficient $a(x,t)$ and a diffusion coefficient $b(x,t)^2$.

The results are precisely analogous to those of Sect. 3.5.2, in which it was shown that the diffusion process could be locally approximated by an equation resembling an Ito stochastic differential equation.

b) Backward Fokker-Planck Equation—the Feynman-Kac Formula: Suppose a function $g(x,t)$ obeys the backward Fokker-Planck equation

$$\partial_t g = -a(x,t)\partial_x g - \tfrac{1}{2}b(x,t)\partial_x^2 g\,, \quad (4.3.31)$$

with the final condition

$$g(x,T) = G(x)\,. \quad (4.3.32)$$

If $x(t)$ obeys the stochastic differential equation (4.3.3), then using Ito's rule (adapted appropriately to account for explicit time dependence), the function $g[x(t),t]$ obeys the stochastic differential equation

$$dg[x(t),t] = \left\{\partial_t g + a[x(t),t]\,\partial_x g[x(t),t] + \tfrac{1}{2}b[x(t),t]^2\,\partial_x^2 g[x(t),t]\right\}dt$$
$$+\, b[x(t),t]\partial_x g[x(t),t]\,dW(t)\,, \quad (4.3.33)$$

and using (4.3.31) this becomes

$$dg[x(t),t] = b[x(t),t]\,\partial_x g[x(t),t]\,dW(t)\,. \quad (4.3.34)$$

Now integrate from t to T, and take the mean

$$\langle g[x(T),T]\rangle - \langle g[x(t),t]\rangle = \left\langle \int_t^T b[x(t'),t']\,\partial_x g[x(t'),t']\,dW(t')\right\rangle = 0. \quad (4.3.35)$$

Let the initial condition of the stochastic differential equation for $x(t')$ and $t' = t$ be

$$x(t) = x\,, \quad (4.3.36)$$

where x is a non-stochastic value, so that

$$\langle g[x(t),t]\rangle = g(x,t)\,. \quad (4.3.37)$$

At the other end of the interval, use the final condition (4.3.32) to write

$$\langle g[x(T),T]\rangle = \langle G[x(T)]\,|\,x(t) = x\rangle\,, \quad (4.3.38)$$

where the notation on the right hand side indicates the mean conditioned on the initial condition (4.3.36).

Putting these two together, the *Feynman-Kac* formula results:

$$\langle G[x(T)]\,|\,x(t) = x\rangle = g(x,t)\,, \quad (4.3.39)$$

where $g(x,t)$ is the solution of the backward Fokker-Planck equation (4.3.31) with initial condition (4.3.32).

This formula is essentially equivalent to the fact that $p(x,t\,|\,x_0,t_0)$ obeys the backward Fokker-Planck equation in the arguments x_0, t_0, as shown in Sect. 3.6, since

$$\langle G[x(T)]\|x(t_0) = x_0\rangle = \int dx\, G(x)p(x,T\,|\,x_0,t_0)\,. \quad (4.3.40)$$

4.3.6 Multivariable Systems

In general, many variable systems of stochastic differential equations can be defined for n variables by

$$d\boldsymbol{x} = \boldsymbol{A}(\boldsymbol{x},t)\,dt + \mathsf{B}(\boldsymbol{x},t)\,d\boldsymbol{W}(t)\,, \tag{4.3.41}$$

where $d\boldsymbol{W}(t)$ is an n variable Wiener process, as defined in Sect. 3.8.1. The many variable version of the reasoning used in Sect. 4.3.5 shows that the Fokker-Planck equation for the conditional probability density $p(\boldsymbol{x},t\,|\,x_0,t_0) \equiv p$ is

$$\partial_t p = -\sum_i \partial_i[A_i(\boldsymbol{x},t)p] + \tfrac{1}{2}\sum_{i,j}\partial_i\partial_j\{[\mathsf{B}(\boldsymbol{x},t)\mathsf{B}^{\mathrm{T}}(\boldsymbol{x},t)]_{ij}p\}\,. \tag{4.3.42}$$

Notice that the same Fokker-Planck equation arises from all matrices B such that $\mathsf{B}\mathsf{B}^{\mathrm{T}}$ is the same. This means that we can obtain the same Fokker-Planck equation by replacing B by BS where S is orthogonal, i.e., $\mathsf{SS}^{\mathrm{T}} = 1$. Notice that S may depend on $\boldsymbol{x}(t)$.

This can be seen directly from the stochastic differential equation. Suppose $\mathsf{S}(t)$ is an orthogonal matrix with an arbitrary *nonanticipating* dependence on t. Then define

$$d\boldsymbol{V}(t) = \mathsf{S}(t)\,d\boldsymbol{W}(t)\,. \tag{4.3.43}$$

Now the vector $d\boldsymbol{V}(t)$ is a linear combination of Gaussian variables $d\boldsymbol{W}(t)$ with coefficients $\mathsf{S}(t)$ which are independent of $d\boldsymbol{W}(t)$, since $\mathsf{S}(t)$ is nonanticipating. For any fixed value of $\mathsf{S}(t)$, the $d\boldsymbol{V}(t)$ are thus Gaussian and their correlation matrix is

$$\begin{aligned}\langle dV_i(t)\,dV_j(t)\rangle &= \sum_{l,m} S_{il}(t)\,S_{jm}(t)\,\langle dW_l(t)\,dW_m(t)\rangle\,,\\ &= \sum_l S_{il}(t)\,S_{jl}(t)\,dt = \delta_{ij}\,dt\,,\end{aligned} \tag{4.3.44}$$

since $\mathsf{S}(t)$ is orthogonal. Hence, all the moments are independent of $\mathsf{S}(t)$ and are the same as those of $d\boldsymbol{W}(t)$, so $d\boldsymbol{V}(t)$ is itself Gaussian with the same correlation matrix as $d\boldsymbol{W}(t)$. Finally, averages at different times factorise, for example, if $t > t'$ in

$$\sum_{i,k}\langle[dW_i(t)S_{ij}(t)]^m[dW_k(t')S_{kl}(t')]^n\rangle\,, \tag{4.3.45}$$

we can factorise out the averages of $dW_i(t)$ to various powers since $dW_i(t)$ is independent of all other terms. Evaluating these we will find that the orthogonal nature of $\mathsf{S}(t)$ gives, after averaging over $dW_i(t)$, simply

$$\sum_k\langle[dW_j(t)]^m\rangle\,\langle[dW_k(t')S_{kl}(t')]^n\rangle\,, \tag{4.3.46}$$

which similarly gives $\langle[dW_j(t)]^m[dW_l(t')]^n\rangle$. Hence, the $d\boldsymbol{V}(t)$ are also increments of a Wiener process. The orthogonal transformation simply mixes up different sample paths of the process, without changing its stochastic nature.

Hence, instead of (4.3.41) we can write

$$d\boldsymbol{x} = \boldsymbol{A}(\boldsymbol{x},t)dt + \mathsf{B}(\boldsymbol{x},t)\mathsf{S}^{\mathrm{T}}(t)\mathsf{S}(t)\,d\boldsymbol{W}(t)\,, \tag{4.3.47}$$

$$= \boldsymbol{A}(\boldsymbol{x},t)dt + \mathsf{B}(\boldsymbol{x},t)\mathsf{S}^{\mathrm{T}}(t)\,d\boldsymbol{V}(t)\,, \tag{4.3.48}$$

and since $V(t)$ is itself simply a Wiener process, this equation is equivalent to

$$dx = A(x,t)dt + \mathsf{B}(x,t)\mathsf{S}^{\mathrm{T}}(t)\,dW(t)\,, \tag{4.3.49}$$

which has exactly the same Fokker-Planck equation (4.3.42).

We will return to some examples in which this identity is relevant in Sect. 4.5.5.

4.4 The Stratonovich Stochastic Integral

The Stratonovich stochastic integral is an alternative to the Ito definition, in which Ito's formula, developed in Sect. 4.3.4, is replaced by the ordinary chain rule for change of variables. This apparent advantage does not come without cost, since in Stratonovich's definition the independence of a non-anticipating integrand $G(t)$ and the increment $dW(t)$ in a stochastic integral no longer holds. This means that increment and the integrand are correlated, and therefore to give a full definition of the Stratonovich integral requires some way of specifying what this correlation is.

This correlation is implicitly specified in the situation of most interest, the case in which the integrand is a function whose stochastic nature arises from its dependence on a variable $x(t)$ which obeys a stochastic differential equation. Since the aim is to recover the chain rule for change of variables in a stochastic differential equation, this seems a reasonable restriction.

4.4.1 Definition of the Stratonovich Stochastic Integral

Stratonovich [4.2] defined a stochastic integral of an integrand which is a function of $x(t)$ and t by

$$(S)\int_{t_0}^{t} G[x(t'),t']dW(t') = \text{ms-lim}_{n\to\infty} \sum_{i=1}^{n} G\left\{\tfrac{1}{2}\left(x(t_i)+x(t_{i-1})\right),t_{i-1}\right\}[W(t_i)-W(t_{i-1})]\,. \tag{4.4.1}$$

The Stratonovich integral is clearly *related* to a mid-point choice of τ_i in the definition of stochastic integration as given in Sect. 4.2.1, but clearly is *not* necessarily equivalent to that definition. Rather, instead of evaluating x at the midpoint $\frac{1}{2}(t_i+t_{i-1})$, the average of the values at the two time points is taken. Furthermore it is only the dependence on $x(t)$ that is averaged in this way, and not the explicit dependence on t. However, if $G(z,t)$ is differentiable in t, the integral can be shown to be independent of the particular choice of value for t in the range $[t_{i-1},t_i]$.

4.4.2 Stratonovich Stochastic Differential Equation

It is possible to write a stochastic differential equation (SDE) using Stratonovich's integral

$$x(t) = x(t_0) + \int_{t_0}^{t} dt'\alpha[x(t'),t'] + (S)\int_{t_0}^{t} dW(t')\beta[x(t'),t']\,. \tag{4.4.2}$$

a) Change of Variables for the Stratonovich SDE: The definition of the Stratonovich integral is such as to make the ordinary rules of calculus valid for change of variables. This means, that for the Stratonovich integral, Ito's formula (4.3.23) is replaced by the simple calculus rule

$$(S)\, df[x(t)] = f'[x(t)]\left\{a[x(t),t]\,dt + b[x(t),t]\,dW(t)\right\}. \tag{4.4.3}$$

This can be proved quite simply from the definition (4.4.1). The essence of the proof can be explained by using the simple SDE

$$(S)\, dx(t) = B[x(t)]\,dW(t). \tag{4.4.4}$$

In discretised form, this can be written

$$x_{i+1} = x_i + B[\tfrac{1}{2}(x_{i+1}+x_i)]\,(W_{i+1}-W_i). \tag{4.4.5}$$

To find the Stratonovich SDE for $f[x(t)]$, we need only use the Taylor series expansion of a function about a midpoint in the form

$$f(x+a) = f(x-a) + \sum_{n=0}^{\infty} \frac{f^{2n+1}(x)\,a^{2n+1}}{(2n+1)!}\,. \tag{4.4.6}$$

In expanding $f(x_{i+1})$ we only need to keep terms up to second order, so we drop all but the first two terms and write

$$f(x_{i+1}) = f(x_i) + f'[\tfrac{1}{2}(x_{i+1}+x_i)](x_{i+1}-x_i)\,, \tag{4.4.7}$$

$$= f'[\tfrac{1}{2}(x_{i+1}+x_i)]\,B[\tfrac{1}{2}(x_{i+1}+x_i)]\,(W_{i+1}-W_i). \tag{4.4.8}$$

This means that the Stratonovich SDE for $f[x(t)]$ is

$$(S)\, df[x(t)] = f'[x(t)]\,B[x(t)]\,dW(t)\,, \tag{4.4.9}$$

which is the ordinary calculus rule. The extension to the general case (4.4.3) is straightforward.

b) Equivalent Ito SDE: We shall show that the Stratonovich SDE is in fact equivalent to an appropriate Ito SDE. Let us assume that $x(t)$ is a solution of the Ito SDE

$$dx(t) = a[x(t),t]dt + b[x(t),t]\,dW(t)\,, \tag{4.4.10}$$

and deduce the α and β for a corresponding Stratonovich equation of the form (4.4.2). In both cases, the solution $x(t)$ is the same function.

We first compute the connection between the Ito integral $\int_{t_0}^{t} dW(t')b[x(t'),t']$ and the Stratonovich integral $(S)\int_{t_0}^{t} dW(t')\beta[x(t'),t']$:

$$(S)\int_{t_0}^{t} dW(t')\beta[x(t'),t'] \simeq \sum_i \beta\left[\tfrac{1}{2}\big(x(t_i)+x(t_{i-1})\big), t_{i-1}\right]\Delta W(t_{i-1})\,. \tag{4.4.11}$$

In (4.4.11) we write

$$x(t_i) = x(t_{i-1}) + \Delta x(t_{i-1})\,, \tag{4.4.12}$$

and use the Ito SDE (4.4.10) to write

$$\Delta x(t_i) = a[x(t_{i-1}), t_{i-1}]\,\Delta t_{i-1} + b[x(t_{i-1}), t_{i-1}]\,\Delta W(t_{i-1})\,. \tag{4.4.13}$$

Then, applying Ito's formula, we can write

$$\begin{aligned}\beta\Big[\tfrac{1}{2}\big(x(t_i) + x(t_{i-1})\big), t_{i-1}\Big] &= \beta\Big[x(t_{i-1}) + \tfrac{1}{2}\Delta x(t_{i-1}), t_{i-1}\Big],\\ &= \beta(t_{i-1}) + \Big[a(t_{i-1})\partial_x\beta(t_{i-1}) + \tfrac{1}{4}b^2(t_{i-1})\Big]\tfrac{1}{2}\Delta t_{i-1}\\ &\quad + \tfrac{1}{2}b(t_{i-1})\partial_x\beta(t_{i-1})\,\Delta W(t_{i-1})\,.\end{aligned} \tag{4.4.14}$$

(For simplicity, we write $\beta(t_i)$ etc, instead of $\beta[x(t_i), t_i]$ wherever possible). Putting all these back in the original equation (4.4.10) and dropping as usual $dt^2, dt\,dW(t)$, and setting $dW(t)^2 = dt$, we find

$$(S)\int = \sum_i \beta(t_{i-1})\{W(t_i) - W(t_{i-1})\} + \tfrac{1}{2}\sum_i b(t_{i-1})\partial_x\beta(t_{i-1})(t_i - t_{i-1})\,.$$

Hence we derive

$$(S)\int_{t_0}^{t}\beta[x(t'), t']\,dW(t') = \int_{t_0}^{t}\beta[x(t'), t']\,dW(t') + \tfrac{1}{2}\int_{t_0}^{t} b[x(t'), t']\partial_x\beta[x(t'), t']\,dt'\,. \tag{4.4.15}$$

This formula gives a connection between the Ito and Stratonovich integrals of functions $\beta[x(t'), t']$, in which $x(t')$ is the solution of the Ito SDE (4.4.2). It does not give a general connection between the Ito and Stratonovich integrals of arbitrary functions.

If we now make the choice

$$\begin{aligned}\alpha(x,t) &= a(x,t) - \tfrac{1}{2}b(x,t)\partial_x b(x,t)\\ \beta(x,t) &= b(x,t)\end{aligned} \tag{4.4.16}$$

we see that:

The Ito SDE

$$dx = a\,dt + b\,dW(t)\,, \tag{4.4.17}$$

is the same as the Stratonovich SDE

$$(S)\,dx = \big(a - \tfrac{1}{2}b\partial_x b\big)\,dt + b\,dW(t)\,, \tag{4.4.18}$$

and conversely,

The Stratonovich SDE

$$(S)\,dx = \alpha\,dt + \beta\,dW(t)\,, \tag{4.4.19}$$

is the same as the Ito SDE

$$dx = \big(\alpha + \tfrac{1}{2}\beta\partial_x\beta\big)\,dt + \beta\,dW(t)\,. \tag{4.4.20}$$

c) Many Variables: If a many variable Ito equation is

$$dx = A(x,t)\,dt + \mathsf{B}(x,t)\,dW(t)\,, \tag{4.4.21}$$

then the corresponding Stratonovich equation can be shown similarly to be given by replacing

$$\begin{aligned} A_i^{\mathrm{s}} &= A_i - \tfrac{1}{2}\sum_{j,k} B_{kj}\partial_k B_{ij} \\ \mathsf{B}^{\mathrm{s}} &= \mathsf{B}\,. \end{aligned} \tag{4.4.22}$$

d) Fokker-Planck Equation: Corresponding to the Stratonovich SDE,

$$(S)\,dx = A^{\mathrm{s}}(x,t)dt + \mathsf{B}^{\mathrm{s}}(x,t)\,dW(t)\,, \tag{4.4.23}$$

we can, by use of (4.4.22) and the known correspondence (Sect. 4.3.6) between the Ito stochastic differential equation and Fokker-Planck equation, show that the equivalent Fokker-Planck equation is

$$\partial_t p = -\sum_i \partial_i\{A_i^{\mathrm{s}} p\} + \tfrac{1}{2}\sum_{i,j,k} \partial_i\{B_{ik}^{\mathrm{s}}\partial_j[B_{jk}^{\mathrm{s}} p]\}\,, \tag{4.4.24}$$

which is often known as the "Stratonovich form" of the Fokker-Planck equation. In contrast to the two forms of the stochastic differential equation, the two forms of Fokker-Planck equation have a different appearance but are (of course) interpreted with the same rules—those of ordinary calculus. We will find later that the Stratonovich form of the Fokker-Planck equation does arise very naturally in certain contexts—see Sect. 8.3.

4.5 Some Examples and Solutions

4.5.1 Coefficients without x Dependence

The simple equation

$$dx = a(t)\,dt + b(t)\,dW(t)\,, \tag{4.5.1}$$

with $a(t)$ and $b(t)$ nonrandom functions of time, is solved simply by integrating

$$x(t) = x_0 + \int_{t_0}^{t} a(t')\,dt' + \int_{t_0}^{t} b(t')\,dW(t')\,. \tag{4.5.2}$$

Here, x_0 can be either a nonrandom initial condition or may be random, but must be independent of $W(t) - W(t_0)$ for $t > t_0$; otherwise, $x(t)$ is not nonanticipating.

As constructed, $x(t)$ is Gaussian, provided x_0 is either nonrandom or itself Gaussian, since

$$\int_{t_0}^{t} b(t')\,dW(t'), \tag{4.5.3}$$

is simply a linear combination of infinitesimal Gaussian variables. Further,

$$\langle x(t)\rangle = \langle x_0\rangle + \int_{t_0}^{t} a(t')\,dt', \tag{4.5.4}$$

(since the mean of the Ito integral vanishes) and

$$\langle [x(t) - \langle x(t)\rangle][x(s) - \langle x(s)\rangle]\rangle \equiv \langle x(t), x(s)\rangle, \tag{4.5.5}$$

$$= \mathrm{var}[x_0] + \left\langle \int_{t_0}^{t} b(t')\,dW(t') \int_{t_0}^{s} b(s')\,dW(s')\right\rangle, \tag{4.5.6}$$

$$= \mathrm{var}[x_0] + \int_{t_0}^{\min(t,s)} [b(t')]^2\,dt', \tag{4.5.7}$$

where we have used the result (4.2.53) with, however,

$$G(t') = \begin{cases} b(t'), & t' < t, \\ 0, & t' \geqslant t, \end{cases} \tag{4.5.8}$$

$$H(t') = \begin{cases} b(t'), & t' < s, \\ 0, & t' \geqslant s. \end{cases} \tag{4.5.9}$$

The process is thus completely determined.

4.5.2 Multiplicative Linear White Noise Process—Geometric Brownian Motion

The equation

$$dx = cx\,dW(t), \tag{4.5.10}$$

is known as a multiplicative white noise process because it is linear in x, but the "noise term" $dW(t)$ multiplies x. It is also commonly know as *geometric Brownian motion.*

We can solve this exactly by using Ito's formula. Let us define a new variable by

$$y = \log x, \tag{4.5.11}$$

so that

$$dy = \frac{1}{x}dx - \frac{1}{2x^2}(dx)^2 = c\,dW(t) - \tfrac{1}{2}c^2\,dt. \tag{4.5.12}$$

This equation can now be directly integrated, so we obtain

$$y(t) = y(t_0) + c[W(t) - W(t_0)] - \tfrac{1}{2}c^2(t - t_0), \tag{4.5.13}$$

and hence,

$$x(t) = x(t_0)\exp\left\{c[W(t) - W(t_0)] - \tfrac{1}{2}c^2(t - t_0)\right\}. \tag{4.5.14}$$

a) Mean value: We can calculate the mean by using the formula for any Gaussian variable z with zero mean

$$\langle \exp z \rangle = \exp\left(\tfrac{1}{2}\langle z^2 \rangle\right), \tag{4.5.15}$$

so that

$$\langle x(t) \rangle = \langle x(t_0) \rangle \exp\left[\tfrac{1}{2}c^2(t-t_0) - \tfrac{1}{2}c^2(t-t_0)\right] = \langle x(t_0) \rangle . \tag{4.5.16}$$

This result is also obvious from definition, since

$$d\langle x(t) \rangle = \langle dx(t) \rangle = \langle cx(t)\, dW(t) \rangle = 0 . \tag{4.5.17}$$

b) Autocorrelation Function: We can also calculate the autocorrelation function

$$\begin{aligned} \langle x(t)x(s) \rangle &= \langle x(t_0)^2 \rangle \left\langle \exp\left\{c[W(t) + W(s) - 2W(t_0)] - \tfrac{1}{2}c^2(t+s-2t_0)\right\}\right\rangle , \\ &= \langle x(t_0)^2 \rangle \exp\left\{\tfrac{1}{2}c^2[\langle [W(t) + W(s) - 2W(t_0)]^2 \rangle - (t+s-2t_0)]\right\} , \\ &= \langle x(t_0)^2 \rangle \exp\left\{\tfrac{1}{2}c^2[t+s-4t_0 + 2\min(t,s) - (t+s-2t_0)]\right\} , \\ &= \langle x(t_0)^2 \rangle \exp\{c^2 \min(t-t_0, s-t_0)] . \end{aligned} \tag{4.5.18}$$

c) Stratonovich Interpretation: The solution of this equation interpreted as a Stratonovich equation can also be obtained, but ordinary calculus would then be valid. Thus, instead of (4.5.12) we would obtain

$$(S)\, dy = c\, dW(t), \tag{4.5.19}$$

and hence,

$$x(t) = x(t_0) \exp\{c[W(t) - W(t_0)]\} . \tag{4.5.20}$$

In this case,

$$\langle x(t) \rangle = \langle x(t_0) \rangle \exp\left[\tfrac{1}{2}c^2(t-t_0)\right], \tag{4.5.21}$$

and

$$\langle x(t)x(s) \rangle = \langle x(t_0)^2 \exp\left\{\tfrac{1}{2}c^2[t+s-2t_0 + 2\min(t-t_0, s-T_0)]\right\} . \tag{4.5.22}$$

One sees that there is a clear difference between these two answers.

4.5.3 Complex Oscillator with Noisy Frequency

This is a simplification of a model due to *Kubo* [4.4] and is a slight generalisation of the previous example for complex variables. We consider

$$\frac{dz}{dt} = \mathrm{i}\left(\omega + \sqrt{2\gamma}\,\xi(t)\right)z, \tag{4.5.23}$$

which formally represents a simple model of an oscillator with a mean frequency ω perturbed by a noise term $\xi(t)$.

Physically, this is best modelled by writing a Stratonovich equation

$$(S)\, dz = \mathrm{i}\left(\omega\, dt + \sqrt{2\gamma}\, dW(t)\right)z, \tag{4.5.24}$$

which is equivalent to the Ito equation (from Sect. 4.4)

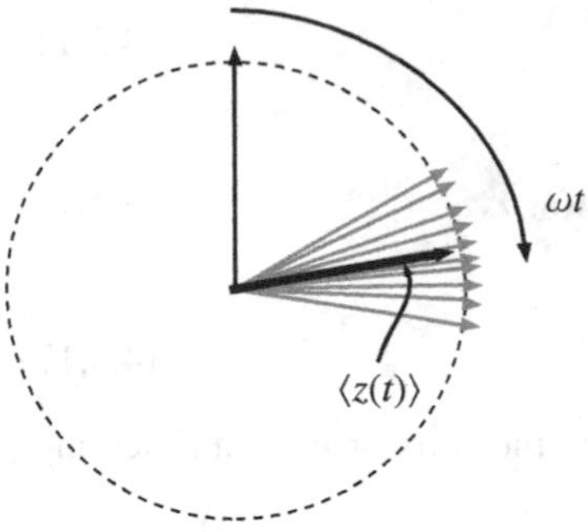

Fig. 4.3. Illustration of the decay of the mean amplitude of a complex oscillator of as a result of dephasing.

$$dz = [(\mathrm{i}\omega - \gamma)\,dt + \mathrm{i}\sqrt{2\gamma}\,dW(t)]z\,. \tag{4.5.25}$$

Taking the mean value, we see immediately that

$$\frac{d\langle z\rangle}{dt} = (\mathrm{i}\omega - \gamma)\langle z\rangle\,, \tag{4.5.26}$$

with the *damped* oscillatory solution

$$\langle z(t)\rangle = \exp[(i\omega - \gamma)t]\langle z(0)\rangle\,. \tag{4.5.27}$$

We shall show fully in Sect. 8.3, why the Stratonovich model is more appropriate. The most obvious way to see this is to note that $\xi(t)$ would, in practice, be somewhat smoother than a white noise and ordinary calculus would apply, as is the case in the Stratonovich interpretation.

Now in this case, the correlation function obtained from solving the original Stratonovich equation is

$$\langle z(t)z(s)\rangle = \langle z(0)^2\rangle \exp[(\mathrm{i}\omega - \gamma)(t+s) - 2\gamma \min(t,s)]\,. \tag{4.5.28}$$

In the limit $t, s \to \infty$, with $t + \tau = s$,

$$\lim_{t\to\infty}\langle z(t+\tau)z(t)\rangle = 0\,. \tag{4.5.29}$$

However, the correlation function of physical interest is the complex correlation

$$\begin{aligned}\langle z(t)z^*(s)\rangle &= \langle |z(0)|^2\rangle\langle \exp\{\mathrm{i}\omega(t-s) + \mathrm{i}\sqrt{2\gamma}[W(t) - W(s)]\}\rangle\,,\\ &= \langle |z(0)|^2\rangle \exp\{\mathrm{i}\omega(t-s) - \gamma[t+s-2\min(t,s)]\}\,,\\ &= \langle |z(0)|^2\rangle \exp[\mathrm{i}\omega(t-s) - \gamma|t-s|]\,.\end{aligned} \tag{4.5.30}$$

Thus, the complex correlation function has a damping term which arises purely from the noise. It may be thought of as a noise induced dephasing effect, whereby the phases of an ensemble of initial states with identical phases diffuse away from the value ωt arising from the deterministic motion, as illustrated in Fig. 4.3. The mean of the ensemble consequently decays, although the amplitude $|z(t)|$ of any member of the ensemble is unchanged. For large time differences, $z(t)$ and $z^*(s)$ become independent.

A realistic oscillator cannot be described by this model of a complex oscillator, as discussed by *van Kampen* [4.5]. However the qualitative behaviour is very similar, and this model may be regarded as a prototype model of oscillators with noisy frequency.

4.5.4 Ornstein-Uhlenbeck Process

Taking the Fokker-Planck equation given for the Ornstein-Uhlenbeck process in Sect. 3.8.4, we can immediately write down the SDE using the result of Sect. 4.3.5:

$$dx = -kx\,dt + \sqrt{D}\,dW(t)\,, \tag{4.5.31}$$

and solve this directly. Putting

$$y = x\,e^{kt}\,, \tag{4.5.32}$$

then

$$\begin{aligned} dy &= (dx)d(e^{kt}) + (dx)e^{kt} + xd(e^{kt}) \\ &= \left[-kx\,dt + \sqrt{D}\,dW(t)\right]k\,e^{kt}\,dt + \left[-kx\,dt + \sqrt{D}\,dW(t)\right]e^{kt} + kx\,e^{kt}\,dt. \end{aligned} \tag{4.5.33}$$

We note that the first product vanishes, involving only dt^2, and $dW(t)\,dt$ (in fact, it can be seen that this will always happen if we simply multiply x by a deterministic function of time). We get

$$dy = \sqrt{D}\,e^{kt}\,dW(t)\,, \tag{4.5.34}$$

so that integrating and resubstituting for y, we get

$$x(t) = x(0)\,e^{-kt} + \sqrt{D}\int_0^t e^{-k(t-t')}\,dW(t')\,. \tag{4.5.35}$$

If the initial condition is deterministic or Gaussian distributed, then $x(t)$ is clearly Gaussian, with mean and variance

$$\langle x(t)\rangle = \langle x(0)\rangle\,e^{-kt}\,, \tag{4.5.36}$$

$$\mathrm{var}[x(t)] = \left\langle\left\{[x(0) - \langle x(0)\rangle]\,e^{-kt} + \sqrt{D}\int_0^t e^{-k(t-t')}\,dW(t')\right\}^2\right\rangle. \tag{4.5.37}$$

Taking the initial condition to be nonanticipating, that is, independent of $dW(t)$ for $t > 0$, we can write using the result of Sect. 4.4f

$$\begin{aligned} \mathrm{var}[x(t)] &= \mathrm{var}[x(0)]\,e^{-2kt} + D\int_0^t e^{-2k(t-t')}\,dt'\,, \\ &= (\mathrm{var}[x(0)] - D/2k)\,e^{-2kt} + D/2k\,. \end{aligned} \tag{4.5.38}$$

These equations are the same as those obtained directly by solving the Fokker-Planck equation in Sect. 3.8.4, with the added generalisation of a nonanticipating random initial condition. Added to the fact that the solution is a Gaussian variable, we also have the correct conditional probability.

The time correlation function can also be calculated directly and is,

$$\begin{aligned}\langle x(t), x(s)\rangle &= \mathrm{var}[x(0)]\,\mathrm{e}^{-k(t+s)} + D\left\langle \int_0^t \mathrm{e}^{-k(t-t')}\,dW(t') \int_0^s \mathrm{e}^{-k(s-s')}\,dW(s')\right\rangle,\\ &= \mathrm{var}\{x(0)\}\,\mathrm{e}^{-k(t+s)} + D \int_0^{\min(t,s)} \mathrm{e}^{-k(t+s-2t')}\,dt',\\ &= \left[\mathrm{var}\{x(0)\} - \frac{D}{2k}\right]\mathrm{e}^{-k(t+s)} + \frac{D}{2k}\,\mathrm{e}^{-k|t-s|}. \end{aligned} \tag{4.5.39}$$

Notice that if $k > 0$, as $t, s \to \infty$ with finite $|t - s|$, the correlation function becomes stationary and of the form deduced in Sect. 3.8.4.

In fact, if we set the initial time at $-\infty$ rather than 0, the solution (4.5.35) becomes

$$x(t) = \sqrt{D} \int_{-\infty}^{t} \mathrm{e}^{-k(t-t)}\,dW(t'). \tag{4.5.40}$$

in which the correlation function and the mean obviously assume their stationary values. Since the process is Gaussian, this makes it stationary.

4.5.5 Conversion from Cartesian to Polar Coordinates

A model often used to describe an optical field is given by a pair of Ornstein-Uhlenbeck processes describing the real and imaginary components of the electric field, i.e.,

$$dE_1(t) = -\gamma E_1(t)\,dt + \varepsilon\,dW_1(t), \tag{4.5.41a}$$

$$dE_2(t) = -\gamma E_2(t)\,dt + \varepsilon\,dW_2(t). \tag{4.5.41b}$$

It is of interest to convert to polar coordinates. We set

$$E_1(t) = a(t)\cos\phi(t), \tag{4.5.42}$$

$$E_2(t) = a(t)\sin\phi(t), \tag{4.5.43}$$

and for simplicity, also define

$$\mu(t) = \log a(t), \tag{4.5.44}$$

so that

$$\mu(t) + \mathrm{i}\phi(t) = \log[E_1(t) + \mathrm{i}E_2(t)]. \tag{4.5.45}$$

We then use the Ito calculus to derive

$$\begin{aligned} d(\mu + \mathrm{i}\phi) &= \frac{d(E_1 + \mathrm{i}E_2)}{E_1 + \mathrm{i}E_2} - \frac{[d(E_1 + \mathrm{i}E_2)]^2}{2(E_1 + \mathrm{i}E_2)^2},\\ &= -\frac{\gamma(E_1 + \mathrm{i}E_2)}{E_1 + \mathrm{i}E_2}\,dt + \frac{\varepsilon[dW_1(t) + \mathrm{i}dW_2(t)]}{(E_1 + \mathrm{i}E_2)} - \frac{\varepsilon^2[dW_1(t) + \mathrm{i}dW_2(t)]^2}{2(E_1 + \mathrm{i}E_2)^2}, \end{aligned} \tag{4.5.46}$$

and noting $dW_1(t)\,dW_2(t) = 0$, and $dW_1(t)^2 = dW_2(t)^2 = dt$, it can be seen that the last term vanishes, so we find

$$d[\mu(t) + \mathrm{i}\phi(t)] = -\gamma\, dt + \varepsilon \exp[-\mu(t) - \mathrm{i}\phi(t)]\{dW_1(t) + \mathrm{i}\, dW_2(t)\}\,. \tag{4.5.47}$$

We now take the real part, set $a(t) = \exp[\mu(t)]$ and using the Ito calculus find

$$da(t) = \left(-\gamma a(t) + \frac{\varepsilon^2}{2a(t)}\right) dt + \varepsilon\left(dW_1(t)\cos\phi(t) + dW_2(t)\sin\phi(t)]\right). \tag{4.5.48}$$

The imaginary part yields

$$d\phi(t) = \frac{\varepsilon}{a(t)}\left(- dW_1(t)\sin\phi(t) + dW_2\cos\phi(t)\right). \tag{4.5.49}$$

We now define

$$\left.\begin{aligned} dW_a(t) &= \ \ dW_1(t)\cos\phi(t) + dW_2(t)\sin\phi(t)\,, \\ dW_\phi(t) &= -dW_1(t)\sin\phi(t) + dW_2(t)\cos\phi(t)\,. \end{aligned}\right\} \tag{4.5.50}$$

We note that this is an orthogonal transformation of the kind mentioned in Sect. 4.3.6, so that we may take $dW_a(t)$ and $dW_\phi(t)$ as increments of independent Wiener processes $W_a(t)$ and $W_\phi(t)$.

Hence, the stochastic differential equations for phase and amplitude are

$$d\phi(t) = \frac{\varepsilon}{a(t)} dW_\phi(t)\,, \tag{4.5.51a}$$

$$da(t) = \left(- \gamma a(t) + \frac{\varepsilon^2}{2a(t)}\right) dt + \varepsilon\, dW_a(t)\,. \tag{4.5.51b}$$

Comment. Using the rules given in Sect. 4.4 (ii), it is possible to convert both the Cartesian equation (4.5.41a, 4.5.41b) and the polar equations (4.5.51a, 4.5.51b) to the Stratonovich form, and to find that both are exactly the same as the Ito form. Nevertheless, a direct conversion using ordinary calculus is not possible. Doing so we would get the same result until (4.5.47) where the term $[\varepsilon^2/2a(t)]\, dt$ would not be found. This must be compensated by an extra term which arises from the fact that the Stratonovich increments $dW_i(t)$ are *correlated* with $\phi(t)$ and thus, $dW_a(t)$ and $dW_\phi(t)$ cannot simply be defined by (4.5.49). We see the advantage of the Ito method which retains the statistical independence of $dW(t)$ and variables evaluated at time t.

Unfortunately, the equations in Polar form are not soluble, as the corresponding Cartesian equations are. There is an advantage, however, in dealing with polar equations in the laser, whose equations are similar, but have an added term proportional to $a(t)^2\, dt$ in (4.5.51b).

4.5.6 Multivariate Ornstein-Uhlenbeck Process

we define the process by the stochastic differential equation

$$d\boldsymbol{x}(t) = -\boldsymbol{A}\boldsymbol{x}(t)\, dt + \boldsymbol{B}\, d\boldsymbol{W}(t), \tag{4.5.52}$$

($\boldsymbol{A}$ and $\boldsymbol{B}$ are constant matrices) for which the solution is easily obtained (as in Sect. 4.5.4):

$$x(t) = \exp(-At)x(0) + \int_0^t \exp[-A(t-t')]B\,dW(t')\,. \tag{4.5.53}$$

The mean is

$$\langle x(t)\rangle = \exp(-At)\langle x(0)\rangle. \tag{4.5.54}$$

The correlation function follows similarly

$$\begin{aligned}\langle x(t), x^{\mathrm{T}}(s)\rangle &\equiv \langle [x(t)-\langle x(t)\rangle][x(s)-\langle x(s)\rangle]^{\mathrm{T}}\rangle\,,\\ &= \exp(-At)\langle x(0), x^{\mathrm{T}}(0)\rangle \exp(-As)\\ &\quad + \int_0^{\min(t,s)} \exp[-A(t-t')]BB^{T}\exp[-A^{\mathrm{T}}(s-t')]\,dt\,.\end{aligned} \tag{4.5.55}$$

The integral can be explicitly evaluated in certain special cases, and for particular low-dimensional problems, it is possible to simply multiply everything out term by term. In the remainder we set $\langle x(0), x^{\mathrm{T}}(0) = 0\rangle$, corresponding to a deterministic initial condition, and evaluate a few special cases.

a) The Case $AA^{\mathrm{T}} = A^{\mathrm{T}}A$: In this case (for real A) we can find a unitary matrix S such that

$$\begin{aligned} SS^{\dagger} &= 1\,,\\ SAS^{\dagger} &= SA^{\mathrm{T}}S^{\dagger} = \mathrm{diag}(\lambda_1, \lambda_2, \ldots \lambda_n)\,.\end{aligned} \tag{4.5.56}$$

For simplicity, assume $t \geqslant s$. Then

$$\langle x(t), x^{\mathrm{T}}(s)\rangle = S^{\dagger}G(t,s)S\,, \tag{4.5.57}$$

where

$$[G(t,s)]_{ij} = \frac{(BB)^{\mathrm{T}}_{ij}}{\lambda_i + \lambda_j}[\exp(-\lambda_i|t-s|) - \exp(-\lambda_i t - \lambda_j s)]\,. \tag{4.5.58}$$

b) Stationary Variance: If A has only eigenvalues with positive real part, a stationary solution exists of the form

$$x_{\mathrm{s}}(t) = \int_{-\infty}^{t} \exp[-A(t-t')]B\,dW(t')\,. \tag{4.5.59}$$

We have of course

$$\langle x_{\mathrm{s}}(t)\rangle = 0\,, \tag{4.5.60}$$

$$\langle x_{\mathrm{s}}(t), x_{\mathrm{s}}^{\mathrm{T}}(s)\rangle = \int_{-\infty}^{\min(t,s)} \exp[-A(t-t')]BB^{\mathrm{T}}\exp[-A^{\mathrm{T}}(s-t')]\,dt'\,. \tag{4.5.61}$$

Let us define the stationary covariance matrix σ by

$$\sigma = \langle x_{\mathrm{s}}(t), x_{\mathrm{s}}^{\mathrm{T}}(t)\rangle\,. \tag{4.5.62}$$

This can be evaluated by means of an algebraic equation thus:

$$
\begin{aligned}
A\sigma + \sigma A^{\mathrm{T}} &= \int_{-\infty}^{t} A \exp[-A(t-t')]BB^{\mathrm{T}} \exp[-A^{\mathrm{T}}(t-t')]\, dt', \\
&\quad + \int_{-\infty}^{t} \exp[-A(t-t')]BB^{\mathrm{T}} \exp[-A^{\mathrm{T}}(t-t')]A^{\mathrm{T}}\, dt', \\
&= \int_{-\infty}^{t} \frac{d}{dt'}\{\exp[-A(t-t')]BB^{\mathrm{T}} \exp[-A^{\mathrm{T}}(t-t')]\}\, dt'.
\end{aligned} \tag{4.5.63}
$$

Carrying out the integral, we find that the lower limit vanishes by the assumed positivity of the eigenvalues of A and hence only the upper limit remains, giving

$$
A\sigma + \sigma A^{\mathrm{T}} = BB^{\mathrm{T}}, \tag{4.5.64}
$$

as an algebraic equation for the stationary covariance matrix.

c) Stationary Variance for Two Dimensions: We note that if A is a 2×2 matrix, it satisfies the characteristic equation

$$
A^2 - (\operatorname{Tr} A)A + (\operatorname{Det} A) = 0, \tag{4.5.65}
$$

and from (4.5.60) and the fact that (4.5.65) implies $\exp(-At)$ is a polynomial of degree 1 in A, we must be able to write

$$
\sigma = \alpha BB^{\mathrm{T}} + \beta(ABB^{\mathrm{T}} + BB^{\mathrm{T}}A^{\mathrm{T}}) + \gamma ABB^{\mathrm{T}}A^{\mathrm{T}}. \tag{4.5.66}
$$

Using (4.5.65), we find (4.5.64) is satisfied if

$$
\alpha + (\operatorname{Tr} A)\beta - (\operatorname{Det} A)\gamma = 0, \tag{4.5.67}
$$

$$
2\beta(\operatorname{Det} A) + 1 = 0, \tag{4.5.68}
$$

$$
\beta + (\operatorname{Tr} A)\gamma = 0. \tag{4.5.69}
$$

From which we have

$$
\sigma = \frac{(\operatorname{Det} A)BB^{\mathrm{T}} + [A - (\operatorname{Tr} A)]BB^{\mathrm{T}}[A - (\operatorname{Tr} A)]^{\mathrm{T}}}{2(\operatorname{Tr} A)(\operatorname{Det} A)}. \tag{4.5.70}
$$

d) Time Correlation Matrix in the Stationary State: From the solution of (4.5.60), we see that if $t > s$,

$$
\begin{aligned}
\langle \boldsymbol{x}_{\mathrm{s}}(t), \boldsymbol{x}_{\mathrm{s}}^{\mathrm{T}}(s)\rangle &= \exp[-A(t-s)] \int_{-\infty}^{s} \exp[-A(s-t')]BB^{\mathrm{T}} \exp[-A^{\mathrm{T}}(s-t')]\, dt', \\
&= \exp[-A(t-s)]\sigma, \qquad t > s,
\end{aligned} \tag{4.5.71a}
$$

and similarly,

$$
= \sigma \exp[-A^{\mathrm{T}}(s-t)], \quad t < s. \tag{4.5.71b}
$$

This depends only on $s - t$, as expected of a stationary solution. Defining then

$$
G_{\mathrm{s}}(t-s) = \langle \boldsymbol{x}_{\mathrm{s}}(t), \boldsymbol{x}_{\mathrm{s}}^{\mathrm{T}}(s)\rangle, \tag{4.5.72}
$$

we see (remembering $\sigma = \sigma^{\mathrm{T}}$) that

$$G_s(t-s) = [G_s(s-t)]^{\mathrm{T}}. \tag{4.5.73}$$

e) Spectrum Matrix in Stationary State: The spectrum matrix turns out to be rather simple. We define similarly to Sect. 1.5.2:

$$S(\omega) = \frac{1}{2\pi}\int_{-\infty}^{\infty} \mathrm{e}^{-i\omega\tau} G_s(\tau)d\tau\,, \tag{4.5.74}$$

$$= \frac{1}{2\pi}\Big\{\int_0^{\infty} \exp[-(i\omega + A)\tau]\sigma\, d\tau + \int_{-\infty}^{0} \sigma \exp[(-\mathrm{i}\omega + A^{\mathrm{T}})\tau]\, d\tau\Big\}, \tag{4.5.75}$$

$$= \frac{1}{2\pi}[(A+\mathrm{i}\omega)^{-1}\sigma + \sigma(A^{\mathrm{T}} - \mathrm{i}\omega)^{-1}]. \tag{4.5.76}$$

Hence,

$$(A+\mathrm{i}\omega)S(\omega)(A^T - \mathrm{i}\omega) = \frac{1}{2\pi}(\sigma A^{\mathrm{T}} + A\sigma), \tag{4.5.77}$$

and using (4.5.64), we get

$$S(\omega) = \frac{1}{2\pi}(A+\mathrm{i}\omega)^{-1}BB^{\mathrm{T}}(A^{\mathrm{T}} - \mathrm{i}\omega)^{-1}. \tag{4.5.78}$$

f) Regression Theorem: The result (4.5.71a) is also known as a regression theorem in that it states that the time development $G_s(\tau)$ is for $\tau > 0$ governed by the same law of time development of the mean, as in (4.5.54). It is a consequence of the Markovian linear nature of the problem. The time derivative of the stationary correlation function is

$$\frac{d}{d\tau}[G_s(\tau)]\, d\tau = \frac{d}{d\tau}\langle \boldsymbol{x}_s(\tau), \boldsymbol{x}_s^{\mathrm{T}}(0)\rangle\, d\tau\,,$$

$$= \langle[-A\boldsymbol{x}_s(\tau)d\tau + B\, d\boldsymbol{W}(\tau)], \boldsymbol{x}_s^{\mathrm{T}}(0)\rangle\,. \tag{4.5.79}$$

Since $\tau > 0$, the increment $d\boldsymbol{W}(\tau)$ is uncorrelated with $\boldsymbol{x}_s^{\mathrm{T}}(0)$, this means that

$$\frac{d}{d\tau}[G_s(\tau)] = -A\, G_s(\tau)\,. \tag{4.5.80}$$

Thus, computation of $G_s(\tau)$ requires the knowledge of $G_s(0) = \sigma$ and the time development equation of the mean. This result is similar to those of Sect. 3.7.4.

4.5.7 The General Single Variable Linear Equation

a) Homogeneous Case: We consider firstly the homogeneous case

$$dx = [b(t)\, dt + g(t)\, dW(t)]x\,, \tag{4.5.81}$$

and using the usual Ito rules, write

$$y = \log x\,, \tag{4.5.82}$$

so that

$$dy = \frac{dx}{x} - \frac{dx^2}{2x^2} = b(t)\,dt + g(t)\,dW(t) - \tfrac{1}{2}g(t^2)\,dt\,, \tag{4.5.83}$$

and integrating and inverting (4.5.82), we get

$$x(t) = x(0)\exp\left\{\int_0^t \left[b(t') - \tfrac{1}{2}g(t')^2\right]dt' + \int_0^t g(t')\,dW(t')\right\}, \tag{4.5.84}$$

$$\equiv x(0)\,\phi(t)\,, \tag{4.5.85}$$

which serves to define $\phi(t)$.

We note that [using (4.5.15)]

$$\langle [x(t)]^n\rangle = \langle [x(0)]^n\rangle \left\langle \exp\left\{n\int_0^t [b(t') - \tfrac{1}{2}g(t')^2]\,dt' + n\int_0^t g(t')\,dW(t')\right\}\right\rangle$$

$$= \langle [x(0)]^n\rangle \exp\left\{n\int_0^t b(t')\,dt' + \tfrac{1}{2}n(n-1)\int_0^t g(t')^2\,dt'\right\}. \tag{4.5.86}$$

b) Inhomogeneous Case: Now consider

$$dx = [a(t) + b(t)x]\,dt + [f(t) + g(t)x]\,dW(t)\,, \tag{4.5.87}$$

and write

$$z(t) = x(t)[\phi(t)]^{-1}\,, \tag{4.5.88}$$

with $\phi(t)$ as defined in (4.5.85) and a solution of the homogeneous equation (4.5.81). Then we write

$$dz = dx[\phi(t)]^{-1} + x\,d[\phi(t)^{-1}] + dx\,d[\phi(t)^{-1}]\,. \tag{4.5.89}$$

Noting that $d[\phi(t)]^{-1} = -d\phi(t)[\phi(t)]^{-2} + [d\phi(t)]^2[\phi(t)]^{-3}$ and using Ito rules, we find

$$dz = \{[a(t) - f(t)g(t)]\,dt + f(t)\,dW(t)\}\phi(t)^{-1} \tag{4.5.90}$$

which is directly integrable. Hence, the solution is

$$x(t) = \phi(t)\left\{x(0) + \int_0^t \phi(t')^{-1}\{[a(t') - f(t')g(t')]\,dt' + f(t')\,dW(t')\}\right\}. \tag{4.5.91}$$

c) Moments and Autocorrelation: It is better to derive equations for the moments from (4.5.87) rather than calculate moments and autocorrelation directly from the solution (4.5.91).

For we have

$$d[x(t)^n] = nx(t)^{n-1}dx(t) + \tfrac{1}{2}n(n-1)x(t)^{n-2}[dx(t)]^2\,,$$

$$= nx(t)^{n-1}dx(t) + \tfrac{1}{2}n(n-1)x(t)^{n-2}[f(t) + g(t)x(t)]^2\,dt\,. \tag{4.5.92}$$

Hence,

$$\frac{d}{dt}\langle x(t)^n\rangle = \langle x(t)^n\rangle[nb(t) + \tfrac{1}{2}n(n-1)g(t)^2], \\ + \langle x(t)^{n-1}\rangle[na(t) + n(n-1)f(t)g(t)], \\ + \langle x(t)^{n-2}\rangle\tfrac{1}{2}n(n-1)f(t)^2 . \tag{4.5.93}$$

These equations from a hierarchy in which the nth equation involves the solutions of the previous two, and can be integrated successively.

4.5.8 Multivariable Linear Equations

a) Homogeneous Case: The equation is

$$d\boldsymbol{x}(t) = \Big[B(t)\,dt + \sum_i G_i(t)dW_i(t)\Big]\boldsymbol{x}(t), \tag{4.5.94}$$

where $B(t), G_i(t)$ are matrices. The equation is not, in general, soluble in closed form unless all the matrices $B(t), G_i(t')$ commute at all times with each other, i.e.

$$\left.\begin{aligned} G_i(t)G_j(t') &= G_j(t')G_i(t), \\ B(t)G_i(t') &= G_i(t')B(t), \\ B(t)B(t') &= B(t')B(t). \end{aligned}\right\} \tag{4.5.95}$$

In this case, the solution is completely analogous to the one variable case and we have

$$\boldsymbol{x}(t) = \Phi(t)\boldsymbol{x}(0), \tag{4.5.96}$$

with

$$\Phi(t) = \exp\Big\{\int_0^t \Big[B(t) - \tfrac{1}{2}\sum_i G_i(t)^2\Big]dt + \int_0^t \sum_i G_i(t)dW_i(t)\Big\}. \tag{4.5.97}$$

b) Inhomogeneous Case: We can reduce the inhomogeneous case to the homogeneous case in exactly the same way as in one dimension. Thus, we consider

$$d\boldsymbol{x}(t) = [\boldsymbol{A}(t) + B(t)\boldsymbol{x}]\,dt, + \sum_i[\boldsymbol{F}_i(t) + G_i(t)\boldsymbol{x}]\,dW(t), \tag{4.5.98}$$

and write

$$\boldsymbol{y}(t) = \psi(t)^{-1}\boldsymbol{x}(t), \tag{4.5.99}$$

where $\psi(t)$ is a matrix solution of the homogeneous equation (4.5.94). We first have to evaluate $d[\psi^{-1}]$. For any matrix M we have $MM^{-1} = 1$, so, expanding to second order, $Md[M^{-1}] + dMM^{-1} + dMd[M^{-1}] = 0$.

Hence, $d[M^{-1}] = -[M + dM]^{-1}dM\,M^{-1}$ and again to second order

$$d[M^{-1}] = -M^{-1}dM\,M^{-1} + M^{-1}dM\,M^{-1}dM\,M^{-1}, \tag{4.5.100}$$

and thus, since $\psi(t)$ satisfies the homogeneous equation,

$$d[\psi(t)^{-1}] = \psi(t)^{-1}\Big\{\Big[-B(t) + \sum_i G_i(t)^2\Big]dt - \sum_i G_i(t)dW_i(t)\Big\}, \tag{4.5.101}$$

and, again taking differentials

$$d\boldsymbol{y}(t) = \psi(t)^{-1}\Big\{\Big[\boldsymbol{A}(t) - \sum_i G_i(t)\boldsymbol{F}_i(t)\Big]dt + \sum_i \boldsymbol{F}_i(t)dW_i(t)\Big\}. \tag{4.5.102}$$

Hence,

$$\boldsymbol{x}(t) = \psi(t)\left\{\boldsymbol{x}(0) + \int_0^t \psi(t')^{-1}\Big\{[\boldsymbol{A}(t') - \sum_i G_i(t')\boldsymbol{F}_i(t')]\,dt' + \sum_i \boldsymbol{F}_i(t')dW_i(t')\Big\}\right\}. \tag{4.5.103}$$

This solution is not very useful for practical purposes, even when the solution for the homogeneous equation is known, because of the difficulty in evaluating means and correlation functions.

4.5.9 Time-Dependent Ornstein-Uhlenbeck Process

This is a particular case of the previous general linear equation which is soluble. It is a generalisation of the multivariate Ornstein-Uhlenbeck process (Sect. 4.5.6) to include time-dependent parameters, namely,

$$d\boldsymbol{x}(t) = -A(t)\boldsymbol{x}(t)\,dt + B(t)d\boldsymbol{W}(t). \tag{4.5.104}$$

This is clearly of the same form as (4.5.98) with the replacements

$$\left.\begin{array}{lcl} A(t) & \to & 0, \\ B(t) & \to & -A(t), \\ \sum_i F_i(t)dW_i(t) & \to & B(t)\,d\boldsymbol{W}(t), \\ G_i(t) & \to & 0. \end{array}\right\} \tag{4.5.105}$$

The corresponding homogeneous equation is simply the deterministic equation

$$d\boldsymbol{x}(t) = -A(t)\boldsymbol{x}(t)\,dt, \tag{4.5.106}$$

which is soluble provided $A(t)A(t') = A(t')A(t)$ and has the solution

$$\boldsymbol{x}(t) = \psi(t)\boldsymbol{x}(0), \tag{4.5.107}$$

with

$$\psi(t) = \exp\Big[-\int_0^t A(t')\,dt'\Big]. \tag{4.5.108}$$

Thus, applying (4.5.103),

$$\boldsymbol{x}(t) = \exp\Big[-\int_0^t A(t')\,dt'\Big]\boldsymbol{x}(0) + \int_0^t \Big\{\exp\Big[-\int_{t'}^t A(s)ds\Big]\Big\}B(t')d\boldsymbol{W}(t'). \tag{4.5.109}$$

This is very similar to the solution of the time-independent Ornstein-Uhlenbeck process, as derived in Sect. 4.5.6, equation (4.5.53).

From this we have

$$\langle \boldsymbol{x}(t)\rangle = \exp\left[-\int_0^t A(t')\,dt'\right]\langle \boldsymbol{x}(0)\rangle\,, \tag{4.5.110}$$

$$\begin{aligned}\langle \boldsymbol{x}(t), \boldsymbol{x}^{\mathrm{T}}(t)\rangle &= \exp\left[-\int_0^t A(t')\,dt'\right]\langle \boldsymbol{x}(0), \boldsymbol{x}(0)^{\mathrm{T}}\rangle \exp\left[-\int_0^t A^{\mathrm{T}}(t')\,dt'\right] \\ &\quad + \int_0^t dt' \exp\left[-\int_{t'}^t A(s)ds\right]B(t')B^{\mathrm{T}}(t')\exp\left[-\int_t^t A^{\mathrm{T}}(s)ds\right].\end{aligned} \tag{4.5.111}$$

The time-dependent Ornstein-Uhlenbeck process will arise very naturally in connection with the development of asymptotic methods in low-noise systems.

5. The Fokker-Planck Equation

In the next two chapters, the theory of continuous Markov processes is developed from the point of view of the corresponding Fokker-Planck equation, which gives the time evolution of the probability density function for the system. This chapter is devoted mainly to single variable systems, since there are a large number of exact results for single variable systems, which makes the separate treatment of such systems appropriate. The next chapter deals with the more general multivariable aspects of many of the same issues treated one-dimensionally in this chapter.

The construction of appropriate boundary conditions is of fundamental importance, and is carried out in Sect. 5.1 in a form applicable to both one-variable and many-variable systems. A corresponding treatment for the boundary conditions on the backward Fokker-Planck equation is given in Sect. 5.1.2. The remaining of the chapter is devoted to a range of exact results, on stationary distribution functions, properties of eigenfunctions, and exit problems, most of which can be explicitly solved in the one variable case.

We have already met the Fokker-Planck equation in several contexts, starting from Einstein's original derivation and use of the diffusion equation (Sect. 1.2), again as a particular case of the differential Chapman-Kolmogorov equation (Sect. 3.5.2), and finally, in connection with stochastic differential equations (Sect. 4.3.5). There are many techniques associated with the use of Fokker-Planck equations which lead to results more directly than by direct use of the corresponding stochastic differential equation; the reverse is also true. To obtain a full picture of the nature of diffusion processes, one must study both points of view.

The origin of the name "Fokker-Planck Equation" is from the work of *Fokker* (1914) [5.1, 5.2] and *Planck* (1917) [5.2] where the former investigated Brownian motion in a radiation field and the latter attempted to build a complete theory of fluctuations based on it. Mathematically oriented works tend to use the term "Kolmogorov's Equation" because of Kolmogorov's work in developing its rigorous basis [5.3]. Yet others use the term "Smoluchowski Equation" because of Smoluchowski's original use of this equation. Without in any way assessing the merits of this terminology, I shall use the term "Fokker-Planck equation" as that most commonly used by the audience to whom this book is addressed.

5.1 Probability Current and Boundary Conditions

The FPE is a second-order parabolic partial differential equation, and for solutions we need an initial condition such as (5.2.5) and boundary conditions at the end of the interval inside which x is constrained. These take on a variety of forms.

It is simpler to derive the boundary conditions in general, than to restrict consideration to the one variable situation. We consider the forward equation

$$\partial_t p(z,t) = -\sum_i \frac{\partial}{\partial z_i} A_i(z,t)p(z,t) + \tfrac{1}{2}\sum_{i,j} \frac{\partial^2}{\partial z_i \partial z_i} B_{ij}(z,t)p(z,t)\,. \tag{5.1.1}$$

We note that this can also be written

$$\frac{\partial p(z,t)}{\partial t} + \sum_i \frac{\partial}{\partial z_i} J_i(z,t) = 0\,, \tag{5.1.2}$$

where we define the *probability current*

$$J_i(z,t) = A_i(z,t)p(z,t) - \tfrac{1}{2}\sum_j \frac{\partial}{\partial z_j} B_{ij}(z,t)p(z,t)\,. \tag{5.1.3}$$

Equation (5.5) has the form of a local conservation equation, and can be written in an integral form as follows. Consider some region R with a boundary $\boldsymbol{S}$ and define

$$P(R,t) = \int_R dz\, p(z,t)\,, \tag{5.1.4}$$

then (5.1.2) is equivalent to

$$\frac{\partial P(R,t)}{\partial t} = -\int_S dS\, \boldsymbol{n}\cdot \boldsymbol{J}(z,t)\,, \tag{5.1.5}$$

where $\boldsymbol{n}$ is the outward pointing normal to $\boldsymbol{S}$. Thus (5.1.5) indicates that the total loss of probability is given by the surface integral of $\boldsymbol{J}$ over the boundary of R.

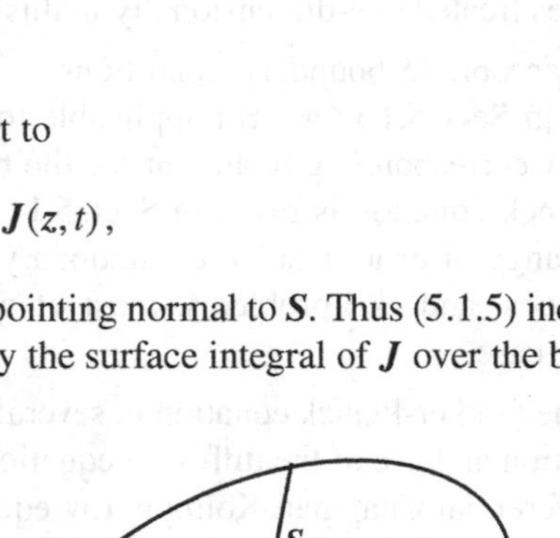

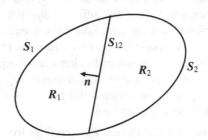

Fig. 5.1. Regions used to demonstrate that the probability current is the flow of probability

We can show as well that the current $\boldsymbol{J}$ does have the somewhat stronger property, that a surface integral over any surface $\boldsymbol{S}$ gives the net flow of probability across that surface. For consider two adjacent regions R_1 and R_2, separated by a surface $\boldsymbol{S}_{12}$. Let $\boldsymbol{S}_1$ and $\boldsymbol{S}_2$ be the surfaces which, together with $\boldsymbol{S}_{12}$, enclose respectively R_1, and R_2 (see Fig. 5.1).

Then the net flow of probability can be computed by noting that we are dealing here with a process with continuous sample paths, so that, in a sufficiently short time Δt, the probability of crossing $\boldsymbol{S}_{12}$ from R_2 to R_1 is the joint probability of being in R_2 at time t and R_1, at time $t+\Delta t$,

$$= \int_{R_1} d\boldsymbol{x} \int_{R_2} d\boldsymbol{y}\, p(\boldsymbol{x}, t+\Delta t; \boldsymbol{y}, t)\,. \tag{5.1.6}$$

The *net flow* of probability from R_2 to R_1 is obtained by subtracting from this the probability of crossing in the reverse direction, and dividing by Δt; i.e.

$$\lim_{\Delta t\to 0} \frac{1}{\Delta t} \int_{R_1} d\boldsymbol{x} \int_{R_2} d\boldsymbol{y}\, [p(\boldsymbol{x}, t+\Delta t; \boldsymbol{y}, t) - p(\boldsymbol{y}, t+\Delta t; \boldsymbol{x}, t)]\,. \tag{5.1.7}$$

Note that

$$\int_{R_1} d\boldsymbol{x} \int_{R_2} d\boldsymbol{y}\, p(\boldsymbol{x}, t; \boldsymbol{y}, t) = 0 \tag{5.1.8}$$

since this is the probability of being in R_1 and R_2 simultaneously. Thus, we can write

$$(5.1.7) = \int_{R_1} d\boldsymbol{x} \int_{R_2} d\boldsymbol{y}\, [\partial_{t'} p(\boldsymbol{x}, t'; \boldsymbol{y}, t) - \partial_{t'} p(\boldsymbol{y}, t'; \boldsymbol{x}, t)]_{t'=t}\,, \tag{5.1.9}$$

and using the Fokker-Planck equation in the form (5.5)

$$= -\int_{R_1} d\boldsymbol{x} \sum_i \frac{\partial}{\partial x_i} J_i(\boldsymbol{x}, t; R_2, t) + \int_{R_2} d\boldsymbol{y} \sum_i \frac{\partial}{\partial y_i} J_i(\boldsymbol{y}, t; R_1, t)\,, \tag{5.1.10}$$

where $J_i(\boldsymbol{x}, t; R_2, t)$ is formed from

$$p(\boldsymbol{x}, t; R_2, t) = \int_{R_2} d\boldsymbol{y}\, p(\boldsymbol{x}, t; \boldsymbol{y}, t)\,, \tag{5.1.11}$$

in the same way as $\boldsymbol{J}(z, t)$ is formed from $p(z, t)$ in (5.1.3) and $J_i(\boldsymbol{y}, t; R_1 t)$ is defined similarly. We now convert the integrals to surface integrals. The integral over $\boldsymbol{S}_2$ vanishes, since it will involve $p(\boldsymbol{x}, t; R_2, t)$, with $\boldsymbol{x}$ not in R_2 or on its boundary (except for a set of measure zero.) Similarly the integral over $\boldsymbol{S}_1$ vanishes, but those over $\boldsymbol{S}_{12}$ do not, since here the integration is simply over part of the boundaries of R_1 and R_2.

Thus we find, the net flow from R_2 to R_1 is

$$\int_{S_{12}} dS\, \boldsymbol{n} \cdot \{\boldsymbol{J}(\boldsymbol{x}, t; R_1, t) + \boldsymbol{J}(\boldsymbol{x}, t; R_2, t)\}\,, \tag{5.1.12}$$

and we finally conclude, since $\boldsymbol{x}$ belongs the union of R_1 and R_2, that the net flow of probability per unit time from R_2 to R_1

$$\equiv \lim_{\Delta t\to 0} \frac{1}{\Delta t} \int_{R_1} d\boldsymbol{x} \int_{R_2} d\boldsymbol{y} \Big[p(\boldsymbol{x}, t+\Delta t; \boldsymbol{y}, t) - p(\boldsymbol{y}, t+\Delta t; \boldsymbol{x}, t)\Big] = \int_{S_{12}} dS\, \boldsymbol{n} \cdot \boldsymbol{J}(\boldsymbol{x}, t)\,,$$

where $\boldsymbol{n}$ points from R_2 to R_1. (5.1.13)

5.1.1 Classification of Boundary Conditions

We can now consider the various kinds of boundary condition separately.

a) Reflecting Barrier: We can consider the situation where the particle cannot leave a region R, hence there is zero net flow of probability across S, the boundary of R. Thus we require

$$\boldsymbol{n} \cdot \boldsymbol{J}(z, t) = 0\,, \quad \text{for } z \in \boldsymbol{S}, \quad \boldsymbol{n} = \text{normal to } \boldsymbol{S}\,, \tag{5.1.14}$$

where $\boldsymbol{J}(z, t)$ is given by (5.5.4).

Since the particle cannot cross S, it must be reflected there, and hence the name *reflecting barrier* for this condition.

b) Absorbing Barrier: Here, one assumes that the moment the particle reaches S, it is removed from the system, thus the barrier absorbs. Consequently, the probability of being on the boundary is zero, i.e.

$$p(z, t) = 0\,, \qquad \text{for } z \in S\,. \tag{5.1.15}$$

c) Boundary Conditions at a Discontinuity: It is possible for both the A_i and B_{ij} coefficients to be discontinuous at a surface S, but for there to be free motion across S. Consequently, the probability and the normal component of the current must both be continuous across S,

$$\boldsymbol{n} \cdot \boldsymbol{J}(z)|_{S_+} = \boldsymbol{n} \cdot \boldsymbol{J}(z)|_{S_-}\,, \tag{5.1.16}$$

$$p(z)|_{S_+} = p(z)|_{S_-}\,, \tag{5.1.17}$$

where S_+, S_-, as subscripts, mean the limits of the quantities from the left and right hand sides of the surface.

The definition (5.1.3) of the current, indicates that the derivatives of $p(z)$ are not necessarily continuous at S.

5.1.2 Boundary Conditions for the Backward Fokker-Planck Equation

We suppose that $p(\boldsymbol{x}, t\,|\,\boldsymbol{x}', t')$ obeys the forward Fokker-Planck equation for a set of $\boldsymbol{x}, t$ and $\boldsymbol{x}', t'$, and that the process is confined to a region R with boundary S. Then, if s is a time between t and t',

$$0 = \frac{\partial}{\partial s} p(\boldsymbol{x}, t\,|\,\boldsymbol{x}', t') = \frac{\partial}{\partial s} \int d\boldsymbol{y}\, p(\boldsymbol{x}, t\,|\,\boldsymbol{y}, s) p(\boldsymbol{y}, s\,|\,\boldsymbol{x}', t')\,, \tag{5.1.18}$$

where we have used the Chapman-Kolmogorov equation. We take the derivative $\partial/\partial s$ inside the integral, use the forward Fokker-Planck equation for the second factor and the backward equation for the first factor. For brevity, let us write

$$\left.\begin{aligned} p(\boldsymbol{y}, s) &= p(\boldsymbol{y}, s\,|\,\boldsymbol{x}', t')\,, \\ \bar{p}(\boldsymbol{y}, s) &= p(\boldsymbol{x}, t\,|\,\boldsymbol{y}, s)\,. \end{aligned}\right\} \tag{5.1.19}$$

Then,

$$0 = \int_R dy \left[-\sum_i \frac{\partial}{\partial y_i}(A_i p) + \sum_{i,j} \frac{\partial^2}{\partial y_i \partial y_j}(B_{ij} p) \right] \bar{p}$$

$$+ \int_R dy \left[-\sum_i A_i \frac{\partial \bar{p}}{\partial y_i} - \sum_{i,j} B_{ij} \frac{\partial^2 \bar{p}}{\partial y_i \partial y_j} \right] p, \tag{5.1.20}$$

and after some manipulation

$$= \int_R dy \sum_i \frac{\partial}{\partial y_i} \left\{ -A_i p \bar{p} + \frac{1}{2} \sum_j \left[\bar{p} \frac{\partial}{\partial y_j}(B_{ij} p) - p B_{ij} \frac{\partial \bar{p}}{\partial y_j} \right] \right\}, \tag{5.1.21}$$

$$= \int_S \sum_i dS_i \left\{ \bar{p} \left[-A_i p + \frac{1}{2} \sum_j \frac{\partial}{\partial y_i}(B_{ij} p) \right] \right\} - \frac{1}{2} \int_S \sum_j dS_i p \left(\sum_j B_{ij} \frac{\partial \bar{p}}{\partial y_j} \right). \tag{5.1.22}$$

We now treat the various cases individually.

a) Absorbing Boundaries: This requires $p = 0$ on the boundary. That it also requires $\bar{p}(y, t) = 0$ on the boundary is easily seen to be consistent with (5.1.22) since on substituting $p = 0$ in that equation, we get

$$0 = \int_S \bar{p} \sum_{i,j} dS_i B_{ij} \frac{\partial p}{\partial y_j}. \tag{5.1.23}$$

However, if the boundary is absorbing, clearly

$$p(x, t \,|\, y, s) = 0, \qquad \text{for} \quad y \in \text{boundary}, \tag{5.1.24}$$

since this merely states that the probability of X re-entering R from the boundary is zero.

b) Reflecting Boundaries: Here the condition on the forward equation makes the first integral vanish in (5.1.22). The final factor vanishes for arbitrary p only if

$$\sum_{i,j} n_i B_{ij}(y) \frac{\partial}{\partial y_j} [p(x, t \,|\, y, s)] = 0\,. \tag{5.1.25}$$

In one dimension this reduces to

$$\frac{\partial}{\partial y} p(x, t \,|\, y, s) = 0\,, \tag{5.1.26}$$

unless B vanishes.

c) Other Boundaries: We shall not consider these this section. For further details see [5.4].

5.2 Fokker-Planck Equation in One Dimension

In one dimension, the Fokker-Planck equation (FPE) takes the simple form

$$\frac{\partial f(x,t)}{\partial t} = -\frac{\partial}{\partial x}[A(x,t) f(x,t)] + \frac{1}{2} \frac{\partial^2}{\partial x^2}[B(x,t) f(x,t)]\,. \tag{5.2.1}$$

In Sects. 3.4, 3.5, the Fokker-Planck equation was shown to be valid for the conditional probability, that is, the choice

$$f(x,t) = p(x,t\,|\,x_0,t_0)\,, \tag{5.2.2}$$

for any initial x_0, t_0, and with the initial condition

$$p(x,t_0\,|\,x_0,t_0) = \delta(x-x_0)\,. \tag{5.2.3}$$

However, using the definition for the one time probability

$$p(x,t) = \int dx_0\, p(x,t;x_0,t_0) \equiv \int dx_0\, p(x,t\,|\,x_0,t_0)p(x_0,t_0)\,, \tag{5.2.4}$$

we see that it is also valid for $p(x,t)$ with the initial condition

$$p(x,t)|_{t=t_0} = p(x,t_0)\,, \tag{5.2.5}$$

which is generally less singular than (5.2.3).

From the result of Sect. 4.3.5, we know that the stochastic process described by a conditional probability satisfying the FPE is equivalent to the Ito stochastic differential equation (SDE)

$$dx(t) = A[x(t),t]\,dt + \sqrt{B[x(t),t]}\,dW(t)\,, \tag{5.2.6}$$

and that the two descriptions are to be regarded as complementary to each other. We will see that perturbation theories based on the FPE are very different from those based on the SDE and both have their uses.

5.2.1 Boundary Conditions in One Dimension

The general formulation of boundary conditions as given in Sect. 5.1.1 can be augmented by some more specific results for the one-dimensional case.

a) Periodic Boundary Condition: We assume that the process takes place on an interval $[a,b]$in which the two end points are identified with each other. (this occurs, for example, if the diffusion is on a circle). Then we impose boundary conditions derived from those for a discontinuity, i.e.,

$$\text{I}: \lim_{x\to b-} p(x,t) = \lim_{x\to a+} p(x,t)\,, \tag{5.2.7}$$

$$\text{II}: \lim_{x\to b-} J(x,t) = \lim_{x\to a+} J(x,t)\,. \tag{5.2.8}$$

Most frequently, periodic boundary conditions are imposed when the functions $A(x,t)$ and $B(x,t)$ are periodic on the same interval so that we have

$$\begin{aligned} A(b,t) &= A(a,t)\,,\\ B(b,t) &= B(a,t)\,, \end{aligned} \tag{5.2.9}$$

and this means that I and II simply reduce to an equality of $p(x,t)$ and its derivatives at the points a and b.

b) Prescribed Boundaries: If the diffusion coefficient vanishes at a boundary, we have a situation in which the kind of boundary may be automatically prescribed. Suppose the motion occurs only for $x > a$. If a Lipschitz condition is obeyed by $A(x,t)$ and $\sqrt{B(x,t)}$ at $x = a$ Sect. 4.3.1b) and $B(x,t)$ is differentiable at $x = a$ then

$$\partial_x B(a,t) = 0\,. \tag{5.2.10}$$

The SDE then has solutions, and we may write

$$dx(t) = A(x,t)dt + \sqrt{B(x,t)}\,dW(t)\,. \tag{5.2.11}$$

In this rather special case, the situation is determined by the sign of $A(x,t)$. Three cases then occur, as follows.

i) *Exit boundary.* In this case, we suppose

$$A(a,t) < 0\,, \tag{5.2.12}$$

so that if the particle reaches the point a, it will certainly proceed out of region to $x < a$. Hence the name "exit boundary"

ii) *Entrance boundary.* Suppose

$$A(a,t) > 0\,. \tag{5.2.13}$$

In this case, if the particle reaches the point a, the sign of $A(a,t)$ is such as to return it to $x > a$; thus a particle placed to the right of a can never leave the region. However, a particle introduced at $x = a$ will certainly enter the region. Hence the name, "entrance boundary".

iii) *Natural boundary.* Finally consider

$$A(a,t) = 0\,. \tag{5.2.14}$$

The particle, once it reaches $x = a$, will remain there. However it can be demonstrated that it cannot ever reach this point. This is a boundary from which we can neither absorb nor at which we can introduce any particles.

c) Feller's Classification of Boundaries: *Feller* [5.4] showed that in general the boundaries can be assigned to one of the four types; regular, entrance, exit and natural. His general criteria for the classification of these boundaries are as follows.

Define

$$f(x) = \exp\left[-2\int_{x_0}^{x} ds\, A(s)/B(s)\right], \tag{5.2.15}$$

$$g(x) = 2/[B(x)f(x)]\,, \tag{5.2.16}$$

$$h_1(x) = f(x)\int_{x_0}^{x} g(s)\,ds\,, \tag{5.2.17}$$

$$h_2(x) = g(x)\int_{x_0}^{x} f(s)\,ds\,. \tag{5.2.18}$$

Here $x_0 \in (a,b)$, and is fixed. Denote by

$$\mathscr{L}(x_1,x_2)\,, \tag{5.2.19}$$

the space of all functions integrable on the interval (x_1,x_2).

Then the boundary at a can be classified as

I:	Regular:	if $f(x) \in \mathscr{L}(a, x_0)$	and	$g(x) \in \mathscr{L}(a, x_0)$
II:	Exit:	if $g(x) \notin \mathscr{L}(a, x_0)$	and	$h_1(x) \in \mathscr{L}(a, x_0)$
III:	Entrance:	if $g(x) \in \mathscr{L}(a, x_0)$	and	$h_2(x) \in \mathscr{L}(a, x_0)$
IV:	Natural:	all other cases.		

It can be seen from the results of Sect. 5.3 that for an exit boundary there is no normalisable stationary solution of the Fokker-Planck equation, and that the mean time to reach the boundary, (5.5.24), is finite. Similarly, if the boundary is exit, a stationary solution can exist, but the mean time to reach the boundary is infinite. In the case of a regular boundary, the mean time to reach the boundary is finite, but a stationary solution with a reflecting boundary at a does exist. The case of natural boundaries is harder to analyse. The reader is referred to [5.4] for a more complete description.

c) Boundaries at Infinity: All of the above kinds of boundary can occur at infinity, provided we can simultaneously guarantee the normalisation of the probability which, if $p(x)$ is reasonably well behaved, requires

$$\lim_{x\to\infty} p(x,t) = 0\,. \tag{5.2.20}$$

If $\partial_x p(x)$ is reasonably well behaved (i.e., does not oscillate infinitely rapidly as $x \to \infty$),

$$\lim_{x\to\infty} \partial_x p(x,t) = 0\,, \tag{5.2.21}$$

so that a nonzero current at infinity will usually require either $A(x,t)$ or $B(x,t)$ to become infinite there. Treatment of such cases is usually best carried out by changing to another variable which is finite at $x = \infty$.

Where there are boundaries at $x = \pm\infty$ and nonzero currents at infinity are permitted, we have two possibilities which do not allow for loss of probability:

$$\text{i)}\quad J(\pm\infty,t) = 0\,, \tag{5.2.22}$$

$$\text{ii)}\quad J(+\infty,t) = J(-\infty,t)\,. \tag{5.2.23}$$

These are the limits of reflecting and periodic boundary conditions, respectively.

5.3 Stationary Solutions for Homogeneous Fokker-Planck Equations

We recall (Sect. 3.7.2) that in a homogeneous process, the drift and diffusion coefficients are time independent. In such a case, the equation satisfied by the stationary distribution is

$$\frac{d}{dx}[A(x)p_s(x)] - \frac{1}{2}\frac{d^2}{dx^2}[B(x)p_s(x)] = 0\,, \tag{5.3.1}$$

which can also be written simply in terms of the current (as defined in Sect. 5.1)

$$\frac{dJ(x)}{dx} = 0\,, \tag{5.3.2}$$

which clearly has the solution

$$J(x) = \text{constant}\,. \tag{5.3.3}$$

Suppose the process takes place on an interval (a, b). Then we must have

$$J(a) = J(x) = J(b) \equiv J\,, \tag{5.3.4}$$

and if one of the boundary conditions is reflecting, this means that both are reflecting, and $J = 0$.

If the boundaries are not reflecting, (5.3.4) requires them to be periodic. We then use the boundary conditions given by (5.2.7) and (5.2.8).

a) Zero Current—Potential Solution:
Setting $J = 0$, we rewrite (5.3.4) as

$$A(x)p_s(x) = \frac{1}{2}\frac{d}{dx}[B(x)p_s(x)] = 0\,, \tag{5.3.5}$$

for which the solution is

$$p_s(x) = \frac{\mathcal{N}}{B(x)}\exp\left[2\int\limits_a^x dx' A(x')/B(x')\right], \tag{5.3.6}$$

where $\mathcal{N}$ is a normalisation constant such that

$$\int\limits_a^b dx\, p_s(x) = 1\,. \tag{5.3.7}$$

Such a solution is known as a *potential solution*, for various historical reasons, but chiefly because the stationary solution is obtained by a single integration (the full significance of this term will be treated in Sect. 6.2.2).

b) Periodic Boundary Condition: Here we have nonzero current J and we rewrite (5.3.3) as

$$A(x)p_s(x) - \frac{1}{2}\frac{d}{dx}[B(x)p_s(x)] = J\,. \tag{5.3.8}$$

However, J is not arbitrary, but is determined by normalisation and the periodic boundary condition

$$p_s(a) = p_s(b)\,, \tag{5.3.9}$$

$$J(a) = J(b)\,. \tag{5.3.10}$$

For convenience, define

$$\psi(x) = \exp\left[2\int\limits_a^x dx' A(x')/B(x')\right]. \tag{5.3.11}$$

Then we can easily integrate (5.3.8) to get

$$\frac{p_s(x)B(x)}{\psi(x)} = \frac{p_s(a)B(a)}{\psi(a)} - 2J\int\limits_a^x \frac{dx'}{\psi(x')}\,. \tag{5.3.12}$$

By imposing the boundary condition (5.3.9) we find that

$$J = \frac{\left[\dfrac{B(b)}{\psi(b)} - \dfrac{B(a)}{\psi(a)}\right] p_s(a)}{\int\limits_a^b \dfrac{dx'}{\psi(x')}}\,, \tag{5.3.13}$$

so that

$$p_s(x) = p_s(a)\left[\frac{\int\limits_a^x \dfrac{dx'}{\psi(x')}\dfrac{B(b)}{\psi(b)} + \int\limits_x^b \dfrac{dx'}{\psi(x')}\dfrac{B(a)}{\psi(a)}}{\dfrac{B(x)}{\psi(x)}\int\limits_a^b \dfrac{dx'}{\psi(x')}}\right]. \tag{5.3.14}$$

c) Infinite Range and Singular Boundaries: In either of these cases, one or the other of the above possibilities may turn out to be forbidden because of divergences, etc. A full enumeration of the possibilities is, in general, very complicated. We shall demonstrate these by means of the examples given in the next section.

5.3.1 Examples of Stationary Solutions

a) Diffusion in a Gravitational Field: A strongly damped Brownian particle moving in a constant gravitational field is often described by the stochastic differential equation (8.2.15)

$$dx = -g\,dt + \sqrt{D}\,dW(t)\,, \tag{5.3.15}$$

for which the Fokker-Planck equation is

$$\frac{\partial p}{\partial t} = \frac{\partial}{\partial x}(gp) + \frac{1}{2}D\frac{\partial^2 p}{\partial x^2}\,. \tag{5.3.16}$$

On the interval (a, b) with reflecting boundary conditions, the stationary solution is given by (5.3.6), i.e.

$$p_s(x) = \mathcal{N}\exp[-2gx/D]\,, \tag{5.3.17}$$

where we have absorbed constant factors into the definition of $\mathcal{N}$.

Clearly this solution is normalisable on (a, b) only if a a is finite, though b may be infinite. The result is no more profound than to say that particles diffusing in a beaker of fluid will fall down, and if the beaker is infinitely deep, they will never stop falling! Diffusion upwards against gravity is possible for any distance but with exponentially small probability.

Now assume periodic boundary conditions on (a, b). Substitution into (5.3.14) yields

$$p_s(x) = p_s(a)\,, \tag{5.3.18}$$

a constant distribution.

The interpretation is that the particles pass freely from a to b and back.

b) Ornstein Uhlenbeck Process: We use the notation of Sect. 3.8.4 where the Fokker-Planck equation was

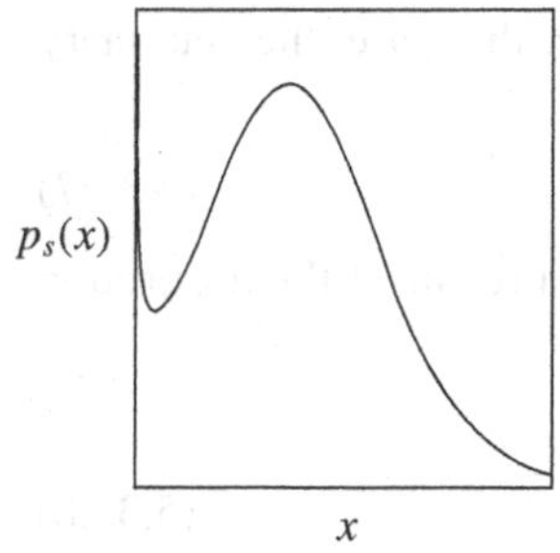

Fig. 5.2. Non-normalisable "stationary" $p_s(x)$ for the reaction $X + A \rightleftharpoons 2X$.

$$\frac{\partial p}{\partial t} = \frac{\partial}{\partial x}(kxp) + \frac{1}{2}D\frac{\partial^2 p}{\partial x^2}, \tag{5.3.19}$$

whose stationary solution on the interval (a, b) with reflecting barriers is

$$p_s(x) = \mathcal{N} \exp(-kx^2/D). \tag{5.3.20}$$

Provided $k > 0$, this is normalisable on $(-\infty, \infty)$.

If $k < 0$, one can only make sense of it on a finite interval. In this case suppose

$$a = -b < 0. \tag{5.3.21}$$

so that from (5.3.11),

$$\psi(x) = \exp\left[-\frac{k}{D}(x^2 - a^2)\right], \tag{5.3.22}$$

and if we consider the periodic boundary condition on this interval, by noting

$$\psi(a) = \psi(-a), \tag{5.3.23}$$

we find that

$$p_s(x) = p_s(a)\frac{\psi(x)}{\psi(a)} = p_s(a)\exp\left[-\frac{k}{D}(x^2 - a^2)\right], \tag{5.3.24}$$

so that the symmetry yields the same solution as in the case of reflecting barriers.

Letting $a \to \infty$, we see that we still have the same solution. The result is also true if $a \to \infty$ independently of $b \to -\infty$, provided $k > 0$.

c) A Chemical Reaction Model: Although chemical reactions are normally best modelled by a birth-death master equation formalism, as in Chap. 11, approximate treatments are often given by means of a Fokker-Planck equation. The reaction

$$X + A \rightleftharpoons 2X, \tag{5.3.25}$$

is of interest since it possesses an exit boundary at $x = 0$ (where x is the number of molecules of X). Clearly if there is no X, a collision between X and A cannot occur so no more X is produced.

The Fokker-Planck equation is derived in Sect. 11.6.1 and is

$$\partial_t p(x,t) = -\partial_x\left[\left(ax - x^2\right)p(x,t)\right] + \tfrac{1}{2}\partial_x^2\left[\left(ax + x^2\right)p(x,t)\right]. \tag{5.3.26}$$

We introduce reflecting boundaries at $x = \alpha$ and $x = \beta$. In this case, the stationary solution is

$$p_s(x) = e^{-2x}(a+x)^{4a-1}x^{-1}\,, \tag{5.3.27}$$

which is not normalisable if $\alpha = 0$. The pole at $x = 0$ is a result of the absorption there. In fact, comparing with (5.2.18), we see that

$$\left.\begin{aligned} B(0,t) &\equiv (ax+x^2)_{x=0} = 0\,, \\ A(0,t) &\equiv (ax-x^2)_{x=0} = 0\,, \\ \partial_x B(0,t) &\equiv (a+2x)_{x=0} > 0\,, \end{aligned}\right\} \tag{5.3.28}$$

so we indeed have an exit boundary. The stationary solution has relevance only if $\alpha > 0$ since it is otherwise not normalisable. The physical meaning of a reflecting barrier is quite simple: whenever a molecule of X disappears, we simply add another one immediately. A plot of $p_s(x)$ is given in Fig. 5.2. The time for all x to disappear is in practice extraordinarily long, and the stationary solution (5.3.27) is, in practice, a good representation of the distribution except near $x = 0$.

5.4 Eigenfunction Methods for Homogeneous Processes

We shall now show how, in the case of homogeneous processes, solutions can most naturally be expressed in terms of eigenfunctions. We consider reflecting and absorbing boundaries.

5.4.1 Eigenfunctions for Reflecting Boundaries

We consider a Fokker-Planck equation for a process on a interval (a, b) with reflecting boundaries. We suppose the Fokker-Planck equation to have a stationary solution $p_s(x)$ and the from

$$\partial_t p(x,t) = -\partial_x[A(x)p(x,t)] + \tfrac{1}{2}\partial_x^2[B(x)p(x,t)]\,. \tag{5.4.1}$$

We define a function $q(x, t)$ by

$$p(x,t) = p_s(x)q(x,t)\,, \tag{5.4.2}$$

and, by direct substitution, find that $q(x, t)$ satisfies the *backward equation*

$$\partial_t q(x,t) = A(x)\partial_x q(x,t) + \tfrac{1}{2}B(x)\partial_x^2 q(x,t)\,. \tag{5.4.3}$$

We now wish to consider solutions of the form

$$p(x,t) = P_\lambda(x)e^{-\lambda t}\,, \tag{5.4.4}$$

$$q(x,t) = Q_\lambda(x)e^{-\lambda t}\,, \tag{5.4.5}$$

which obey the eigenfunction equations

$$-\partial_x[A(x)P_\lambda(x)] + \tfrac{1}{2}\partial_x^2[B(x)P_\lambda(x)] = -\lambda P_\lambda(x)\,, \tag{5.4.6}$$

$$A(x)\partial_x Q_{\lambda'}(x) + \tfrac{1}{2}B(x)\partial_x^2 Q_{\lambda'}(x) = -\lambda' Q_{\lambda'}(x)\,. \tag{5.4.7}$$

i) *Relationship between P_λ and Q_λ :* From (5.4.2) and (5.4.3) it follows that

$$P_\lambda(x) = p_s(x)Q_\lambda(x). \tag{5.4.8}$$

This simple result does not generalise completely to many dimensional situations, which are treated in Sect. 6.5.

ii) *Orthogonality of eigenfunctions:* We can straightforwardly show by partial integration that

$$(\lambda' - \lambda)\int_a^b dx P_\lambda(x)Q_{\lambda'}(x)$$
$$= \left[Q_{\lambda'}(x)\left\{-A(x)P_\lambda(x) + \tfrac{1}{2}\partial_x[B(x)P_\lambda(x)]\right\} - \tfrac{1}{2}B(x)P_\lambda(x)\partial_x Q_{\lambda'}(x)\right]_a^b. \tag{5.4.9}$$

Using the reflecting boundary condition on the coefficient of $Q_{\lambda'}(x)$, we see that this coefficient vanishes. Further, using the definition of $q(x,t)$ in terms of the stationary solution (5.4.2), it is simple to show that

$$\tfrac{1}{2}B(x)\partial_x Q_{\lambda'}(x) = -A(x)P_{\lambda'}(x) + \tfrac{1}{2}\partial_x[B(x)P_{\lambda'}(x)], \tag{5.4.10}$$

so that term vanishes also. Hence, the $Q_\lambda(x)$ and $P_\lambda(x)$ form a bi-orthogonal system

$$\int_a^b dx\, P_\lambda(x)Q_{\lambda'}(x) = \delta_{\lambda\lambda'}. \tag{5.4.11}$$

There are thus are two alternative *orthogonality* systems,

$$\int_a^b dx\, p_s(x)Q_\lambda(x)Q_{\lambda'}(x) = \delta_{\lambda\lambda'}, \tag{5.4.12}$$

$$\int_a^b dx[p_s(x)]^{-1}P_\lambda(x)P_{\lambda'}(x) = \delta_{\lambda\lambda'}. \tag{5.4.13}$$

It should be noted that setting $\lambda = \lambda' = 0$ gives the normalisation of the stationary solution $p_s(x)$ since

$$P_0(x) = p_s(x), \tag{5.4.14}$$

$$Q_0(x) = 1. \tag{5.4.15}$$

iii) *Expansion in eigenfunctions:* Using this orthogonality (and assuming completeness) we can write any solution in terms of eigenfunctions. For if

$$p(x,t) = \sum_\lambda A_\lambda P_\lambda(x)e^{-\lambda t}, \tag{5.4.16}$$

then

$$\int_a^b dx\, Q_\lambda(x)p(x,0) = A_\lambda. \tag{5.4.17}$$

iv) *Conditional probability:* For example, the conditional probability $p(x, t \mid x_0, 0)$ is given by the initial condition

$$p(x, 0 \mid x_0, 0) = \delta(x - x_0)\,, \tag{5.4.18}$$

so that

$$A_\lambda = \int_a^b dx\, Q_\lambda(x)\delta(x - x_0) = Q_\lambda(x_0)\,, \tag{5.4.19}$$

and hence,

$$p(x, t \mid x_0, 0) = \sum_\lambda P_\lambda(x) Q_\lambda(x_0) \mathrm{e}^{-\lambda t}\,. \tag{5.4.20}$$

v) *Autocorrelation function:* We can write the autocorrelation function quite elegantly as

$$\langle x(t)x(0)\rangle = \int dx \int dx_0\, x\, x_0\, p(x, t \mid x_0, 0) p_\mathrm{s}(x)\,, \tag{5.4.21}$$

$$= \sum_\lambda \left[\int dx\, x P_\lambda(x)\right]^2 \mathrm{e}^{-\lambda t}\,, \tag{5.4.22}$$

where we have used the definition of $Q_\lambda(x)$ by (5.4.5).

5.4.2 Eigenfunctions for Absorbing Boundaries

These are treated similarly.

We define P_λ and Q_λ as above, except that $p_\mathrm{s}(x)$ is still the stationary solution of the Fokker-Planck equation with *reflecting boundary conditions*. With this definition, we find that we must have

$$P_\lambda(a) = Q_\lambda(a) = P_\lambda(b) = Q_\lambda(b) = 0\,, \tag{5.4.23}$$

and the orthogonality proof still follows through. Eigenfunctions are then computed using this condition and the eigenfunction equations (5.4.6) and (5.4.7) and all other results look the same. However, the range of λ does not include $\lambda = 0$, and hence $p(x, t \mid x_0, 0) \to 0$ as $t \to \infty$.

5.4.3 Examples

a) A Wiener Process with Absorbing Boundaries: The Fokker-Planck equation

$$\partial_t p = \tfrac{1}{2}\partial_x^2 p\,, \tag{5.4.24}$$

is treated on the interval (0, 1). The absorbing boundary condition requires

$$p(0, t) = p(1, t) = 0\,, \tag{5.4.25}$$

and the appropriate eigenfunctions are $\sin(n\pi x)$ so we expand in a Fourier sine series

$$p(x, t) = \sum_{n=1}^{\infty} b_n(t) \sin(n\pi x)\,, \tag{5.4.26}$$

which automatically satisfies (5.4.25). The initial condition is chosen so that

$$p(x, 0) = \delta(x - x_0), \tag{5.4.27}$$

for which the Fourier coefficients are

$$b_n(0) = 2 \int_0^1 dx\, \delta(x - x_0) \sin(n\pi x) = 2 \sin(n\pi x_0). \tag{5.4.28}$$

Substituting the Fourier expansion (5.4.26) into (5.4.24) gives

$$\frac{db_n(t)}{dt} = -\lambda_n b_n(t), \tag{5.4.29}$$

with

$$\lambda_n = n^2\pi^2/2, \tag{5.4.30}$$

and the solution

$$b_n(t) = b_n(0) \exp(-\lambda_n t). \tag{5.4.31}$$

So we have the solution [which by the initial condition (5.4.27) is for the conditional probability $p(x, t \,|\, x_0, 0)$]

$$p(x, t \,|\, x_0, 0) = 2 \sum_{n=1}^{\infty} \exp(-\lambda_n t) \sin(n\pi x_0) \sin(n\pi x). \tag{5.4.32}$$

b) Wiener Process with Reflecting Boundaries: Here the boundary condition reduces to [on the interval (0, 1)]

$$\partial_x p(0, t) = \partial_x p(1, t) = 0, \tag{5.4.33}$$

and the eigenfunctions are now $\cos(n\pi x)$, so we make a Fourier cosine expansion

$$p(x, t) = \tfrac{1}{2} a_0 + \sum_{n=1}^{\infty} a_n(t) \cos(n\pi x), \tag{5.4.34}$$

with the same initial condition

$$p(x, 0) = \delta(x - x_0), \tag{5.4.35}$$

so that

$$a_n(0) = 2 \int_0^1 dx\, \cos(n\pi x)\delta(x - x_0) = 2 \cos(n\pi x_0). \tag{5.4.36}$$

In the same way as before, we find

$$a_n(t) = a_n(0) \exp(-\lambda_n t), \tag{5.4.37}$$

with

$$\lambda_n = n^2\pi^2/2, \tag{5.4.38}$$

so that

$$p(x, t \,|\, x_0, 0) = 1 + 2 \sum_{n=1}^{\infty} \cos(n\pi x_0) \cos(n\pi x) \exp(-\lambda_n t). \tag{5.4.39}$$

As $t \to \infty$, the process becomes stationary, with stationary distribution

$$p_s(x) = \lim_{t\to\infty} p(x,t\,|\,x_0,0) = 1. \tag{5.4.40}$$

We can compute the stationary autocorrelation function by

$$\langle x(t)x(0)\rangle_s = \int_0^1\int_0^1 dx\, dx_0\, x\, x_0\, p(x,t\,|\,x_0,0)p_s(x)\,, \tag{5.4.41}$$

and carrying out the integrals explicitly,

$$\langle x(t)x(0)\rangle_s = \frac{1}{4} + \frac{8}{\pi^4}\sum_{n=0}^{\infty} \exp(-\lambda_{2n+1}t)(2n+1)^{-4}\,. \tag{5.4.42}$$

We see that as $t \to \infty$, all the exponentials vanish and

$$\langle x(t)x(0)\rangle_s \to \frac{1}{4} = [\langle x\rangle_s]^2\,, \tag{5.4.43}$$

and as $t \to 0$,

$$\langle x(t)x(0)\rangle_s \to \frac{1}{4} + \frac{8}{\pi^4}\sum_{n=0}^{\infty}(2n+1)^{-4} = \frac{1}{3} = \langle x_s^2\rangle\,, \tag{5.4.44}$$

when one takes account of the identity (from the theory of the Riemann zeta-function)

$$\sum_{n=0}^{\infty}(2n+1)^{-4} = \frac{\pi^4}{96}\,. \tag{5.4.45}$$

c) Ornstein-Uhlenbeck Process: As in Sect. 3.8.4 the Fokker-Planck equation is

$$\partial_t p(x,t) = \partial_x[kxp(x,t)] + \tfrac{1}{2}D\partial_x^2 p(x,t)\,. \tag{5.4.46}$$

The eigenfunction equation for Q_λ is

$$d_x^2 Q_\lambda - \frac{2kx}{d}d_x Q_\lambda + \frac{2\lambda}{D}Q_\lambda = 0\,, \tag{5.4.47}$$

and this becomes the differential equation for *Hermite polynomials* $\mathrm{H}_n(y)$ [5.5] on making the replacement $y = x\sqrt{k/D}$:

$$d_y^2 Q_\lambda - 2y d_y Q_\lambda + (2\lambda/k)Q_\lambda = 0\,. \tag{5.4.48}$$

We can write

$$Q_\lambda = (2^n n!)^{-1/2}\mathrm{H}_n\left(x\sqrt{k/D}\right), \tag{5.4.49}$$

where

$$\lambda = nk\,, \tag{5.4.50}$$

and these solutions are normalised as in (5.4.11–5.4.13).

The stationary solution is, as previously found,

$$p_s(x) = (k/\pi D)^{1/2}\exp(-kx^2/D)\,, \tag{5.4.51}$$

and a general solution can be written as

$$p(x,t) = \sum_n \sqrt{[k/(2^n n!\pi D)]}\exp(-kx^2/D)\mathrm{H}_n\left(x\sqrt{k/D}\right)\mathrm{e}^{-nkt}A_n\,, \tag{5.4.52}$$

with

$$A_n = \int_{-\infty}^{\infty} dx\, p(x,0) \mathrm{H}_n\left(x\sqrt{k/D}\right)(2^n n!)^{-1/2} . \tag{5.4.53}$$

The result can also be obtained directly from the explicit solution for the conditional probability given in Sect. 3.8.4 by using generating functions for Hermite polynomials. One sees that the time scale of relaxation to the stationary state is given by the eigenvalues

$$\lambda_n = nk . \tag{5.4.54}$$

Here, k is the rate constant for deterministic relaxation, and it thus determines the slowest time in the relaxation. One can also compute the autocorrelation function directly using (5.4.22). We use the result [5.5] that

$$\mathrm{H}_1(y) = 2y , \tag{5.4.55}$$

so that the orthogonality property means that only the eigenfunction corresponding to $n = 1$ has a nonzero coefficient. We must compute

$$\int x P_{\lambda_1}(x) dx = \int_{-\infty}^{\infty} dx\, \sqrt{k/(2\pi D)} \exp(-kx^2/D)\left(2x\sqrt{k/D}\right)x = \sqrt{D/2k} , \tag{5.4.56}$$

so that

$$\langle x(t)x(0)\rangle_s = \frac{D}{2k}\mathrm{e}^{-kt} , \tag{5.4.57}$$

as found previously in Sects. 3.8.4 and 4.5.4.

d) Rayleigh Process: We take the model of amplitude fluctuations developed in Sect. 4.5.5. The Fokker-Planck equation is

$$\partial_t p(x,t) = \partial_x[(\gamma x - \mu/x)p(x,t)] + \mu\partial_x^2 p(x,t) , \tag{5.4.58}$$

where

$$\mu = \varepsilon^2/2 . \tag{5.4.59}$$

The range here is $(0,\infty)$ and the stationary solution (normalised)

$$p_s(x) = (\gamma x/\mu)\exp(-\gamma x^2/2\mu) . \tag{5.4.60}$$

The eigenfunction equation for the $Q_\lambda(x)$ is

$$d_x^2 Q_\lambda + (1/x - \gamma x/\mu) d_x Q_\lambda + (\lambda/\mu) Q_\lambda = 0 . \tag{5.4.61}$$

By setting

$$z = x^2\gamma/2\mu , \tag{5.4.62}$$

we obtain

$$z d_z^2 Q_\lambda + (1-z) d_z Q_\lambda + (\lambda/2\gamma) Q_\lambda = 0 . \tag{5.4.63}$$

This is the differential equation for the *Laguerre polynomials* $\mathrm{L}_n(y)$ [5.5] provided

$$\lambda = 2n\gamma . \tag{5.4.64}$$

We can write

$$Q_\lambda(x) = \mathrm{L}_n(\gamma x^2/2\mu)\,, \tag{5.4.65}$$

which is normalised. Hence, the conditional probability is

$$p(x,t\,|\,x_0,0) = \sum_{n=0}^{\infty} \frac{\gamma x}{\mu} \exp\left(-\frac{\gamma x^2}{2\mu}\right) \mathrm{L}_n\left(\frac{\gamma x_0^2}{2\mu}\right) \mathrm{L}_n\left(\frac{\gamma x^2}{2\mu}\right) \mathrm{e}^{-2n\gamma t}\,. \tag{5.4.66}$$

We can compute the autocorrelation function by the method of (5.4.22):

$$\langle x(t)x(0)\rangle = \sum_{n=0}^{\infty}\left[\int_0^\infty x\,dx\frac{\gamma x}{\mu}\exp\left(\frac{-\gamma x^2}{2\mu}\right)\mathrm{L}_n\left(\frac{\gamma x^2}{2\mu}\right)\right]^2 \exp(-2n\gamma t)\,, \tag{5.4.67}$$

and using

$$\int_0^\infty dz\, z^\alpha \mathrm{e}^{-z}\mathrm{L}_n(z) = (-1)^n \Gamma(\alpha+1)\binom{\alpha}{n}, \tag{5.4.68}$$

we find for the autocorrelation function

$$\langle x(t)x(0)\rangle = \frac{2\mu}{\gamma}\sum_{n=0}^{\infty}\frac{\pi}{4}\binom{\frac{1}{2}}{n}^2 \exp(-2n\gamma t)\,. \tag{5.4.69}$$

5.5 First Passage Times for Homogeneous Processes

It is often of interest to know how long a particle whose position is described by a Fokker-Planck equation remains in a certain region of x. The solution of this problem can be achieved by use of the *backward Fokker-Planck equation*, as described in Sect. 3.6.

5.5.1 Two Absorbing Barriers

Let the particle be initially at x at time $t = 0$ and let us ask how long it remains in the interval (a, b) which is assumed to contain x:

$$a \leqslant x \leqslant b\,. \tag{5.5.1}$$

We erect absorbing barriers at a and b so that the particle is removed from the system when it reaches a or b. Hence, if it is still in the interval (a, b), it has never left that interval.

i) *Distribution of exit times:* Under these conditions, the probability that at time t the particle is still in (a, b) is

$$\int_a^b dx'\, p(x',t\,|\,x,0) \equiv G(x,t)\,. \tag{5.5.2}$$

Let the time that the particle leaves (a, b) be T. Then we can rewrite (5.5.2) as

$$\mathrm{Prob}(T \geqslant t) = \int_a^b dx'\, p(x',t\,|\,x,0)\,, \tag{5.5.3}$$

which means that $G(x,t)$ is the same as $\text{Prob}(T \geqslant t)$. Since the system is time homogeneous, we can write

$$p(x',t\,|\,x,0) = p(x',0\,|\,x,-t)\,, \tag{5.5.4}$$

and the backward Fokker-Planck equation can be written

$$\partial_t p(x',t\,|\,x,0) = A(x)\partial_x p(x',t\,|\,x,0) + \tfrac{1}{2}B(x)\partial_x^2 p(x',t\,|\,x,0)\,, \tag{5.5.5}$$

and hence, $G(x,t)$ obeys the equation

$$\partial_t G(x,t) = A(x)\partial_x G(x,t) + \tfrac{1}{2}B(x)\partial_x^2 G(x,t)\,. \tag{5.5.6}$$

ii) *Initial condition:* Clearly that

$$p(x',0\,|\,x,0) = \delta(x-x')\,, \tag{5.5.7}$$

and hence,

$$G(x,0) = \begin{cases} 1\,, & a \leqslant x \leqslant b, \\ 0\,. & \text{elsewhere.} \end{cases} \tag{5.5.8}$$

iii) *Boundary conditions:* If $x = a$ or b, the particle is absorbed immediately, so $\text{Prob}(T \geqslant t) = 0$ when $x = a$ or $x = b$, i.e.,

$$G(a,t) = G(b,t) = 0\,. \tag{5.5.9}$$

iv) *Moments of the exit time:* Since $G(x,t)$ is the probability that $T \geqslant t$, the mean of any function of T is

$$\langle f(T)\rangle = -\int_0^\infty f(t)\,dG(x,t)\,. \tag{5.5.10}$$

Thus, the *mean exit time* (or *mean first passage time*)

$$T(x) = \langle T\rangle\,, \tag{5.5.11}$$

is given by

$$T(x) = -\int_0^\infty t\,\partial_t\,G(x,t)dt \tag{5.5.12}$$

$$= \int_0^\infty G(x,t)dt\,, \tag{5.5.13}$$

after integrating by parts. Similarly, defining

$$T_n(x) = \langle T^n\rangle\,, \tag{5.5.14}$$

we find

$$T_n(x) = \int_0^\infty t^{n-1}G(x,t)dt\,. \tag{5.5.15}$$

v) *Differential equation for the mean exit time:* We can derive a simple ordinary differential equation for $T(x)$ by using (5.5.13) and integrating (5.5.6) over $(0,\infty)$. Noting that

$$\int_0^\infty \partial_t G(x,t)dt = G(x,\infty) - G(x,0) = -1\,, \tag{5.5.16}$$

we derive

$$A(x)\partial_x T(x) + \tfrac{1}{2}B(x)\partial_x^2 T(x) = -1\,, \tag{5.5.17}$$

with the boundary condition

$$T(a) = T(b) = 0\,. \tag{5.5.18}$$

Similarly, we see that

$$-nT_{n-1}(x) = A(x)\partial_x T_n(x) + \tfrac{1}{2}B(x)\partial_x^2 T_n(x)\,. \tag{5.5.19}$$

vi) *Solutions of the Equations:* Equation (5.5.17) can be solved directly by integration. The solution, after some manipulation, can be written in terms of

$$\psi(x) = \exp\left\{\int_0^x dx'[2A(x')/B(x')]\right\}. \tag{5.5.20}$$

We find

$$T(x) = \frac{2\left[\left(\int_a^x \frac{dy}{\psi(y)}\right)\int_x^b \frac{dy'}{\psi(y')}\int_a^{y'} \frac{dz\,\psi(z)}{B(z)} - \left(\int_x^b \frac{dy}{\psi(y)}\right)\int_a^x \frac{dy'}{\psi(y')}\int_a^{y'} \frac{dz\,\psi(z)}{B(z)}\right]}{\int_a^b \frac{dy}{\psi(y)}}\,. \tag{5.5.21}$$

5.5.2 One Absorbing Barrier

We consider motion still in the interval (a, b) but suppose the barrier at a to be reflecting. The boundary conditions then become

$$\partial_x G(a,t) = 0\,, \tag{5.5.22a}$$

$$G(b,t) = 0\,, \tag{5.5.22b}$$

which follow from the conditions on the backward Fokker-Planck equation derived in Sect. 5.1.2. We solve (5.5.17) with the corresponding boundary condition and obtain

$$T(x) = 2\int_x^b \frac{dy}{\psi(y)}\int_a^y \frac{\psi(z)}{B(z)}dz \qquad \begin{array}{l} a \text{ reflecting,} \\ b \text{ absorbing,} \\ a < b\,. \end{array} \tag{5.5.23}$$

Similarly, one finds

$$T(x) = 2\int_a^x \frac{dy}{\psi(y)}\int_y^b \frac{\psi(z)}{B(z)}dz \qquad \begin{array}{l} b \text{ reflecting,} \\ a \text{ absorbing,} \\ a < b\,. \end{array} \tag{5.5.24}$$

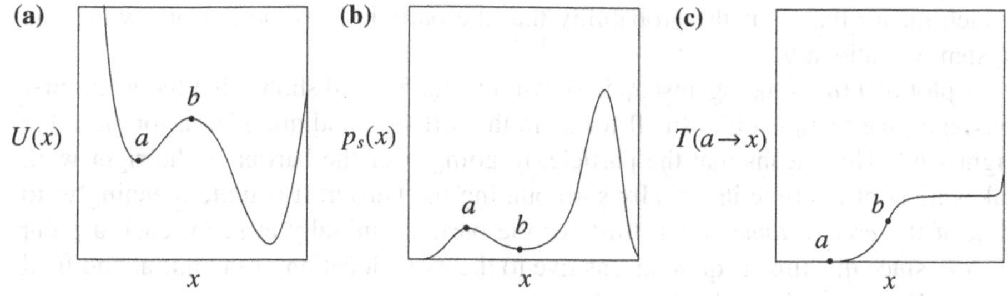

Fig. 5.3. (**a**) Double well potential $U(x)$; (**b**) Stationary distribution $p_s(x)$; (**c**) Mean first passage time from a to x, $T(a \to x)$

5.5.3 Application—Escape Over a Potential Barrier

We suppose that a point moves according to the Fokker-Planck equation

$$\partial_t p(x,t) = \partial_x[U'(x)p(x,t)] + D\partial_x^2 p(x,t). \tag{5.5.25}$$

The potential has maxima and minima, as shown in Fig. 5.3. We suppose that motion is on an infinite range, which means the stationary solution is

$$p_s(x) = \mathcal{N} \exp[-U(x)/D], \tag{5.5.26}$$

which is bimodal (as shown in Fig. 5.3) so that there is a relatively high probability of being on the left or the right of b, but not near b. What is the mean escape time from the left hand well? By this we mean, what is the mean first passage time from a to x, where x is in the vicinity of b? We use (5.5.23) with the substitutions

$$b \to x, \quad a \to -\infty, \quad x \to a, \tag{5.5.27}$$

so that

$$T(a \to x) = \frac{1}{D} \int_a^x dy \exp[U(y)/D] \int_{-\infty}^{y} \exp[-U(z)/D]dz. \tag{5.5.28}$$

If the central maximum of $U(x)$ is large and D is small, then exp $[U(y)/D]$ is sharply peaked at $x = b$, while $\exp[-U(z)/D]$ is very small near $z = b$. Therefore, $\int_{-\infty}^{y} \exp[-U(z)/D]dz$ is a very slowly varying function of y near $y = b$. This means that the value of the integral $\int_{-\infty}^{y} \exp[-U(z)/D]dz$ will be approximately constant for those values of y which yield a value of $\exp[U(y)/D]$ which is significantly different from zero. Hence, in the inner integral, we can set $y = b$ and remove the resulting constant factor from inside the integral with respect to y. Thus, we can approximate (5.5.28) by

$$T(a \to x) \approx \left\{ \frac{1}{D} \int_{-\infty}^{b} dy \exp[-U(z)/D] \right\} \int_a^x dy \exp[U(y)/D]. \tag{5.5.29}$$

Notice that by the definition of $p_s(x)$ in (5.5.26), we can say that

$$\int_{-\infty}^{b} dy \exp[-U(z)/D] = n_a/\mathcal{N}, \tag{5.5.30}$$

which means that n_a is the probability that the particle is to the left of b when the system is stationary.

A plot of $T(a \to x_0)$ against x_0 is shown in Fig. 5.3 and shows that the mean first passage time to x_0 is quite small for x_0 in the left well and quite large for x_0 in the right well. This means that the particle, in going over the barrier to the right well, takes most of the time in actually surmounting the barrier. It is quite meaningful to talk of the *escape time* as that time for the particle, initially at a, to reach a point near c, since this time is quite insensitive to the exact location of the initial and final points. We can evaluate this by further assuming that near b we can write

$$U(x) \approx U(b) - \tfrac{1}{2}\left(\frac{x-b}{\delta}\right)^2, \tag{5.5.31}$$

and near a

$$U(x) \approx U(a) + \tfrac{1}{2}\left(\frac{x-a}{\alpha}\right)^2. \tag{5.5.32}$$

The constant factor in (5.5.29) is evaluated as

$$\int_{-\infty}^{b} dz \exp[-U(z)/D] \approx \int_{-\infty}^{\infty} dz \exp\left[-\frac{U(a)}{D} - \frac{(z-a)^2}{2D\alpha^2}\right], \tag{5.5.33}$$

$$\approx \alpha\sqrt{2\pi D}\exp[-U(a)/D], \tag{5.5.34}$$

and the inner factor becomes, on assuming x_0 is well to the right of the central point b,

$$\int_{a}^{x} dy \exp U(y)/D] \approx \int_{-\infty}^{\infty} dy \exp\left[\frac{U(b)}{D} - \frac{(y-b)^2}{2D\delta^2}\right], \tag{5.5.35}$$

$$= \delta\sqrt{2\pi D}\exp[U(b)/D]. \tag{5.5.36}$$

Putting both of these in (5.5.29), we get

$$T(a \to x) \approx 2\alpha\delta\pi \exp\{[U(b) - U(a)]/D\}. \tag{5.5.37}$$

This is the classical *Arrhenius formula* of chemical reaction theory. In a chemical reaction, we can model the reaction by introducing a coordinate such that $x = a$ is species A and $x = c$ is species C. The reaction is modelled by the above diffusion process and the two distinct chemical species are separated by the potential barrier at b. In the chemical reaction, statistical mechanics gives the value

$$D = kT, \tag{5.5.38}$$

where k is Boltzmann's constant and T is the absolute temperature. We see that the most important dependence on temperature comes from the exponential factor which is often written

$$\exp(\Delta E/kT), \tag{5.5.39}$$

and predicts a very characteristic dependence on temperature. Intuitively, the answer is obvious. The exponential factor represents the probability that the energy will exceed that of the barrier when the system is in thermal equilibrium. Those molecules that reach this energy then react, with a certain finite probability.

We will come back to problems like this in great detail in Chap. 14.

5.5.4 Probability of Exit Through a Particular End of the Interval

What is the probability that the particle, initially at x in (a, b) exits through a, and what is the mean exit time?

The total probability that the particle exits through a after time t is given by the time integral of the probability current at a. We thus define this probability by

$$g_a(x,t) = -\int_t^\infty dt'\, J(a,t'\,|\,x,0)\,, \tag{5.5.40}$$

$$= \int_t^\infty dt' \left\{-A(a)p(a,t'\,|\,x,0) + \tfrac{1}{2}\partial_a[B(a)p(a,t'\,|\,x,0)]\right\}. \tag{5.5.41}$$

the negative sign being chosen since we need the current pointing to the left. Similarly we define

$$g_b(x,t) = \int_t^\infty dt' \left\{A(b)p(b,t'\,|\,x,0) - \tfrac{1}{2}\partial_b[B(b)p(b,t'\,|\,x,0)]\right\}. \tag{5.5.42}$$

These two quantities give the probabilities that the particle exits through a or b after time t, respectively. The probability that (given that it exits through a) it exits after time t is

$$\text{Prob}(T_a > t) = g_a(x,t)/g_a(x,0)\,. \tag{5.5.43}$$

We now find an equation for $g_a(x,t)$, using the fact that $p(a,t\,|\,x,0)$ satisfies a backward Fokker-Planck equation. Thus,

$$\begin{aligned} A(x)\partial_x g_a(x,t) + \tfrac{1}{2}B(x)\partial_x^2 g_a(x,t) &= -\int_t^\infty dt'\partial_{t'} J(a,t'\,|\,x,0)\,,\\ &= J(a,t\,|\,x,0)\,,\\ &= \partial_t g_a(x,t)\,. \end{aligned} \tag{5.5.44}$$

The mean exit time, given that exit is through a, is

$$T(a,x) = -\int_0^\infty t\,\partial_t \text{Prob}\,(T_a > t)\,dt = \int_0^\infty g_a(x,t)\,dt/g_a(x,\infty)\,. \tag{5.5.45}$$

Simply integrating (5.5.44) with respect to t, we get

$$A(x)\partial_x[\pi_a(x)T(a,x)] + \tfrac{1}{2}B(x)\partial_x^2[\pi_a(x)T(a,x)] = -\pi_a(x)\,, \tag{5.5.46}$$

where we define

$$\pi_a(x) = (\text{probability of exit through } a) = g_a(x,0)\,. \tag{5.5.47}$$

The boundary conditions on (5.5.46) are quite straightforward since they follow from those for the backward Fokker-Planck equation, namely,

$$\pi_a(a)T(a,a) = \pi_a(b)T(a,b) = 0\,. \tag{5.5.48}$$

In the first of these clearly $T(a,a)$ is zero (the time to reach a from a is zero) and in the second, $\pi_a(b)$ is zero (the probability of exiting through a, starting from b, is zero).

By letting $t \to 0$ in (5.5.44), we see that $J(a,0|x,0)$ must vanish if $a \neq x$, since. $p(a,0\,|\,x,0) = \delta(x-a)$. Hence, the right-hand side tends to zero and we get

$$A(x)\partial_x \pi_a(x) + \tfrac{1}{2}B(x)\partial_x^2 \pi_a(x) = 0\,, \tag{5.5.49}$$

the boundary condition this time being

$$\pi_a(a) = 1, \quad \pi_a(b) = 0\,. \tag{5.5.50}$$

The solution of (5.5.49) subject to this boundary condition and the condition

$$\pi_a(x) + \pi_b(x) = 1\,, \tag{5.5.51}$$

is

$$\pi_a(x) = \left[\int_x^b dy\,\psi(y)\right] \Big/ \int_a^b dy\,\psi(y), \tag{5.5.52}$$

$$\pi_b(x) = \left[\int_a^x dy\,\psi(y)\right] \Big/ \int_a^b dy\,\psi(y)\,. \tag{5.5.53}$$

with $\psi(x)$ as defined in (5.5.20).

These formulae find application in the problem of relaxation of a distribution initially concentrated at an unstable stationary point (Sect. 14.1.4).

Example—Diffusive Traversal Time of a One-Dimensional medium: A particle diffuses in a one-dimension according to the diffusion equation $\partial_t p = \frac{1}{2}D\partial_x^2 p$. What is the mean time for the particle to diffuse from b to a under the condition that it does not leave the interval (a,b) before reaching a? [5.6]

In the case that the particle starts at x within (a,b), we find from (5.5.52) and (5.5.46)

$$\pi_a(x) = \frac{b-x}{b-a}\,, \tag{5.5.54}$$

$$\tfrac{1}{2}D\partial_x^2[\pi_a(x)T_a(x)] = -\pi_a(x)\,. \tag{5.5.55}$$

Using the boundary conditions (5.5.48) the second equation is easily integrated to give

$$\pi_a(x)T_a(x) = \frac{(x-b)(x-a)(x+a-2b)}{3D(b-a)}\,, \tag{5.5.56}$$

and hence

$$T_a(x) = \frac{(x-a)(2b-x-a)}{3D}\,. \tag{5.5.57}$$

In the limiting case $x \to b$, the probability of exit through a as given by (5.5.54), is zero. Nevertheless, in the limit that x is approaches b, the mean time to make the exit given that the exit is at a is quite well defined and is

$$T_a(b) = \frac{(b-a)^2}{3D}\,. \tag{5.5.58}$$

This is also clearly the time to exit at a without ever leaving the interval (a, b) before exiting at a.

Semi-infinite interval: Notice that if we fix x and let $b \to \infty$, we find

$$\pi_a(x) \to 1, \tag{5.5.59}$$

$$T_a(x) \to \infty. \tag{5.5.60}$$

The result is that the particle is certain to escape at a, but the average time to escape is infinite. This arises because the particle can spend a great deal of time exploring the infinite half of the interval, giving rise to an escape time distribution which is normalisable, but decays so slowly that all moments diverge.

6. The Fokker-Planck Equation in Several Dimensions

In many variable situations, Fokker-Planck equations take on an essentially more complex range of behaviour than is possible in the case of one variable. Boundaries are no longer simple end points of a line but rather curves or surfaces, and the nature of the boundary can change from place to place. Stationary solutions even with reflecting boundaries can correspond to nonzero probability currents and eigenfunction methods are no longer so simple.

Nevertheless, the analogies between one and many dimensions are useful, and this chapter will follow the same general outline as that on one-variable situations.

The subject matter of this chapter covers a number of exact Fokker-Planck equation results for many variable systems. These results are not as explicit as for the one variable case. An extra feature which is included is the concept of *detailed balance* in multivariable systems, which is almost trivial in one variable systems, but leads to very interesting conclusions in multivariable systems.

The chapter concludes with a treatment of exact results in exit problems for multivariable Fokker-Planck equations.

6.1 Change of Variables

Suppose we have a Fokker-Planck equation in variable x_i,

$$\partial_t p(\boldsymbol{x},t) = -\sum_i \partial_i[A_i(\boldsymbol{x})p(\boldsymbol{x},t] + \tfrac{1}{2}\sum_{i,j}\partial_i\partial_j[B_{ij}(\boldsymbol{x})p(\boldsymbol{x},t)]\,, \tag{6.1.1}$$

and we want to know the corresponding equation for the variables

$$y_i = f_i(\boldsymbol{x})\,, \tag{6.1.2}$$

where f_i are certain differentiable independent functions. Let us denote by $\tilde{p}(\boldsymbol{y},t)$ the probability density for the new variable, which is given by

$$\tilde{p}(\boldsymbol{y},t) = p(\boldsymbol{x},t)\left|\frac{\partial(x_1,x_2\ldots)}{\partial(y_1,y_2\ldots)}\right|\,. \tag{6.1.3}$$

The simplest way to effect the change of variables is to use Ito's formula on a corresponding stochastic differential equation

$$d\boldsymbol{x}(t) \quad = \boldsymbol{A}(\boldsymbol{x})\,dt + \mathsf{b}(\boldsymbol{x})\,d\boldsymbol{W}(t)\,, \tag{6.1.4}$$

where

$$\mathsf{b}(\boldsymbol{x})^{\mathrm{T}}\mathsf{b}(\boldsymbol{x}) = \mathsf{B}(\boldsymbol{x}), \tag{6.1.5}$$

and then recompute the corresponding Fokker-Planck equation for $\tilde{p}(\boldsymbol{y}, t)$ from the resulting stochastic differential equation as derived in Sect. 4.3.5.

The result is rather complicated. In specific situations, direct implementation (6.1.3) may be preferable. There is no way of avoiding a rather messy calculation unless full use of symmetries and simplifications is made.

Example—Cartesian to Polar Coordinates: As an example, one can consider the transformation to polar coordinates of the Rayleigh process, previously done by the stochastic differential equation method in Sect. 4.5.5. Thus, the Fokker-Planck equation is

$$\partial_t p(E_1, E_2, t) = \gamma \frac{\partial}{\partial E_1} E_1 p + \gamma \frac{\partial}{\partial E_2} E_2 p + \tfrac{1}{2}\varepsilon^2 \left(\frac{\partial^2 p}{\partial E_1^2} + \frac{\partial^2 p}{\partial E_2^2} \right), \tag{6.1.6}$$

and we want to find the Fokker-Planck equation for a and ϕ defined by

$$E_1 = a\cos\phi, \quad E_2 = a\sin\phi. \tag{6.1.7}$$

The Jacobian is

$$|J| = \frac{\partial(E_1, E_2)}{\partial(a, \phi)} = \begin{vmatrix} \cos\phi & \sin\phi \\ -\sin\phi & a\cos\phi \end{vmatrix} = a. \tag{6.1.8}$$

We use the polar form of the Laplacian to write

$$\frac{\partial^2}{\partial E_1^2} + \frac{\partial^2}{\partial E_2^2} = \frac{1}{a^2}\frac{\partial^2}{\partial \phi^2} + \frac{1}{a}\frac{\partial}{\partial a}\left(a \frac{\partial}{\partial a} \right), \tag{6.1.9}$$

and inverting (6.1.7)

$$a = \sqrt{E_1^2 + E_2^2}, \quad \phi = \tan^{-1}(E_2/E_1). \tag{6.1.10}$$

We note

$$\left.\begin{aligned} \frac{\partial a}{\partial E_1} &= \frac{E_1}{\sqrt{E_1^2 + E_2^2}} = \cos\phi, \\ \frac{\partial a}{\partial E_2} &= \frac{E_2}{\sqrt{E_1^2 + E_2^2}} = \sin\phi, \end{aligned}\right\} \tag{6.1.11}$$

and

$$\left.\begin{aligned} \frac{\partial \phi}{\partial E_2} &= \frac{E_1}{E_1^2 + E_2^2} = \frac{\cos\phi}{a}. \\ \frac{\partial \phi}{\partial E_1} &= -\frac{E_2}{E_1^2 + E_2^2} = -\frac{\sin\phi}{a}. \end{aligned}\right\} \tag{6.1.12}$$

Hence,

$$\frac{\partial}{\partial E_1}E_1p+\frac{\partial}{\partial E_2}E_2p,$$
$$=2p+E_1\left(\frac{\partial p}{\partial a}\frac{\partial a}{\partial E_1}+\frac{\partial p}{\partial \phi}\frac{\partial \phi}{\partial E_2}\right)+E_2\left(\frac{\partial p}{\partial a}\frac{\partial a}{\partial E_2}+\frac{\partial p}{\partial \phi}\frac{\partial \phi}{\partial E_2}\right),$$
$$=2p+a\frac{\partial p}{\partial a}=\frac{1}{a}\frac{\partial}{\partial a}(a^2p)\,. \tag{6.1.13}$$

Let us use the symbol $\tilde{p}(a,\phi)$ for the density function in terms of a and ϕ. The Jacobian formula (6.1.8) us that

$$\tilde{p}(a,\phi)=\left|\frac{\partial(E_1,E_2)}{\partial(a,\phi)}\right|p(E_1,E_2)=a\,p(E_1,E_2)\,. \tag{6.1.14}$$

Putting together (6.1.6), (6.1.9), (6.1.13) and (6.1.14), we get

$$\frac{\partial \tilde{p}}{\partial t}=-\frac{\partial}{\partial a}\left[\left(-\gamma a+\frac{\varepsilon^2}{2a}\right)\tilde{p}\right]+\frac{\varepsilon^2}{2}\left(\frac{1}{a^2}\frac{\partial^2\tilde{p}}{\partial\phi^2}+\frac{\partial^2\tilde{p}}{\partial a^2}\right), \tag{6.1.15}$$

which (of course) is the Fokker-Planck equation, corresponding to the two stochastic differential equations in Sect. 4.5.5, which were derived by changing variables according to Ito's formula.

6.2 Stationary Solutions of Many Variable Fokker-Planck Equations

6.2.1 Boundary Conditions

We have already touched on boundary conditions in general in Sect. 5.1 where they were considered in terms of probability current. The full range of boundary conditions for an arbitrary multidimensional Fokker-Planck equation is very much broader than for the one dimensional case, and probably has never been completely specified. In this book we shall therefore consider mostly *reflecting barrier* boundary conditions at a surface $\boldsymbol{S}$, namely,

$$\boldsymbol{n}\cdot\boldsymbol{J}=0\quad\text{for }\boldsymbol{x}\in\boldsymbol{S}\,, \tag{6.2.1}$$

where $\boldsymbol{n}$ is the normal to the surface and

$$J_i(\boldsymbol{x},t)=A_i(\boldsymbol{x},t)p(\boldsymbol{x},t)-\tfrac{1}{2}\sum_j\frac{\partial}{\partial x_j}B_{ij}(\boldsymbol{x},t)p(\boldsymbol{x},t)\,, \tag{6.2.2}$$

and *absorbing barrier* boundary conditions

$$p(\boldsymbol{x},t)=0\quad\text{for }\boldsymbol{x}\in\boldsymbol{S}\,. \tag{6.2.3}$$

In practice, some part of the surface may be reflecting and another absorbing. At a surface $\boldsymbol{S}$ on which A_i or B_{ij} are discontinuous, we enforce

$$\left.\begin{aligned}\boldsymbol{n}\cdot\boldsymbol{J}(\boldsymbol{x})_1&=\boldsymbol{n}\cdot\boldsymbol{J}(\boldsymbol{x})_2\,,\\ p_1(\boldsymbol{x})&=p_2(\boldsymbol{x})\,,\end{aligned}\right\}\quad\boldsymbol{x}\text{ on }\boldsymbol{S}\,. \tag{6.2.4}$$

The tangential current component is permitted to be discontinuous.

The boundary conditions on the backward equation have already been derived in Sect. 5.1.2 For completeness, they are

$$\text{Absorbing Boundary:}\quad p(\boldsymbol{x},t\,|\,\boldsymbol{y},t')=0,\qquad \boldsymbol{y}\in\boldsymbol{S}, \tag{6.2.5}$$

$$\text{Reflecting Boundary:}\quad \sum_{i,j} n_i B_{ij}(\boldsymbol{y})\frac{\partial}{\partial y_j}p(\boldsymbol{x},t\,|\,\boldsymbol{y},t')=0,\qquad \boldsymbol{y}\in\boldsymbol{S}. \tag{6.2.6}$$

6.2.2 Potential Conditions

A large class of interesting systems is described by Fokker-Planck equations which permit a stationary distribution for which the probability current vanishes for all $\boldsymbol{x}$ in R. Assuming this to be the case, by rearranging the definition of $\boldsymbol{J}$ (6.2.2), we obtain a completely equivalent equation

$$\tfrac{1}{2}\sum_j B_{ij}(\boldsymbol{x})\frac{\partial p_s(\boldsymbol{x})}{\partial x_j} = p_s(\boldsymbol{x})\left[A_i(\boldsymbol{x})-\tfrac{1}{2}\sum_j\frac{\partial}{\partial x_j}B_{ij}(\boldsymbol{x})\right]. \tag{6.2.7}$$

If the matrix $B_{ij}(\boldsymbol{x})$ has an inverse for all $\boldsymbol{x}$, we can write (6.2.7)

$$\frac{\partial}{\partial x_i}\log[p_s(\boldsymbol{x})] = \sum_k B_{ik}^{-1}(\boldsymbol{x})\left[2A_k(\boldsymbol{x})-\sum_j\frac{\partial}{\partial x_j}\boldsymbol{B}_{kj}(\boldsymbol{x})\right], \tag{6.2.8}$$

$$\equiv Z_i[A,B,\boldsymbol{x}]\,. \tag{6.2.9}$$

This equation cannot be satisfied for arbitrary $B_{ij}(\boldsymbol{x})$ and $A_i(\boldsymbol{x})$ since the left-hand side is explicitly a gradient. Hence, Z_i must also be a gradient, and a necessary and sufficient condition for that is the vanishing of the curl, i.e.,

$$\frac{\partial Z_i}{\partial x_j}=\frac{\partial Z_j}{\partial x_i}\,. \tag{6.2.10}$$

If this condition is satisfied, the stationary solution can be obtained by simple integration of (6.2.8):

$$p_s(\boldsymbol{x})=\exp\left\{\int^{x} d\boldsymbol{x}'\cdot\boldsymbol{Z}[A,B,\boldsymbol{x}']\right\}. \tag{6.2.11}$$

The conditions (6.2.10) are known as *potential conditions* since we derive the quantities Z_i from derivatives of log $[p_s(\boldsymbol{x})]$, which, therefore, is often thought of as a potential $-\phi(\boldsymbol{x})$ so that more precisely,

$$p_s(\boldsymbol{x})=\exp[-\phi(\boldsymbol{x})]\,, \tag{6.2.12}$$

and

$$\phi(\boldsymbol{x})=-\int^{x} d\boldsymbol{x}'\cdot\boldsymbol{Z}[A,B,\boldsymbol{x}']\,. \tag{6.2.13}$$

Example—Rayleigh Process in Polar Coordinates: From (6.1.15) we find

$$A = \begin{pmatrix} -\gamma a + \varepsilon^2/2a \\ 0 \end{pmatrix}, \tag{6.2.14}$$

$$\mathsf{B} = \begin{pmatrix} \varepsilon^2 & 0 \\ 0 & \varepsilon^2/a^2 \end{pmatrix}, \tag{6.2.15}$$

from which

$$\sum_j \frac{\partial}{\partial x_j} B_{a,j} = \sum_j \frac{\partial}{\partial x_j} B_{\phi,j} = 0\,, \tag{6.2.16}$$

so that

$$Z = 2\mathsf{B}^{-1} A = \begin{pmatrix} -2\gamma a/\varepsilon^2 + 1/a \\ 0 \end{pmatrix}, \tag{6.2.17}$$

and clearly

$$\frac{\partial Z_a}{\partial \phi} = \frac{\partial Z_\phi}{\partial a} = 0\,. \tag{6.2.18}$$

The stationary solution is then

$$p_s(a,\phi) = \exp\left[\int^{(a,\phi)} (d\phi\, Z_\phi + da\, Z_a)\right], \tag{6.2.19}$$

$$= \mathcal{N} \exp\left(-\frac{\gamma a^2}{\varepsilon^2} + \log a\right), \tag{6.2.20}$$

$$= \mathcal{N} a \exp\left(-\frac{\gamma a^2}{\varepsilon^2}\right). \tag{6.2.21}$$

6.3 Detailed Balance

6.3.1 Definition of Detailed Balance

The fact that the stationary solution of certain Fokker-Planck equations corresponds to a vanishing probability current is a particular version of the physical phenomenon of *detailed balance*. A Markov process satisfies detailed balance if, roughly speaking, in the stationary situation each possible transition balances with the reversed transition. The concept of detailed balance comes from physics, so let us explain more precisely with a physical example. We consider a gas of particles with positions $\boldsymbol{r}$ and velocities $\boldsymbol{v}$. Then a transition corresponds to a particle at some time t with position velocity $(\boldsymbol{r}, \boldsymbol{v})$ having acquired by a later time $t+\tau$ position and velocity $(\boldsymbol{r}',\boldsymbol{v}')$. The probability density of this transition is the *joint probability density* $p(\boldsymbol{r}', \boldsymbol{v}', t+\tau\,;\, \boldsymbol{r}, \boldsymbol{v}, t)$.

We may symbolically write this transition as

$$(\boldsymbol{r}, \boldsymbol{v}, t) \to (\boldsymbol{r}', \boldsymbol{v}', t+\tau)\,. \tag{6.3.1}$$

The reversed transition is not given simply by interchanging primed and unprimed quantities Rather, it is

$$(\boldsymbol{r}', -\boldsymbol{v}', t) \to (\boldsymbol{r}, -\boldsymbol{v}, t+\tau)\,. \tag{6.3.2}$$

It corresponds to the *time reversed transition* and requires the velocities to be reversed because the motion from $\boldsymbol{r}'$ to $\boldsymbol{r}$ is in the opposite direction from that from $\boldsymbol{r}$ to $\boldsymbol{r}'$.

The probability density for the reversed transition is thus the joint probability density

$$p(\boldsymbol{r}, -\boldsymbol{v}, t+\tau; \boldsymbol{r}', -\boldsymbol{v}', t)\,. \tag{6.3.3}$$

The *principle of detailed balance* requires the equality of these two joint probabilities when the system is in a stationary state. Thus, we may write

$$p_s(\boldsymbol{r}', \boldsymbol{v}', \tau; \boldsymbol{r}, \boldsymbol{v}, 0) = p_s(\boldsymbol{r}, -\boldsymbol{v}, \tau; \boldsymbol{r}', -\boldsymbol{v}', 0)\,. \tag{6.3.4}$$

The principle can be derived under certain conditions from the laws of physics—see [6.1] and Sect. 6.4.2.

6.3.2 Detailed Balance for a Markov Process

When the probabilities correspond to those of a Markov process we can rewrite (6.3.4) as

$$p(\boldsymbol{r}', \boldsymbol{v}', \tau \,|\, \boldsymbol{r}, \boldsymbol{v}, 0) p_s(\boldsymbol{r}, \boldsymbol{v}) = p(\boldsymbol{r}, -\boldsymbol{v}, \tau \,|\, \boldsymbol{r}', -\boldsymbol{v}', 0) p_s(\boldsymbol{r}', -\boldsymbol{v}')\,, \tag{6.3.5}$$

where the conditional probabilities now apply to the corresponding homogeneous Markov process (if the process was not Markov, the conditional probabilities would be for the stationary system only).

In its general form, detailed balance is formulated in terms of arbitrary variables x_i which, under time reversal, transform to the reversed variables according to the rule

$$x_i \;\to\; \varepsilon_i x_i\,, \tag{6.3.6}$$

$$\varepsilon_i \;=\; \pm 1\,, \tag{6.3.7}$$

depending on whether the variable is odd or even under time reversal. In the above, $\boldsymbol{r}$ is even, $\boldsymbol{v}$ is odd.

Then by detailed balance we require

$$p_s(\boldsymbol{x}, t+\tau; \boldsymbol{x}', t) = p_s(\boldsymbol{\varepsilon x}', t+\tau; \boldsymbol{\varepsilon x}, t)\,. \tag{6.3.8}$$

By $\boldsymbol{\varepsilon x}$, we mean $(\varepsilon_1 x_1, \varepsilon_2 x_2 \ldots)$.

Notice that setting $\tau = 0$ in (6.3.8) we obtain

$$\delta(\boldsymbol{x} - \boldsymbol{x}') p_s(\boldsymbol{x}') = \delta(\boldsymbol{\varepsilon x} - \boldsymbol{\varepsilon x}') p_s(\boldsymbol{\varepsilon x})\,. \tag{6.3.9}$$

The two delta functions are equal since only sign changes are involved. Hence,

$$p_s(\boldsymbol{x}) = p_s(\boldsymbol{\varepsilon x})\,, \tag{6.3.10}$$

is a consequence of the formulation of detailed balance by (6.3.8). Rewriting now in terms of conditional probabilities, we have

$$p(\boldsymbol{x},\tau\,|\,\boldsymbol{x}',0)\,p_{\mathrm{s}}(\boldsymbol{x}') = p(\boldsymbol{\varepsilon x}',\tau\,|\,\boldsymbol{\varepsilon x},0)\,p_{\mathrm{s}}(\boldsymbol{x})\,. \tag{6.3.11}$$

6.3.3 Consequences of Detailed Balance for Stationary Mean, Autocorrelation Function and Spectrum

An important consequence of (6.3.10) is that

$$\langle \boldsymbol{x}\rangle_{\mathrm{s}} = \boldsymbol{\varepsilon}\,\langle \boldsymbol{x}\rangle_{\mathrm{s}}\,, \tag{6.3.12}$$

hence all odd variables have zero stationary mean, and for the autocorrelation function

$$G(\tau) \equiv \langle \boldsymbol{x}(\tau)\boldsymbol{x}^{\mathrm{T}}(0)\rangle_{\mathrm{s}}\,, \tag{6.3.13}$$

we have

$$G(\tau) = \boldsymbol{\varepsilon}\,\langle \boldsymbol{x}(0)\boldsymbol{x}^{\mathrm{T}}(\tau)\rangle_{\mathrm{s}}\,\boldsymbol{\varepsilon}\,, \tag{6.3.14}$$

hence,

$$G(\tau) = \boldsymbol{\varepsilon}\,G^{\mathrm{T}}(\tau)\,\boldsymbol{\varepsilon}\,, \tag{6.3.15}$$

and setting $\tau = 0$ and noting that the covariance matrix σ satisfies $\sigma = \sigma^{\mathrm{T}}$,

$$\sigma\boldsymbol{\varepsilon} = \boldsymbol{\varepsilon}\sigma\,. \tag{6.3.16}$$

For the spectrum matrix

$$S(\omega) = \frac{1}{2\pi}\int_{-\infty}^{\infty} e^{-\mathrm{i}\omega\tau}G(\tau)\,, \tag{6.3.17}$$

we find from (6.3.15) that

$$S(\omega) = \boldsymbol{\varepsilon}S^{T}(\omega)\boldsymbol{\varepsilon}\,. \tag{6.3.18}$$

6.3.4 Situations in Which Detailed Balance must be Generalised

It is possible that there exist several stationary solutions to a Markov process, and in this situation, a weaker form of detailed balance may hold, namely, instead of (6.3.8), we have

$$p_{\mathrm{s}}^{1}(\boldsymbol{x},t+\tau\,;\,\boldsymbol{x}',t) = p_{\mathrm{s}}^{2}(\boldsymbol{\varepsilon x}',t+\tau\,;\,\boldsymbol{\varepsilon x},t)\,, \tag{6.3.19}$$

where the superscripts 1 and 2 refer to two different stationary solutions. Such a situation can exist if one of the variables is odd under time reversal, but does not change with time; for example, in a centrifuge the total angular momentum has this property. A constant magnetic field acts the same way.

Mostly, one writes the detailed balance conditions in such situations as

$$p_s^{\lambda}(\boldsymbol{x}, t+\tau\,;\, \boldsymbol{x}^1, t) = p_s^{\varepsilon\lambda}(\boldsymbol{\varepsilon x}', t+\tau\,;\, \boldsymbol{\varepsilon x}, t)\,, \tag{6.3.20}$$

where λ is a vector of such constant quantities, which change to $\boldsymbol{\varepsilon}\lambda$ under time reversal. According to one point of view, such a situation does not represent detailed balance; since in a given stationary situation, the transitions do *not* balance in detail. It is perhaps better to call the property (6.3.20) *time reversal invariance*.

In the remainder of our considerations, we shall mean by detailed balance the situation (6.3.10), since no strong consequences arise from the form (6.3.20).

6.3.5 Implementation of Detailed Balance in the Differential Chapman-Kolmogorov Equation

The formulation of detailed balance for the Fokker-Planck equation was done by *van Kampen* [6.1], and *Graham* and *Haken* [6.2]. We will formulate the conditions in a slightly more direct and more general way. We want necessary and sufficient conditions on the drift and diffusion coefficients and the jump probabilities for a homogeneous Markov process to have stationary solutions which satisfy detailed balance. We shall show that necessary and sufficient conditions are given by

$$\begin{aligned}
&\text{(i)} \quad W(\boldsymbol{x}\,|\,\boldsymbol{x}')p_s(\boldsymbol{x}') = W(\boldsymbol{\varepsilon x}'\,|\,\boldsymbol{\varepsilon x})p_s(\boldsymbol{x})\,,\\
&\text{(ii)} \quad \varepsilon_i A_i(\boldsymbol{\varepsilon x})p_s(x) = -A_i(\boldsymbol{x})p_s(\boldsymbol{x}) + \sum_j \frac{\partial}{\partial x_j}[\boldsymbol{B}_{ij}(\boldsymbol{x})p_s(\boldsymbol{x})]\,,\\
&\text{(iii)} \quad \varepsilon_i\varepsilon_j B_{ij}(\boldsymbol{\varepsilon x}) = B_{ij}(\boldsymbol{x})\,.
\end{aligned}$$

The specialisation to a Fokker-Planck equation is simply done by setting the jump probabilities $W(\boldsymbol{x}\,|\,\boldsymbol{x}')$ equal to zero.

a) Necessary Conditions: It is simpler to formulate conditions for the differential Chapman-Kolmogorov equation than to restrict ourselves to the Fokker-Planck equation. According to Sect. 3.4, which defines the quantities $W(\boldsymbol{x}\,|\,\boldsymbol{x}')$, $A_i(\boldsymbol{x})$ and $B_{ij}(\boldsymbol{x})$ (all of course being time independent, since we are considering homogeneous process), we have the trivial result that detailed balance requires, from (6.3.11)

$$W(\boldsymbol{x}\,|\,\boldsymbol{x}')p_s(\boldsymbol{x}') = W(\boldsymbol{\varepsilon x}'\,|\,\boldsymbol{\varepsilon x})p_s(\boldsymbol{x})\,. \tag{6.3.21}$$

Consider now the drift coefficient. For simplicity write

$$\boldsymbol{x}' = \boldsymbol{x} + \boldsymbol{\delta}\,. \tag{6.3.22}$$

Then from (6.3.11) we have

$$\int_{|\delta|<K} d\delta\, \delta_i p(\boldsymbol{\varepsilon x} + \boldsymbol{\varepsilon\delta}, \Delta t\,|\,\boldsymbol{\varepsilon x}, 0)p_s(\boldsymbol{x}) = \int_{|\delta|<K} d\delta\, \delta_i p(\boldsymbol{x}, \Delta t\,|\,\boldsymbol{x}+\boldsymbol{\delta}, 0)p_s(\boldsymbol{x}+\boldsymbol{\delta})\,, \tag{6.3.23}$$

(we use K instead of $\boldsymbol{\varepsilon}$ in the range of integration to avoid confusion with ε_i); divide by Δt and take the limit $\Delta t \to 0$, and the left-hand side yields

$$\varepsilon_i A_i(\boldsymbol{\varepsilon x})p_s(\boldsymbol{x}) + O(K)\,. \tag{6.3.24}$$

On the right-hand side we write

$$p(\boldsymbol{x}+\boldsymbol{\delta}-\boldsymbol{\delta},\Delta t\,|\,\boldsymbol{x}+\boldsymbol{\delta},0)p_s(\boldsymbol{x}+\boldsymbol{\delta}) = p(\boldsymbol{x}-\boldsymbol{\delta},\Delta t\,|\,\boldsymbol{x},0)p_s(\boldsymbol{x}) + \sum_j \delta_j \frac{\partial}{\partial x_j}[p(\boldsymbol{x}-\boldsymbol{\delta},\Delta t\,|\,\boldsymbol{x},0)p_s(\boldsymbol{x})] + O(\delta^2)\,, \tag{6.3.25}$$

so that the right-hand side is

$$\lim_{\Delta t\to 0}\frac{1}{\Delta t}\int d\boldsymbol{\delta}\left\{\delta_i p(\boldsymbol{x}-\boldsymbol{\delta},\Delta t\,|\,\boldsymbol{x},0)p_s(\boldsymbol{x}) + \sum_j \delta_i\delta_j \frac{\partial}{\partial x_j}[p(\boldsymbol{x}-\boldsymbol{\delta},\Delta t\,|\,\boldsymbol{x},0)p_s(\boldsymbol{x})]\right\} + O(K)\,,$$

$$= -A_i(\boldsymbol{x})p_s(\boldsymbol{x}) + \sum_j \frac{\partial}{\partial x_j}[\boldsymbol{B}_{ij}(\boldsymbol{x})p_s(\boldsymbol{x})] + O(K)\,, \tag{6.3.26}$$

where we have used the fact demonstrated in Sect. 3.4 that terms involving higher powers of δ than δ^2 are of order K. Letting $K \to 0$, we find

$$\varepsilon_i A_i(\boldsymbol{\varepsilon x})p_s(\boldsymbol{x}) = -A_i(\boldsymbol{x})p_s(\boldsymbol{x}) + \sum_j \frac{\partial}{\partial x_j}[B_{ij}(\boldsymbol{x})p_s(\boldsymbol{x})]\,. \tag{6.3.27}$$

The condition on $B_{ij}(\boldsymbol{x})$ is obtained similarly, but in this case no term like the second on the right of (6.3.27) arises, since the principal term is $O(\delta^2)$. We find

$$\varepsilon_i\varepsilon_j B_{ij}(\boldsymbol{\varepsilon x}) = B_{ij}(\boldsymbol{x})\,. \tag{6.3.28}$$

A third condition is, of course, that $p_s(\boldsymbol{x})$ be a stationary solution of the differential Chapman-Kolmogorov equation. This is not a trivial condition, and is, in general, independent of the others.

b) Sufficient Conditions: We now show that (6.3.21) are sufficient. Assume that these conditions are satisfied, that $p_s(\boldsymbol{x})$ is a stationary solution of the differential Chapman-Kolmogorov equation, and that $p(\boldsymbol{x},t\,|\,\boldsymbol{x}',0)$ is a solution of the differential Chapman-Kolmogorov equation. We now consider a quantity

$$\hat{p}(\boldsymbol{x},t\,|\,\boldsymbol{x}',0) \equiv p(\boldsymbol{\varepsilon x}',t\,|\,\boldsymbol{\varepsilon x},0)p_s(\boldsymbol{x})p_s(\boldsymbol{x}')\,. \tag{6.3.29}$$

Clearly

$$\hat{p}(\boldsymbol{x},0\,|\,\boldsymbol{x}',0) = \delta(\boldsymbol{x}-\boldsymbol{x}') = p(\boldsymbol{x},0\,|\,\boldsymbol{x}',0)\,. \tag{6.3.30}$$

We substitute $\hat{p}$ into the differential Chapman-Kolmogorov equation and show that because $p(\boldsymbol{x}',t\,|\,\boldsymbol{x},0)$ obeys the *backward differential Chapman-Kolmogorov equation* in the variable $\boldsymbol{x}$, the quantity $\hat{p}$ is a solution of the *forward differential Chapman-Kolmogorov equation.*

We do this explicitly. The notation is abbreviated for clarity, so that we write

$$\left.\begin{array}{lll} \hat{p} & \text{for} & \hat{p}(\boldsymbol{x},t\,|\,\boldsymbol{x}',0)\,, \\ p_s & \text{for} & p_s(\boldsymbol{x})\,, \\ p_s' & \text{for} & p_s(\boldsymbol{x}')\,, \\ p(\boldsymbol{x}) & \text{for} & p(\boldsymbol{x}',t\,|\,\boldsymbol{x},0)\,. \end{array}\right\} \tag{6.3.31}$$

We proceed term by term.

i) *Drift Term:*

$$-\sum_i \frac{\partial}{\partial x_i}(A_i\hat{p}) = -\sum_i \frac{\partial}{\partial x_i}(A_i p(\boldsymbol{\varepsilon x})p_s/p_s'), \tag{6.3.32}$$

$$= -\sum_i \left\{\frac{\partial}{\partial x_i}[A_i p_s]p(\boldsymbol{\varepsilon x}) + A_i p_s \frac{\partial}{\partial x_i} p(\boldsymbol{\varepsilon x})\right\}/p_s'.$$

ii) *Diffusion Term:*

$$\tfrac{1}{2}\sum_{i,j}\frac{\partial^2}{\partial x_i \partial x_j}(B_{ij}\hat{p}) = \tfrac{1}{2}\sum_{i,j}\left[\frac{\partial^2}{\partial x_i \partial x_j}(B_{ij}p_s)p(\boldsymbol{\varepsilon x}) + 2\frac{\partial}{\partial x_i}(B_{ij}p_s)\frac{\partial}{\partial x_j}p(\boldsymbol{\varepsilon x}) + B_{ij}p_s\frac{\partial^2}{\partial x_i \partial x_j}p(\boldsymbol{\varepsilon x})\right]/p_s'. \tag{6.3.33}$$

iii) *Jump Term:*

$$\int dz[W(\boldsymbol{x}\,|\,\boldsymbol{z})\hat{p}(\boldsymbol{z},t\,|\,\boldsymbol{x}',0) - W(\boldsymbol{z}\,|\,\boldsymbol{x})\hat{p}(\boldsymbol{x},t\,|\,\boldsymbol{x}',0)]$$
$$= \int dz[W(\boldsymbol{x}\,|\,\boldsymbol{z})p_s(\boldsymbol{z})p(\boldsymbol{\varepsilon x}',t\,|\,\boldsymbol{\varepsilon z},0) - W(\boldsymbol{z}\,|\,\boldsymbol{x})p_s(\boldsymbol{x})p(\boldsymbol{\varepsilon x}',t\,|\,\boldsymbol{\varepsilon x},0)]p_s. \tag{6.3.34}$$

We now use the fact that $p_s(\boldsymbol{x})$ is a solution of the stationary differential Chapman-Kolmogorov equation to write

$$-\sum_i\left[\frac{\partial}{\partial x_i}(A_i p_s) + \tfrac{1}{2}\sum_{i,j}\frac{\partial^2}{\partial x_i \partial x_i}(\boldsymbol{B}_{ij}p_s)\right] - \int dz\, W(\boldsymbol{z}\,|\,\boldsymbol{x})p_s(\boldsymbol{x})$$
$$= -\int dz\, W(\boldsymbol{x}\,|\,\boldsymbol{z})p_s(\boldsymbol{z}), \tag{6.3.35}$$

and using the detailed balance condition (6.3.21(i)) for W

$$= -\int dz\, W(\boldsymbol{\varepsilon z}\,|\,\boldsymbol{\varepsilon x})p_s(\boldsymbol{x}). \tag{6.3.36}$$

iv) *Combining all terms :* Now substitute

$$\boldsymbol{y} = \boldsymbol{\varepsilon x}, \quad \boldsymbol{y}' = \boldsymbol{\varepsilon x}', \tag{6.3.37}$$

and all up all three contributions, taking care of (6.3.35), (6.3.36):

$$= \Big\{-\sum_i \varepsilon_i A_i(\boldsymbol{\varepsilon y})p_s(\boldsymbol{y})\left[\frac{\partial}{\partial y_i}p(\boldsymbol{y})\right] + \sum_{i,j}\varepsilon_i\varepsilon_j\frac{\partial}{\partial y_j}[B_{ij}(\boldsymbol{\varepsilon y})p_s(\boldsymbol{y})]\left[\frac{\partial}{\partial y_i}p(\boldsymbol{y})\right]$$
$$+\tfrac{1}{2}\sum_{i,j}\varepsilon_i\varepsilon_j B_{ij}(\boldsymbol{\varepsilon y})p_s(\boldsymbol{y})\left[\frac{\partial^2}{\partial y_i \partial y_j}p(\boldsymbol{y})\right] + \int dz[W(\boldsymbol{\varepsilon y}\,|\,\boldsymbol{z})p_s(\boldsymbol{z})p(\boldsymbol{y}',t\,|\,\boldsymbol{\varepsilon z},0)$$
$$- W(\boldsymbol{\varepsilon y}\,|\,\boldsymbol{z})p_s(\boldsymbol{z})p(\boldsymbol{y}',t\,|\,\boldsymbol{y},0)]\Big\}/p_s(\boldsymbol{y}'). \tag{6.3.38}$$

We now substitute the detailed balance conditions (6.3.21).

$$= \Big\{\sum_i A_i(\boldsymbol{y})\frac{\partial}{\partial y_i}p(\boldsymbol{y}',t\,|\,\boldsymbol{y},0) + \tfrac{1}{2}\sum_{i,j}\boldsymbol{B}_{ij}(\boldsymbol{y})\frac{\partial^2}{\partial y_i \partial y_j}p(\boldsymbol{y}',\boldsymbol{t}\,|\,\boldsymbol{y},0)$$
$$+ \int dz[W(\boldsymbol{z}\,|\,\boldsymbol{y})p(\boldsymbol{y}',t\,|\,\boldsymbol{z},0) - W(\boldsymbol{z}\,|\,\boldsymbol{y})p(\boldsymbol{y}',t\,|\,\boldsymbol{y},0)\Big\}p_s(\boldsymbol{y})/p_s(\boldsymbol{y}'). \tag{6.3.39}$$

The term in the large curly brackets is now recognisable as the *backward differential Chapman-Kolmogorov operator* [Sect. 3.6, (3.6.4)]. Note that the process is homogeneous, so that

$$p(\boldsymbol{y}',t\,|\,\boldsymbol{y},0) = p(\boldsymbol{y},0\,|\,\boldsymbol{y},-t)\,. \tag{6.3.40}$$

We see that

$$(6.3.39) = \frac{\partial}{\partial t}[p(\boldsymbol{y}',t\,|\,,0)p_s(\boldsymbol{y})/p_s(\boldsymbol{y}')] = \frac{\partial}{\partial t}\hat{p}(\boldsymbol{x},t\,|\,\boldsymbol{x}',0)\,, \tag{6.3.41}$$

which means that $\hat{p}(\boldsymbol{x},t\,|\,x',0)$, defined in (6.3.29), satisfies the forward differential Chapman-Kolmogorov equation. Since the initial condition of $p(\boldsymbol{x},t\,|\,\boldsymbol{x}',0)$ and $\hat{p}(\boldsymbol{x},t\,|\,\boldsymbol{x}',0)$ at $t=0$ are the same (6.3.30) and the solutions are unique, we have shown that provided that detailed balance conditions (6.3.21) are satisfied, detailed balance is satisfied. Hence, sufficiency is shown.

c) Comments:

i) *Even variables only:* the conditions are considerably simpler if all ε_i are $+1$. In this case, the conditions reduce to

$$W(\boldsymbol{x}|\boldsymbol{x}')p_s(\boldsymbol{x}') = W(\boldsymbol{x}'|\boldsymbol{x})p_s(\boldsymbol{x})\,, \tag{6.3.42}$$

$$A_i(\boldsymbol{x})p_s(\boldsymbol{x}) = \tfrac{1}{2}\sum_j \frac{\partial}{\partial x_j}[B_{ij}(\boldsymbol{x})p_s(\boldsymbol{x})]\,, \tag{6.3.43}$$

$$B_{ij}(\boldsymbol{x}) = B_{ij}(\boldsymbol{x}), \tag{6.3.44}$$

the last of which is trivial. The condition (6.3.43) is exactly the same as the potential condition (6.2.7) which expresses the vanishing of $\boldsymbol{J}$, the probability current in the stationary state.

The conditions (6.3.42) and(6.3.43) taken together imply that $p_s(\boldsymbol{x})$ satisfies the stationary differential Chapman-Kolmogorov equation, which is not the case for the general conditions (6.3.21).

ii) *Fokker-Planck equations:* The concept of reversible and irreversible drift parts was introduced by *van Kampen*, [6.1], and *Graham* and *Haken* [6.2]. The irreversible drift is

$$D_i(\boldsymbol{x}) = \tfrac{1}{2}[A_i(\boldsymbol{x}) + \varepsilon_i A_i(\boldsymbol{\varepsilon x})]\,, \tag{6.3.45}$$

and the reversible drift

$$I_i(\boldsymbol{x}) = \tfrac{1}{2}[A_i(\boldsymbol{x}) - \varepsilon_i A_i(\boldsymbol{\varepsilon x}]\,. \tag{6.3.46}$$

Using again the potential defined by

$$p_s(\boldsymbol{x}) = \exp[-\phi(\boldsymbol{x})]\,, \tag{6.3.47}$$

we see that in the case of a Fokker-Planck equation, we can write the conditions for detailed balance as

$$\varepsilon_i \varepsilon_j B_{ij}(\boldsymbol{\varepsilon x}) = B_{ij}(\boldsymbol{x}), \tag{6.3.48}$$

$$D_i(\boldsymbol{x}) - \tfrac{1}{2} \sum_j \frac{\partial}{\partial x_j}[B_{ij}(\boldsymbol{x})] = -\tfrac{1}{2} \sum_i B_{ij}(\boldsymbol{x}) \frac{\partial \phi(\boldsymbol{x})}{\partial x_j}, \tag{6.3.49}$$

$$\sum_i \left[\frac{\partial}{\partial x_i} I_i(\boldsymbol{x}) - I_i(\boldsymbol{x}) \frac{\partial \phi(\boldsymbol{x})}{\partial x_i} \right] = 0, \tag{6.3.50}$$

where the last equation is simply the stationary Fokker-Planck equation for $p_s(\boldsymbol{x})$, after substituting (6.3.21 ii). As was the case for the potential conditions, it can be seen that (6.3.49) gives an equation for $\partial\phi/\partial x_j$ which can only be satisfied provided certain conditions on $D_i(\boldsymbol{x})$ and $B_{ij}(\boldsymbol{x})$ are satisfied. If $B_{ij}(\boldsymbol{x})$ has an inverse, these take the form

$$\frac{\partial \hat{Z}_i}{\partial x_j} = \frac{\partial \hat{Z}_j}{\partial x_i}, \tag{6.3.51}$$

where

$$\hat{Z}_i = \sum_k B_{ik}^{-1}(\boldsymbol{x}) \Big[2D_k(\boldsymbol{x}) - \sum_j \frac{\partial}{\partial x_j} B_{kj}(\boldsymbol{x}) \Big], \tag{6.3.52}$$

and we have

$$p_s(\boldsymbol{x}) = \exp[-\phi(\boldsymbol{x})] = \exp\left(\int^{\boldsymbol{x}} d\boldsymbol{x}' \cdot \hat{\boldsymbol{Z}} \right). \tag{6.3.53}$$

Thus, as in the case of a vanishing probability current, $p_s(\boldsymbol{x})$ can be determined explicitly as an integral.

iii) *Connection between backward and forward operators* of differential Chapman-Kolmogorov equations is provided by the detailed balance. The proof of sufficient conditions amounts to showing that if $f(\boldsymbol{x}, t)$ is a solution of the forward differential Chapman-Kolmogorov equation, then

$$\tilde{f}(\boldsymbol{x}, t) = f(\boldsymbol{\varepsilon x}, -t)/p_s(\boldsymbol{x}), \tag{6.3.54}$$

is a solution of the backward differential Chapman-Kolmogorov equation. This relationship will be used in Sect. 6.5 for the construction of eigenfunctions.

6.4 Examples of Detailed Balance in Fokker-Planck Equations

6.4.1 Kramers' Equation for Brownian Motion in a Potential

This problem was given its definitive formulation by Kramers [6.3], when considering a model of molecular dissociation.

We take the motion of a particle in a fluctuating environment. The motion is in one dimension and the state of the particle is described by its position x and velocity v. This gives the differential equations

$$\frac{dx}{dt} = v\,, \tag{6.4.1}$$

$$m\frac{dv}{dt} = -V'(x) - \beta v + \sqrt{2\beta kT}\,\xi(t)\,, \tag{6.4.2}$$

which are essentially Langevin's equations (1.2.14) in which, for brevity, we write

$$6\pi\eta a = \beta\,, \tag{6.4.3}$$

and $V(x)$ is a potential whose gradient $V'(x)$ gives rise to a force on the particle. By making the assumption that the physical fluctuating force $\xi(t)$ is to be interpreted as

$$\xi(t)dt = dW(t)\,, \tag{6.4.4}$$

as explained in Sect. 4.1, we obtain stochastic differential equations

$$dx = v\,dt\,, \tag{6.4.5}$$

$$m\,dv = -[V'(x) + \beta v]dt + \sqrt{2\beta kT}\,dW(t)\,, \tag{6.4.6}$$

for which the corresponding Fokker-Planck equation is

$$\frac{\partial p}{\partial t} = -\frac{\partial}{\partial x}(vp) + \frac{1}{m}\frac{\partial}{\partial v}\Big([V'(x) + \beta v]p\Big) + \frac{\beta kT}{m^2}\frac{\partial^2 p}{\partial v^2}\,. \tag{6.4.7}$$

The equation can be slightly simplified by introducing new scaled variables

$$y = x\sqrt{m/kT}\,, \tag{6.4.8}$$

$$u = v\sqrt{m/kT}\,, \tag{6.4.9}$$

$$U(y) = V(x)/kT\,, \tag{6.4.10}$$

$$\gamma = \beta/m\,, \tag{6.4.11}$$

so that the Fokker-Planck equation takes the form

$$\frac{\partial p}{\partial t} = -\frac{\partial}{\partial y}(up) + \frac{\partial}{\partial u}\Big([U'(y) + \gamma u]p\Big) + \gamma\frac{\partial}{\partial u}\left(up + \frac{\partial p}{\partial u}\right), \tag{6.4.12}$$

which we shall call *Kramers' equation*.

Here, y (the position) is an even variable and u(the velocity) an odd variable, as explained in Sect. 6.3. The drift and diffusion can be written

$$\mathsf{A}(y,u) = \begin{bmatrix} u \\ -U'(y) - \gamma u \end{bmatrix}, \tag{6.4.13}$$

$$\mathsf{B}(y,u) = \begin{bmatrix} 0 & 0 \\ 0 & 2\gamma \end{bmatrix}. \tag{6.4.14}$$

The detailed balance transformation is

$$\boldsymbol{\varepsilon}\begin{bmatrix} y \\ u \end{bmatrix} = \begin{bmatrix} y \\ -u \end{bmatrix}. \tag{6.4.15}$$

We can check the conditions one by one.

The condition (6.3.21 iii) is trivially satisfied. The condition (6.3.21 i) is somewhat degenerate, since B is not invertible. It can be written

$$\varepsilon A(y,-u)p_s(y,u) = -A(y,u)p_s(y,u) + \begin{bmatrix} 0 \\ 2\gamma \dfrac{\partial p_s}{\partial u} \end{bmatrix}, \tag{6.4.16}$$

or, more fully

$$\begin{bmatrix} -u \\ U'(y) - \gamma u \end{bmatrix} p_s = \begin{bmatrix} -u \\ U'(y) + \gamma u \end{bmatrix} p_s + \begin{bmatrix} 0 \\ 2\gamma \dfrac{\partial p_s}{\partial u} \end{bmatrix}. \tag{6.4.17}$$

The first line is an identity and the second states

$$-u p_s(y,u) = \frac{\partial p_s}{\partial u}, \tag{6.4.18}$$

i.e.,

$$p_s(y,u) = \exp\left(-\tfrac{1}{2}u^2\right) f(y). \tag{6.4.19}$$

This means that if $p_s(y,u)$ is written in the form (6.4.19), then the detailed balance conditions are satisfied. One must now check whether (6.4.19) indeed gives a stationary solution of Kramers' equation (6.4.12) by substitution. The final bracket vanishes, leaving

$$0 = -u\frac{\partial f}{\partial y} - U'(y)uf, \tag{6.4.20}$$

which means

$$f(y) = \mathscr{N} \exp[-U(y)], \tag{6.4.21}$$

and

$$p_s(y,u) = \mathscr{N} \exp\left[-U(y) - \tfrac{1}{2}u^2\right]. \tag{6.4.22}$$

In terms of the original (x, v) variables,

$$p_s(x,v) = \mathscr{N} \exp\left[-\frac{V(x)}{kT} - \frac{mv^2}{2kT}\right], \tag{6.4.23}$$

which is the familiar Boltzmann distribution of statistical mechanics. Notice that the denominators kT arise from the assumed coefficient $\sqrt{2\beta kT}$ of the fluctuating force in (6.4.2). Thus, we take the macroscopic equations and add a fluctuating force, whose magnitude is fixed by the requirement that the solution be the Boltzmann distribution corresponding to the temperature T.

But we have also achieved exactly the right distribution function. This means that the assumption that Brownian motion is described by a *Markov* process of the form (6.4.1), (6.4.2) must have considerable validity.

6.4.2 Deterministic Motion

Here we have $B_{ij}(\boldsymbol{x})$ and $W(\boldsymbol{x}|\boldsymbol{x}')$ equal to zero, so the detailed balance conditions are simply

$$\varepsilon_i A_i(\boldsymbol{\varepsilon x}) = -A_i(\boldsymbol{x}) . \tag{6.4.24}$$

Since we are now dealing with a Liouville equation (Sect. 3.5.3), the motion of a point whose coordinates are $\boldsymbol{x}$ is described by the ordinary differential equation

$$\frac{d}{dt}\boldsymbol{x}(t) = A[\boldsymbol{x}(t)] . \tag{6.4.25}$$

Suppose a solution of (6.4.25) which passes through the point $\boldsymbol{y}$ at $t = 0$ is

$$\boldsymbol{q}[t, \boldsymbol{y}] , \tag{6.4.26}$$

which therefore satisfies

$$\boldsymbol{q}[0, \boldsymbol{y}] = \boldsymbol{y} . \tag{6.4.27}$$

Then the relation (6.4.24) implies that the reversed solution

$$\boldsymbol{\varepsilon q}(-t, \boldsymbol{\varepsilon y}) , \tag{6.4.28}$$

is also a solution of (6.4.25), and since

$$\boldsymbol{\varepsilon q}(0, \boldsymbol{\varepsilon y}) = \boldsymbol{\varepsilon\varepsilon y} = \boldsymbol{y} , \tag{6.4.29}$$

i.e., the initial conditions are the same, these solutions must be identical, i.e.,

$$\boldsymbol{\varepsilon q}(-t, \boldsymbol{\varepsilon y}) = \boldsymbol{q}(t, \boldsymbol{y}) . \tag{6.4.30}$$

Now the joint probability in the stationary state can be written as

$$\begin{aligned} p_s(\boldsymbol{x}, t; \boldsymbol{x}', t') &= \int d\boldsymbol{y}\, p_s(\boldsymbol{x}, t; \boldsymbol{x}', t'; \boldsymbol{y}, 0) \\ &= \int d\boldsymbol{y}\, \delta[\boldsymbol{x} - \boldsymbol{q}(t, \boldsymbol{y})]\delta[\boldsymbol{x}' - \boldsymbol{q}(t, \boldsymbol{y})] p_s(\boldsymbol{y}) , \end{aligned} \tag{6.4.31}$$

and

$$p_s(\boldsymbol{\varepsilon x}', -t'; \boldsymbol{\varepsilon x}, -t) = \int dy\, \delta[\boldsymbol{\varepsilon x} - \boldsymbol{q}(-t, \boldsymbol{y})]\delta[\boldsymbol{\varepsilon x}' - \boldsymbol{q}(-t', \boldsymbol{y})] p_s(\boldsymbol{y}) . \tag{6.4.32}$$

Change the variables from $\boldsymbol{y}$ to $\boldsymbol{\varepsilon y}$ and note that $p_s(\boldsymbol{y}) = p_s(\boldsymbol{\varepsilon y})$, and $d\boldsymbol{\varepsilon y} = d\boldsymbol{y}$, so that

$$(6.4.32) = \int dy\, \delta[\boldsymbol{x} - \boldsymbol{\varepsilon q}(-t, \boldsymbol{\varepsilon y})]\delta[\boldsymbol{x}' - \boldsymbol{\varepsilon q}(-t', \boldsymbol{\varepsilon y})] p_s(\boldsymbol{y}) , \tag{6.4.33}$$

and using (6.4.30),

$$= \int dy\, \delta[\boldsymbol{x} - \boldsymbol{q}(t, \boldsymbol{y})]\delta[\boldsymbol{x}' - \boldsymbol{q}(t', \boldsymbol{y})] p_s(\boldsymbol{y}) , \tag{6.4.34}$$

$$= p_s(\boldsymbol{x}, t; \boldsymbol{x}', t') . \tag{6.4.35}$$

Using the stationarity property, that p_s depends only on the time difference, we see that detailed balance is satisfied. This direct proof is, of course, unnecessary since the original general proof is valid for this deterministic system.

6.4.3 Detailed Balance in Markovian Physical Systems

In physical systems, which are where detailed balance is important, we often have an unbelievably large number of variables, of the order of 10^{20} at least. These variables (say, momentum and velocity of the particles in a gas) are those which occur in the distribution function which obeys a Liouville equation for they follow deterministic equations of motion, like Newton's laws of motion.

It can be shown directly that, for appropriate forms of interaction, Newton's laws obey the principle of *microscopic reversibility* which means that they can be put in the form (6.4.25), where $\boldsymbol{A}(\boldsymbol{x})$ obeys the reversibility condition (6.4.24). The macroscopically observable quantities in such a system are functions of these variables (for example, pressure, temperature, density of particles) and, by appropriate changes of variable, can be represented by the first few components of the vector $\boldsymbol{x}$.

Thus, we assume $\boldsymbol{x}$ can be written

$$\boldsymbol{x} = (\boldsymbol{a}, \hat{\boldsymbol{x}}) \tag{6.4.36}$$

where the vector $\boldsymbol{a}$ represents the macroscopically observable quantities and $\hat{\boldsymbol{x}}$ is all the others. Then, in practice, we are interested in

$$\begin{aligned}&\tilde{p}(\boldsymbol{a}_1, t_1; \boldsymbol{a}_2, t_2; \boldsymbol{a}_3, t_3; \dots)\\ &\quad = \int\int \dots \int d\hat{\boldsymbol{x}}_1, d\hat{\boldsymbol{x}}_2 \dots p(\boldsymbol{x}_1, t_1; \boldsymbol{x}_2, t_2; \boldsymbol{x}_3, t_3; \dots)\,.\end{aligned} \tag{6.4.37}$$

From the microscopic reversibility, it follows from our reasoning above that p, and thus also $\tilde{p}$, both obey the detailed balance conditions but, of course, $\tilde{p}$ does not obey a Liouville equation. *If it turns out or can be proven that $\tilde{p}$ obeys, to some degree approximation, a Markov equation of motion*, then we must preserve the detailed balance property, which takes the same form for $\tilde{p}$ as for p. In this sense, the condition (6.3.8) for detailed balance may be said to be derived from microscopic reversibility of the equations of motion.

6.4.4 Ornstein-Uhlenbeck Process

Most systems in which detailed balance is of interest can be approximated by an Ornstein-Uhlenbeck process, i.e., this means we assume

$$A_i(\boldsymbol{x}) = \sum_j A_{ij} x_j\,, \tag{6.4.38}$$

$$B_{ij}(\boldsymbol{x}) = B_{ij}\,. \tag{6.4.39}$$

The detailed balance conditions are not trivial, even for the case of such a linear system. They take the form

$$\sum_j (\varepsilon_i \varepsilon_j A_{ij} + A_{ij}) x_j = \sum_j B_{ij} \frac{\partial}{\partial x_j} \log p_s(\boldsymbol{x})\,, \tag{6.4.40}$$

and

$$\varepsilon_i \varepsilon_j B_{ij} = B_{ij}\,. \tag{6.4.41}$$

Equation (6.4.40) has the qualitative implication that $p_s(\boldsymbol{x})$ is a Gaussian since derivative of $\log p_s(\boldsymbol{x})$ log is linear in $\boldsymbol{x}$. Furthermore, since the left-hand side contains no constant term, this Gaussian must have zero mean, hence, we can write

$$p_s(\boldsymbol{x}) = \mathcal{N} \exp\left(-\tfrac{1}{2}\boldsymbol{x}^{\mathrm{T}}\sigma^{-1}\boldsymbol{x}\right). \tag{6.4.42}$$

One can now substitute (6.4.42) in the stationary Fokker-Planck equation and rearrange to obtain

$$-\sum_i A_{ii} - \tfrac{1}{2}\sum_{i,j} B_{ij}\sigma_{ij}^{-1} + \sum_{k,j}\left(\sum_i \sigma_{ki}^{-1}A_{ij} + \tfrac{1}{2}\sum_{i,l}\sigma_{ij}^{-1}B_{il}\sigma_{ij}^{-1}\right)x_k x_j = 0\,, \tag{6.4.43}$$

where we have used the symmetry of the matrix σ. The quadratic term vanishes if the symmetric part of its coefficient is zero. This condition may be written in matrix form as

$$\sigma^{-1}A + A^{\mathrm{T}}\sigma^{-1} = -\sigma^{-1}B\sigma^{-1}\,, \tag{6.4.44}$$

or

$$A\sigma + \sigma A^{\mathrm{T}} = -B\,. \tag{6.4.45}$$

The constant term also vanishes if (6.4.44) is satisfied. Equation (6.4.45) is, of course, exactly that derived by stochastic differential equation techniques in Sect. 4.5.6 (4.5.64) with the substitutions

$$A \to -A\,, \tag{6.4.46}$$

$$BB^{\mathrm{T}} \to B. \tag{6.4.47}$$

We can now write the detailed balance conditions in their most elegant form. We define the matrix $\boldsymbol{\varepsilon}$ by

$$\boldsymbol{\varepsilon} = \mathrm{diag}(\varepsilon_1, \varepsilon_2, \varepsilon_3, \dots)\,, \tag{6.4.48}$$

and clearly

$$\boldsymbol{\varepsilon}^2 = 1\,. \tag{6.4.49}$$

Then the conditions (6.4.40) and (6.4.41) become in matrix notation

$$\boldsymbol{\varepsilon}A\boldsymbol{\varepsilon} + A = -B\sigma^{-1} \tag{6.4.50}$$

$$\boldsymbol{\varepsilon}B\boldsymbol{\varepsilon} = B\,. \tag{6.4.51}$$

The potential condition (6.3.51) is simply equivalent to the symmetry of σ.

As noted in Sect. 6.3 (6.3.16), detailed balance requires

$$\boldsymbol{\varepsilon}\sigma = \sigma\boldsymbol{\varepsilon}\,. \tag{6.4.52}$$

Bearing this in mind, we take (6.4.45)

$$A\sigma + \sigma A^{\mathrm{T}} = -B\,, \tag{6.4.53}$$

and from (6.4.50)

$$\boldsymbol{\varepsilon}A\boldsymbol{\varepsilon}\sigma + A\sigma = -B\,, \tag{6.4.54}$$

which yield

$$\varepsilon A \varepsilon \sigma = \sigma A^{\mathrm{T}}, \tag{6.4.55}$$

and with (6.4.52)

$$\varepsilon (A\sigma) = (A\sigma)^{\mathrm{T}} \varepsilon. \tag{6.4.56}$$

These are equivalent to the celebrated Onsager relations; *Onsager*, [6.4]; *Casimir*, [6.5]. The derivation closely follows *van Kampen's* [6.1] work.

6.4.5 The Onsager Relations

The physical form of the Onsager relations arises when we introduce *phenomenological forces* defined as the gradient of the potential $\phi = \log[p_s(\boldsymbol{x})]$:

$$\boldsymbol{F}(\boldsymbol{x}) = -\nabla\phi(\boldsymbol{x}) = \sigma^{-1}\boldsymbol{x}, \tag{6.4.57}$$

(in physics, ϕ/kT is the entropy of the system). Because of the linear form of the $A_i(\boldsymbol{x})$ [(6.4.38)], the exact equations of motion for $\langle \boldsymbol{x} \rangle$ are

$$\frac{d\langle \boldsymbol{x} \rangle}{dt} = A\langle \boldsymbol{x} \rangle = A\sigma \boldsymbol{F}(\langle \boldsymbol{x} \rangle). \tag{6.4.58}$$

Thus, if the *fluxes* $d\langle \boldsymbol{x} \rangle/dt$ are related linearly to the *forces* $\boldsymbol{F}(\langle \boldsymbol{x} \rangle)$ by a matrix L defined by

$$L = A\sigma, \tag{6.4.59}$$

then (6.4.56) says

$$\varepsilon L \varepsilon = L^{\mathrm{T}}, \tag{6.4.60}$$

or

$$\begin{aligned} L_{ij} &= L_{ji}, \quad && \varepsilon_i \text{ and } \varepsilon_j \text{ of the same sign,} \\ L_{ij} &= -L_{ji}, \quad && \varepsilon_i \text{ and } \varepsilon_j \text{ of different sign.} \end{aligned} \tag{6.4.61}$$

Notice also that the relations

$$\varepsilon B \varepsilon = B, \qquad \varepsilon \sigma \varepsilon = \sigma, \tag{6.4.62}$$

imply that B_{ij} and σ_{ij} vanish if ε_i and ε_j have opposite signs.

In the special case that all the ε_i have the same sign, we find that

$$L = L^{\mathrm{T}}, \tag{6.4.63}$$

and noting that, since σ is symmetric and positive definite it has a *real* square root $\sigma^{1/2}$, we find that

$$\hat{A} \equiv \sigma^{-1/2} A \sigma^{1/2}. \tag{6.4.64}$$

is symmetric, so that A is similar to a symmetric matrix. Hence, all the eigenvalues of A are real.

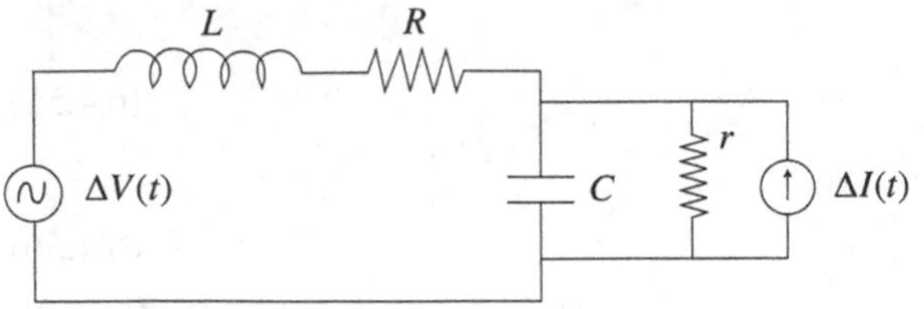

Fig. 6.1. Electric circuit used in the derivation of Nyquist's formula

6.4.6 Significance of the Onsager Relations—Fluctuation-Dissipation Theorem

The Onsager relations are for a set of macroscopically observable quantities and thus provide an easily observed consequence of detailed balance, which is itself a consequence of the reversibility of microscopic equations of motion, as outlined in Sect. 6.4.2 above. However, to check the validity in a given situation requires a knowledge of the covariance matrix σ.

Fortunately, in such situations, statistical mechanics gives us the form of the stationary distribution, provided this is thermodynamic equilibrium in which detailed balance is always satisfied. The principle is similar to that used by Langevin (Sect. 1.2.2).

Example—Derivation of Nyquists's Formula: As an example of the use of the Onsager relations, we can give a derivation of Nyquist's formula

We assume an electric circuit as in Fig. 6.1 in which there is assumed to be a fluctuating voltage and a fluctuating charge, both of which arise from the system having a nonzero temperature, and which we will show have their origin in the resistors R and r. The electrical equations arise from conservation of an electric charge q and Kirchoff's voltage law. The charge equation takes the form

$$\frac{dq}{dt} = i - \gamma q + \Delta I(t), \qquad \gamma \equiv 1/rC, \tag{6.4.65}$$

in which we equate the rate of gain of charge on the capacitor to the current i, less a leakage term $\gamma q \equiv q/rC$ plus a possible fluctuating term $\Delta I(t)$, which arises from leakage through the capacitor, and whose magnitude will be shortly calculated.

Kirchoff's voltage law is obtained by adding up all the voltages around the circuit, including a possible fluctuating voltage $\Delta V(t)$:

$$\frac{di}{dt} = \frac{1}{L}\left[-\frac{q}{c} - iR + \Delta V(t)\right]. \tag{6.4.66}$$

We now assume that $\Delta I(t)$ and $\Delta V(t)$ are white noise. We can write in the most general case,

$$\Delta I(t) = b_{11}\xi_1(t) + b_{12}\xi_2(t), \tag{6.4.67}$$

$$\frac{1}{L}\Delta V(t) = b_{21}\xi_1(t) + b_{22}\xi_2(t), \tag{6.4.68}$$

in which $\xi_1(t)$ and $\xi_2(t)$ are uncorrelated Langevin sources, i.e.,

$$\xi_1(t)\,dt = dW_1(t)\,, \tag{6.4.69}$$
$$\xi_2(t)\,dt = dW_2(t)\,, \tag{6.4.70}$$

where $W_1(t)$ and $W_2(t)$ are independent Wiener processes.

Thus, here

$$A = -\begin{bmatrix} \gamma & -1 \\ \dfrac{1}{LC} & \dfrac{R}{L} \end{bmatrix}. \tag{6.4.71}$$

The total energy in the system is

$$E = \frac{L}{2}i^2 + \frac{1}{2C}q^2\,. \tag{6.4.72}$$

From statistical mechanics, we know that $p_s(q,i)$ is given by the *Boltzmann Distribution* at temperature T, i.e.,

$$\begin{aligned} p_s(q,i) &= \mathcal{N}\exp(-E/kT) \\ &= \mathcal{N}\exp\left(-\frac{Li^2}{2kT} - \frac{q^2}{2CkT}\right), \end{aligned} \tag{6.4.73}$$

where k is the Boltzmann constant, so that the covariance matrix is

$$\sigma = \begin{bmatrix} kTC & 0 \\ 0 & kT/L \end{bmatrix}. \tag{6.4.74}$$

The Onsager symmetry can now be checked:

$$A\sigma = -\begin{bmatrix} \gamma kTC & -kT/L \\ kT/L & RkT/L^2 \end{bmatrix}. \tag{6.4.75}$$

For this system q, the total charge is even under time inversion and i, the current, is odd Thus,

$$\boldsymbol{\varepsilon} = \mathrm{diag}(1,-1)\,, \tag{6.4.76}$$

and it is clear that

$$(A\sigma)_{12} = -(A\sigma)_{21}\,, \tag{6.4.77}$$

is the only consequence of the symmetry and is satisfied by (6.4.75).

Here, also,

$$B = -(A\sigma + \sigma A^{\mathrm{T}}) = 2kT\begin{bmatrix} \gamma C & 0 \\ 0 & R/L^2 \end{bmatrix}, \tag{6.4.78}$$

so that

$$b_{12} = b_{21} = 0\,, \tag{6.4.79}$$
$$b_{11} = \sqrt{2kT\gamma C} = \sqrt{2kT/r}\,, \tag{6.4.80}$$
$$b_{22} = \sqrt{2kTR}\Big/L\,, \tag{6.4.81}$$

and we see that $B_{12} = B_{21} = 0$, as required by physical intuition, which suggests that the two sources of fluctuations arise from different causes and should be independent.

These results are *fluctuation dissipation* results. The magnitudes of the fluctuations b_{ij} are determined by the *dissipative terms* r and R. In fact, the result (6.4.80) is precisely Nyquist's theorem, which we discussed in Sect. 1.5.4. The noise voltage and current in the circuit are given by

$$\Delta V(t) = \sqrt{2kTR}\,\xi_2(t)\,, \qquad \Delta I(t) = \sqrt{2kT/r}\,\xi_1(t)\,, \tag{6.4.82}$$

so that

$$\left.\begin{aligned} \langle \Delta V(t+\tau)\Delta V(t)\rangle &= 2kTR\,\delta(\tau)\,,\\ \langle \Delta I(t+\tau)\Delta I(t)\rangle &= 2kT/r\,\delta(\tau)\,,\\ \langle \Delta V(t+\tau)\Delta I(t)\rangle &= 0\,. \end{aligned}\right\} \tag{6.4.83}$$

The first of these is Nyquists' theorem in the form quoted in (1.5.48–1.5.50), the second is the corresponding result for the noise regarded as arising from a fluctuation currrent from the resistor r, and these fluctuations are independent.

The terms r and R are called dissipative because they give rise to energy dissipation; in fact, deterministically (i.e., setting noise equal to 0),

$$\frac{dE}{dt} = -i^2R - \frac{\gamma q^2}{C}\,, \tag{6.4.84}$$

which explicitly exhibits the dissipation.

6.5 Eigenfunction Methods in Many Variables

Here we shall proceed similarly to Sect. 5.4, treating only homogeneous processes. We *assume* the existence of a complete set of eigenfunctions $P_\lambda(\boldsymbol{x})$ of the forward Fokker-Planck and a set $Q_\lambda(\boldsymbol{x})$ of the backward Fokker-Planck equation. Thus,

$$-\sum_i \partial_i[A_i(\boldsymbol{x})P_\lambda(\boldsymbol{x})] + \tfrac{1}{2}\sum_{i,j}\partial_i\partial_j[B_{ij}(\boldsymbol{x})P_\lambda(\boldsymbol{x})] = -\lambda P_\lambda(\boldsymbol{x})\,, \tag{6.5.1}$$

$$\sum_i A_i(\boldsymbol{x})\partial_i Q_{\lambda'}(\boldsymbol{x}) + \tfrac{1}{2}\sum_{i,j}B_{ij}(\boldsymbol{x})\partial_i\partial_j Q_{\lambda'}(\boldsymbol{x}) = -\lambda' Q_{\lambda'}(\boldsymbol{x})\,. \tag{6.5.2}$$

Whether $Q_{\lambda'}(\boldsymbol{x})$ and $P_\lambda(\boldsymbol{x})$ satisfy absorbing or reflecting boundary conditions, one can show, in a manner very similar to that used in Sect. 5.1.2, that

$$-(\lambda - \lambda')\int_R d\boldsymbol{x}\, P_\lambda(\boldsymbol{x})Q_{\lambda'}(\boldsymbol{x}) = 0\,, \tag{6.5.3}$$

so that the $P_\lambda(\boldsymbol{x})$ and $Q_{\lambda'}(\boldsymbol{x})$ from a bi-orthogonal set, which we normalise as

$$\int d\boldsymbol{x}\, P_\lambda(\boldsymbol{x})Q_\lambda(\boldsymbol{x}) = \delta_{\lambda\lambda'}\,, \tag{6.5.4}$$

if the spectrum of eigenvalues λ is discrete. If the spectrum is continuous, the Kronecker $\delta_{\lambda\lambda'}$, is to be replaced by $\delta(\lambda - \lambda')$, except where we have reflecting boundary conditions, and thus $\lambda = 0$ corresponds to the stationary state. The normalisation of $p_s(\boldsymbol{x})$ then gives

$$\int d\boldsymbol{x}\, P_0(\boldsymbol{x})Q_0(\boldsymbol{x}) = \int d\boldsymbol{x}\, P_0(\boldsymbol{x}) = 1\,, \tag{6.5.5}$$

so that there is also a discrete point with zero eigenvalue in the spectrum then.

6.5.1 Relationship between Forward and Backward Eigenfunctions

The functional relationship (5.4.8) between the P_λ and Q_λ, which is always true in one dimension, only pertains in many dimensions if detailed balance with all $\varepsilon_i = 1$ is valid. If some $\varepsilon_i = -1$, there is a modified relationship.

To show this, note that if detailed balance is valid, we have already seen in Sect. 6.3.5c(iii) that $p(\boldsymbol{\varepsilon x}, -t)/p_s(x)$ is a solution of the backward Fokker-Planck equation so that, from the uniqueness of solutions, we can say

$$Q_\lambda(\boldsymbol{x}) = \eta_\lambda P_\lambda(\boldsymbol{\varepsilon x})/p_s(\boldsymbol{x}) . \tag{6.5.6}$$

Here, $\eta_\lambda = \pm 1$ but is otherwise undetermined. For, if $\lambda = \lambda'$, we can then write (6.5.4) in the form

$$\int dx \frac{P_\lambda(\boldsymbol{x}) P_\lambda(\boldsymbol{\varepsilon x})}{P_s(\boldsymbol{x})} = \eta_\lambda . \tag{6.5.7}$$

We cannot determine η_λ *a priori*, but by suitable normalisation it may be chosen ± 1. If, for example, all ε_i are — 1 and $P_\lambda(\boldsymbol{x})$ is an odd function of x, it is clear that η_λ must be -1.

6.5.2 Even Variables Only—Negativity of Eigenvalues

If all the ε_i are equal to one, then we can write

$$Q_\lambda(\boldsymbol{x}) = P_\lambda(\boldsymbol{x})/p_s(\boldsymbol{x}) , \tag{6.5.8}$$

and η_λ can always be set equal to one.

Hence, the expansions in eigenfunctions will be much the same as for the onevariable case.

The *completeness* of the eigenfunctions is a matter for proof in each individual case. If all the ε_i are equal to one, then we can show that the Fokker-Planck operator is self adjoint and negative semi-definite in a certain Hilbert space.

To be more precise, let us write the forward Fokker-Planck operator as

$$\mathscr{L}(\boldsymbol{x}) = -\sum_i \partial_i A_i(\boldsymbol{x}) + \tfrac{1}{2} \sum_{i,j} \partial_i \partial_j B_{ij}(\boldsymbol{x}) , \tag{6.5.9}$$

and the backward Fokker-Planck operator as

$$\mathscr{L}^*(\boldsymbol{x}) = \sum_i A_i(\boldsymbol{x}) \partial_i + \tfrac{1}{2} \sum_{i,j} B_{ij}(\boldsymbol{x}) \partial_i \partial_j . \tag{6.5.10}$$

Then the fact that, if all $\varepsilon_i = 1$, we can transform a solution of the forward Fokker-Planck equation to a solution of the backward Fokker-Planck equation by dividing by $p_s(\boldsymbol{x})$ arises from the fact that for any $f(\boldsymbol{x})$,

$$\mathscr{L}(\boldsymbol{x})[f(\boldsymbol{x}) p_s(\boldsymbol{x}))] = p_s(\boldsymbol{x}) \mathscr{L}^*(\boldsymbol{x})[f(\boldsymbol{x})] . \tag{6.5.11}$$

Define a Hilbert space of functions $f(\boldsymbol{x}), g(\boldsymbol{x}) \dots$ by the scalar product

$$(f, g) = \int_R dx \frac{f(\boldsymbol{x}) g(\boldsymbol{x})}{p_s(\boldsymbol{x})} . \tag{6.5.12}$$

Then from (6.5.11),

$$(f, \mathscr{L}g) = \int_R dx \frac{f(\boldsymbol{x})}{p_s(\boldsymbol{x})} \mathscr{L}(\boldsymbol{x}) \left[\frac{g(\boldsymbol{x})}{p_s(\boldsymbol{x})} p_s(\boldsymbol{x}) \right], \tag{6.5.13}$$

$$= \int_R dx\, f(\boldsymbol{x}) \mathscr{L}^*(\boldsymbol{x}) \left[\frac{g(\boldsymbol{x})}{p_s(\boldsymbol{x})} \right], \tag{6.5.14}$$

and integrating by parts, discarding surface terms by use of either reflecting or absorbing boundary conditions

$$= \int_R dx\, \frac{g(\boldsymbol{x})}{p_s(\boldsymbol{x})} \mathscr{L}(\boldsymbol{x})[f(\boldsymbol{x})]. \tag{6.5.15}$$

Thus, in this Hilbert space,

$$(f, \mathscr{L}g) = (g, \mathscr{L}f). \tag{6.5.16}$$

This condition is sufficient to ensure that the eigenvalues of $\mathscr{L}(\boldsymbol{x})$ are real. To prove negative semi-definiteness, notice that for even variables only, [see (6.3.43)]

$$A_i(\boldsymbol{x}) = \sum_j \partial_j [B_{ij}(\boldsymbol{x}) p_s(\boldsymbol{x})] / 2 p_s(\boldsymbol{x}), \tag{6.5.17}$$

so that for any $p(\boldsymbol{x})$,

$$\mathscr{L}(\boldsymbol{x}) p(\boldsymbol{x}) = \sum_{i,j} \partial_i \left\{ - \frac{\partial_j [B_{ij}(\boldsymbol{x}) p_s(\boldsymbol{x})] p(\boldsymbol{x})}{2 p_s(\boldsymbol{x})} + \tfrac{1}{2} \partial_j [B_{ij}(\boldsymbol{x}) p(\boldsymbol{x})] \right\},$$

$$= \tfrac{1}{2} \sum_{i,j} \partial_i \left\{ B_{ij}(\boldsymbol{x}) p_s(\boldsymbol{x}) \partial_j [p(\boldsymbol{x}) / p_s(\boldsymbol{x})] \right\}. \tag{6.5.18}$$

Hence,

$$(p, \mathscr{L}p) = \tfrac{1}{2} \int dx\, p(\boldsymbol{x}) / p_s(\boldsymbol{x}) \sum_{i,j} \partial_i \{ B_{ij}(\boldsymbol{x}) p_s(\boldsymbol{x}) \partial_j [p(\boldsymbol{x}) / p_s(\boldsymbol{x})] \}, \tag{6.5.19}$$

and integrating by parts (discarding surface terms),

$$(p, \mathscr{L}p) = - \int dx B_{ij}(\boldsymbol{x}) p_s(\boldsymbol{x}) \partial_i [p(\boldsymbol{x}) / p_s(\boldsymbol{x})] \partial_j [p(\boldsymbol{x}) / p_s(\boldsymbol{x})] \leqslant 0, \tag{6.5.20}$$

since $B_{ij}(\boldsymbol{x})$ is positive semi-definite.

Hence, we conclude for any eigenfunction $P_\lambda(\boldsymbol{x})$ that λ is real, and

$$\lambda \geqslant 0, \tag{6.5.21}$$

(remember that $-\lambda$ is the eigenvalue).

6.5.3 A Variational Principle

Combined with this property is a variational principle. For, suppose we choose any real function $f(\boldsymbol{x})$ and expand it in eigenfunctions

$$f(\boldsymbol{x}) = \sum_\lambda \alpha_\lambda P_\lambda(\boldsymbol{x}). \tag{6.5.22}$$

Further, let us fix

$$(f, f) = \int dx\, f(x)^2 / p_s(x) = 1. \tag{6.5.23}$$

Then,

$$-(f, \mathscr{L} f) = \sum_\lambda \lambda \alpha_\lambda^2, \tag{6.5.24}$$

and

$$(f, f) = \sum_\lambda \alpha_\lambda^2. \tag{6.5.25}$$

Clearly $-(f, \mathscr{L} f)$ has its minimum of zero only if only the term $\lambda = 0$ exists, i.e.,

$$\begin{aligned} &\alpha_0 \neq 0 \\ &\alpha_\lambda = 0 \quad \text{for} \quad \lambda \neq 0. \end{aligned} \tag{6.5.26}$$

Now choose $f(x)$ orthogonal to $P_0(x)$ so $\alpha_0 = 0$, and we see that the minimum of $(f, \mathscr{L} f)$ occurs when

$$\begin{aligned} &\alpha_{\lambda_1} = 1, \\ &\text{and}, \\ &\alpha_\lambda = 0 \qquad \text{for all other } \lambda, \end{aligned} \tag{6.5.27}$$

where λ_1 is the next smallest eigenvalue. This means that $P_{\lambda_1}(x)$ is obtained by minimising $-(p, \mathscr{L} p)$ [which can be put in the form of (6.5.20)], subject to the condition that

$$\left.\begin{aligned} (p, p) &= 1, \\ (p, P_0) &= 0. \end{aligned}\right\} \tag{6.5.28}$$

This method can yield a numerical way of estimating eigenfunctions in terms of trial solutions and is useful particularly in bistability situations. It is, however, limited to situations in which detailed balance is valid.

6.5.4 Conditional Probability

Assuming completeness, we find that the conditional probability can be written

$$p(x, t \mid x_0, 0) = \sum_\lambda P_\lambda(x) Q_\lambda(x_0) e^{-\lambda t}. \tag{6.5.29}$$

6.5.5 Autocorrelation Matrix

If detailed balance is valid, for the stationary autocorrelation matrix [using (6.5.6)] we have

$$G(t) = \langle x(t) x^T(0) \rangle = \sum \eta_\lambda e^{-\lambda t} \left[\int_R dx\, x P_\lambda(x) \right] \left[\int_R dx\, x P_\lambda(x) \right]^T \varepsilon, \tag{6.5.30}$$

which explicitly satisfies the condition

$$G(t) = \varepsilon G^T(t) \varepsilon, \tag{6.5.31}$$

derived in Sect. 6.3.3.

6.5.6 Spectrum Matrix

The spectrum matrix is

$$S(\omega) = \frac{1}{2\pi}\int_{-\infty}^{\infty} \mathrm{e}^{-\mathrm{i}\omega t} G(t)\,dt\,,$$
$$= \frac{1}{2\pi}\left\{\int_{0}^{\infty} \mathrm{e}^{-\mathrm{i}\omega t} G(t)\,dt + \int_{0}^{\infty} \mathrm{e}^{\mathrm{i}\omega t} G^{\mathrm{T}}(t)\,dt\right\}. \tag{6.5.32}$$

If we define, for convenience,

$$\boldsymbol{U}_\lambda = \int_R d\boldsymbol{x}\,\boldsymbol{x} P_\lambda(\boldsymbol{x})\,, \tag{6.5.33}$$

then

$$S(\omega) = \frac{1}{2\pi}\sum_\lambda\left[\frac{\eta_\lambda}{\lambda + \mathrm{i}\omega}\boldsymbol{U}_\lambda \boldsymbol{U}_\lambda^{\mathrm{T}}\boldsymbol{\varepsilon} + \frac{\eta_\lambda}{\lambda - \mathrm{i}\omega}\boldsymbol{\varepsilon}\boldsymbol{U}_\lambda \boldsymbol{U}_\lambda^{\mathrm{T}}\right]. \tag{6.5.34}$$

If any of the λ are complex, from the reality of $\mathscr{L}(\boldsymbol{x})$ we find that the eigenfunction belonging to λ^* is $[P_\lambda(\boldsymbol{x})]^*$, $\eta_\lambda = \eta_\lambda^*$ and $[\boldsymbol{U}_\lambda]^* = \boldsymbol{U}_{\lambda^*}$. The spectrum is then obtained by adding the complex conjugate of those terms involving complex eigenvalues to (6.5.34).

In the case where $\boldsymbol{\varepsilon} = 1$ and hence $\eta_\lambda = 1$, the spectrum matrix has the simpler form

$$S(\omega) = \frac{1}{\pi}\sum_\lambda \frac{\lambda \boldsymbol{U}_\lambda \boldsymbol{U}_\lambda^{\mathrm{T}}}{\lambda^2 + \omega^2}\,, \tag{6.5.35}$$

which is explicitly a positive definite matrix. The spectrum of a single variable q made up as a linear combination of the $\boldsymbol{x}$ by a formula such as

$$q = \boldsymbol{m}\cdot\boldsymbol{x}\,, \tag{6.5.36}$$

is given by

$$S_q(\omega) = \boldsymbol{m}^{\mathrm{T}} S(\omega)\boldsymbol{m} = \frac{1}{\pi}\sum_\lambda \frac{\lambda(\boldsymbol{m}\cdot\boldsymbol{U}_\lambda)^2}{\lambda^2 + \omega^2}\,, \tag{6.5.37}$$

and is a strictly decreasing function of ω.

In the case where $\boldsymbol{\varepsilon} \neq 1$, the positivity of the spectrum is no longer obvious, though general considerations such as the result (3.7.20) show that it must be. An important difference arises because the λ now need not all be real, and this means that denominators of the form

$$1/[\lambda_R^2 + (\omega - \lambda_I)^2]\,, \tag{6.5.38}$$

can occur, giving rise to peaks in the spectrum away from $\omega = 0$.

6.6 First Exit Time from a Region (Homogeneous Processes)

We wish to treat here the multidimensional analogue of the first passage time problem in one dimension, treated in Sect. 5.5. As in that section, we will restrict ourselves to homogeneous processes.

The analogous problem here is to compute the earliest time at which a particle, initially inside a region R with boundary S, leaves that region.

As in the one-dimensional case, we consider the problem of solving the backward Fokker-Planck equation with an absorbing boundary condition on S, namely,

$$p(\boldsymbol{x}', t \,|\, \boldsymbol{x}, 0) = 0 \quad (\boldsymbol{x} \in S). \tag{6.6.1}$$

The probability that the particle, initially at $\boldsymbol{x}$, is somewhere within R after a time t is

$$G(\boldsymbol{x}, t) = \int_R d\boldsymbol{x}' \, p(\boldsymbol{x}', t \,|\, \boldsymbol{x}, 0), \tag{6.6.2}$$

and if T is the time at which the particle leaves R, then

$$\text{Prob}\,(T \geqslant t) = G(\boldsymbol{x}, t). \tag{6.6.3}$$

Since the process is homogeneous, we find that $G(\boldsymbol{x}, t)$ obeys the backward Fokker-Planck equation

$$\partial_t G(\boldsymbol{x}, t) = \sum_i A_i(\boldsymbol{x}) \partial_i G(\boldsymbol{x}, t) + 2 \sum_{i,j} B_{ij}(\boldsymbol{x}) \partial_i \partial_j G(\boldsymbol{x}, t). \tag{6.6.4}$$

The initial conditions on (6.6.4) will arise from:

i) The initial condition

$$p(\boldsymbol{x}', 0 \,|\, \boldsymbol{x}, 0) = \delta(\boldsymbol{x} - \boldsymbol{x}'), \tag{6.6.5}$$

so that

$$G(\boldsymbol{x}, 0) = \begin{cases} 1, & \boldsymbol{x} \in R, \\ 0, & \text{elsewhere.} \end{cases} \tag{6.6.6}$$

ii) The boundary condition (6.6.1) requires

$$G(\boldsymbol{x}, t) = 0, \quad \boldsymbol{x} \in S. \tag{6.6.7}$$

As in Sect. 5.5, we find that these imply that the *mean exit time* from R starting at $\boldsymbol{x}$, for which we shall use the symbol $T(\boldsymbol{x})$, satisfies

$$\sum_i A_i(\boldsymbol{x}) \partial_i T(\boldsymbol{x}) + 2 \sum_{i,j} B_{ij}(\boldsymbol{x}) \partial_i \partial_j T(\boldsymbol{x}) = -1, \tag{6.6.8}$$

with the boundary condition

$$T(\boldsymbol{x}) = 0, \quad \boldsymbol{x} \in S, \tag{6.6.9}$$

and the nth moments

$$T_n(\boldsymbol{x}) = \langle T^n \rangle = \int_0^\infty t^{n-1} G(\boldsymbol{x}, t)\, dt, \tag{6.6.10}$$

satisfy

$$-nT_{n-1}(\boldsymbol{x}) = \sum_i A_i(\boldsymbol{x})\partial_i T_n(\boldsymbol{x}) + \sum_{i,j} B_{ij}(\boldsymbol{x})\partial_i\partial_j T_n(\boldsymbol{x})\,, \tag{6.6.11}$$

with the boundary conditions

$$T_n(\boldsymbol{x}) = 0 \quad \boldsymbol{x} \in S\,. \tag{6.6.12}$$

Inclusion of Reflecting Regions: It is possible to consider that S, the boundary of R, is divided into two regions S_{r} and S_{a} such that the particle is reflected when it meets S_{r} and is absorbed when it meets S_{a}. The boundary conditions on $G(\boldsymbol{x},t)$ are then, from those derived in Sect. 5.1.2,

$$\sum_{i,j} n_i B_{ij}(\boldsymbol{x})\partial_j G(\boldsymbol{x},t) = 0\,, \quad (\boldsymbol{x} \in S_{\mathrm{r}})\,, \tag{6.6.13}$$

$$G(\boldsymbol{x},t) = 0\,, \quad (\boldsymbol{x} \in S_{\mathrm{a}})\,, \tag{6.6.14}$$

and hence,

$$\sum_{i,j} n_i B_{ij}(\boldsymbol{x})\partial_j T_n(\boldsymbol{x}) = 0\,, \quad (\boldsymbol{x} \in S_{\mathrm{r}})\,, \tag{6.6.15}$$

$$T_n(\boldsymbol{x}) = 0\,, \quad (\boldsymbol{x} \in S_{\mathrm{a}})\,. \tag{6.6.16}$$

6.6.1 Solutions of Mean Exit Time Problems

The basic partial differential equation for the mean first passage time is only simple to solve in one dimension or in situations where there is a particular symmetry available. Asymptotic approximations can provide very powerful results, but these will be dealt with in Chap. 9. We will illustrate some methods here with some examples.

a) Ornstein-Uhlenbeck Process in Two Dimensions (Rotationally Symmetric): We suppose that a particle moves according to

$$\left.\begin{aligned} dx &= -kx\,dt + \sqrt{D}\,dW_1(t)\,, \\ dy &= -ky\,dt + \sqrt{D}\,dW_2(t)\,, \end{aligned}\right\} \tag{6.6.17}$$

and want to know the mean exit time from the region

$$x^2 + y^2 \leqslant a^2\,, \tag{6.6.18}$$

given the initial position is at x_0, y_0.

This is easily reduced to the one-dimensional problem for the variable

$$r = \sqrt{x^2 + y^2}\,, \tag{6.6.19}$$

by changing variables as in Sect. 4.5.5, namely,

$$dr = \left(-kr + \tfrac{1}{2}D/r\right)dt + \sqrt{D}\,dW(t)\,, \tag{6.6.20}$$

and we want to know the mean exit time from the region $(0, a)$. This can be solved by using (5.5.28) with the replacements

$$\left.\begin{aligned} U(x) &\to \tfrac{1}{2}kr^2 - \tfrac{1}{2}D\log r \\ D &\to \tfrac{1}{2}D \\ x_0 &\to a \\ a &\to \sqrt{x_0^2 + y_0^2} \equiv r_0 \\ -\infty &\to 0. \end{aligned}\right\}, \tag{6.6.21}$$

Thus, $T(r_0 \to a)$

$$= \frac{1}{D}\int_{r_0}^{a} y^{-1} \exp[ky^2/D]dy \int_0^y z\exp(-kz^2/D)\,dz\,. \tag{6.6.22}$$

The problem is thus essentially one dimensional. This does not often happen.

b) Application of Eigenfunctions: Suppose we use the eigenfunctions $Q_\lambda(\boldsymbol{x})$ and $P_\lambda(\boldsymbol{x})$ to expand the mean first passage time as

$$T(\boldsymbol{x}) = \sum t_\lambda Q_\lambda(\boldsymbol{x})\,. \tag{6.6.23}$$

We suppose that the $P_\lambda(\boldsymbol{x})$ and $Q_\lambda(\boldsymbol{x})$ satisfy the boundary conditions required for the particular exit problem being studied, so that $T(\boldsymbol{x})$ as written in (6.6.23) satisfies the appropriate boundary conditions.

We then expand

$$-1 = \sum_\lambda I_\lambda Q_\lambda(\boldsymbol{x})\,, \tag{6.6.24}$$

where

$$I_\lambda = -\int_R d\boldsymbol{x}\, P_\lambda(\boldsymbol{x})\,. \tag{6.6.25}$$

Inserting (6.6.23) into (6.6.8) yields

$$\lambda t_\lambda = -T_\lambda\,, \tag{6.6.26}$$

so that

$$T(\boldsymbol{x}) = \sum_\lambda \frac{1}{\lambda} Q_\lambda(\boldsymbol{x}) \int_R d\boldsymbol{x}' P_\lambda(\boldsymbol{x}')\,. \tag{6.6.27}$$

The success of the method depends on the knowledge of the eigenvalues satisfying the correct boundary conditions on $\boldsymbol{S}$ and normalised on R.

c) Asymptotic Result: If the first eigenvalue λ_1 is very much less than all other eigenvalues, the series may be approximated by its first term. This will mean that the eigenfunction Q_1 will be very close to a solution of

$$\sum_i A_i(\boldsymbol{x})\partial_i f(\boldsymbol{x}) + \tfrac{1}{2}\sum_{i,j} B_{ij}(\boldsymbol{x})\partial_i\partial_j f(\boldsymbol{x}) = 0\,, \tag{6.6.28}$$

since λ_1 is very small. Hence,

$$Q_1(\boldsymbol{x}) \sim K\,, \tag{6.6.29}$$

where K is a constant. Taking account of the bi-orthonormality of P_λ and Q_λ, we see

$$1 = \int dx\, P_1(\boldsymbol{x})Q_1(\boldsymbol{x}) \sim K \int dx\, P_1(\boldsymbol{x}), \tag{6.6.30}$$

so that

$$T(\boldsymbol{x}) \sim 1/\lambda_1 . \tag{6.6.31}$$

The reasoning given here is rather crude. It can be refined by the asymptotic methods of Chap. 14.

d) Application of the Eigenfunction Method to Two-dimensional Brownian motion: The particle moves in the xy plane within a square whose corners are $(0,0)$, $(0,1)$, $(1,0)$, $(1,1)$. The sides of this square are absorbing barriers. $T(x,y)$ obeys

$$\frac{D}{2}\left(\frac{\partial^2 T}{\partial x^2} + \frac{\partial^2 T}{\partial y^2}\right) = -1 . \tag{6.6.32}$$

The eigenfunctions satisfying the boundary condition $T = 0$ on the edges of the square are

$$P_{n,m}(x,y) = \sin(n\pi x)\sin(m\pi x), \tag{6.6.33}$$

with

$$Q_{n,m}(x,y) = 4\sin(n\pi x)\sin(m\pi x), \tag{6.6.34}$$

and n, m positive and integral. The eigenvalues are

$$\lambda_{n,m} = \frac{\pi^2 D}{2}(n^2 + m^2) . \tag{6.6.35}$$

The coefficient

$$\begin{aligned} \int_R dx\, dy P_{n,m}(x,y) &= 0 && (\text{either } n \text{ or } m \text{ even}), \\ &= \frac{4}{mn\pi^2} && (m \text{ and } n \text{ both odd}) . \end{aligned} \tag{6.6.36}$$

Hence,

$$T(x,y) = \frac{1}{D} \sum_{\substack{n,m \\ \text{odd}}} \frac{32}{\pi^2 nm(m^2+n^2)} \sin(n\pi x)\sin(m\pi y) . \tag{6.6.37}$$

6.6.2 Distribution of Exit Points

This problem is the multidimensional analogue of that treated in Sect. 5.5.4. Namely, what is the probability of exiting through an element $d\boldsymbol{S}(\boldsymbol{a})$ at $\boldsymbol{a}$ of the boundary $\boldsymbol{S}$ of the region R. We assume absorption on all $\boldsymbol{S}$.

The probability that the particle exits through $d\boldsymbol{S}(\boldsymbol{a})$ after time t is

$$g(\boldsymbol{a},\boldsymbol{x},t)\,|\,d\boldsymbol{S}(\boldsymbol{a})| = -\int_t^\infty dt' \boldsymbol{J}(\boldsymbol{a},t'\,|\,\boldsymbol{x},0)\cdot d\boldsymbol{S}(\boldsymbol{a}) . \tag{6.6.38}$$

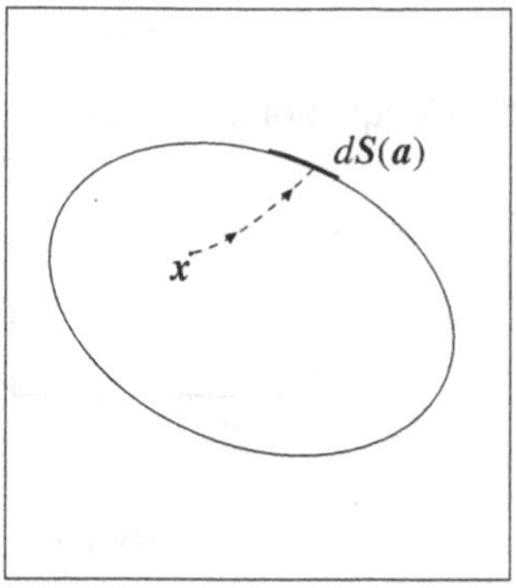

Fig. 6.2. Region and surface considered in Sect. 6.6.2

Similar reasoning to that in Sect. 5.5.4 shows that $g(\boldsymbol{a},\boldsymbol{x},t)$ obeys the backward Fokker-Planck equation

$$\partial_t g(\boldsymbol{a},\boldsymbol{x},t) = \sum_i A_i(\boldsymbol{x})\partial_i g(\boldsymbol{a},\boldsymbol{x},t) + \tfrac{1}{2}\sum_{i,j} B_{ij}(\boldsymbol{x})\partial_i\partial_j g(\boldsymbol{a},\boldsymbol{x},t)\,. \tag{6.6.39}$$

The boundary conditions follow by definition. Initially we have

$$g(\boldsymbol{a},\boldsymbol{x},0) = 0 \quad \text{for } \boldsymbol{x} \neq \boldsymbol{a},\ \boldsymbol{x} \in R, \tag{6.6.40}$$

and at all times

$$g(\boldsymbol{a},\boldsymbol{x},t) = 0 \quad \text{for } \boldsymbol{x} \neq \boldsymbol{a},\ \boldsymbol{x} \in S\,. \tag{6.6.41}$$

If $\boldsymbol{x} = \boldsymbol{a}$, then exit through $d\boldsymbol{S}(\boldsymbol{a})$ is certain, hence,

$$g(\boldsymbol{a},\boldsymbol{a},t)d\boldsymbol{S}(\boldsymbol{a}) = 1 \quad \text{for all } t, \tag{6.6.42}$$

or effectively

$$g(\boldsymbol{a},\boldsymbol{x},t) = \delta_s(\boldsymbol{a}-\boldsymbol{x}) \quad \boldsymbol{x} \in \boldsymbol{S} \text{ for all } t, \tag{6.6.43}$$

where $\delta_s(\boldsymbol{a}-\boldsymbol{x})$ is an appropriate surface delta function such that

$$\int |d\boldsymbol{S}(\boldsymbol{a})|\,\delta_s(\boldsymbol{a}-\boldsymbol{x}) = 1\,. \tag{6.6.44}$$

The probability of ultimate exit through $d\boldsymbol{S}(\boldsymbol{a})$ is

$$\pi(\boldsymbol{a},\boldsymbol{x})\,|d\boldsymbol{S}(\boldsymbol{a})| = g(\boldsymbol{a},\boldsymbol{x},0)\,|d\boldsymbol{S}(\boldsymbol{a})|\,. \tag{6.6.45}$$

The mean exit time given that exit occurred at $\boldsymbol{a}$ is

$$T(\boldsymbol{a},\boldsymbol{x}) = \int_0^\infty dt\, g(\boldsymbol{a},\boldsymbol{x},t)/\pi(\boldsymbol{a},\boldsymbol{x})\,, \tag{6.6.46}$$

and in the same way as in Sect. 5.5.4, we show that this satisfies

$$\boxed{\sum_i A_i(\boldsymbol{x})\partial_i[\pi(\boldsymbol{a},\boldsymbol{x})T(\boldsymbol{a},\boldsymbol{x})] + \tfrac{1}{2}\sum_{i,i} B_{ij}(\boldsymbol{x})\partial_i\partial_j[\pi(\boldsymbol{a},\boldsymbol{x})T(\boldsymbol{a},\boldsymbol{x})] = -\pi(\boldsymbol{a},\boldsymbol{x})\,,} \tag{6.6.47}$$

and the boundary conditions is

$$\pi(\boldsymbol{a},\boldsymbol{x})T(\boldsymbol{a},\boldsymbol{x}) = 0, \qquad \boldsymbol{x} \in S\,. \tag{6.6.48}$$

Further, by letting $t \to \infty$ in the corresponding Fokker-Planck equation for $g(\boldsymbol{a},\boldsymbol{x},t)$, we obtain the equation for $\pi(\boldsymbol{a},\boldsymbol{x})$:

$$\sum A_i(\boldsymbol{x})\partial_i[\pi(\boldsymbol{a},\boldsymbol{x})] + \tfrac{1}{2}\sum_{i,j} B_{ij}(\boldsymbol{x})\partial_i\partial_j[\pi(\boldsymbol{a},\boldsymbol{x})] = 0\,. \tag{6.6.49}$$

The boundary condition for (6.6.49) is

$$\pi(\boldsymbol{a},\boldsymbol{x}) = 0 \quad \boldsymbol{a} \neq \boldsymbol{x}, \quad \boldsymbol{x} \in S\,, \tag{6.6.50}$$

and

$$\int |dS\,(\boldsymbol{a})|\,\pi(\boldsymbol{a},\boldsymbol{x}) = 1\,. \tag{6.6.51}$$

Thus, we can summarise this as

$$\pi(\boldsymbol{a},\boldsymbol{x}) = \delta_{\mathrm{s}}(\boldsymbol{a}-\boldsymbol{x})\,, \quad \boldsymbol{x} \in \boldsymbol{S}, \tag{6.6.52}$$

where $\delta_{\mathrm{s}}(\boldsymbol{a}-\boldsymbol{x})$ is the surface delta function for the boundary $\boldsymbol{S}$.

7. Small Noise Approximations for Diffusion Processes

The methods described in the previous two chapters have concentrated on exact results, and of course the results available are limited in their usefulness. Approximation methods are the essence of most applications, where some way of reducing a problem to an exactly soluble one is always sought. It could even be said that most work on applications is concerned with the development of various approximations.

There are two major approximation methods of great significance. The first is the small noise expansion theory which gives solutions linearised about a deterministic equation. Since noise is often small, this is a method of wide practical application, and is the subject of this chapter. The equations are reduced into a sequence of time-dependent Ornstein-Uhlenbeck processes—mostly the first order is used.

Another large class of methods is given by adiabatic elimination, in which different time scales are identified and fast variables are eliminated completely. This forms the basis of the Chap. 8.

7.1 Comparison of Small Noise Expansions for Stochastic Differential Equations and Fokker-Planck Equations

In many physical and chemical problems, the stochastic element in a dynamical system arises from thermal fluctuations, which are always very small. Unless one measures very carefully, it is difficult to detect the existence of fluctuations. In such a case, the time development of the system will be almost deterministic and the fluctuations will be a small perturbation.

a) Ornstein-Uhlenbeck Process with Small Noise: With this in mind, we consider a simple linear example which is exactly soluble: a one-variable Ornstein-Uhlenbeck process described by the stochastic differential equation:

$$dx = -kx\,dt + \varepsilon\,dW(t)\,, \tag{7.1.1}$$

for which the Fokker-Planck equation is

$$\partial_t p = \partial_x(kx\,p) + \tfrac{1}{2}\varepsilon^2\partial_x^2 p\,. \tag{7.1.2}$$

The solutions of these have been previously investigated in Sects. 3.8.4, 4.5.4. Here ε is a small parameter which is zero in the deterministic limit. However, the nature of the limit $\varepsilon \to 0$ is essentially different in the two cases.

In the stochastic differential equation (7.1.1), as $\varepsilon \to 0$, the differential equation becomes nonstochastic but remains of first order in t, and the limit $\varepsilon \to 0$ is therefore not singular. In contrast, in the Fokker-Planck equation (7.1.2), the limit $\varepsilon \to 0$ reduces a second-order differential equation to one of first order. This limit is singular and any perturbation theory is a singular perturbation theory.

The solution to (7.1.1) is known exactly—it is (4.5.35), i.e.,

$$x_\varepsilon(t) = c\,\mathrm{e}^{-kt} + \varepsilon \int_0^t \mathrm{e}^{-k(t-t')}\, dW(t'), \tag{7.1.3}$$

which can be written

$$x_\varepsilon(t) = x_0(t) + \varepsilon x_1(t), \tag{7.1.4}$$

and this is generic; that is, we can normally solve in a power series in the small parameter ε. Furthermore, the zero-order term $x_0(t)$ is the solution of the equation obtained by setting $\varepsilon = 0$, i.e., of

$$dx = -kx\,dt. \tag{7.1.5}$$

The situation is by no means so simple for the Fokker-Planck Equation (7.1.2). Assuming the initial condition c is a nonstochastic variable, the exact solution is the Gaussian with mean and variance given by

$$\langle x(t)\rangle \quad = \alpha(t) \quad \equiv c\,\mathrm{e}^{-kt}, \tag{7.1.6}$$

$$\mathrm{var}\{x(t)\} = \varepsilon^2\beta(t) \equiv \varepsilon^2(1-\mathrm{e}^{-2kt})/2k, \tag{7.1.7}$$

so that

$$p_\varepsilon(x,t|c,0) = \frac{1}{\varepsilon}\frac{1}{\sqrt{2\pi\beta(t)}}\exp\left\{-\frac{1}{\varepsilon^2}\frac{[x-\alpha(t)]^2}{2\beta(t)}\right\}. \tag{7.1.8}$$

The solution for the conditional probability has the limiting form as $\varepsilon \to 0$ of

$$p_\varepsilon(x,t\,|\,c,0) \to \delta[x-\alpha(t)] \tag{7.1.9}$$

which corresponds exactly to the first-order solution of the stochastic differential equation. This is a deterministic trajectory along the path $x(t) = c\exp(-kt)$. However, p_ε cannot be expanded as a simple power series in ε. To carry out a power series expansion, one must define *a scaled variable* at each time;

$$y = [x-\alpha(t)]/\varepsilon \tag{7.1.10}$$

so that a probability density for y is

$$\bar{p}_\varepsilon(y,t\,|\,0,0) = p_\varepsilon(x,t\,|\,c,0)\,\frac{dx}{dy} = \frac{1}{\sqrt{2\pi\beta(t)}}\exp\left[-\frac{y^2}{2\beta(t)}\right]. \tag{7.1.11}$$

The probability density for the scaled variable y has no singularity; indeed, we have no ε dependence.

The transformation (7.1.10) can be rewritten as

$$x = \alpha(t) + \varepsilon y, \tag{7.1.12}$$

and is to be interpreted as follows. From (7.1.11) we see that the distribution of y is Gaussian with mean zero, variance $\beta(t)$. Equation (7.1.12) says that the deviation of x from $\alpha(t)$, the deterministic path, is of order ε as $\varepsilon \to 0$, and the coefficient is the Gaussian random variable y. This is essentially the same conclusion as was reached by the stochastic differential equation method.

b) General Expansion for Small Noise: The general form of these results is the following. We have a system described by the stochastic differential equation

$$dx = a(x)dt + \varepsilon b(x)dW(t)\,. \tag{7.1.13}$$

Then we can write the solution as

$$x_\varepsilon(t) = x_0(t) + \varepsilon x_1(t) + \varepsilon^2 x_2(t) + \cdots \tag{7.1.14}$$

and solve successively for the $x_n(t)$. In particular, $x_0(t)$ is the solution of the deterministic equation

$$dx = a(x)dt. \tag{7.1.15}$$

Alternatively, we consider the Fokker-Planck equation

$$\partial_t p = -\partial_x[a(x)p] + \tfrac{1}{2}\varepsilon^2\partial_x^2[b(x)^2 p]\,. \tag{7.1.16}$$

Then by changing the variable to the *scaled variable* and thus writing

$$y = [x - x_0(t)]/\varepsilon\,, \tag{7.1.17}$$

$$\bar{p}_\varepsilon(y,t) = \varepsilon p(x,t\,|\,c,0)\,, \tag{7.1.18}$$

we can write the perturbation expansion

$$\bar{p}_\varepsilon(y,t) = \bar{p}_0(y,t) + \varepsilon\bar{p}_1(y,t) + \varepsilon^2\bar{p}_2(y,t) + \ldots. \tag{7.1.19}$$

Here we will find that $\bar{p}_0(y,t)$ is indeed a genuine probability density, i.e., is positive and normalised, while the higher-order terms are negative in some regions. Thus, it can be said that the Fokker-Planck perturbation theory is not probabilistic. In contrast, the stochastic differential equation theory expands in a series of random variables $x_n(t)$, each of which has its own probability distribution. At every stage, the system is probabilistic.

And finally, the most noticeable difference. The first term in the stochastic differential equation perturbation theory is $x_0(t)$ which is the solution of the stochastic differential equation obtained by setting $\varepsilon = 0$ in (7.1.1). In contrast, the first term $\bar{p}_0(y,t)$ in (7.1.19) is *not* the solution of the equation obtained by setting $\varepsilon = 0$ in (7.1.2). In general, it is a limiting form of a Fokker-Planck equation for $\bar{p}_\varepsilon(y,t)$ obtained by setting $\varepsilon = 0$ in the Fokker-Planck equation for the scaled variable, y.

7.2 Small Noise Expansions for Stochastic Differential Equations

We consider a stochastic differential equation of the form

$$dx = a(x)\,dt + \varepsilon b(x)\,dW(t)\,, \tag{7.2.1}$$

in which ε is a small parameter. At this stage we exclude time dependence from $a(x)$ and $b(x)$, which is only necessary in order to simplify the algebra. The results and methods are exactly the same if there is a time dependence. We then assume that the solution $x(t)$ of (7.2.1) can be written

$$x(t) = x_0(t) + \varepsilon x_1(t) + \varepsilon^2 x_2(t) + \dots . \tag{7.2.2}$$

We also assume that we can write

$$a(x) = a(x_0 + \varepsilon x_1 + \varepsilon^2 x_2 + \dots) \tag{7.2.3}$$

$$= a_0(x_0) + \varepsilon a_1(x_0, x_1) + \varepsilon^2 a_2(x_0, x_1, x_2) + \dots . \tag{7.2.4}$$

The particular functional dependence in (7.2.4) is important and is easy to demonstrate, for

$$\begin{aligned} a(x) &= a\Big(x_0 + \sum_{m=1}^{\infty} \varepsilon^m x_m\Big), \\ &= a(x_0) + \sum_{p=1}^{\infty} \frac{1}{p!} \frac{d^p a(x_0)}{dx_0^p} \Big(\sum_{m=1}^{\infty} \varepsilon^m x_m \Big)^p . \end{aligned} \tag{7.2.5}$$

Formally resumming is not easy, but it is straightforward to compute the first few powers of ε and to obtain

$$\left.\begin{aligned} a_0(x_0) &= a(x_0), \\ a_1(x_0, x_1) &= x_1 \frac{da(x_0)}{dx_0}, \\ a_2(x_0, x_1, x_2) &= x_2 \frac{da(x_0)}{dx_0} + \tfrac{1}{2} x_1^2 \frac{d^2 a(x_0)}{dx_0^2}, \\ a_3(x_0, x_1, x_2, x_3) &= x_3 \frac{da(x_0)}{dx_0} + x_1 x_2 \frac{d^2 a(x_0)}{dx_0^2} + \frac{1}{6} x_1^3 \frac{d^3 a(x_0)}{dx_0^3} . \end{aligned}\right\} \tag{7.2.6}$$

Although it is not easy to write explicitly the full set of terms in general, it is easy to see that we can always write for $n \geqslant 1$,

$$a_n(x_0, x_1, \dots x_n) = x_n \frac{da(x_0)}{dx_0} + A_n(x_0, x_1, \dots x_{n-1}), \tag{7.2.7}$$

where it is seen that A_n is independent of x_n. In fact, it is easy to see directly from (7.2.5) that the coefficient of ε^n can only involve x_n if this is contributed from the term $p = 1$, and the only other possibility for ε^n to arise is from terms with $m < n$, which do not involve x_n.

The form (7.2.7) for a_n is very important. It is clear that if we also require

$$b(x) = b_0(x_0) + \varepsilon b_1(x_0, x_1) + \varepsilon^2 b_2(x_0, x_1, x_2) + \dots, \tag{7.2.8}$$

then all the same conclusions apply. However, they are not so important in the perturbation expansion.

We now substitute the expansions (7.2.2, 7.2.7, 7.2.8) in the stochastic differential equation (7.2.1) and equate coefficients of like powers of ε. We then obtain an infinite set of stochastic differential equation's. In these we use the notation

$$k(x_0) = -\frac{da(x_0)}{dx_0}\,, \tag{7.2.9}$$

for simplicity.

We obtain

$$dx_0 = a(x_0)\,dt\,, \tag{7.2.10a}$$

$$dx_n = [-k(x_0)x_n + A_n(x_0, \dots x_{n-1})]\,dt + b_{n-1}(x_0, \dots x_{n-1})\,dW(t). \tag{7.2.10b}$$

Solutions: These equations can now be solved sequentially. Equation (7.2.10a) is a (possibly non-linear) ordinary differential equation whose solution is assumed to be known, subject to an initial condition. It is of course possible to assume independent nontrivial initial conditions for all the x_n, but this is unnecessarily complicated. It is simplest to write (setting $t = 0$ as the initial time)

$$\left.\begin{aligned} x_0(0) &= x(0)\,, \\ x_n(0) &= 0\,, \qquad n \geqslant 1\,. \end{aligned}\right\} \tag{7.2.11}$$

Assuming that the solution of (7.2.9) is given by

$$x_0(t) = \alpha(t)\,, \tag{7.2.12}$$

the equation (7.2.10a) for $n = 1$ can be written as

$$dx_1 = -k[\alpha(t)]x_1\,dt + b[\alpha(t)]\,dW(t)\,, \tag{7.2.13}$$

where we have noted from (7.2.5) that A_0 vanishes and $b_0 = b$.

This equation, the first of a perturbation theory, is a time-dependent Ornstein-Uhlenbeck process whose solution can be obtained straightforwardly by the methods of Sect. 4.5.9, reduced to one dimension. The solution is obtained simply by multiplying by the integrating factor

$$\exp\{\textstyle\int dt'\,k[\alpha(t')]\}\,, \tag{7.2.14}$$

and is

$$x_1(t) = \int_0^t b[\alpha(t')]\exp\left\{-\int_{t'}^t k[\alpha(s)\,ds\right\}\,dW(t')\,, \tag{7.2.15}$$

where the initial condition $x_1(0) = 0$ has been included.

For many purposes this form is quite adequate and amounts to a linearisation of the original equation about the deterministic solution. Higher-order terms are more complex because of the more complicated form of (7.2.10b) but are, in essence, treated in exactly the same way. In order to solve the equation for $x_N(t)$, we assume we know all the $x_n(t)$ for $n < N$ so that A_n and b_{n-1} become known (stochastic) functions of t after substituting these solutions. Then (7.2.10b) becomes

$$dx_N = \{-k[\alpha(t)]x_N + A_N(t)\}dt + b_{N-1}(t)dW(t) \tag{7.2.16}$$

whose solution is obtained directly, or from Sect. 4.5.9, as

$$x_N(t) = \int_0^t [A_N(t')dt' + b_{N-1}(t')dW(t')]\exp\left\{-\int_{t'}^t k[\alpha(s)]ds\right\}\,. \tag{7.2.17}$$

Formally, the procedure is now complete. The range of validity of the method and its practicability are yet unanswered. Like all perturbation theories, terms rapidly become unwieldy with increasing order.

7.2.1 Validity of the Expansion

The expansion will not normally be a convergent power series. For (7.2.15) shows that $x_1(t)$ is a Gaussian variable, being simply an Ito integral with nonstochastic coefficients, and hence $x_1(t)$ can with finite probability assume a value greater than any fixed value. Thus, only if all power series involved in the derivation of (7.2.5) are convergent for all arguments, no matter how large, can we expect the method to yield a convergent expansion.

We can, in fact, show that the expansion is *asymptotic* by using the results on dependence on a parameter given in Sect. 4.3.2. We define a remainder by

$$y_n(\varepsilon, t) = \left[x(t) - \sum_{r=0}^{n} \varepsilon^r x_r(t)\right] / \varepsilon^{n+1}, \tag{7.2.18}$$

where the $x_r(t)$ are solutions of the set of stochastic differential equations (7.2.10a, 7.2.10b) with initial conditions (7.2.11).

We then derive an equation for $y_n(t)$. We can write

$$a[x(t)] = a\left[\sum_{r=0}^{n} \varepsilon^r x_r(t) + y_n(\varepsilon, t)\varepsilon^{n+1}\right], \tag{7.2.19}$$

and we define a function $\hat{a}_{n+1}[X_0, X_1, X_2 \ldots X_n, Y, \varepsilon]$ by

$$\begin{aligned} &\hat{a}_{n+1}[X_0, X_1, \ldots X, Y, \varepsilon] \\ &\quad = \varepsilon^{-n-1}\left\{a\left[\sum_{r=0}^{n} \varepsilon^r X_r + \varepsilon^{n+1} Y\right] - \sum_{r=0}^{n} \varepsilon^r a_r(X_0, X_1, \ldots X_r)\right\}. \end{aligned} \tag{7.2.20}$$

We require that for all fixed $X_0, X_1, \ldots X_N, Y$,

$$\lim_{\varepsilon \to 0} \hat{a}_{n+1}[X_0, X_1, \ldots X_N, Y, \varepsilon], \tag{7.2.21}$$

exists. We similarly define $\hat{b}_n[X_0, X_1, \ldots X_n, Y, \varepsilon]$ and impose the same condition on it.

This condition is not probabilistic, but expresses required analytic properties of the functions $a(x)$ and $b(x)$; it requires, in fact, that the expansions (7.2.4, 7.2.8) be merely *asymptotic* expansions.

Now we can write the differential equation for $y_n(\varepsilon, t)$ as

$$\begin{aligned} dy_n &= \hat{a}_{n+1}[x_0(t), x_1(t), \ldots x_n(t), y_n, \varepsilon]\, dt \\ &\quad + \hat{b}_n[x_0(t), x_1(t), \ldots x_{n-1}(t), y_n, \varepsilon]\, dW(t). \end{aligned} \tag{7.2.22}$$

The coefficients of dt and $dW(t)$ are now *stochastic* functions because the $x_r(t)$ are stochastic. However, the requirement (7.2.21) is now an almost certain limit, and hence implies the existence of the stochastic limits

$$\text{st-}\lim_{\varepsilon\to 0} \hat{a}_{n+1}[x_0(t), x_1(t), \dots x_n(t), y_n, \varepsilon] \equiv \tilde{a}_{n+1}(t, y_n), \tag{7.2.23}$$

$$\text{st-}\lim_{\varepsilon\to 0} \hat{b}_n[x_0(t), x_1(t), \dots x_{n-1}(t), y_n, \varepsilon] \equiv \tilde{b}_n(t, y_n), \tag{7.2.24}$$

which is sufficient to satisfy the result of Sect. 4.3.2 on the continuity of solutions of the stochastic differential equation (7.2.22) with respect to the parameter ε, provided the appropriate Lipschitz conditions (ii) and (iii) of Sect. 4.3.2 are satisfied. Thus, $y_n(0,t)$ exists as a solution of the stochastic differential equation

$$dy_n(0,t) = \tilde{a}_{n+1}[t, y_n(0,t)] + \tilde{b}_n[t, y_n(0,t)]\, dW(t), \tag{7.2.25}$$

which, from the definition (7.2.18) shows that

$$x(t) - \sum_{r=0}^{t} \varepsilon^r x_r(t) \sim \varepsilon^{n+1}. \tag{7.2.26}$$

Hence, the expansion in power of ε is an *asymptotic expansion.*

7.2.2 Stationary Solutions (Homogeneous Processes)

A stationary solution is obtained by letting $t \to \infty$. If the process is, as written, homogeneous and ergodic, it does not matter what the initial condition is. In this case, one chooses $x_0(0)$ so that $a[x_0(0)]$ vanishes and the solution to (7.2.10a) is

$$x_0(t) = x_0(0) \equiv \alpha, \tag{7.2.27}$$

[where we write simply α instead of $\alpha(t)$].

Because of the initial condition at $t = 0$ the solution (7.2.15) to the equation of order one is not a stationary process. One must either let $t \to \infty$ or set the initial condition not at $t = 0$, but at $t = -\infty$. Choosing the latter, we have

$$x_1^s(t) = \int_{-\infty}^{t} b(\alpha) \exp[-(t-t')k(\alpha)] dW(t'). \tag{7.2.28}$$

Similarly,

$$x_n^s(t) = \int_{-\infty}^{t} [A_n^s(t')dt' + b_{n-1}^s(t')dW(t')] \exp[-(t-t')k(\alpha)], \tag{7.2.29}$$

where by A_n^s and b_{n-1}^s we mean the values of A_n and b_{n-1} obtained by inserting the stationary values of all arguments. From (7.2.29) it is clear that $x_n^s(t)$ is, by construction, stationary. Clearly the integrals in (7.2.28, 7.2.29) converge only if $k(\alpha) > 0$, which will mean that only a *stable* stationary solution of the deterministic process generates a stationary solution by this method. This is rather obvious—the addition of fluctuations to an unstable state derives the system away from that state.

7.2.3 Mean, Variance, and Time Correlation Function

If the series expansion in ε is valid in some sense, it is useful to know the expansion for mean and variance. Clearly

$$\langle x(t)\rangle = \sum_{n=0}^{\infty} \varepsilon^n \langle x_n(t)\rangle\,, \tag{7.2.30}$$

$$\mathrm{var}\{x(t)\} = \sum_{n=0}^{\infty} \varepsilon^n \sum_{m=0}^{\infty} \langle x_m(t)x_{n-m}(t)\rangle - \langle x_m(t)\rangle\langle x_{n-m}(t)\rangle]\,. \tag{7.2.31}$$

Since, however, we assume a deterministic initial condition and $x_0(t)$ is hence deterministic, all terms involving $x_0(t)$ vanish. We can then work out that

$$\begin{aligned}\mathrm{var}\{x(t)\} &= \varepsilon^2 \mathrm{var}\{x_1(t)\} + 2\varepsilon^3 \langle x_1(t), x_2(t)\rangle \\ &\quad + \varepsilon^4 [2\langle x_1(t), x_3(t)\rangle + \mathrm{var}\{x_2(t)\}] + \dots\,,\end{aligned} \tag{7.2.32}$$

and similarly,

$$\begin{aligned}\langle x(t), x(s)\rangle &= \varepsilon^2 \langle x_1(t), x_1(s)\rangle + \varepsilon^3 [\langle x_1(t), x_2(s)\rangle + \langle x_1(s), x_2(t)\rangle] \\ &\quad + \varepsilon^4 [\langle x_1(t), x_3(s)\rangle + \langle x_1(s), x_3(t)\rangle + \langle x_2(t), x_2(s)\rangle] + \dots\,.\end{aligned} \tag{7.2.33}$$

7.2.4 Failure of Small Noise Perturbation Theories

a) Example—Cubic Process and Related Behaviour: Consider the stochastic differential equation

$$dx = -x^3 dt + \varepsilon dW(t)\,. \tag{7.2.34}$$

It is not difficult to see that the expansion conditions (7.2.21) are trivially satisfied for the coefficients of both dt and $dW(t)$, and in fact for any *finite* t, an asymptotic expansion with terms $x_N(t)$ given by (7.2.17) is valid. However, at $x = 0$, it is clear that

$$\left.\frac{d}{dx}(x^3)\right|_{x=0} = k(0) = 0\,, \tag{7.2.35}$$

and because $x = 0$ is the stationary solution of the deterministic equation, the perturbation series for stationary solutions is not likely to converge since the exponential time factors are all constant. For example, the first-order term in the stationary expansion is, from (7.2.28),

$$x_1^s(t) = \int_{-\infty}^{t} dW(t') = W(t) - W(-\infty)\,, \tag{7.2.36}$$

which is infinite with probability one (being a Gaussian variable with infinite variance).

The problem is rather obvious. Near $x = 0$, the motion described by (7.2.34) is simply not able to be approximated by an Ornstein-Uhlenbeck process. For example, the stationary probability distribution, which is the stationary solution of the Fokker-Planck equation

$$\partial_t p = \partial_x (x^3 p) + \tfrac{1}{2}\varepsilon^2 \partial_x^2 p\,, \tag{7.2.37}$$

is given by

$$p_s(x) = \mathcal{N} \exp(-x^4/2\varepsilon^2)\,, \tag{7.2.38}$$

and the moments are

$$\langle x^n \rangle = \begin{cases} (2\varepsilon^2)^{n/4}\, \Gamma\left(\tfrac{1}{4}(n+1)\right) \big/ \Gamma\left(\tfrac{1}{4}\right), & n \text{ even}, \\ 0, & n \text{ odd}. \end{cases} \tag{7.2.39}$$

The lowest-order term of the expansion of the variance is proportional to ε to the first power, not ε^2 as in (7.2.32).

In this case, we must simply regard the *cubic process* described by (7.2.34) as a fundamental process. If we introduce the new scaled variables through the definitions

$$x = \sqrt{\varepsilon}\, y \qquad t = \tau/\varepsilon, \tag{7.2.40}$$

and use

$$dW(\tau/\varepsilon) = dW(\tau)/\sqrt{\varepsilon}, \tag{7.2.41}$$

then the cubic process can be reduced to a parameterless form

$$dy = -y^3 d\tau + dW(\tau). \tag{7.2.42}$$

Regarding the solution of (7.2.42) as a known quantity, we can write

$$x(t) = \sqrt{\varepsilon}\, y(\varepsilon t), \tag{7.2.43}$$

so that the limit $\varepsilon \to 0$ is approached like $\sqrt{\varepsilon}$, and also with a slower time scale. This kind of scaling result is the basis of many critical phenomena.

A successful perturbation theory in the case where $a(x)$ behaves like x^3 near $x = 0$ must involve firstly the change of variables (7.2.40) then a similar kind of perturbation theory to that already outlined—but in which the zero-order solution is the cubic process. Thus, let us assume that we can write

$$a(x) = -x^3 c(x), \tag{7.2.44}$$

where $c(x)$ is a smooth function with $c(0) \neq 0$. Then, using the transformations (7.2.40, 7.2.41), we can rewrite the stochastic differential equation as

$$dy = -y^3 c\left(y\sqrt{\varepsilon}\right) d\tau + b\left(y\sqrt{\varepsilon}\right) dW(\tau). \tag{7.2.45}$$

If we expand $y(t)$, $c\left(y\sqrt{\varepsilon}\right)$, $b\left(y\sqrt{\varepsilon}\right)$ as series in $\sqrt{\varepsilon}$, we obtain a perturbation theory. By writing

$$y(t) = \sum_{n=0}^{\infty} \varepsilon^{n/2} y_n(t), \tag{7.2.46}$$

we get for the first two terms

$$dy_0 = -y_0^3 c(0)\, d\tau + b(0)\, dW(\tau), \tag{7.2.47}$$

$$dy_1 = -y_1 \left[3y_0^2 c(0) + y_0^3 \frac{dc}{dx}(0)\right] d\tau + \left[\frac{db}{dx}(0) y_0\right] dW(\tau). \tag{7.2.48}$$

We see that the equation for y_1 is in fact that of a time-dependent Ornstein-Uhlenbeck process with stochastic coefficients. Thus, *in principle*, as long as the cubic process is known, the rest is easily computed.

b) Order ν Processes (ν odd): If instead we have the Stochastic differential equation

$$dx = -x^{\nu}dt + \varepsilon\, dW(t)\,, \tag{7.2.49}$$

we find a stable deterministic state only if ν is an odd integer. In this case we make the transformation to scaled variables

$$x = y\varepsilon^{2/(1+\nu)}\,, \qquad t = \tau\,\varepsilon^{2(1-\nu)/(1+\nu)}\,, \tag{7.2.50}$$

and follow very similar procedures to those used for the cubic process.

c) Bistable Systems: Suppose we have a system described by

$$dx = (x - x^3)dt + \varepsilon\, dW(t)\,, \tag{7.2.51}$$

for which there are three deterministic stationary states, at $x = 0, \pm 1$. The state at $x = 0$ is deterministically unstable, and we can see directly from the perturbation theory that no stationary process arises from it, since the exponentials in the perturbation series integrals (7.2.28, 7.2.29) have increasing arguments.

The solutions $x_0(t)$ of the deterministic differential equation

$$dx/dt = x - x^3\,, \tag{7.2.52}$$

divide into three classes depending on their behaviour as $t \to \infty$. Namely,

$$\left.\begin{array}{lll} \text{i)}\ x_0(0) < 0 \ \Rightarrow\ x_0(t) \to -1\,, & & \\ \text{ii)}\ x_0(0) = 0 \ \Rightarrow\ x_0(t) = \ \ 0 & \text{for all } t\,, & \\ \text{iii)}\ x_0(0) > 0 \ \Rightarrow\ x_0(t) \to \ \ 1\,. & & \end{array}\right\} \tag{7.2.53}$$

Thus, depending on the initial condition, we get two different asymptotic expansions, whose stationary limits represent the fluctuations about the two deterministic stationary states. There is no information in these solutions about the possible jump from the branch $x = 1$ to the branch $x = -1$, or conversely – at least not in any obvious form. In this sense the asymptotic expansion fails, since it does not give a picture of the overall behaviour of the stationary state. We will see in Chap. 14 that this results because an asymptotic expansion of behaviour characteristic of jumps from one branch to the other is typically of the order of magnitude of $\exp(-1/\varepsilon^2)$, which approaches zero faster than any power as $\varepsilon \to 0$, and thus is not represented in an expansion in powers of ε.

7.3 Small Noise Expansion of the Fokker-Planck Equation

As mentioned in Sect. 7.1, a small noise expansion of a Fokker-Planck equation is a singular expansion involving the introduction of scaled variables. Let us consider how this is done.

We consider the Fokker-Planck equation

$$\partial_t p = -\partial_x[A(x)p] + \tfrac{1}{2}\varepsilon^2\partial_x^2[B(x)p]\,. \tag{7.3.1}$$

We assume the solution of the deterministic equation to be $\alpha(t)$ so that

$$d_t\alpha(t) = A[\alpha(t)], \tag{7.3.2}$$

and introduce new variables (y, s) by

$$y = [x - \alpha(t)]/\varepsilon, \tag{7.3.3}$$

$$s = t, \tag{7.3.4}$$

and we set

$$\hat{p}(y, s) = \varepsilon p(x, t). \tag{7.3.5}$$

We note that

$$\partial_t \hat{p}(y, s) = \frac{\partial \hat{p}}{\partial y}\frac{\partial y}{\partial t} + \frac{\partial \hat{p}}{\partial s}\frac{\partial s}{\partial t} = -\frac{\dot{\alpha}(t)}{\varepsilon}\frac{\partial \hat{p}}{\partial y} + \frac{\partial \hat{p}}{\partial s}, \tag{7.3.6}$$

$$\partial_x \hat{p}(y, s) = \frac{1}{\varepsilon}\frac{\partial \hat{p}}{\partial y}, \tag{7.3.7}$$

so that substituting into (7.3.1) we get, with the help of the equation of motion (7.3.2) for $\alpha(t)$

$$\frac{\partial \hat{p}}{\partial s} = -\frac{\partial}{\partial y}\left\{\frac{A[\alpha(s) + \varepsilon y] - A[\alpha(s)]}{\varepsilon}\hat{p}\right\} + \tfrac{1}{2}\frac{\partial^2}{\partial y^2}\left\{B[\alpha(s) + \varepsilon y]\hat{p}\right\}. \tag{7.3.8}$$

We are now in a position to make an expansion in powers of ε. We assume that A and B have an expansion in powers of ε of the form

$$A[\alpha(s) + \varepsilon y] = \sum_{n=0}^{\infty} \tilde{A}_n(s)\varepsilon^n y^n, \tag{7.3.9}$$

$$B[\alpha(s) + \varepsilon y] = \sum_{n=0}^{\infty} \tilde{B}_n(s)\varepsilon^n y^n, \tag{7.3.10}$$

and expand $\hat{p}$ in powers of ε:

$$\hat{p} = \sum_{n=0}^{\infty} \hat{p}_n \varepsilon^n. \tag{7.3.11}$$

Substituting these expansions into the Fokker-Planck equation (7.3.8), we get by equating coefficients

$$\frac{\partial \hat{p}_0}{\partial s} = -\tilde{A}_1(s)\frac{\partial}{\partial y}(y\hat{p}_0) + \tfrac{1}{2}B_0(s)\frac{\partial^2 \hat{p}_0}{\partial y^2}, \tag{7.3.12}$$

$$\frac{\partial \hat{p}_r}{\partial s} = -\frac{\partial}{\partial y}[\tilde{A}_1(s)y\hat{p}_1 + \tilde{A}_2(s)y^2\hat{p}_0] + \tfrac{1}{2}\frac{\partial^2}{\partial y^2}[\tilde{B}_0(s)\hat{p}_1 + \tilde{B}_1(s)y\hat{p}_0], \tag{7.3.13}$$

and, in general,

$$\frac{\partial \hat{p}_r}{\partial s} = -\frac{\partial}{\partial y}\left[\sum_{m=0}^{r} y^{r-m+1}\tilde{A}_{r-m+1}(s)\hat{p}_m\right] + \tfrac{1}{2}\frac{\partial^2}{\partial y^2}\left[\sum_{m=0}^{r} y^{r-m}\tilde{B}_{r-m}(s)\hat{p}_m\right]. \tag{7.3.14}$$

Only the equation for $\hat{p}_0$ is a Fokker-Planck equation and, as mentioned in Sect. 7.1, only $\hat{p}_0$ is a probability. The first equation in the hierarchy, (7.3.12), is a time-dependent Ornstein-Uhlenbeck process which corresponds exactly to (7.2.13), the first equation in the hierarchy for the stochastic differential equation. Thereafter the correspondence ceases.

The boundary conditions on the $\hat{p}_r$ do present technical difficulties since the transformation from x to y is time dependent, and a boundary at a fixed position in the x variable corresponds to a moving boundary in the y variable. Further, a boundary at $x = a$ corresponds to one at

$$y = \frac{[a - \alpha(s)]}{\varepsilon}, \tag{7.3.15}$$

which approaches $\pm\infty$ as $\varepsilon \to 0$. There does not seem to be any known technique of treating such boundaries, except when $a = \pm\infty$, so that the y boundary is also at $\pm\infty$ and hence constant. Boundary conditions then assume the same form in the y variable as in the x variable.

In the case where the boundaries are at infinity, the result of the transformation (7.3.3) is to change a singular perturbation problem (7.3.1), in which the limit $\varepsilon \to 0$ yields an equation of lower order, into an ordinary perturbation problem (7.3.8) in which the coefficients of the equation depend smoothly on ε, and the limit $\varepsilon \to 0$ is an equation of second order. The validity of the expansion method will depend on the form of the coefficients.

7.3.1 Equations for Moments and Autocorrelation Functions

The hierarchy (7.3.14) is not very tractable, but yields a relatively straightforward procedure for computing the moments perturbatively. We assume that the boundaries are at $\pm\infty$ so that we can integrate by parts and discard surface terms. Then we define

$$\langle [y(t)]^n \rangle = \sum_{r=0}^{\infty} \varepsilon^r M_r^n(t). \tag{7.3.16}$$

Then clearly

$$M_r^n(t) = \int dy\, y^n \hat{p}_r(y, t), \tag{7.3.17}$$

and using (7.3.12–7.3.14), we easily derive by integrating by parts

$$\frac{dM_r^n(t)}{dt} = \sum_{m=0}^{r} \left[n\tilde{A}_{r-m+1}(t) M_m^{n+r-m}(t) + \frac{n(n-1)}{2} \tilde{B}_{r-m}(t) M_m^{n+r-m-2}(t) \right], \tag{7.3.18}$$

which is a closed hierarchy of equations, since the equation for $M_r^n(t)$ can be solved if all $M_m^p(t)$ are known for $m < r$ and $p < n + r$.

Writing out the first few for the mean $M_r^1(t)$ and mean square $M_r^2(t)$, we find

$$\frac{dM_0^1(t)}{dt} = \tilde{A}_1(t) M_0^1(t), \tag{7.3.19}$$

$$\frac{dM_1^1(t)}{dt} = \tilde{A}_1(t) M_1^1(t) + \tilde{A}_2(t) M_0^2(t), \tag{7.3.20}$$

$$\frac{dM_2^1(t)}{dt} = \tilde{A}_1(t) M_2^1(t) + \tilde{A}_2(t) M_1^2(t) + \tilde{A}_3(t) M_0^3(t), \tag{7.3.21}$$

$$\frac{dM_0^2(t)}{dt} = 2\tilde{A}_1(t)M_0^2(t) + \tilde{B}_0(t)\,, \tag{7.3.22}$$

$$\frac{dM_1^2(t)}{dt} = 2\tilde{A}_1(t)M_1^2(t) + 2\tilde{A}_2(t)M_0^3(t) + \tilde{B}_1(t)M_0^1(t)\,, \tag{7.3.23}$$

$$\frac{dM_0^3(t)}{dt} = 3\tilde{A}_1(t)M_0^3(t) + 3\tilde{B}_0(t)M_0^1(t)\,. \tag{7.3.24}$$

In deriving the last two equations we note that

$$M_r^0(t) = \int dy\, \hat{p}_r(y,t)\,, \tag{7.3.25}$$

and using

$$\int dy\, \hat{p}(y,t) = 1 = \sum_r \varepsilon^r M_r^0(t)\,, \tag{7.3.26}$$

we see that

$$M_0^0(t) = 1\,, \tag{7.3.27}$$

$$M_r^0(t) = 0\,, \qquad r \neq 1\,. \tag{7.3.28}$$

The equations are linear ordinary differential equations with inhomogeneities that are computed from lower equations in the hierarchy.

a) Stationary Moments: These are obtained by letting $t \to \infty$ and setting the left-hand side of (7.3.18) equal to zero. (All coefficients, $\tilde{A}, \tilde{B}$, etc. are taken time independent.)

From (7.3.19–7.3.22) we find

$$\left.\begin{aligned}
&M_0^1(\infty) = 0\,,\\
&M_0^2(\infty) = -\tfrac{1}{2}\tilde{B}_0/\tilde{A}_1\,,\\
&M_0^3(\infty) = 0\,,\\
&M_1^1(\infty) = -\tilde{A}_2 M_0^2(\infty)/\tilde{A}_1 = \tfrac{1}{2}\,\tilde{A}_2\tilde{B}_0/(\tilde{A}_1)^2\,,\\
&M_1^2(\infty) = -\tilde{A}_2 M_0^3(\infty)/\tilde{A}_1 = 0\,,\\
&M_2^1(\infty) = -[\tilde{A}_2 M_1^2(\infty) + \tilde{A}_3 M_0^3(\infty)]/\tilde{A}_1 = 0\,.
\end{aligned}\right\} \tag{7.3.29}$$

Thus, the stationary mean, variance, etc., are

$$\begin{aligned}
\langle x\rangle_s &= \alpha + \varepsilon[M_0^1(\infty) + \varepsilon M_1^1(\infty) + \varepsilon^2 M_2^1(\infty)]\,,\\
&= \alpha + \tfrac{1}{2}\varepsilon^2 \tilde{A}_2\tilde{B}_0/(\tilde{A}_1)^2\,,
\end{aligned} \tag{7.3.30}$$

$$\begin{aligned}
\mathrm{var}\{x\}_s &= \langle x^2\rangle_s - \langle x_s\rangle^2\,,\\
&= \langle(\alpha+\varepsilon y)^2\rangle_s - \langle \alpha + \varepsilon y\rangle_s^2 = \varepsilon^2 \mathrm{var}\{y\}_s\,,\\
&= -\frac{\varepsilon^2}{2}\tilde{B}_0/\tilde{A}_1 \qquad \text{to order } \varepsilon^2\,.
\end{aligned} \tag{7.3.31}$$

The procedure can clearly be carried on to arbitrarily high order. Of course in a one-variable system, the stationary distribution can be evaluated exactly and the moments found by integration. But in many variable systems this is not always possible, whereas the multivariate extension of this method is always able to be carried out.

b) Stationary Autocorrelation Function: The autocorrelation function of x is simply related to that of y in a stationary state by

$$\langle x(t)x(0)\rangle_s = \alpha^2 + \varepsilon^2\langle y(t)y(0)\rangle_s\,, \tag{7.3.32}$$

and a hierarchy of equations for $\langle y(t)y(0)\rangle$ is easily developed. Notice that

$$\frac{d}{dt}\langle y(t)^n y(0)\rangle_s = \left\langle\left\{\frac{A[\alpha+\varepsilon y(t)]-A(\alpha)}{\varepsilon}ny(t)^{n-1} + \tfrac{1}{2}n(n-1)B[\alpha+\varepsilon y(t)]y(t)^{n-2}\right\}y(0)\right\rangle_s\,, \tag{7.3.33}$$

which can be derived by using the Fokker-Planck equation (7.3.1) for $p(y,t\,|\,y_0,t_0)$ and integrating by parts, or by using Ito's formula for the corresponding stochastic differential equation.

Using the definition of $\tilde{A}_r$, $\tilde{B}_r$ in (7.3.9, 7.3.10) and expanding A and B in a power series, we get

$$\frac{d}{dt}\langle y(t)^n y(0)\rangle_s = \sum_{q=0}^{\infty}\varepsilon^q\Big[n\tilde{A}_{q+1}\langle y(t)^{q+n}y(0)\rangle_s + \tfrac{1}{2}n(n-1)\tilde{B}_q\langle y(t)^{q+n-2}y(0)\rangle_s\Big]. \tag{7.3.34}$$

These equations themselves form a hierarchy which can be simply solved in a power series in ε. Normally one is most interested in $\langle y(t)y(0\rangle_s$, which can be calculated to order ε^q, provided one knows $\langle y(t)^p y(0)_s\rangle$ for $p \leqslant q+1$. We have the initial condition

$$\langle y(0)^n y(0)\rangle_s = \langle y^{n+1}\rangle_s\,, \tag{7.3.35}$$

and the stationary moments can be evaluated from the stationary distribution, or as has just been described.

7.3.2 Example

We consider the Fokker-Planck equation

$$\frac{\partial p}{\partial t} = \frac{\partial}{\partial x}[(x+x^3)p] + \frac{\varepsilon^2}{6}\frac{\partial^2 p}{\partial x^2}\,, \tag{7.3.36}$$

for which we have [in the stationary state $\alpha(t) = 0$]

$$\left.\begin{aligned}&\tilde{A}_1 = -1\,, \qquad \tilde{A}_2 = 0\,, \qquad \tilde{A}_3 = -1\,,\\ &\tilde{A}_q = 0 \quad (q>3)\,,\ \tilde{B}_q = \tfrac{1}{3}\delta_{q,0}\,,\ \alpha = 0\,.\end{aligned}\right\} \tag{7.3.37}$$

Using (7.3.30, 7.3.31) we have

$$\left.\begin{aligned}\langle x\rangle_s &= 0\,,\\ \mathrm{var}[x]_s &= \varepsilon^2/6\,.\end{aligned}\right\} \tag{7.3.38}$$

For convenience, let us use the notation

$$c_n(t) = \langle y^n(t)y(0)\rangle_s \tag{7.3.39}$$

so that the equations for the c_1 and c_3 are

$$\begin{bmatrix} \dot{c}_1 \\ \dot{c}_3 \end{bmatrix} = \begin{bmatrix} -1 & -\varepsilon^2 \\ 1 & -3 \end{bmatrix} \begin{bmatrix} c_1 \\ c_3 \end{bmatrix} . \tag{7.3.40}$$

The equations for the c_{2n} decouple from those for c_{2n+1} because $B(x)$ is constant and $A(x)$ is an odd function of x.

It is simpler to solve (7.3.40) exactly than to perturb. The eigenvalues of the matrix are

$$\left.\begin{aligned} \lambda_1 &= -2 + \sqrt{1-\varepsilon^2} , \\ \lambda_2 &= -2 - \sqrt{1-\varepsilon^2} , \end{aligned}\right\} \tag{7.3.41}$$

and the corresponding eigenvectors

$$\left.\begin{aligned} \boldsymbol{u}_1 &= \begin{bmatrix} 1+ \sqrt{1-\varepsilon^2} \\ 1 \end{bmatrix} , \\ \boldsymbol{u}_2 &= \begin{bmatrix} 1- \sqrt{1-\varepsilon^2} , \\ 1 \end{bmatrix} . \end{aligned}\right\} \tag{7.3.42}$$

The solution of (7.3.40) can then be written

$$\begin{bmatrix} c_1(t) \\ c_3(t) \end{bmatrix} = \alpha_1 \mathrm{e}^{-\lambda_1 t} \boldsymbol{u}_1 + \alpha_2 \mathrm{e}^{-\lambda_2 t} \boldsymbol{u}_2 , \qquad (t > 0) . \tag{7.3.43}$$

The initial condition is

$$\begin{aligned} c_1(0) &= \langle y^2 \rangle_\mathrm{s} , \\ c_3(0) &= \langle y^4 \rangle_\mathrm{s} . \end{aligned} \tag{7.3.44}$$

We can compute $\langle y^4 \rangle_\mathrm{s}$ using the moment hierarchy (7.3.10) extended to M_0^4: we find

$$M_0^4 = -\frac{3}{2} \frac{\tilde{B}_0 M_0^2}{\tilde{A}_1} = \frac{1}{12} , \tag{7.3.45}$$

then $\langle y^4 \rangle = 1/12$.

Hence, we obtain

$$\begin{bmatrix} 1/6 \\ 1/12 \end{bmatrix} = \alpha_1 \boldsymbol{u}_1 + \alpha_2 \boldsymbol{u}_2 . \tag{7.3.46}$$

which have the solutions

$$\begin{aligned} \alpha_1 &= \frac{1}{24} (1 + \sqrt{1-\varepsilon^2}) / \sqrt{1-\varepsilon^2} , \\ \alpha_2 &= \frac{1}{24} (-1 + \sqrt{1-\varepsilon^2}) / \sqrt{1-\varepsilon^2} . \end{aligned} \tag{7.3.47}$$

The correlation function is, to 2nd order in ε (many terms cancel)

$$c_1(t) = \frac{1}{6} \mathrm{e}^{-|\lambda_1| t} . \tag{7.3.48}$$

Notice that the eigenvalues λ_1 and λ_2 depend on ε^2. Any attempt to solve the system (7.3.40) perturbatively would involve expanding $\exp(\lambda_1 t)$ and $\exp(\lambda_2 t)$ in powers of ε^2 and would yield terms like $t^N \exp(-2t)$ which would not be an accurate representation of the long-time behaviour of the autocorrelation function.

The x correlation function is

$$\langle x(t)x(0)\rangle_s = \varepsilon^2 c_1(t)\,, \tag{7.3.49}$$

and the spectrum

$$S(\omega) = \frac{\varepsilon^2}{2\pi} \int_{-\infty}^{\infty} dt\, \mathrm{e}^{-i\omega t} c_1(t)\,, \tag{7.3.50}$$

$$= \frac{\varepsilon^2}{12\pi\left(\lambda_1^2 + \omega^2\right)}\,. \tag{7.3.51}$$

7.3.3 Asymptotic Method for Stationary Distribution

For an arbitrary Fokker-Planck equation

$$\partial_t p = -\sum_i \partial_i A_i(\boldsymbol{x})p + \tfrac{1}{2}\varepsilon^2 \sum_{i,j} \partial_i \partial_j B_{ij}(\boldsymbol{x})p\,, \tag{7.3.52}$$

one can generate an asymptotic expansion for the stationary solution by setting

$$p_s(\boldsymbol{x}) = \exp[-\phi(\boldsymbol{x})/\varepsilon^2]\,, \tag{7.3.53}$$

in terms of which we find

$$\sum_i A_i(\boldsymbol{x})\partial_i\phi + \tfrac{1}{2}\sum_{i,j} B_{ij}(\boldsymbol{x})\partial_i\phi\partial_j\phi$$
$$+\,\varepsilon^2\left[-\sum_i \partial_i A_i(\boldsymbol{x}) + \sum_{ij}\partial_i B_{ij}\partial_j\phi + \sum_{i,j}\partial_i\partial_i B_{ij}(\boldsymbol{x})\right] = 0\,. \tag{7.3.54}$$

The first term, which is of order ε^0, is a Hamilton Jacobi equation. The main significance of the result is that an asymptotic expansion for $\phi(\boldsymbol{x})$ can be, in principle, developed:

$$\phi(\boldsymbol{x}) = \sum_n \varepsilon^n \phi_n(\boldsymbol{x})\,, \tag{7.3.55}$$

where $\phi_0(x)$ satisfies

$$\sum_i A_i(\boldsymbol{x})\partial_i\phi_0 + \tfrac{1}{2}\sum_{i,j} B_{ij}(\boldsymbol{x})\partial_i\phi_0\partial_j\phi_0 = 0\,, \tag{7.3.56}$$

Graham and *Tel* [7.1, 7.2] have shown how equation (7.3.56) may be solved in the general case. Their main result is that solutions, though continuous, in general have infinitely many discontinuities in their derivatives, except in certain special cases, which are closely related to the situation in which the Fokker-Planck equation satisfies potential conditions.

8. The White Noise Limit

There is a wide range of situations in theoretical science in which a well-defined separation of time scales arises, that is, a class of variables varies on a time scale which is characteristically more rapid than the time scale of the remaining variables. When there is dissipation, it is possible for the fast variable to relax to a quasistationary state, in which the values of the fast variable follow those of the slow variables. The subject of this chapter is the formulation of the treatment of this problem for stochastic systems, a formulation in which the fast variable may be eliminated form the equations of motion in some well-defined limit. In deterministic systems the procedure is comparatively simple compared to that needed for stochastic systems, and there is a variety of limits available, with the results depending on what limit is taken. The choice of the limit to be taken is not purely mathematical, depending rather on the actual system being modelled.

8.1 White Noise Process as a Limit of Nonwhite Process

The relationship between real noise and white noise has been mentioned previously in Sect. 1.5.4, and in Sect. 4.1 it was discussed how this limit might be implemented. Let us consider a noise $\zeta_\gamma(t)$ which approaches a delta-correlated noise in some well-defined limit. We can implement this by writing

$$\zeta_\gamma(t) = \gamma\alpha(\gamma^2 t)\,, \tag{8.1.1}$$

where $\alpha(t)$ is a stochastic variable with zero mean and some nonzero correlation time; explicitly

$$\langle\alpha(t)\rangle = 0\,, \tag{8.1.2}$$

$$\langle\alpha(t)\alpha(0)\rangle_s = g(t)\,. \tag{8.1.3}$$

We will assume that the correlation time is finite, and normalise the correlation function of $\alpha(t)$ to 1—thus

$$\int_{-\infty}^{\infty} g(t)\,dt = 1\,, \tag{8.1.4}$$

$$\int_{-\infty}^{\infty} |t| g(t)\,dt = \tau_c\,. \tag{8.1.5}$$

Thus, we can say that

$$\lim_{\gamma\to\infty}\langle\zeta_\gamma(t)\zeta_\gamma(t')\rangle = \lim_{\gamma\to\infty}\gamma^2 g\left(\gamma^2(t-t')\right) \to \delta(t-t')\,. \tag{8.1.6}$$

We are interested in the $\gamma \to \infty$ limit of a differential equation in which the noise source is given by $\zeta_\gamma(t')$. We will show that if this differential equation is written

$$\frac{dx}{dt} = a(x) + \gamma b(x)\,\zeta_\gamma(t)\,, \tag{8.1.7}$$

then if $\alpha(t)$ is a *Markov process*, in the limit that $\gamma \to 0$ that $\zeta_\gamma(t)$ becomes a delta correlated process, the differential equation becomes a *Stratonovich stochastic differential equation* with the same coefficients, that is, it becomes

$$(S)\,dx = a(x)\,dt + b(x)\,dW(t)\,, \tag{8.1.8}$$

which is equivalent to the Ito equation

$$dx = \left[a(x) + \tfrac{1}{2}b(x)b'(x)\right]\,dt + b(x)\,dW(t)\,. \tag{8.1.9}$$

8.1.1 Formulation of the Limit

We will give a demonstration in the case where $\alpha(t)$ is a Markov diffusion process. This means that the pair of variables (x, α) is governed by a conditional probability density $p(x, \alpha, t \,|\, x_0, \alpha_0, 0)$, which satisfies a Fokker-Planck equation of the form

$$\frac{\partial p(x, \alpha \,|\, x_0, \alpha_0, 0)}{\partial t} = (\gamma^2 L_1 + \gamma L_2 + L_3)p(x, \alpha \,|\, x_0, \alpha_0, 0)\,, \tag{8.1.10}$$

with

$$L_1 = -\frac{\partial}{\partial \alpha}A(\alpha) + \tfrac{1}{2}\frac{\partial^2}{\partial \alpha^2}B(\alpha)\,, \tag{8.1.11}$$

$$L_2 = -\frac{\partial}{\partial x}b(x)\alpha\,, \tag{8.1.12}$$

$$L_3 = -\frac{\partial}{\partial x}a(x)\,. \tag{8.1.13}$$

a) Definition of a Projector: We now define a projector P on the space of functions of x and α by

$$(Pf)(x, \alpha) = p_s(\alpha) \int d\alpha f(x, \alpha)\,, \tag{8.1.14}$$

where $p_s(\alpha)$ is the stationary distribution function for α, which is a solution of

$$L_1 p_s(\alpha) = 0\,. \tag{8.1.15}$$

The projector satisfies the following conditions, which are essential for the derivation.

$$P^2 = P\,, \tag{8.1.16}$$

$$PL_1 = L_1 P = 0\,, \tag{8.1.17}$$

$$P = \lim_{t\to\infty} \exp(L_1 t)\,, \tag{8.1.18}$$

$$PL_3 = L_3 P\,, \tag{8.1.19}$$

$$PL_2 P = 0\,. \tag{8.1.20}$$

Properties (8.1.16, 8.1.17) are trivial to show. The property (8.1.18) arises because L_1 is a genuine Fokker-Planck operator, and from the fact that $\exp(L_1 t) f(x,\alpha)$ is a solution of the Fokker-Planck equation $\partial_t p = L_1 p$, which therefore approaches the stationary solution as $t \to \infty$.

Finally, (8.1.20) is true because $\langle \alpha \rangle_s$, the stationary mean of α, is zero, and therefore

$$(PL_2Pf)(x,\alpha) = \big(p_s(\alpha) \int d\alpha\big)\Big(-\frac{\partial}{\partial x} b(x)\alpha\Big)\big(p_s(\alpha) \int d\alpha'\big) f(x,\alpha'),$$
$$= -p_s(\alpha)\langle\alpha\rangle_s \frac{\partial}{\partial x} b(x) \int d\alpha' f(x,\alpha') = 0. \tag{8.1.21}$$

b) Application of the Laplace Transform and the Projector: Because the Fokker-Planck equation (8.1.10) is a linear differential equation which is first order in time, a treatment in terms of the Laplace transform is very appropriate, and readily yields a perturbation expansion.

The Laplace transform of any function of time $f(t)$ is defined by

$$\tilde{f}(s) = \int_0^\infty e^{-st} f(t), \tag{8.1.22}$$

and satisfies

$$\int_0^\infty e^{-st} \frac{df}{dt} = s\tilde{f}(s) - f(0). \tag{8.1.23}$$

It may quite readily be defined for operators and abstract vectors.

The Laplace transform of the Fokker-Planck equation (8.1.10) is

$$s\,\tilde{p}(s) - p(0) = (\gamma^2 L_1 + \gamma L_2 + L_3)\tilde{p}(s). \tag{8.1.24}$$

To generate a perturbation method define,

$$\tilde{v}(s) = P\tilde{p}(s), \tag{8.1.25}$$
$$Q = 1 - P, \tag{8.1.26}$$
$$\tilde{w}(s) = Q\tilde{p}(s). \tag{8.1.27}$$

Applying the projector P to (8.1.24) we get

$$s\,\tilde{v}(s) = P(\gamma^2 L_1 + \gamma L_2 + L_3)\tilde{p}(s) + v(0),$$
$$= \gamma P L_2\big(P\tilde{p}(s) + Q\tilde{p}(s)\big) + L_3 P\tilde{p}(s) + v(0),$$
$$= \gamma P L_2 \tilde{w}(s) + L_3\tilde{v}(s) + v(0), \tag{8.1.28}$$

where we have used (8.1.17) (8.1.19) and (8.1.20) in going from the second line to the third.

Similarly, applying the projector Q to (8.1.24) we get

$$s\,\tilde{w}(s) = Q(\gamma^2 L_1 + \gamma L_2 + L_3)\tilde{p}(s) + v(0),$$
$$= \gamma^2 L_1 Q\tilde{p}(s) + \gamma Q L_2\big(P\tilde{p}(s) + Q\tilde{p}(s)\big) + L_3 Q\tilde{p}(s) + w(0),$$
$$= \big(\gamma^2 L_1 + \gamma Q L_2 + L_3\big)\tilde{w}(s) + \gamma L_2\tilde{v}(s) + w(0) \tag{8.1.29}$$

We will assume $w(0) = 0$, which means that $\alpha(t)$ is a *stationary* Markov process, so that

$$s\tilde{v}(s) = L_3\tilde{v}(s) - \gamma PL_2[-s + \gamma^2 L_1 + \gamma QL_2 + L_3]^{-1}\gamma L_2\tilde{v}(s) + v(0)\,. \tag{8.1.30}$$

Now the limit $\gamma \to \infty$ gives

$$s\tilde{v}(s) \simeq (L_3 - PL_2L_1^{-1}L_2)\tilde{v}(s) + v(0)\,. \tag{8.1.31}$$

We now compute $PL_2L_1^{-1}L_2\tilde{v}$. We write

$$\tilde{v}(s) = \tilde{p}(x)p_s(\alpha)\,, \tag{8.1.32}$$

$$PL_2L_1^{-1}L_2\tilde{v} = p_s(\alpha)\int d\alpha' \left(-\frac{\partial}{\partial x}b(x)\alpha'\right) L_1^{-1}\left(-\frac{\partial}{\partial x}b(x)\alpha'\right) p_s(\alpha')\tilde{p}(x)\,. \tag{8.1.33}$$

c) Relationship to the Correlation Function: We now need to evaluate

$$\int d\alpha\, \alpha L_1^{-1}\alpha p_s(\alpha) \equiv -D\,, \tag{8.1.34}$$

and to do this we need a convenient expression for L_1^{-1}. Consider

$$\int_0^\infty \exp(L_1 t)\,dt = L_1^{-1}\left(\lim_{t\to\infty}\exp(L_1 t) - 1\right), \tag{8.1.35}$$

$$= -L_1^{-1}Q\,, \tag{8.1.36}$$

by using (8.1.18).

Since

$$P\alpha p_s(\alpha) = p_s(\alpha')\langle\alpha\rangle_s = 0\,, \tag{8.1.37}$$

we have

$$D = \int d\alpha\, \alpha \int_0^\infty \exp(L_1 t)\alpha p_s(\alpha)\,dt\,. \tag{8.1.38}$$

We note that $\exp(L_1 t)\alpha p_s(\alpha)$ is the solution of the Fokker-Planck equation $\partial_t f = L_1 f$ with initial condition $f(\alpha, 0) = \alpha p_s(\alpha)$. Hence,

$$\exp(L_1 t)\alpha p_s(\alpha) = \int d\alpha'\, p(\alpha, t\,|\,\alpha', 0)\alpha' p_s(\alpha')\,, \tag{8.1.39}$$

and substituting in (8.1.38), we find

$$D = \int_0^\infty dt \int d\alpha\, d\alpha'\, \alpha\alpha'\, p(\alpha', t\,|\,\alpha, 0)p_s(\alpha)\,, \tag{8.1.40}$$

$$= \int_0^\infty dt\, \langle\alpha(t)\alpha(0)\rangle_s = \tfrac{1}{2}\,. \tag{8.1.41}$$

after using (8.1.4) and the symmetry of the correlation function.

d) Reduced Stochastic Equations: Using this value of D, we find

$$-PL_2L_1^{-1}L_2\tilde{v} = \tfrac{1}{2}p_s(\alpha)\frac{\partial}{\partial x}b(x)\frac{\partial}{\partial x}b(x)\hat{p}(x)\,, \tag{8.1.42}$$

so that the Fokker-Planck equation equation corresponding to (8.1.31) for

$$\hat{p}(x,t) = \int d\alpha\, p(x,\alpha)\,, \tag{8.1.43}$$

is

$$\frac{\partial \hat{p}}{\partial t} = -\frac{\partial}{\partial x}a(x)\hat{p}(x) + \tfrac{1}{2}\frac{\partial}{\partial x}b(x)\frac{\partial}{\partial x}b(x)\hat{p}(x)\,. \tag{8.1.44}$$

This is, of course, the Fokker-Planck equation in the Stratonovich form which corresponds to

$$(S)\,dx = a(x)\,dt + b(x)\,dW(t)\,, \tag{8.1.45}$$

or which has the Ito form

$$dx = \left[a(x) + \tfrac{1}{2}b'(x)b(x)\right]\,dt + b(x)\,dW(t)\,, \tag{8.1.46}$$

as originally asserted.

e) Generality of the Result: A glance at the proof shows that all we needed was for $\alpha(t)$ to form a stationary Markov process with zero mean and with an evolution equation of the form

$$\frac{\partial p(\alpha)}{\partial t} = L_1 p(\alpha)\,, \tag{8.1.47}$$

where L_1 is a linear operator. This is possible for any kind of Markov process, in particular, for example, the random telegraph process in which $\alpha(t)$ takes on values $\pm a$. In the limit $\gamma \to \infty$, the result is still a Fokker-Planck equation. This is a reflection of the central limit theorem. For, the effective Gaussian white noise is made up of the sum of many individual components, as $\gamma \to \infty$, and the net result is still effectively Gaussian. In fact, *Papanicolaou* and *Kohler* [8.1] have rigorously shown that the result is valid even if $\alpha(t)$ is a non-Markov process, provided it is "strongly mixing" which, loosely speaking, means that all its correlation functions decay rapidly for large time differences.

8.1.2 Generalisations of the Method

a) Noise Terms Dependent on x: Notice that in (8.1.1), instead of defining $\zeta_\gamma(t)$ as simply $\gamma\alpha(t/\gamma^2)$, we can use the more general form

$$\zeta_\gamma(t,x) = \gamma\psi[x,\alpha(t/\gamma^2)]\,, \tag{8.1.48}$$

and now consider only $b(x) = 1$, since all x dependence can be included in ψ. We assume that

$$\int d\alpha\,\psi(x,\alpha)p_s(\alpha) = 0\,, \tag{8.1.49}$$

in analogy to the previous assumption $\langle\alpha\rangle_s = 0$.

Then D becomes x dependent, and we have to use

$$D(x) = \int_0^\infty dt\,\big\langle\psi\big(x,\alpha(t)\big)\psi\big(x,\alpha(0)\big)\big\rangle\,, \tag{8.1.50}$$

and

$$E(x) = \int_0^\infty dt \left\langle \frac{\partial \psi}{\partial x}\big(x, \alpha(t)\big)\psi\big(x, \alpha(0)\big)\right\rangle, \tag{8.1.51}$$

and the Fokker-Planck equation becomes

$$\frac{\partial \hat{p}}{\partial t} = -\frac{\partial}{\partial x}\left\{\big(a(x) + E(x)\big)\hat{p}\right\} + \frac{\partial^2}{\partial x^2}[D(x)\hat{p}]. \tag{8.1.52}$$

In this form we have agreement with the form derived by *Stratonovich* [8.2], Eq. (4.4.39).

b) Time Nonhomogeneous Systems: If instead of (8.1.7) we have

$$\frac{dx}{dt} = a(x,t) + b(x,t)\zeta_\gamma(t), \tag{8.1.53}$$

the Laplace transform method cannot be used simply. We can evade this difficulty by the following trick. Introduce the extra variable τ so that the equations become

$$dx = [a(x,\tau) + \gamma b(x,\tau)\alpha]\, dt, \tag{8.1.54}$$

$$d\alpha = \gamma^2 A(\alpha)dt + \gamma\sqrt{B(\alpha)}\, dW(t), \tag{8.1.55}$$

$$d\tau = dt. \tag{8.1.56}$$

The final equation constrains t to be the same as τ, but the system now forms a homogeneous Markov process in the variables (x, α, τ). Indeed, any nonhomogeneous Markov process can be written as a homogeneous Markov process using this trick.

The Fokker-Planck equation is now

$$\frac{\partial p}{\partial t} = \gamma^2 L_1 + \gamma L_2 + L_3, \tag{8.1.57}$$

with

$$L_1 = -\frac{\partial}{\partial \alpha}A(\alpha) + \tfrac{1}{2}\frac{\partial^2}{\partial \alpha^2}B(\alpha), \tag{8.1.58}$$

$$L_2 = -\frac{\partial}{\partial x}b(x,\tau)\alpha, \tag{8.1.59}$$

$$L_3 = -\frac{\partial}{\partial \tau} - \frac{\partial}{\partial x}a(x,\tau). \tag{8.1.60}$$

Using the same procedure as before, we obtain

$$\frac{\partial \hat{p}}{\partial t} = \left[-\frac{\partial}{\partial \tau} - \frac{\partial}{\partial x}a(x,\tau) + \tfrac{1}{2}\frac{\partial}{\partial x}b(x,\tau)\frac{\partial}{\partial x}b(x,\tau)\right]\hat{p}, \tag{8.1.61}$$

which yields

$$d\tau = dt, \tag{8.1.62}$$

so that we have, after eliminating τ in terms of t,

$$\frac{\partial \hat{p}}{\partial t} = \left[-\frac{\partial}{\partial x} a(x,t) + \tfrac{1}{2} \frac{\partial}{\partial x} b(x,t) \frac{\partial}{\partial x} b(x,t) \right] \hat{p} \,, \tag{8.1.63}$$

in exact analogy to (8.1.64).

c) Effect of Time Dependence in L_1: Suppose, in addition, that A and B depend on time as well, so that

$$L_1 = -\frac{\partial}{\partial \alpha} A(\alpha, \tau) + \tfrac{1}{2} \frac{\partial^2}{\partial \alpha^2} B(\alpha, \tau) \,. \tag{8.1.64}$$

In this case, we find P is a function of τ and hence does *not* commute with L_3. Thus,

$$P L_3 \neq L_3 P \,. \tag{8.1.65}$$

Nevertheless, we can take care of this. Defining $\tilde{v}(s)$ and $\tilde{w}(s)$ as before, we have

$$s\,\tilde{v}(s) = P(\gamma L_2 + L_3)\tilde{w}(s) + P L_3 \tilde{v}(s) + v(0) \,, \tag{8.1.66}$$

$$s\,\tilde{w}(s) = [\gamma^2 L_1 + \gamma Q L_2 + Q L_3]\tilde{w}(s) + \gamma L_2 \tilde{v}(s) + Q L_3 \tilde{v}(s) \,, \tag{8.1.67}$$

so that

$$s\,\tilde{v}(s) = P L_3 \tilde{v}(s) + P(\gamma L_2 + L_3)[s - \gamma^2 L_1 - \gamma Q L_2 - Q L_3]^{-1}[\gamma L_2 + Q L_3]\tilde{v}(s) + v(0) \,. \tag{8.1.68}$$

We see that because L_2 is multiplied by γ and L_3 is not, we get in the limit of large γ

$$s\,\tilde{v}(s) \simeq (P L_3 - P L_2 L_1^{-1} L_2)\tilde{v}(s) + v(0) \,. \tag{8.1.69}$$

In this case we will not assume that we can normalise the autocorrelation function to a constant. The term $-P L_2 L_1^{-1} L_2$ gives

$$\frac{\partial}{\partial x} b(x,\tau) \frac{\partial}{\partial x} b(x,\tau) \int_0^\infty dt \langle \alpha_\tau(t) \alpha_\tau(0) \rangle \,, \tag{8.1.70}$$

where by $\alpha_\tau(t)$ we mean the random variable whose Fokker-Planck equation is

$$\frac{\partial p}{\partial t} = \left[\frac{\partial}{\partial \alpha} A(\alpha, \tau) + \tfrac{1}{2} \frac{\partial^2}{\partial \alpha^2} B(\alpha, \tau) \right] p \,. \tag{8.1.71}$$

Thus, the limit $\gamma \to \infty$ effectively makes the random motion of α infinitely faster than the motion due to the time dependence of α arising from the time dependence of A and B. Defining

$$D(\tau) = \int_0^\infty dt \langle \alpha_\tau(t) \alpha_\tau(0) \rangle \,, \tag{8.1.72}$$

we find, by eliminating τ as before,

$$\frac{\partial \hat{p}}{\partial t} = \left[-\frac{\partial}{\partial x} a(x,t) + D(t) \frac{\partial}{\partial x} b(x,t) \frac{\partial}{\partial x} b(x,t) \right] \hat{p} \,. \tag{8.1.73}$$

8.2 Brownian Motion and the Smoluchowski Equation

The original example of adiabatic elimination, and the most straightforward to illustrate the technique, is found in the case of Brownian motion. Here it is normal to observe only the position of the Brownian particle, but the fundamental equations involve the velocity as well, which is normally unobservable, because the extreme rapidity of the changes in velocity. Thus, Langevin's equation (1.2.14) can be regarded as two equations, one for position and the other for velocity

$$\frac{dx}{dt} = v\,, \tag{8.2.1}$$

$$m\frac{dv}{dt} = -\beta v + \sqrt{2k\beta T}\,\xi(t)\,. \tag{8.2.2}$$

If we interpret the equations as Ito stochastic differential equations, the method of solution has already been given in Sect. 4.5.6. However, it is simpler to integrate (8.2.2) first to give the solution

$$v(t) = v(0)\exp(-\beta t/m) + \frac{\sqrt{2kT\beta}}{m}\int_0^t \exp[-\beta(t-t')/m]\xi(t')dt'\,. \tag{8.2.3}$$

We now want to consider the situation in which the friction coefficient β is not small but the mass m is very small. Then for times t such that

$$t \gg m/\beta \equiv \tau\,, \tag{8.2.4}$$

the exponential in the first term will be negligible and the lower limit in the integral will be able to be extended to $-\infty$, without significant error. Hence,

$$v(t) \to \frac{\sqrt{2kT\beta}}{m}\int_{-\infty}^t \exp\left(-(t-t')/\tau\right)\,\xi(t')dt'\,. \tag{8.2.5}$$

Here τ will be called *relaxation time* since it determines the time scale of relaxation to (8.2.5).

Let us define

$$\eta(t,\tau) = \tau^{-1}\int_{-\infty}^t \exp\left(-(t-t')/\tau\right)\,dW(t')\,, \tag{8.2.6}$$

which is, from Sect. 4.5.4, a stationary Ornstein-Uhlenbeck process. The correlation function is

$$\langle \eta(t,\tau)\eta(t',\tau)\rangle = \frac{1}{2\tau}\exp(-|t-t'|/\tau)\,, \tag{8.2.7}$$

$$\underset{\tau\to 0}{\longrightarrow} \delta(t-t')\,. \tag{8.2.8}$$

We see that the limit $\tau \to 0$ corresponds to a white noise limit in which the correlation function becomes a delta function.

Thus, we can write (8.2.1) as

$$\frac{dx}{dt} = \sqrt{\frac{2kT}{\beta}}\eta(t,\tau)\,, \tag{8.2.9}$$

and in the limit $\tau \to 0$, this should become

$$\frac{dx}{dt} = \sqrt{\frac{2kT}{\beta}}\xi(t). \tag{8.2.10}$$

An alternative, and much more transparent way of looking at this is to say that in (8.2.2) the limit $m \to 0$ corresponds to setting the left-hand side equal to zero, so that

$$v(t) = \sqrt{\frac{2kT}{\beta}}\xi(t). \tag{8.2.11}$$

More generally, we can consider Brownian motion in a potential for which the Langevin equations are (Sect. 6.4)

$$\frac{dx}{dt} = v, \tag{8.2.12}$$

$$m\frac{dv}{dt} = -\beta v - V'(x) + \sqrt{2\beta kT}\,\xi(t). \tag{8.2.13}$$

The limit of large β should result in very rapid relaxation of the second equation to a quasistationary state in which $dv/dt \to 0$. Hence, we assume that for large enough β,

$$v = -\beta^{-1}V'(x) + \sqrt{\frac{2kT}{\beta}}\,\xi(t), \tag{8.2.14}$$

and substituting in (8.2.12) we get

$$\frac{dx}{dt} = -\beta^{-1}V'(x) + \sqrt{\frac{2kT}{\beta}}\,\xi(t), \tag{8.2.15}$$

corresponding to a Fokker-Planck equation for $\hat{p}(x)$ known as the *Smoluchowski equation:*

$$\frac{\partial \hat{p}}{\partial t} = \beta^{-1}\frac{\partial}{\partial x}\left\{V'(x)\hat{p} + kT\frac{\partial \hat{p}}{\partial x}\right\}. \tag{8.2.16}$$

In this case we have eliminated the *fast variable* v, which is assumed to relax very rapidly to the value given by (8.2.14).

This procedure is the prototype of all *adiabatic elimination* procedures which have been used as the basis of Haken's *slaving principle* [8.3]. The basic physical assumption is that large β (or, in general, short relaxation times) force the variables governed by equations involving large β (e.g., v) to relax to a value given by assuming the slow variable (in this case x) to be constant. Such fast variables are then effectively *slaved* by the slow variables.

The reasoning here is very suggestive but completely nonrigorous and gives no idea of any systematic approximation method, which should presumably be some asymptotic expansion in a small dimensionless parameter. Furthermore, there does not seem to be any way of implementing such an expansion directly on the stochastic differential equation—at least to the author's knowledge no one has successfully developed such a scheme.

Surprisingly, the problem of a rigorous derivation of the Smoluchowski equation and an estimation of corrections to it, was only solved half a century after the time of Smoluchowski himself. The first treatment was by *Brinkman* [8.4] who only estimated the order of magnitude of corrections to (8.2.16) but did not give all the correction terms to lowest order. The first correct solution was by *Stratonovich* [8.2] (Chap. 4, Sect. 11.1). Independently, *Wilemski* [8.5] and *Titulaer* [8.6] have also given correct treatments.

In the following sections we will present a systematic and reasonably general theory of the problem of the derivation of the Smoluchowski equation and corrections to it, and will then proceed to more general adiabatic elimination problems. The procedure is the same as that used in deriving the white noise limit in the earlier part of this chapter.

8.2.1 Systematic Formulation in Terms of Operators and Projectors

Let us consider the rescaled form of the Fokker-Planck equation corresponding to (8.2.1,8.2.2) derived as (6.4.12) in Sect. 6.4, which we can write in the form

$$\frac{\partial p}{\partial t} = (\gamma L_1 + L_2)p\,, \tag{8.2.17}$$

where L_1 and L_2 are differential operators given by

$$L_1 = \frac{\partial}{\partial u}\left(u + \frac{\partial}{\partial u}\right), \tag{8.2.18}$$

$$L_2 = -\frac{\partial}{\partial y}u + U'(y)\frac{\partial}{\partial u}\,. \tag{8.2.19}$$

The procedure we follow is exactly that of Sect. 8.1, with the substitution $\alpha \to u$. The only differences are the *form* of the operator L_2, and the fact that $L_3 = 0$. The eliminated equation is

$$\frac{\partial v}{\partial t} = -\gamma^{-1}PL_2L_1^{-1}L_2v\,, \tag{8.2.20}$$

a) Use of the Eigenfunctions of the Ornstein-Uhlenbeck Process: For this problem the eigenfunctions given in Sect. 5.4.3c are to be implemented with the parameters,

$$\tfrac{1}{2}D = k = 1\,, \tag{8.2.21}$$

and take the form

$$P_n(u) = (2\pi)^{-1/2}\exp\left(-\tfrac{1}{2}u^2\right)Q_n(u)\,, \tag{8.2.22}$$

with

$$Q_n(u) = (2^n n!)^{-1/2}\mathrm{H}_n(u/\sqrt{2})\,. \tag{8.2.23}$$

We will use

$$L_1 P_n(u) = -n P_n(u)\,, \tag{8.2.24}$$

and the recursion formulae for Hermite polynomials

$$x\mathrm{H}_n(x) = \tfrac{1}{2}\mathrm{H}_{n+1}(x) + n\mathrm{H}_{n-1}(x)\,, \tag{8.2.25}$$

$$\frac{d}{dx}[\mathrm{e}^{-x^2}\mathrm{H}_n(x)] = -\mathrm{e}^{-x^2}\mathrm{H}_{n+1}(x)\,. \tag{8.2.26}$$

b) Derivation of the Smoluchowski Equation: In terms of these eigenfunctions

$$L_2 v = \left[-\frac{\partial}{\partial y}u + U'(y)\frac{\partial}{\partial u}\right] P_0(u) \int du'\, p(u', y, t)\,, \tag{8.2.27}$$

and we see that

$$L_2 v = -\left[U'(y) + \frac{\partial}{\partial y}\right] P_1(u)\hat{p}(y)\,, \tag{8.2.28}$$

so that, using (8.2.24),

$$L_1^{-1} L_2 v = \left[U'(y) + \frac{\partial}{\partial y}\right] P_1(u)\hat{p}(y)\,. \tag{8.2.29}$$

We now apply L_2 once more and use the relations (8.2.25, 8.2.26) again, finding

$$L_2 P_1(u) = -\left[\sqrt{2}\, P_2(u) + P_0(u)\right]\frac{\partial}{\partial y} - \sqrt{2}\, P_2(u) U'(y)\,, \tag{8.2.30}$$

$$P L_2 L_1^{-1} L_2 v = -\frac{\partial}{\partial y}\left[U'(y) + \frac{\partial}{\partial y}\right]\hat{p}(y) P_0(u)\,, \tag{8.2.31}$$

and the equation of motion (8.2.20) takes the form, after cancelling the factor $P_0(u)$,

$$\frac{\partial \hat{p}}{\partial t} = \gamma^{-1}\frac{\partial}{\partial y}\left[U'(y)\hat{p} + \frac{\partial \hat{p}}{\partial y}\right], \tag{8.2.32}$$

which is exactly the Smoluchowski equation, derived using the naive elimination given in Sect. 8.2, page 193.

8.2.2 Short-Time Behaviour

One should note that the limit $\gamma \to \infty$ in implies that s is finite, or more precisely speaking $s \ll \gamma$. This means that the solution in the time domain is valid on a *time-scale* $t \gg \gamma^{-1}$, and that finer details of time development will not be given accurately. To get a better idea of the finer details, let us define a scaled Laplace transform variable

$$s_1 = s\gamma^{-1}\,, \tag{8.2.33}$$

so that the limit $\gamma \to \infty$ now gives

$$\gamma s_1 \tilde{v} \simeq \gamma^{-1} P L_2 (s_1 - L_1)^{-1} L_2 \tilde{v} + v(0)\,. \tag{8.2.34}$$

Using the fact that $L_2\tilde{v}$ is proportional to $P_1(u)$ (8.2.28), we see that

$$\gamma s_1 \tilde{v} = \gamma^{-1}(s_1+1)^{-1} P L_2^2 \tilde{v} + v(0) . \tag{8.2.35}$$

Changing back to the variable s again and rearranging, we find

$$s\tilde{v} = \gamma^{-1}\left(\frac{s}{\gamma}+1\right)^{-1} P L_2^2 \tilde{v} + v(0) , \tag{8.2.36}$$

which is equivalent to

$$\frac{\partial v}{\partial t} = \int_0^t dt' \exp[\gamma(t'-t)] P L_2^2 v(t')\, dt' . \tag{8.2.37}$$

Alternatively, we can rewrite (8.2.36) in the form

$$\frac{1}{\gamma}[s^2\tilde{v} - sv(0)] + [s\tilde{v} - v(0)] = \gamma^{-1} P L_2^2 v , \tag{8.2.38}$$

which, using (8.1.23) and the same working out as in Sect. 8.2.1, is equivalent to the equation for $\hat{p}$:

$$\frac{1}{\gamma}\frac{\partial^2 p}{\partial t^2} + \frac{\partial \hat{p}}{\partial t} = \gamma^{-1}\frac{\partial}{\partial y}\left[U'(y)\hat{p} + \frac{\partial \hat{p}}{\partial y}\right] , \tag{8.2.39}$$

in which the initial condition

$$\frac{\partial \hat{p}}{\partial t}(0) = 0 , \tag{8.2.40}$$

is implied because

$$\int_0^\infty \mathrm{e}^{-st} f''(t) = s^2 \tilde{f}(s) - s\, f(0) - f'(0) , \tag{8.2.41}$$

and no constant term appears in the first bracket in (8.2.38).

We may similarly rewrite (8.2.37) or integrate (8.2.39) to get

$$\frac{\partial \hat{p}}{\partial t} = \frac{\partial}{\partial y}\left[U'(y) + \frac{\partial}{\partial y}\right] \int_0^t dt' \exp[\gamma(t'-t)]\hat{p}(t') . \tag{8.2.42}$$

Equations (8.2.39, 8.2.42) demonstrate a non-Markov nature, seen explicitly in (8.2.42) which indicates that the prediction of $\hat{p}(t+\Delta t)$ requires the knowledge of $\hat{p}(t')$ for $0 \leqslant t' \leqslant t$. However, the kernel $\exp[\gamma(t'-t)]$ is significantly different from zero only for $|t'-t| \sim \gamma^{-1}$ and on a time scale much longer than this, (8.2.42) is approximated by the Smoluchowski equation (8.2.32). We achieve this by integrating by parts in (8.2.42):

$$\int_0^t dt' \exp[\gamma(t'-t)]\hat{p}(t') = \frac{\hat{p}(t) - \mathrm{e}^{-\gamma t}\hat{p}(0)}{\gamma} - \gamma^{-1}\int_0^t \exp[\gamma(t'-t)]\frac{\partial \hat{p}}{\partial t} dt' . \tag{8.2.43}$$

Neglecting the last term as being of order γ^{-2}, we find the Smoluchowski equation replaced by

$$\frac{\partial \hat{p}}{\partial t} = \frac{\partial}{\partial y}\left[U'(y) + \frac{\partial}{\partial y}\right]\left[\hat{p}(t) - \mathrm{e}^{-\gamma t}\hat{p}(0)\right] . \tag{8.2.44}$$

This equation is to lowest order in γ equivalent to (8.2.42) for *all times*, that is, very short ($< \gamma^{-1}$) and very long ($\gg \gamma^{-1}$) times. It shows the characteristic *memory time*, γ^{-1}, which elapses before the equation approaches the Smoluchowski form.

To this order then, the process can be approximated by a Markov process, but to the same order, there are the alternative expressions (8.2.39, 8.2.42) which are not Markov processes. Clearly, to any higher order, we must have a non-Markov process.

8.2.3 Boundary Conditions

Let us consider a barrier to be erected at $y = a$ with the particle confined to $y \leqslant a$. The behaviors for $u > 0$ and $u < 0$ are distinct. From the stochastic differential equations

$$\left.\begin{aligned} dy &= u\,dt\,, \\ du &= -[U'(y) + \gamma u]dt + \sqrt{2\gamma}\,dW(t)\,. \end{aligned}\right\} \tag{8.2.45}$$

We see that:

For $u > 0$, $y = a$ is an exit boundary;

For $u < 0$, $y = a$ is an entrance boundary;

since a particle with $u > 0$ at $y = a$ must proceed to $y > a$ or be absorbed. Similarly, particles to the left of the boundary can never reach $y = a$ if $u < 0$. Conventionally, we describe $y = a$ as absorbing or reflecting as follows:

i) *Absorbing barrier:* particle absorbed for $u > 0$, no particles with $u < 0$; this means

$$p(u, a, t) = \begin{cases} 0, & u > 0, \\ 0, & u < 0. \end{cases} \tag{8.2.46}$$

These two apparently similar conditions are in fact quite distinct from each other. The first condition is the usual absorbing boundary condition, as derived in Sect. 5.1. The second condition expresses the fact that any particle placed with $u < 0$ at $y = a$ *immediately* proceeds to $y < a$, and no further particles are introduced.

The absorbing barrier condition clearly implies

$$\hat{p}(a, t) = 0\,, \tag{8.2.47}$$

which is the usual absorbing barrier condition in a one variable Fokker-Planck equation.

ii) *Reflecting barrier:* physically, a reflection at $y = a$ implies that the particle reaches $y = a$ with the velocity u, and is immediately reflected with a different but negative velocity. If we *assume* that

$$u \to -u\,, \tag{8.2.48}$$

then we have a *periodic boundary condition* as in Sects. 5.1 and 6.2.1.

This means that

$$p(u, a, t) = p(-u, a, t)\,, \tag{8.2.49}$$

and that the normal component of the current leaving at (u, a) is equal to that entering at $(-u, a)$. However, since for Kramers' equation the current is

$$J = \left\{-up, [\gamma u + U'(y)]p + \gamma \frac{\partial p}{\partial u}\right\}, \tag{8.2.50}$$

and the normal component is the component along the direction y, we see that the vanishing of the normal component of the current is equivalent to (8.2.49).

This boundary condition leads very naturally to that for the Smoluchowski equation, as follows. Equation (8.2.49) implies that only even-order eigenfunctions $P_n(u)$ can occur in the expansion of $p(u, a, t)$. Hence,

$$Qp(u, a, t) = w(u, a, t)\,, \tag{8.2.51}$$

contains only *even* eigenfunctions, and the same is true of $\tilde{w}(u, a, s)$, the Laplace transform. But, from (8.1.29) we see that to lowest order in γ^{-1}

$$\tilde{w}(u, a, s) \simeq -\gamma^{-1} L_1^{-1} L_2\, \tilde{u}(a, s)\,, \tag{8.2.52}$$

and using (8.2.29)

$$= -P_1(u)\gamma^{-1}\left[U'(y)\hat{p}(y) + \frac{\partial \hat{p}(y)}{\partial y}\right]_{y=a}. \tag{8.2.53}$$

Since this is proportional to the *odd* eigenfunction $P_1(u)$, it must vanish. Hence we derive

$$\left[U'(y)\hat{p}(y) + \frac{\partial \hat{p}(y)}{\partial y}\right]_{y=a} = 0\,, \tag{8.2.54}$$

which is the appropriate reflecting barrier boundary condition for the Smoluchowski equation.

It is not difficult to show similarly that the same boundary conditions can be derived for the equations derived in Sect. 8.2.2.

8.2.4 Evaluation of Higher Order Corrections

Let us consider the general expressions (8.1.28, 8.1.29) for $\tilde{w}(s)$ and $\tilde{v}(s)$, setting $w(0) = 0$ for simplicity, which give in this case (since here $L_3 = 0$)

$$\tilde{w}(s) = [s - \gamma L_1 - QL_2]^{-1} L_2 \tilde{v}(s)\,, \tag{8.2.55}$$

and

$$s\,\tilde{v}(s) = PL_2[s - \gamma L_1 - QL_2]^{-1} L_2 \tilde{v}(s) + v(0)\,. \tag{8.2.56}$$

We can straightforwardly expand the inverse in (8.2.56) in powers of γ. However, the order of magnitude of s in this expansion must be decided. From the previous sections we see that there is a possibility of defining

$$s_1 = s\gamma^{-1}, \quad \tilde{v} \to \gamma^{-1}\tilde{v}, \tag{8.2.57}$$

which gives rise to an expansion in which the initial time behaviour in times of order γ^{-1} is explicitly exhibited.

An alternative is to define

$$s_2 = s\gamma, \quad \tilde{v} \to \gamma\tilde{v}, \tag{8.2.58}$$

which we will find yields an expansion in which only the long-time behaviour is exhibited.

By substituting (8.2.58) we find that (8.2.56) takes the form

$$s_2\tilde{v} = -PL_2[L_1 + QL_2\gamma^{-1} - s_2\gamma^{-2}]^{-1}L_2\tilde{v} + v(0), \tag{8.2.59}$$

and we may now write

$$s_2\tilde{v} = -PL_2L_1^{-1}L_2\tilde{v} + v(0), \quad (\gamma \to \infty), \tag{8.2.60}$$

whereas, without this substitution, the limit $\gamma \to \infty$ yields simply

$$s\tilde{v} = v(0). \tag{8.2.61}$$

The substitution of $s_1 = s\gamma^{-1}$,

$$s_1\tilde{v} = \gamma^{-2}PL_2[s_1 - L_1 - QL_2\gamma^{-1}]^{-1}L_2\tilde{v}(s) + v(0), \tag{8.2.62}$$

does not lead to a limit as $\gamma \to \infty$. However, it does yield an expansion which is similar to an ordinary perturbation problem, and, as we have seen, exhibits short-time behaviour.

a) Long-Time Perturbation Theory: We can expand (8.2.59) to order γ^{-2} in the form

$$s_2 v = [A + B\gamma^{-1} + (C + Ds_2)\gamma^{-2}]\tilde{v} + v(0), \tag{8.2.63}$$

with

$$\left.\begin{aligned} A &= -\,PL_2L_1^{-1}L_2\,, \\ B &= \,PL_2L_1^{-1}QL_2L_1^{-1}L_2\,, \\ C &= -\,PL_2L_1^{-1}QL_2L_1^{-1}QL_2L_1^{-1}L_2\,, \\ D &= -\,PL_2L_1^{-2}L_2\,. \end{aligned}\right\} \tag{8.2.64}$$

Rearranging (8.2.63) we have

$$s_2(1 - \gamma^{-2}D)\tilde{v} = [A + B\gamma^{-1} + C\gamma^{-2}]\tilde{v} + v(0), \tag{8.2.65}$$

or to order γ^{-2} this is equivalent to

$$s_2\tilde{v} = [A + B\gamma^{-1} + (C + DA)\gamma^{-2}]\tilde{v} + (1 + \gamma^{-2}D)v(0). \tag{8.2.66}$$

This gives the Laplace transform of an equation of motion for v, *in which the initial condition is not* $v(0)$, *but* $(1 + \gamma^{-2}D)v(0)$. Notice that this equation will be of first order in time, since s_2^n for $n > 1$ does not occur. This is not possible, of course, for an expansion to higher powers of γ.

b) Application to Brownian Motion: For Brownian motion we have already computed the operator A of (8.2.64); it is given by (8.2.31)

$$A = -PL_2\, L_1^{-1}\, L_2 = \frac{\partial}{\partial y}\left[U'(y) + \frac{\partial}{\partial y}\right]. \tag{8.2.67}$$

The other operators can be similarly calculated.

For example, from (8.2.29, 8.2.30),

$$L_2 L_1^{-1} L_2 = -\left\{P_0(u)\frac{\partial}{\partial y} + \sqrt{2}\, P_2(u)\left[U'(y) + \frac{\partial}{\partial y}\right]\right\}\left\{U'(y) + \frac{\partial}{\partial y}\right\}. \tag{8.2.68}$$

Multiplying by $(1 - P)$ simply removes the term involving P_0, and multiplying by L_1^{-1} afterwards multiplies by $-\frac{1}{2}$. Hence,

$$L_1^{-1} Q L_2 L_1^{-1} L_2 = \sqrt{2}\, P_2(u)\left[U'(y) + \frac{\partial}{\partial y}\right]^2. \tag{8.2.69}$$

We now use the Hermite polynomial recursion relations (8.2.25, 8.2.26) when multiplying by L_2: we derive

$$\begin{aligned} L_2 P_2(u) &= \left[-\frac{\partial}{\partial y} u + U'(y)\frac{\partial}{\partial u}\right] P_2(u), \\ &= -\sqrt{3}\, U'(y) P_3(u) - \frac{\partial}{\partial y}\left[\sqrt{3}\, P_3(u) + \sqrt{2}\, P_1(u)\right]. \end{aligned} \tag{8.2.70}$$

Finally, multiplying by P annihilates all terms since $P_0(u)$ does not occur. Hence,

$$B = P L_2 L_1^{-1} Q L_2 L_1^{-1} L_2 = 0\,. \tag{8.2.71}$$

The computation of C and DA follow similarly.

One finds that

$$C = \frac{\partial^2}{\partial y^2}\left[U'(y) + \frac{\partial}{\partial y}\right]\left[U'(y) + \frac{\partial}{\partial y}\right], \tag{8.2.72}$$

and

$$DA = -\frac{\partial}{\partial y}\left[U'(y) + \frac{\partial}{\partial y}\right]\frac{\partial}{\partial y}\left[U'(y) + \frac{\partial}{\partial y}\right], \tag{8.2.73}$$

so that

$$C + DA = \frac{\partial}{\partial y} U''(y)\left[U'(y) + \frac{\partial}{\partial y}\right], \tag{8.2.74}$$

and (8.2.66) is equivalent to the differential equation

$$\frac{\partial \hat{p}}{\partial t} = \gamma^{-1}\frac{\partial}{\partial y}[1 + \gamma^{-2} U''(y)]\left[U'(y)\hat{p} + \frac{\partial \hat{p}}{\partial y}\right]. \tag{8.2.75}$$

The initial condition becomes

$$\lim_{t\to\infty} \hat{p}(y,t) = \left\{1 - \gamma^{-2}\frac{\partial}{\partial y}\left[U'(y) + \frac{\partial}{\partial y}\right]\right\} \int d\alpha\, p(y,\alpha,0)\,. \tag{8.2.76}$$

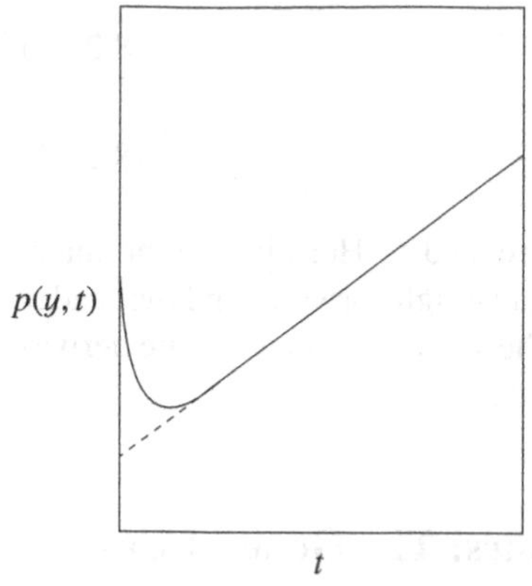

Fig. 8.1. Formation of a layer at a boundary. The exact solution (——) changes rapidly near the boundary on the left. The approximation (- - - - - -) is good except near the boundary. The appropriate boundary condition for the approximation is thus given by the smaller value, where the dashed line meets the boundary.

This alteration of the initial condition is a reflection of a "layer" phenomenon. Equation (8.2.75) is valid for $t \gg \gamma^{-1}$ and is known as the *corrected Smoluchowski equation.*

The exact solution would take account of the behaviour in the time up to $t \sim \gamma^{-1}$ in which terms like $\exp(-\gamma t)$ occur. Graphically, the situation is as in Fig. 8.1. The changed initial condition accounts for the effect of the initial layer near $t \sim \gamma^{-1}$, which means that

$$\hat{p}(y,0) \neq \int d\alpha\, p(y,\alpha,0)\,, \tag{8.2.77}$$

but that for $t \gg \gamma^{-1}$

$$\hat{p}(y,t) \to \int d\alpha\, p(y,\alpha,t). \tag{8.2.78}$$

c) Boundary Conditions: The higher order implementation of boundary conditions cannot be carried out by the methods of this section, since a rapidly varying layer occurs in the variable y near the boundary, and the assumption that the operator $\partial/\partial y$ is bounded becomes unjustified. Significant progress has been made by *Titulaer* and co-workers [8.8, 8.9]. Suppose the boundary is at $y = 0$, then we can substitute $z = \gamma y$ and $\tau = \gamma t$ into the Kramers' equation (8.2.17) to obtain

$$\frac{\partial p}{\partial \tau} = \left[\frac{\partial}{\partial u}\left(u + \frac{\partial}{\partial u}\right) - u\frac{\partial}{\partial z}\right]p + \frac{1}{\gamma}U'(z/y)\frac{\partial p}{\partial u}\,, \tag{8.2.79}$$

so that the zero order problem is the solution of that part of the equation independent of γ: to this order the potential is irrelevant. Only the stationary solution of (8.2.79) has so far been amenable to treatment. It can be shown that the stationary solution to the $\gamma \to \infty$ limit of (8.2.79) can be written in the form

$$P(u,z) = \psi_0'(u,z) + d_0^s\psi(u,z) + \sum_{n=1}^{\infty} d_n^s\psi_n(u,z)\,, \tag{8.2.80}$$

where

$$\psi_0(u,z) = (2\pi)^{-1/2}\exp\left(-\frac{1}{2}u^2\right), \tag{8.2.81}$$

$$\psi_0'(u,z) = (2\pi)^{-1/2}(z-u)\exp\left(-\frac{1}{2}u^2\right), \tag{8.2.82}$$

and the $\psi_n(u,z)$ are certain complicated functions related to Hermite polynomials. The problem of determining the coefficients d_n^s is not straightforward, and the reader is referred to [8.10] for a treatment. It is found that the solution has an infinite derivative at $z = 0$, and for small z is of the form $a + bz^{1/2}$.

8.3 Adiabatic Elimination of Fast Variables: The General Case

We now want to consider the general case of two variables x and α which are coupled together in such a way that each affects the other. This is now a problem analogous to the derivation of the Smoluchowski equation with nonvanishing $V'(x)$, whereas the previous section was a generalization of the same equation with $V'(x)$ set equal to zero.

The most general problem of this kind would be so complex and unwieldy as to be incomprehensible. In order to introduce the concepts involved, we will first consider an example of a linear chemical system and then develop a generalised theory.

8.3.1 Example: Elimination of Short-Lived Chemical Intermediates

We consider the example of a chemically reacting system

$$X \underset{k}{\overset{\gamma}{\rightleftharpoons}} Y \underset{\gamma}{\overset{k}{\rightleftharpoons}} A, \tag{8.3.1}$$

where X and Y are chemical species whose quantities vary, but A is by some means held fixed. The deterministic rate equations for this system are

$$\left.\begin{aligned} \frac{dx}{dt} &= -x + \gamma y, \\ \frac{dy}{dt} &= -2\gamma y + x + a. \end{aligned}\right\} \tag{8.3.2}$$

Here x, y, a, are the concentrations of X, Y, A. The rate constants have been chosen so that $k = 1$, for simplicity.

Physically, Y is often a very short-lived intermediate state which can decay to X or A, with time constant γ^{-1}. Thus, the situation of interest is the limit of large γ in which the short-lived intermediate Y becomes even more short lived, and its concentration negligible. This results in the situation where we solve the first equation of (8.3.2) with $dy/dt = 0$ so that

$$y = (x+a)/2\gamma, \tag{8.3.3}$$

and substitute this in the second equation to get

$$\frac{dx}{dt} = \frac{x}{2} - \frac{a}{2}. \tag{8.3.4}$$

The stochastic analogue of this procedure is complicated by the fact that the white noises to be added to (8.3.2) are correlated, and the stationary distribution of y depends on γ. More precisely, the stochastic differential equations corresponding to (8.3.2) are usually chosen to be (Sect. 11.6.1)

$$\left.\begin{aligned} dx &= (-x + \gamma y)\,dt + \varepsilon B_{11}\,dW_1(t) + \varepsilon B_{12}\,dW_2(t)\,, \\ dy &= (-2\gamma y + x + a)\,dt + \varepsilon B_{21}\,dW_1(t) + \varepsilon B_{22}\,dW_2(t)\,, \end{aligned}\right\} \tag{8.3.5}$$

where the matrix B satisfies

$$BB^{\mathrm{T}} = \begin{bmatrix} 2a & -2a \\ -2a & 4a \end{bmatrix}. \tag{8.3.6}$$

Here ε is a parameter, which is essentially the square root of the inverse volume of the reacting system, and is usually small, though we shall not make use of this fact in what follows.

We wish to eliminate the variable y, whose mean value would be given by (8.3.3) and becomes vanishingly small in the limit. It is only possible to apply the ideas we have been developing if the variable being eliminated has a distribution function in the stationary state which is independent of γ. We will thus have to define a new variable as a function of y and x which possesses this desirable property.

The Fokker-Planck equation corresponding to (8.3.5) is

$$\frac{\partial p}{\partial t} = \left[\frac{\partial}{\partial x}(x - \gamma y) + \frac{\partial^2}{\partial x^2}\varepsilon^2 a - 2\frac{\partial^2}{\partial x \partial y}\varepsilon^2 a + \frac{\partial}{\partial y}(2\gamma y - x - a) + 2\varepsilon^2 a \frac{\partial^2}{\partial y^2}\right] p. \tag{8.3.7}$$

It seems reasonable to define a new variable z by

$$z = 2\gamma y - x - a\,, \tag{8.3.8}$$

which is proportional to the difference between y and its stationary value. Thus, we formally define a pair of new variables (x_1, z) by

$$\left.\begin{aligned} x_1 &= x\,, & x &= x_1\,, \\ z &= 2\gamma y - x - a\,, & y &= (z + x_1 + a)/2\gamma\,, \end{aligned}\right\} \tag{8.3.9}$$

so that we can transform the Fokker-Planck equation using

$$\left.\begin{aligned} \frac{\partial}{\partial x} &= \frac{\partial}{\partial x_1} - \frac{\partial}{\partial z}\,, \\ \frac{\partial}{\partial y} &= 2\gamma\frac{\partial}{\partial z}\,, \end{aligned}\right\} \tag{8.3.10}$$

to obtain

$$\frac{\partial p}{\partial t} = \left[\frac{\partial}{\partial x_1}\left(\frac{x_1 - a}{2} - \frac{z}{2}\right) + \varepsilon^2 a \frac{\partial^2}{\partial x_1^2} + \frac{\partial^2}{\partial x_1 \partial z}(-2\varepsilon^2 a - 4\gamma\varepsilon^2 a)\right.$$
$$\left. + \frac{\partial}{\partial z}\left(2\gamma z - \frac{x - a}{2} + \frac{z}{z}\right) + \frac{\partial^2}{\partial z^2}(8\varepsilon^2\gamma^2 a + \varepsilon^2 a + 4\gamma\varepsilon^2 a)\right] p. \tag{8.3.11}$$

The limit of $\gamma \to \infty$ does not yet give a Fokker-Planck operator in z, which is simply proportional to a fixed operator: we see that the drift and diffusion terms for z are proportional to γ and γ^2, respectively.

However, the substitution

$$\alpha = z\gamma^{-1/2}, \tag{8.3.12}$$

changes this. In terms of α, the drift and diffusion coefficients become proportional to γ and we can see (now writing x instead of x_1)

$$\frac{\partial p}{\partial t} = [\gamma L_1 + \gamma^{1/2} L_2(\gamma) + L_3] p, \tag{8.3.13}$$

in which

$$L_1 = 2\frac{\partial}{\partial \alpha}\alpha + 8\varepsilon^2 a \frac{\partial^2}{\partial \alpha^2}, \tag{8.3.14}$$

$$L_2(\gamma) = \frac{\partial}{\partial x}\left[-4\varepsilon^2 a \frac{\partial}{\partial \alpha} - \frac{1}{2}\alpha\right] + 4\gamma^{-1/2}\varepsilon^2 a \frac{\partial^2}{\partial \alpha^2} - \gamma^{-1}\frac{\partial}{\partial \alpha}\left[\frac{x - a}{2}\right]$$
$$+ \left\{-\gamma^{-1/2}\frac{\partial}{\partial \alpha}\alpha - \gamma^{-1} 2\varepsilon^2 a \frac{\partial^2}{\partial x \partial \alpha} + \gamma^{-3/2}\varepsilon^2 a \frac{\partial^2}{\partial \alpha^2}\right\}, \tag{8.3.15}$$

$$L_3 = \frac{\partial}{\partial x}\left(\frac{x - a}{2}\right) + \varepsilon^2 a \frac{\partial^2}{\partial x^2}. \tag{8.3.16}$$

Notice that $L_2(\gamma)$ has a large γ limit given by the first term of the first line of (8.3.15). The only important property of L_2 is that $PL_2P = 0$. Defining P as usual

$$Pf(x, \alpha) = p_s(\alpha) \int d\alpha' \, f(x, \alpha'), \tag{8.3.17}$$

where $p_s(\alpha)$ is the stationary solution for L_1, we see that for any operator beginning with $\partial/\partial\alpha$ [such as the γ dependent part of $L_2(\gamma)$] we have

$$P\frac{\partial}{\partial \alpha} f = p_s(\alpha) \int d\alpha' \frac{\partial}{\partial \alpha'} f(x, \alpha') = 0, \tag{8.3.18}$$

provided we can drop boundary terms. Hence, the γ dependent part of $L_2(\gamma)$ satisfies $PL_2P = 0$. Further, it is clear from (8.3.14) that $\langle\alpha\rangle_s = 0$, so we find that the first, γ independent part, also satisfies this condition. Thus

$$PL_2(\gamma)P = 0. \tag{8.3.19}$$

Nevertheless, it is worth commenting that the γ dependent part of $L_2(\gamma)$ contains terms which look more appropriate to L_1, that is, terms not involving any x derivatives. However, by moving these terms into L_2, we arrange for L_1 to be independent of γ. Thus, P is independent of γ and the limits are clearer.

The procedure is now quite straightforward. Defining, as usual

$$\left.\begin{aligned} P\tilde{p}(s) &= \tilde{v}(s)\,, \\ (1-P)\tilde{p}(s) &= \tilde{w}(s) \end{aligned}\right\} \tag{8.3.20}$$

and assuming, as usual, $w(0) = 0$, we find

$$s\,\tilde{v}(s) = P[\gamma L_1 + \gamma^{1/2} L_2(\gamma) + L_3][\tilde{v}(s) + \tilde{w}(s)] + v(0)\,, \tag{8.3.21}$$

and using

$$\left.\begin{aligned} PL_1 &= L_1 P = 0\,, \\ PL_2 P &= 0\,, \\ PL_3 &= L_3 P\,, \end{aligned}\right\} \tag{8.3.22}$$

we obtain

$$s\,\tilde{v}(s) = P\gamma^{1/2} L_2(\gamma)\tilde{w}(s) + L_3\tilde{v}(s) + v(0)\,, \tag{8.3.23}$$

and similarly,

$$s\,\tilde{w}(s) = [\gamma L_1 + \gamma^{1/2}(1-P)L_2(\gamma) + L_3]\tilde{w}(s) + \gamma^{1/2} L_2(\gamma)\tilde{v}(s)\,, \tag{8.3.24}$$

so that

$$s\,\tilde{v}(s) = \{L_3 + \gamma P L_2(\gamma)[s - \gamma L_1 - \gamma^{1/2}(1-P)L_2(\gamma) - L_3]^{-1} L_2\}\tilde{v}(s) + v(0)\,. \tag{8.3.25}$$

Now taking the large γ limit, we get

$$s\tilde{v}(s) \simeq (L_3 - PL_2 L_1^{-1} L_2)\tilde{v}(s) + v(0)\,, \tag{8.3.26}$$

where

$$L_2 = \lim_{\gamma\to\infty} L_2(\gamma) = \frac{\partial}{\partial x}\left(-4\varepsilon^2 a \frac{\partial}{\partial \alpha} - \frac{1}{2}\alpha\right). \tag{8.3.27}$$

Equation (8.3.26) is of exactly the same form as (8.1.31) and indeed the formal derivation from (8.3.7) is almost identical. The evaluation of $PL_2L_1^{-1}L_2$ is, however, slightly different because of the existence of terms involving $\partial/\partial\alpha$. Firstly, notice that since $P\,\partial/\partial\alpha = 0$, we can write

$$-PL_2L_1^{-1}L_2 v = -p_s(\alpha) \int d\alpha' \left(-\frac{1}{2}\alpha' \frac{\partial}{\partial x}\right) L_1^{-1} \frac{\partial}{\partial x}\left(-4\varepsilon^2 a \frac{\partial}{\partial \alpha'} - \frac{1}{2}\alpha'\right) p_s(\alpha')\hat{p}(x)\,, \tag{8.3.28}$$

and from the definition of $p_s(\alpha)$ as satisfying $L_1 p_s(\alpha) = 0$, we see from (8.3.14) that

$$\frac{\partial}{\partial \alpha} p_s(\alpha) = -\alpha p_s(\alpha)/4\varepsilon^2 a\,, \tag{8.3.29}$$

and hence that

$$-PL_2L_1^{-1}L_2 v = \frac{1}{4} p_s(\alpha) \frac{\partial^2}{\partial x^2} \int d\alpha' \alpha' L_1^{-1} \alpha'\, p_s(\alpha')\hat{p}(x)\,, \tag{8.3.30}$$

$$= -\frac{1}{4} p_s(\alpha) \frac{\partial^2 \hat{p}(x)}{\partial x^2} \int_0^\infty dt \langle \alpha(t)\alpha(0)\rangle_s\,, \tag{8.3.31}$$

where we have used the reasoning given in Sect. 8.1 to write the answer in terms of the correlation function. Here, L_1 is the generator of an Ornstein-Uhlenbeck process (Sect. 3.8.4) with $k = 2, D = 16\varepsilon^2 a$, so that from (3.8.2),

$$-PL_2L_1^{-1}L_2 v = -\frac{1}{4}p_s(\alpha)\frac{\partial^2 \hat{p}(x)}{\partial x^2}4\varepsilon^2 a\int_0^\infty dt\, e^{-2t}\,, \tag{8.3.32}$$

$$= -\frac{1}{2}\varepsilon^2 a\frac{\partial^2 \hat{p}(x)}{\partial x^2}p_s(\alpha)\,. \tag{8.3.33}$$

Hence, from (8.3.26), the effective Fokker-Planck equation is

$$\frac{\partial \hat{p}}{\partial t} = \frac{\partial}{\partial x}\frac{x-a}{2}\hat{p}(x) + \frac{1}{2}\varepsilon^2 a\frac{\partial^2 \hat{p}}{\partial x^2}\,. \tag{8.3.34}$$

Comments

i) This is exactly the equation expected from the reaction

$$X \underset{k/2}{\overset{k/2}{\rightleftharpoons}} A\,, \tag{8.3.35}$$

(with $k = 1$) (Sect. 11.5.3). It is expected because general principles tell us that the stationary variance of the concentration fluctuations is given by

$$\mathrm{var}[x(t)]_s = \varepsilon^2\langle x\rangle_s\,. \tag{8.3.36}$$

ii) Notice that the net effect of the adiabatic elimination is to *reduce* the coefficient of $\partial^2/\partial x^2$, a result of the correlation between the noise terms in x and y in the original equations.

iii) This result differs from the usual adiabatic elimination in that the noise term in the eliminated variable is important. There are cases where this is not so; they will be treated shortly.

8.3.2 Adiabatic Elimination in Haken's Model

Haken introduced a simple model for demonstrating adiabatic elimination [Ref. 6.1, Sect. 7.2]. The deterministic version of the model is a pair of coupled equations which may be written

$$\dot{x} = -\varepsilon x - ax\alpha\,, \tag{8.3.37}$$

$$\dot{\alpha} = -\kappa\alpha + bx^2\,. \tag{8.3.38}$$

One assumes that if κ is sufficiently large, we may, as before, replace α by the stationary solution (8.3.38) in terms of x to obtain

$$\alpha = \frac{b}{\kappa}x^2\,, \tag{8.3.39}$$

$$\dot{x} = -\varepsilon x - \frac{ab}{\kappa}x^3\,. \tag{8.3.40}$$

The essential aim of the model is to obtain the cubic form on the right-hand side of (8.3.40).

In making the transition to a stochastic system, we find that there are various possibilities available. The usual condition for the validity of adiabatic elimination is

$$\varepsilon \ll \kappa \,. \tag{8.3.41}$$

In a stochastic version, all other parameters come into play as well, and the condition (8.3.41) is, in fact, able to be realised in at least three distinct ways with characteristically different answers.

Let us write stochastic versions of (8.3.37, 8.3.38):

$$\left.\begin{aligned} dx &= -(\varepsilon x + ax\alpha)\,dt + C\,dW_1(t)\,, \\ d\alpha &= -(\kappa\alpha - bx^2)\,dt + D\,dW_2(t)\,, \end{aligned}\right\} \tag{8.3.42}$$

and we assume here, for simplicity, that C and D are constants and $W_1(t)$ and $W_2(t)$ are independent of each other.

The Fokker-Planck equation is

$$\frac{\partial p}{\partial t} = \left[\frac{\partial}{\partial x}(\varepsilon x + ax\alpha) + \frac{1}{2}C^2\frac{\partial^2}{\partial x^2} + \frac{\partial}{\partial \alpha}(\kappa\alpha - bx^2) + \frac{1}{2}D^2\frac{\partial^2}{\partial \alpha^2}\right]p\,. \tag{8.3.43}$$

We wish to eliminate α. It is convenient to define a new variable β by

$$\beta = \alpha - \frac{b}{\kappa}x^2\,, \tag{8.3.44}$$

so that, for fixed x, the quantity β has zero mean. In terms of this variable, we can write a Fokker-Planck equation:

$$\frac{\partial p}{\partial t} = (L_1^0 + L_2^0 + L_3^0)p\,, \tag{8.3.45}$$

$$L_1^0 = \frac{\partial}{\partial \beta}\kappa\beta + \frac{D^2}{2}\frac{\partial^2}{\partial \beta^2}\,, \tag{8.3.46}$$

$$\begin{aligned} L_2^0 = {} & \frac{\partial}{\partial \beta}a\beta x - \frac{2bx}{\kappa}\frac{\partial}{\partial \beta}\left(\varepsilon x + \frac{ab}{\kappa}x^3 + ax\beta\right) \\ & - C^2\left(\frac{bx}{\kappa}\frac{\partial^2}{\partial x \partial \beta} + \frac{\partial^2}{\partial x \partial \beta}\frac{bx}{\kappa}\right) + \frac{2b^2x^2C^2}{\kappa^2}\frac{\partial^2}{\partial \beta^2}\,, \end{aligned} \tag{8.3.47}$$

$$L_3^0 = \left[\frac{\partial}{\partial x}\left(\varepsilon x + \frac{ab}{\kappa}x^3\right) + \frac{C^2}{2}\frac{\partial^2}{\partial x^2}\right]. \tag{8.3.48}$$

In terms of these variables, the limit $\varepsilon \to 0$ is not interesting since we simply get the same system with $\varepsilon = 0$. No elimination is possible since L_1 is not multiplied by a large parameter.

In order for the limit $\varepsilon \to 0$ to have the meaning deterministically that (8.3.40) is a valid *limiting form*, there must exist an A such that

$$\frac{ab}{\kappa} = \varepsilon A, \qquad \text{as} \quad \varepsilon \to 0\,. \tag{8.3.49}$$

For this limit to be recognisable deterministically, it must not be swamped by noise so one must also have

$$\tfrac{1}{2}C^2 = \varepsilon B\,, \quad \text{as}\, \varepsilon \to 0\,, \tag{8.3.50}$$

which means, as $\varepsilon \to 0$

$$L_3^0 \to \varepsilon\left[\frac{\partial}{\partial x}(x + Ax^3) + B\frac{\partial^2}{\partial x^2}\right]. \tag{8.3.51}$$

However, there are two distinct possibilities for L_2^0. In order for L_1^0 to be independent of ε, we must have κ independent of ε, which is reasonable. Thus, the limit (8.3.49) must be achieved by the product ab being proportional to ε. We consider various possibilities.

a) The Silent Slave—*a* Proportional to ε : We assume we can write

$$a = \varepsilon\tilde{a}\,. \tag{8.3.52}$$

We see that L_1^0 is independent of ε while L_2^0 and L_3^0 are proportional to ε. If we rescale time by

$$\tau = \varepsilon t\,, \tag{8.3.53}$$

then

$$\frac{\partial p}{\partial \tau} = \left(\frac{1}{\varepsilon}L_1 + L_2 + L_3\right)p\,, \tag{8.3.54}$$

where

$$\left.\begin{aligned} L_1 &= L_1^0\,,\\ L_2 &= L_2^0/\varepsilon\,,\\ L_3 &= L_3^0/\varepsilon\,. \end{aligned}\right\} \tag{8.3.55}$$

Clearly, the usual elimination procedure gives to lowest order

$$\frac{\partial \hat{p}}{\partial t} = L_3\hat{p} = \left[\frac{\partial}{\partial x}(x + Ax^3) + B\frac{\partial^2}{\partial x^2}\right]\hat{p}\,, \tag{8.3.56}$$

since L_2 does not become infinite as $\varepsilon \to 0$.

This corresponds exactly to eliminating α adiabatically, ignoring the fluctuations in α and simply setting the deterministic value in the x equation. I call it the "silent slave", since (in Haken's terminology) α is slaved by x and makes no contribution to the noise in the x equation. This is the usual form of slaving, as considered by Haken.

b) The Noisy Slave—*a* Proportional to $\varepsilon^{1/2}$: If we alternatively assume that both a and b are proportional to $\varepsilon^{1/2}$, we can write

$$a = \tilde{a}\varepsilon^{1/2}\,, \qquad b = \tilde{b}\varepsilon^{1/2}\,, \tag{8.3.57}$$

where

$$\tilde{a}\tilde{b} = \kappa A\,. \tag{8.3.58}$$

L_1^0 stays constant, L_3^0 is proportional to ε and

$$L_2^0 = \varepsilon^{1/2}L_2 + \text{ higher order terms in } \varepsilon\,, \tag{8.3.59}$$

where

$$L_2 = \tilde{a}\beta \frac{\partial}{\partial x} x \,. \tag{8.3.60}$$

Thus, the limiting equation is

$$\frac{\partial \hat{p}}{\partial \tau} = (L_3 - PL_2L_1^{-1}L_2)\hat{p} \,. \tag{8.3.61}$$

The term $PL_2L_1^{-1}L_2$ can be worked out as previously; we find

$$-PL_2L_1^{-1}L_2 = \tilde{a}^2 \frac{D^2}{2\kappa^2} \frac{\partial}{\partial x} x \frac{\partial}{\partial x} x \,, \tag{8.3.62}$$

so

$$\frac{\partial \hat{p}}{\partial \tau} = \left\{ \frac{\partial}{\partial x} \left[x \left[1 - \frac{\tilde{a}^2 D^2}{2\kappa^2} \right] + Ax^3 \right] + \frac{\partial^2}{\partial x^2} \left[B + \frac{\tilde{a}^2 D^2 x^2}{2\kappa^2} \right] \right\} \hat{p} \,. \tag{8.3.63}$$

I call this the "noisy slave", since the slave makes his presence felt in the final equation by adding noise (and affecting the drift, though this appears only in the Ito form as written; as a Stratonovich form, there would be no extra drift).

c) The General Case: Because we assume $ab \propto \varepsilon$, it can be seen that the second two terms in (8.3.47) are always proportional to ε^p, where $p > 1$, and hence are negligible (provided b is bounded). Thus, the only term of significance in L_2^0 is the first. Then it follows that if

$$a = \varepsilon^r \tilde{a} \,, \tag{8.3.64}$$

we have the following possibilities:

$r > \frac{1}{2}$: no effect from L_2 : limiting equation is (8.3.56),

$r = \frac{1}{2}$: limiting equation is (8.3.63)—a noisy slave,

$r < \frac{1}{2}$: the term $PL_2L_1^{-1}L_2$ becomes of order $\varepsilon^{2r-1} \to \infty$ and is dominant.

The equation is asymptotically (for $r < \frac{1}{2}$)

$$\frac{\partial \hat{p}}{\partial \tau} = \varepsilon^{2r-1} \tilde{a}^2 \frac{D^2}{2\kappa^2} \left(\frac{\partial}{\partial x} x \frac{\partial}{\partial x} x \right) \hat{p} \,. \tag{8.3.65}$$

These are quite distinct differences, all of which can be incorporated in the one formula, namely, in general

$$\frac{\partial \hat{p}}{\partial \tau} = \left[\frac{\partial}{\partial x}(x + Ax^3) + \frac{\partial^2}{\partial x^2} B + \varepsilon^{2r-1} \tilde{a}^2 \frac{D^2}{2\kappa^2} \frac{\partial}{\partial k} x \frac{\partial}{\partial x} x \right] \hat{p} \,. \tag{8.3.66}$$

In applying adiabatic elimination techniques, in general, one simply must take particular care to ensure that the correct dependence on small parameters of all constants in the system has been taken.

8.3.3 Adiabatic Elimination of Fast Variables: A Nonlinear Case

We want to consider the general case of two variables x and α which are coupled together in such a way that each affects the other, though the time scale of α is considerably faster than that of x.

Let us consider a system described by a pair of stochastic differential equations:

$$dx = [a(x) + b(x)\alpha]\, dt + c(x)\, dW_1(t)\,, \tag{8.3.67}$$

$$d\alpha = \gamma^2[A(\alpha) - f(x)]\, dt + \gamma \sqrt{2B(\alpha)}\, dW_2(t)\,. \tag{8.3.68}$$

If we naively follow the reasoning of Sect. 8.2, we immediately meet trouble. For in this limit, one would put

$$A(\alpha) - f(x) = -\sqrt{\frac{2B(\alpha)}{\gamma^2}}\frac{dW_2(t)}{dt}\,, \tag{8.3.69}$$

on the assumption that for large α, (8.3.68) is always such that $d\alpha/dt = 0$. But then to solve (8.3.69) for α in terms x yields, in general, some complicated nonlinear function of x and $dW_2(t)/dt$ whose behaviour is inscrutable.

If, however, $B(\alpha)$ is zero, then we can define $u_0(x)$ to be

$$A[u_0(x)] = f(x)\,, \tag{8.3.70}$$

and substitute in (8.3.67) to obtain

$$dx = [a(x) + b(x)u_0(x)]\, dt + c(x)\, dW_1(t)\,. \tag{8.3.71}$$

We shall devise a somewhat better procedure based on our previous methodology which can also take the effect of fluctuations in (8.3.68) into account.

The Fokker-Planck equation equivalent to (8.3.67, 8.3.68) is

$$\frac{\partial p}{\partial t} = (\gamma^2 L_1 + L_2 + L_3)p\,, \tag{8.3.72}$$

where

$$L_1 = \frac{\partial}{\partial \alpha}[f(x) - A(\alpha)] + \frac{\partial^2}{\partial \alpha^2}B(\alpha)\,, \tag{8.3.73}$$

and L_2 and L_3 are chosen with hindsight in order for the requirement $PL_2P = 0$ to be satisfied. Firstly, we choose P, as usual, to be the projector into the null space of L_1. We write $p_x(\alpha)$ for the stationary solution i.e., the solution of

$$L_1 P_x(\alpha) = 0\,, \tag{8.3.74}$$

so that $p_x(\alpha)$ explicitly depends on x, because L_1 explicitly depends on x through the function $f(x)$. The projector P is defined by

$$(PF)(x, \alpha) = p_x(\alpha) \int d\alpha' F(x, \alpha')\,, \tag{8.3.75}$$

for any function $F(x, \alpha)$.

We now define the function $u(x)$ as

$$u(x) = \int d\alpha\, \alpha p_x(\alpha) \equiv \langle \alpha \rangle_x\,. \tag{8.3.76}$$

Then we define

$$L_2 = -\frac{\partial}{\partial x}\{b(x)[\alpha - u(x)]\}\,, \tag{8.3.77}$$

$$L_3 = -\frac{\partial}{\partial x}[a(x) + b(x)u(x)] + \frac{1}{2}\frac{\partial^2}{\partial x^2}[c(x)]^2\,, \tag{8.3.78}$$

so that the term $\{\partial/\partial x\}b(x)u(x)$ cancels when these are added. Thus (8.3.72) is the correct Fokker-Planck equation corresponding to (8.3.67, 8.3.68).

Now we have

$$\begin{aligned} PL_2PF &= -p_x(\alpha)\int d\alpha' \frac{\partial}{\partial x}\{b(x)[\alpha' - u(x)]\}p_x(\alpha')\int d\alpha'' F(x,\alpha'')\,,\\ &= 0\,, \end{aligned} \tag{8.3.79}$$

since $\int \alpha p_x(\alpha)d\alpha = u(x)$.

It is of course true that

$$PL_1 = L_1P = 0\,, \tag{8.3.80}$$

but

$$PL_3 \neq L_3P\,. \tag{8.3.81}$$

We now carry out the normal procedure. Writing, as usual,

$$P\,\tilde{p}(s) = \tilde{v}(s)\,, \tag{8.3.82}$$

$$(1 - P)\tilde{p}(s) = \tilde{w}(s)\,, \tag{8.3.83}$$

and assuming $w(0) = 0$, we find

$$s\,\tilde{v}(s) = P(L_2 + L_3)\tilde{w}(s) + PL_3\tilde{v}(s) + v(0)\,, \tag{8.3.84}$$

$$\begin{aligned} s\,\tilde{w}(s) &= [\gamma^2 L_1 + (1-P)L_2 + (1-P)L_3]\tilde{w}(s) + L_2\tilde{v}(s)\\ &\quad + (1-P)L_3\tilde{v}(s)\,, \end{aligned} \tag{8.3.85}$$

so that

$$\begin{aligned} s\,\tilde{v}(s) &= PL_3\tilde{v}(s) + P(L_2 + L_3)\left[s - \gamma^2 L_1 - (1-P)L_2 - (1-P)L_3\right]^{-1}\\ &\quad \times [L_2 + (1-P)L_3]\tilde{v}(s) + v(0)\,. \end{aligned} \tag{8.3.86}$$

To second order we have simply

$$s\,\tilde{v}(s) \simeq \{(PL_3 - \gamma^{-2}P(L_2 + L_3)L_1^{-1}[L_2 + (1-P)L_3]\}\tilde{v}(s) + v(0)\,. \tag{8.3.87}$$

The term $PL_3\tilde{v}(s)$ is the most important term and yields the deterministic adiabatic elimination result. Writing

$$v(t) = p_x(\alpha)\hat{p}(x)\,, \tag{8.3.88}$$

we find

$$PL_3v(t) = p_x(\alpha)\int d\alpha' \left\{-\frac{\partial}{\partial x}[a(x) + b(x)u(x)] + \frac{1}{2}\frac{\partial^2}{\partial x^2}[c(x)^2]\right\} p_x(\alpha)\hat{p}(x)\,. \tag{8.3.89}$$

Since

$$\int d\alpha\, p_x(\alpha) = 1,$$

$$PL_3 v(t) = p_x(\alpha)\left\{-\frac{\partial}{\partial x}[a(x) + b(x)u(x)] + \frac{1}{2}\frac{\partial^2}{\partial x^2}[c(x)^2]\right\}\hat{p}(x), \tag{8.3.90}$$

the lowest-order differential equation is

$$\frac{\partial \hat{p}(x)}{\partial t} = -\frac{\partial}{\partial x}[a(x) + b(x)u(x)]\hat{p}(x) + \frac{1}{2}\frac{\partial^2}{\partial x^2}c(x)^2\hat{p}(x), \tag{8.3.91}$$

which is equivalent to the stochastic differential equation

$$dx = [a(x) + b(x)u(x)]\,dt + c(x)\,dW(t). \tag{8.3.92}$$

To this order, the equation of motion contains no fluctuating term whose origin is in the equation for α. However, the effect of completely neglecting fluctuations is given by (8.3.71) which is very similar to (8.3.92) but has $u_0(x)$ instead of $u(x)$. While it is expected that the average value of α in the stationary state would be similar to $u_0(x)$, it is not the same, and the similarity would only be close when the noise term $B(\alpha)$ was small.

Second-Order Corrections: It is possible to evaluate (8.3.87) to order γ^{-2}. At first glance, the occurrence of second derivatives in L_3 would seem to indicate that to this order, fourth derivatives occur since L_3 occurs twice. However, we can show that the terms involving fourth-order derivatives vanish.

Consider the expression

$$P(L_2 + L_3)L_1^{-1}[L_2 + (1 - P)L_3]\tilde{v}(s). \tag{8.3.93}$$

We know

$$\left.\begin{aligned} &\text{i)} \quad P\tilde{v}(s) = \tilde{v}(s),\\ &\text{ii)} \quad (1 - P)L_2P\tilde{v}(s) = L_2P\tilde{v}(s), \end{aligned}\right\} \tag{8.3.94}$$

where we have used $PL_2P = 0$.

Thus, (8.3.93) becomes

$$\begin{aligned} P(L_2 + L_3)L_1^{-1}(1 - P)(L_2 + L_3)P\tilde{v}(s) &= P\{PL_2 + [P, L_3] + L_3P\}\\ \times(1 - P)L_1^{-1}(1 - P)\{L_2P + [L_3, P] + PL_3\}\tilde{v}(s) \end{aligned} \tag{8.3.95}$$

where the *commutator* $[A, B]$ is defined by

$$[A, B] = AB - BA. \tag{8.3.96}$$

We have noted that L_1^{-1} commutes with $(1-P)$ and used $(1-P)^2 = (1-P)$ in (8.3.95) to insert another $(1 - P)$ before L_1^{-1}. We have also inserted another P in front of the whole expression, since $P^2 = P$. Using now

$$P(1 - P) = (1 - P)P = 0,$$

(8.3.95) becomes $P\{PL_2 + [P, L_3]\}L_1^{-1}(1 - P)\{L_2 + [L_3, P]\}\tilde{v}(s)$. (8.3.97)

We will now compute $[P, L_3]$:

$$(PL_3f)(x,\alpha) = P_x(\alpha)\left\{-\frac{\partial}{\partial x}[a(x)+b(x)u(x)] + \frac{1}{2}\frac{\partial^2}{\partial x^2}[c(x)]^2\right\}\int f(x,\alpha')d\alpha' . \tag{8.3.98}$$

and

$$(L_3P)f(x,\alpha) = \left\{-\frac{\partial}{\partial x}[a(x)+b(x)u(x)]\frac{1}{2}\frac{\partial^2}{\partial x^2}[a(x)]^2\right\} p_x(\alpha)\int d\alpha' f(x,\alpha') . \tag{8.3.99}$$

Subtracting these, and defining,

$$r_x(\alpha) = \frac{1}{p_x(\alpha)}\frac{\partial p_x(\alpha)}{\partial x}, \tag{8.3.100}$$

$$s_x(\alpha) = \frac{1}{p_x(\alpha)}\frac{\partial^2 p_x(\alpha)}{\partial x^2} . \tag{8.3.101}$$

One finds

$$([P,L_3]f)(x,\alpha) = r_x(\alpha)[a(x)+b(x)u(x)]Pf(x,\alpha) - \tfrac{1}{2}s_x(\alpha)c(x)^2Pf(x,\alpha)$$
$$-\, r_x(\alpha)P\frac{\partial}{\partial x}[c(x)^2f(x,\alpha)] . \tag{8.3.102}$$

The last term can be simplified even further since we are only interested in the case where $f(x,\alpha)$ is v, i.e.,

$$f(x,\alpha) = p_x(\alpha)\hat{p}(x) . \tag{8.3.103}$$

Then,

$$P\frac{\partial}{\partial x}c(x)^2p_x(\alpha)\hat{p}(x) = p_x(\alpha)\frac{\partial}{\partial x}c(x)^2\int d\alpha' p_x(\alpha')\hat{p}(x), \tag{8.3.104}$$

$$= p_x(\alpha)\frac{\partial}{\partial x}c(x)^2\hat{p}(x) . \tag{8.3.105}$$

We can further show that

$$P[P,L_3] = 0 . \tag{8.3.106}$$

For since

$$\int d\alpha\, p_x(\alpha) = 1 , \tag{8.3.107}$$

it follows that

$$\int d\alpha\, r_x(\alpha)p_x(\alpha) = \int d\alpha\, s_x(\alpha)p_x(\alpha) = 0 , \tag{8.3.108}$$

which is sufficient to demonstrate (8.3.106). Hence, instead of (8.3.97) we may write (8.3.97) in the form

$$PL_2L_1^{-1}\{L_2 + [L_3,P]\}\tilde{v}(s) , \tag{8.3.109}$$

and

$$[L_3,P]\tilde{v}(s) = -p_x(\alpha)\left\{r_x(\alpha)[a(x)+b(x)u(x)] - \tfrac{1}{2}s_x(\alpha)c(x)^2\right\}\hat{p}(x)$$
$$+\, p_x(\alpha)r_x(\alpha)\frac{\partial}{\partial x}[c(x)^2\hat{p}(x)] . \tag{8.3.110}$$

8.3.4 An Example with Arbitrary Nonlinear Coupling

We consider the pair of equations

$$\left.\begin{aligned} dx &= \gamma b(x)\alpha\, dt\,, \\ d\alpha &= -\gamma^2 A(x,\alpha,\gamma)\, dt + \gamma\sqrt{2B(x,\alpha,\gamma)}\, dW(t)\,, \end{aligned}\right\} \tag{8.3.111}$$

and assume the existence of the following limits and asymptotic expansions

$$\left.\begin{aligned} A(x,\alpha,\gamma) &\sim \sum_{n=0}^{\infty} A_n(x,\alpha)\gamma^{-n}\,, \\ B(x,\alpha,\gamma) &\sim \sum_{n=0}^{\infty} B_n(x,\alpha)\gamma^{-n}\,. \end{aligned}\right\} \tag{8.3.112}$$

These expansions imply that there is an asymptotic stationary distribution of α at fixed x given by

$$p_s(\alpha, x) = \lim_{\gamma\to\infty} p_s(\alpha, x, \gamma)\,, \tag{8.3.113}$$

$$p_s(\alpha, x) \propto B_0(x,\alpha)^{-1} \exp\left\{\int d\alpha [A_0(x,\alpha)/B_0(x,\alpha)]\right\}\,. \tag{8.3.114}$$

We assume that $A_0(x,\alpha)$ and $B_0(x,\alpha)$ are such that

$$\langle\alpha(x)\rangle_s = \int d\alpha\, \alpha p_s(\alpha, x) = 0\,, \tag{8.3.115}$$

so that we deduce from (8.3.112) that, for finite γ

$$\langle\alpha(x,\gamma)\rangle_s = \int d\alpha\, \alpha p_s(\alpha, x, \gamma) \sim \alpha_0(x)\gamma^{-1}\,, \tag{8.3.116}$$

where $\alpha_0(x)$ can be determined from (8.3.112).

We define the new variables

$$\left.\begin{aligned} \beta &= \alpha - \frac{1}{\gamma}\alpha_0(x)\,, \\ x_1 &= x\,. \end{aligned}\right\} \tag{8.3.117}$$

In terms of these variables the Fokker-Planck operator becomes (the Jacobian is a constant, as usual) on changing x_1 back to x

$$\begin{aligned} L = &-\frac{\partial}{\partial x}\alpha_0(x)b(x) \\ &-\gamma\beta\frac{\partial}{\partial x}b(x) + \frac{1}{\gamma}\alpha_0{}'(x)\alpha_0(x)b(x)\frac{\partial}{\partial\beta} + \frac{\partial}{\partial\beta}\beta\alpha_0{}'(x)b(x) \\ &+\gamma^2\left[\frac{\partial}{\partial\beta}A\left(\frac{\alpha_0(x)}{\gamma}+\beta, x, \gamma\right) + \frac{\partial^2}{\partial\beta^2}B\left(\frac{\alpha_0(x)}{\gamma}+\beta, x, \gamma\right)\right]. \end{aligned} \tag{8.3.118}$$

By using the asymptotic expansions (8.3.112), we can write this as

$$L = L_3 + \gamma L_2(\gamma) + \gamma^2 L_1\,, \tag{8.3.119}$$

with

$$L_3 \quad = \frac{\partial}{\partial x}\alpha_0(x)b(x)\,, \tag{8.3.120}$$

$$L_1 \quad = \frac{\partial}{\partial \beta}A_0(\beta,x) + \frac{\partial^2}{\partial \beta^2}B_0(\beta,x)\,, \tag{8.3.121}$$

$$L_2(\gamma) = L_2 + O(\gamma^{-1})\,, \tag{8.3.122}$$

$$L_2 \quad = -\beta\frac{\partial}{\partial x}b(x) - \frac{\partial}{\partial \beta}\left[\frac{\partial A_0(\beta,x)}{\partial \beta}\alpha_0(x) + A_1(\beta,x)\right] - \frac{\partial^2}{\partial \beta^2}\left[\frac{\partial B_0(\beta,x)}{\partial \beta}\alpha_0(x) + B_1(\beta,x)\right]. \tag{8.3.123}$$

We note that L_3 and L_1 do not commute, but, as in Sect. 8.1.2c, this does not affect the limiting result,

$$\frac{\partial \hat{p}}{\partial t} = (L_3 - PL_2L_1^{-1}L_2)\hat{p}\,. \tag{8.3.124}$$

The evaluation of the $PL_2L_1^{-1}L_2$ term is straightforward, but messy. We note that the terms involving $\partial/\partial\beta$ vanish after being operated on by P. From the explicit form of $p_s(\alpha, x)$ one can define $G(\beta, x)$ by

$$G(\beta,x)p_s(\beta,x) = \frac{\partial}{\partial \beta}\left\{\left[\frac{\partial A_0(\beta,x)}{\alpha\beta}\alpha_0(x) + A_1(\beta,x)\right]p_s(\beta,x)\right\} + p_s(\beta,x)\left\{\frac{\partial^2}{\partial \beta^2}\left[\frac{\partial B_0(\beta,x)}{\alpha\beta}\alpha_0(x) + B_1(\beta,x)\right]\right\}, \tag{8.3.125}$$

and one finds that

$$PL_2L_1^{-1}L_2\hat{p} = \left\{\frac{\partial}{\partial x}b(x)D(x)\frac{\partial}{\partial x}b(x) + \frac{\partial}{\partial x}b(x)E(x)\right\}\hat{p} \tag{8.3.126}$$

with

$$\left.\begin{aligned} D(x) &= \int_0^\infty dt\,\langle\beta(t),\beta(0)\,|\,x\rangle\,,\\ E(x) &= \int_0^\infty dt\,\langle\beta(t),G(\beta,x)\,|\,x\rangle\,, \end{aligned}\right\} \tag{8.3.127}$$

where $\langle\ldots|x\rangle$ indicates an average over $p_s(\beta, x)$. This is a rather strong adiabatic elimination result, in which an arbitrary nonlinear elimination can be handled and a finite resulting noise dealt with. The calculation is simpler than that in the previous section, since the terms involving L_3 are of lower order here than those involving L_2.

9. Beyond the White Noise Limit

In the previous chapter we considered a stochastic system, expressible as a differential equation such as

$$\frac{dx}{dt} = a(x) + b(x)\xi_\gamma(t),$$

in which x is a variable in which we have an interest, and $\xi_\gamma(t)$ represents some fluctuating function, whose properties are only statistically known. The formalism aimed to develop an approximate equation for this equation under the condition that $\xi_\gamma(t)$ is approximately white noise. The result was that provided we can write $\xi_\gamma(t) = \gamma\alpha(\gamma^2 t)$, where $\alpha_0(t)$ is a stationary stochastic process, then an approximate equation for $x(t)$ in the limit of very large γ is the Stratonovich stochastic differential equation

$$(S)\, dx = a(x)\, dt + \sqrt{2D}b(x)\, dW(t),$$

in which $D = \int dt\, \langle \alpha_0(t)\alpha_0(0)\rangle$.

In Chap. 8 we extended this result to several more complicated systems, but the aim was always to obtain a stochastic differential equation of the form in which, apart from D, all reference to any specific properties of $\xi_\gamma(t)$ was eliminated.

There are situations in which this complete elimination of $\xi_\gamma(t)$ can be considered a disadvantage. For example, the noise $\xi_\gamma(t)$ may be a physically measurable quantity, such as a fluctuating laser field, or a noisy electric voltage, and it may be both possible and useful to measure correlations, such as $\langle x(t)\xi_\gamma(t')\rangle$, for which the stochastic differential equation in the white noise limit provides no clues.

Indeed, there is a result that comes from electrical engineering, that in a linear system driven with white noise, the correlation function of the system variable with the applied white noise is the impulse response function of the system, i.e., the response of the system to a delta function input. In this chapter we shall show that essentially this result is true even for a nonlinear system, and that it has generalisations to the case of non-white noise, which are however dependent on the exact structure of the noise. The method of solution uses a kind of perturbation theory developed by *C. Bloch* [9.1] which was introduced into this kind of problem by Titulaer [9.2, 9.3], who showed that a large class of elimination methods (including the Chapman-Enskog solution of the Boltzmann equation) could advantageously be considered as the kind of degenerate perturbation theory developed by Bloch.

The structure of the chapter is as follows: In Sect. 9.1, we define the kind of problem being studied, and outline the eigenfunction structure of the problem. Sect. 9.2 outlines Bloch's perturbation theory in a form adapted to this problem, and Sect. 9.2.2 applies it to the expansion, to first order, of the conditional probability.

We give some specific examples, namely, when the eliminated variable corresponds to the Ornstein-Uhlenbeck process, or to the random telegraph process. The results are somewhat different from each other. We also show that an arbitrary number of independent eliminated variables can be treated, which enables the results to be extended to situations in which the eliminated variable can be treated as the sum of a number of independent Markov variables. In particular, this allows results to be derived for almost arbitrary Gaussian eliminated variables. Sect. 9.3 shows how to compute the correlations between input noise and the variable x. The results take a simple, almost intuitive form, valid for almost arbitrary noise inputs, including non-Markovian Gaussian noise. When we observe on a coarse time scale we obtain particularly simple white noise results, which generalise the electrical engineering result involving the impulse response function.

9.1 Specification of the Problem

As in Sect. 8.1.1 we want to consider a system governed by a differential equation of the form

$$\frac{dx}{dt} = a(x) + b(x)\gamma p(\gamma^2 t) . \tag{9.1.1}$$

For definiteness, we assume that $p(t)$ is a Markov diffusion process, so that we can write a Fokker-Planck equation for the conditional probability of the two variables, x, and p, in the form

$$\partial_t P(x, p, t \,|\, x_0, p_0, t_0) = (\gamma^2 L_1 + \gamma L_2 + L_3) P(x, p, t \,|\, x_0, p_0, t_0) , \tag{9.1.2}$$

with

$$L_1 = -\frac{\partial}{\partial p} A(p) + \frac{\partial^2}{\partial p^2} B(p) , \tag{9.1.3a}$$

$$L_2 = -\frac{\partial}{\partial x} b(x) p , \tag{9.1.3b}$$

$$L_3 = -\frac{\partial}{\partial x} a(x) . \tag{9.1.3c}$$

In fact L_1 can be any evolution operator such that the evolution of p is a Markov process independent of x—for example p could be the random telegraph process, with p taking on values ± 1—the derivation which follows can easily be adapted appropriately.

The operators L_1, L_2 and L_3 are independent of t, so the conditional probability $P(x, p, t \,|\, x_0, p_0, t_0)$ will be a function of $t - t_0$ only, and hence t_0 will be set equal to zero in most of what follows.

The initial condition for the conditional probability is

$$P(x, p, t \,|\, x_0, p_0, 0) = \delta(x - x_0)\delta(p - p_0) , \tag{9.1.4}$$

and we wish to find an asymptotic expansion for $P(x, p, t \,|\, x_0, p_0, to)$ valid for large γ. Using this we will be able to construct asymptotic expansions of all correlation functions of x and p.

9.1.1 Eigenfunctions of L_1

We will need to know the eigenfunctions of L_1. We call these $P_\lambda(p)$, such that

$$L_1 P_\lambda(p) = -\lambda P_\lambda(p), \tag{9.1.5}$$

where $\lambda > 0$ is required for a Markov process with a stationary state. The eigenfunctions of the backward Fokker-Planck operator L_1^*, defined by

$$L_1^* = A(p)\frac{\partial}{\partial p} + B(p)\frac{\partial^2}{\partial p^2} \tag{9.1.6}$$

are $Q_\lambda(p)$, and satisfy

$$L_1^* Q_\lambda(p) = -\lambda Q_\lambda(p), \tag{9.1.7}$$

These are related, as in (5.2.5), by

$$P_\lambda(p) = P_0(p) Q_\lambda(p), \tag{9.1.8}$$

and $P_0(p)$ is of course the stationary distribution function for p. As shown in Sect. 5.4, the system is bi-orthogonal

$$\int dp\, Q_\lambda(p) P_\mu(p) = \delta_{\lambda,\mu}. \tag{9.1.9}$$

The normalisation is such that

$$Q_0(p) = 1. \tag{9.1.10}$$

9.1.2 Projectors

We define two kinds of projectors, firstly from the eigenspace λ to itself

$$\mathscr{P}_\lambda f(p) = P_\lambda(p) \int dp'\, Q_\lambda(p') f(p'), \tag{9.1.11}$$

and from the eigenspace μ to the eigenspace λ,

$$\mathscr{P}_{\lambda\mu} f(p) = P_\lambda(p) \int dp'\, Q_\mu(p') f(p'). \tag{9.1.12}$$

Clearly

$$\mathscr{P}_{\lambda\mu}\mathscr{P}_{\alpha\beta} = \delta_{\mu\alpha}\mathscr{P}_{\lambda\beta}, \tag{9.1.13}$$

and

$$\mathscr{P}_{\lambda\lambda} = \mathscr{P}_\lambda. \tag{9.1.14}$$

It is not conventional to consider operators like $\mathscr{P}_{\lambda\mu}$ for $\mu \neq \lambda$ to be projectors, since $\mathscr{P}_{\lambda\mu}^2 \neq \mathscr{P}_{\lambda\mu}$. On the other hand, a purely geometric point of view will find no difficulty in viewing a linear mapping of one space to another as a projection. For this reason, and for linguistic simplicity, one can view $\mathscr{P}_{\lambda\mu}$ as a non-diagonal projector, called simply a "projector" for brevity.

a) An expression for $pP_\lambda(p)$: Because the operator L_2 contains a factor p it will be necessary to define some constants, $c_{\lambda,\mu}$ by

$$pP_\lambda(p) = \sum_\mu c_{\lambda\mu} P_\mu(p), \tag{9.1.15}$$

It follows from the definition of the projector, that

$$\mathscr{P}_\mu p \mathscr{P}_\lambda = c_{\lambda\mu} \mathscr{P}_{\mu\lambda}. \tag{9.1.16}$$

b) Solution of the Problem as perturbation of Eigenvalues: If we set

$$\tau = \gamma^2 t, \quad \epsilon = 1/\gamma, \tag{9.1.17}$$

the equation of motion (9.1.2) can be written

$$\frac{\partial P}{\partial \tau} = (L_1 + \epsilon L_2 + \epsilon^2 L_3)P. \tag{9.1.18}$$

We shall solve this equation by obtaining the eigenvalues and eigenfunctions for the operator $L_1 + \epsilon L_2 + \epsilon^2 L_3$ as a perturbation theory in ϵ. The unperturbed eigenvalue problem has the eigenvalues λ, of L_1, and these are infinitely degenerate. Including L_2 and L_3 will split the degeneracy: What we do is to reduce the eigenvalue problem for (9.1.18) to a separate eigenvalue problem for each λ. The eigenvalues for $\lambda = 0$ will be the same as those of the Fokker-Planck operator corresponding to the limiting stochastic differential equation $(S)\, dx = a(x)\, dt + \sqrt{2D}b(x)\, dW(t)$.

9.2 Bloch's Perturbation Theory

The problem we have to treat is very similar to degenerate perturbation theory in quantum mechanics, but differs in that the perturbation theory is not expressed in terms of the eigenvalues of a Hermitian Hamiltonian. This kind of problem is most compactly solved by the methods of *Bloch* [9.1]—however, since this theory is not widely used, we shall now give a brief outline of it.

We wish to calculate the eigenfunctions and eigenvalues of $L_1 + \epsilon L_2 + \epsilon^2 L_3$. For simplicity we write

$$L = L_1 + \epsilon L_2 + \epsilon^2 L_3 = L_1 + V. \tag{9.2.1}$$

As $\epsilon \to 0$, the eigenvalues of L approach the infinitely degenerate set of eigenfunctions of L_1, which can be written as $P_\lambda(p)F(x)$, where $F(x)$ is an arbitrary function. Abstractly, we denote an eigenfunction which approaches the eigenspace λ as $\epsilon \to 0$ by

$$|\lambda, \alpha\rangle, \tag{9.2.2}$$

and its eigenvalue by

$$-E_{\lambda\alpha} - \lambda, \tag{9.2.3}$$

and it is assumed that all the $E_{\lambda\alpha}$ are different. It is also necessary to consider two other sets of vectors as follows.

The projections of $|\lambda\alpha\rangle$ on the eigenspace λ, these are

$$|\lambda\alpha) = \mathscr{P}_\lambda |\lambda\alpha\rangle, \tag{9.2.4}$$

The vectors $|\lambda\alpha)$ can be written in the form

$$|\lambda\alpha) = P_\lambda(p) f_{\alpha\lambda}(x) . \tag{9.2.5}$$

There is no scalar product yet defined on the space spanned by the vectors $|\lambda\alpha)$, except in so far as is already defined on the $P_\lambda(p)$, $Q_\lambda(p)$. We choose the adjoint vectors as

$$(\lambda\alpha| = Q_\lambda(p) g_{\alpha\lambda}(x) , \tag{9.2.6}$$

where

$$\int dx\, g_{\alpha\lambda}(x) f_{\beta\lambda}(x) = \delta_{\alpha\beta} . \tag{9.2.7}$$

We also assume completeness so that

$$\sum_\alpha g_{\alpha\lambda}(x) f_{\alpha\lambda}(x') = \delta(x - x'). \tag{9.2.8}$$

In the vectorial notation, this means that

$$(\lambda\alpha\,|\,\lambda'\alpha') = \delta_{\lambda\lambda'}\delta_{\alpha\alpha'} , \tag{9.2.9}$$

$$\sum_{\alpha\lambda} |\lambda\alpha)(\lambda\alpha| = 1. \tag{9.2.10}$$

9.2.1 Formalism for the Perturbation Theory

a) Fundamental Operators: We need two fundamental operators in terms of which the perturbation theory is to be formulated.

i) We define U_λ as the operator which transforms the *projected vectors* $|\lambda\alpha)$ in the subspace λ into the corresponding exact *eigenvectors* of L, $|\lambda\alpha\rangle$, and which is null on all other projected vectors $|\mu\beta)$ with $\mu \neq \lambda$; thus

$$U_\lambda|\mu\beta) = \delta_{\lambda\mu}|\lambda\beta\rangle . \tag{9.2.11}$$

The operator U_λ can obviously be written

$$U_\lambda = \sum_\alpha |\lambda\alpha\rangle(\lambda\alpha| , \tag{9.2.12}$$

ii) There are two identities which are easily verified

$$\mathcal{P}_\lambda \mathcal{U}_\lambda = \mathcal{P}_\lambda , \tag{9.2.13}$$

$$\mathcal{U}_\lambda \mathcal{P}_\lambda = \mathcal{U}_\lambda . \tag{9.2.14}$$

iii) The second operator we need is formed from U_λ thus

$$A_\lambda \equiv \mathcal{P}_\lambda V U_\lambda , \tag{9.2.15}$$

where V is defined by (9.2.1). This is also an operator which is non-null only in the eigenspace λ.

The significance of this operator is that although it acts only within the subspace λ, its eigenvalues are the exact eigenvalues we need to determine, $-E_{\lambda\alpha}$, and its eigenfunctions are $|\lambda\alpha)$. This can be proved as follows: Firstly

$$A_\lambda|\lambda,\alpha) = \mathscr{P}_\lambda V U_\lambda|\lambda,\alpha) = \mathscr{P}_\lambda(-L_1 + L)|\lambda\alpha\rangle. \tag{9.2.16}$$

The operator $-L_1$ is equivalent to $-L_1^*$ acting leftwards on $\mathscr{P}_\lambda$, and gives the eigenvalue λ; on the other hand the operator L acting on the eigenstate $|\lambda\alpha\rangle$ by definition gives the eigenvalue $-\lambda - E_{\lambda\alpha}$. Hence we find

$$A_\lambda|\lambda,\alpha) = \mathscr{P}_\lambda(-E_{\lambda\alpha})|\lambda\alpha\rangle = -E_{\lambda\alpha}|\lambda,\alpha) \,. \tag{9.2.17}$$

Consequently, we can write

$$A_\lambda = -\sum_\alpha E_{\lambda\alpha}|\lambda\alpha)(\lambda\alpha| \,. \tag{9.2.18}$$

b) Nonlinear equation for U_λ: Following Bloch, a nonlinear equation for U_λ is developed, which enables a perturbative expansion to be made. We start from the eigenvalue equation for A_λ, which yields

$$-E_{\lambda\alpha}|\lambda\alpha) = A_\lambda|\lambda\alpha) = \mathscr{P}_\lambda V\mathscr{U}_\lambda|\lambda\alpha) = \mathscr{P}_\lambda V|\lambda\alpha\rangle \,, \tag{9.2.19}$$

so that multiplying by U_λ from the left, and using (9.2.14), one obtains

$$U_\lambda V|\lambda\alpha\rangle = -E_{\lambda\alpha}|\lambda\alpha\rangle \,. \tag{9.2.20}$$

However, by definition

$$(L + \lambda)|\lambda\alpha\rangle = -E_{\lambda\alpha}|\lambda\alpha\rangle \,, \tag{9.2.21}$$

so that equating with (9.2.20) we can find an equation which does not explicitly involve the eigenvalue $E_{\lambda\alpha}$

$$(L + \lambda - U_\lambda V)|\lambda\alpha\rangle = 0 \,. \tag{9.2.22}$$

Multiply this equation on the right with $(\lambda\alpha|$ and sum over α to obtain from (9.2.12), the definition of $\mathscr{U}_\lambda$,

$$(L + \lambda - U_\lambda V)U_\lambda = 0 \,. \tag{9.2.23}$$

This can be rewritten

$$(L_1 + \lambda)U_\lambda = VU_\lambda - U_\lambda VU_\lambda \,. \tag{9.2.24}$$

Define the projector complementary to $\mathscr{U}_\lambda$

$$\mathscr{Q}_\lambda = 1 - \mathscr{P}_\lambda = \sum_{\mu\neq\lambda} \mathscr{P}_\mu \,. \tag{9.2.25}$$

Then, using the formulae (9.2.13, 9.2.14) that we can write

$$U_\lambda = \mathscr{P}_\lambda U_\lambda + \mathscr{Q}_\lambda U_\lambda = \mathscr{P}_\lambda + \mathscr{Q}_\lambda U_\lambda \,. \tag{9.2.26}$$

Substitute (9.2.26) into (9.2.24), and note $L_1\mathscr{P}_\lambda = -\lambda\mathscr{P}_\lambda$, to obtain

$$-(L_1 + \lambda)\mathscr{Q}_\lambda U_\lambda = VU_\lambda - U_\lambda VU_\lambda = \mathscr{Q}_\lambda(VU_\lambda - U_\lambda VU_\lambda) \,, \tag{9.2.27}$$

by use of (9.2.26) and (9.2.13, 9.2.14). In the space into which $\mathscr{Q}_\lambda$ projects, $L_1 + \lambda$ has an inverse, since the eigenvalue λ does not occur.

$$\mathscr{Q}_\lambda U_\lambda = -(L_1 + \lambda)^{-1}\mathscr{Q}_\lambda(VU_\lambda - U_\lambda VU_\lambda) \,. \tag{9.2.28}$$

Finally, we can now use (9.2.26) to write the equation for $\mathscr{U}_\lambda$

$$U_\lambda = \mathscr{P}_\lambda - (L_1 + \lambda)^{-1}\mathscr{Q}_\lambda(VU_\lambda - U_\lambda VU_\lambda). \tag{9.2.29}$$

c) Perturbation Theory for U_λ and A_λ: We can solve (9.2.29) perturbatively in powers of ϵ. We will solve this only to second order;

$$U_\lambda = U_0^\lambda + \epsilon U_1^\lambda + \epsilon^2 U_2^\lambda + \dots \tag{9.2.30}$$

and correspondingly from (9.2.15) we can write

$$A_\lambda = \epsilon A_0^\lambda + \epsilon^2 A_1^\lambda + \epsilon^3 A_2^\lambda + \dots \tag{9.2.31}$$

Then it is easy to show that

$$\left.\begin{aligned} U_0^\lambda &= \mathscr{P}_\lambda\,, \\ U_1^\lambda &= -(\lambda + L_1)^{-1}\mathscr{Q}_\lambda L_2 U_0^\lambda\,, \\ U_2^\lambda &= -(\lambda + L_1)^{-1}\mathscr{Q}_\lambda(L_2 U_1^\lambda + L_3 U_0^\lambda - U_1^\lambda L_2 U_0^\lambda)\,, \end{aligned}\right\} \tag{9.2.32}$$

and

$$\left.\begin{aligned} A_0^\lambda &= \mathscr{P}_\lambda L_2 U_0^\lambda\,, \\ A_1^\lambda &= \mathscr{P}_\lambda(L_2 U_1^\lambda + L_3 U_0^\lambda)\,, \\ A_2^\lambda &= \mathscr{P}_\lambda(L_2 U_2^\lambda + L_3 U_1^\lambda)\,. \end{aligned}\right\} \tag{9.2.33}$$

9.2.2 Application of Bloch's Perturbation Theory

The results of the procedure summarised by the two sets of equations (9.2.32, 9.2.33) contain the complete solution to the problem of constructing the perturbation series for the conditional probability, which we will now demonstrate.

a) The Eigenvalue Spectrum to Order Zero: From (9.2.32)

$$U_0^\lambda = \mathscr{P}_\lambda\,, \tag{9.2.34}$$

and from (9.2.33)

$$A_0^\lambda = \mathscr{P}_\lambda L_2 \mathscr{P}_\lambda = \mathscr{P}_\lambda \hat{L}_2 p \mathscr{P}_\lambda = L_2 c_{\lambda\lambda} \mathscr{P}_\lambda\,, \tag{9.2.35}$$

where we used (9.1.16), and have defined

$$\hat{L}_2 = -\frac{\partial}{\partial x} b(x)\,. \tag{9.2.36}$$

i) In many situations, it is found that $c_{\lambda\lambda} = 0$ This is related to the condition $PL_2P = 0$ given in (8.1.20) as a requirement for the adiabatic elimination, which would be equivalent to $c_{00} = 0$. However, using Bloch's method, neither condition is mandatory for the validity of the perturbation procedure.

ii) It is not difficult to see that $c_{\lambda\lambda}$ will occur only in the evaluation of A_0^λ. If it vanishes, then $A_0^\lambda = 0$, and the perturbation expansion commences only at the next order.

iii) If it does not vanish, we then find that

$$A_0^\lambda = -c_{\lambda\lambda}\frac{\partial}{\partial x}b(x)\,, \tag{9.2.37}$$

and this is the most significant term in the perturbation expansion. It corresponds to a non-stochastic driving force given by the mean value of $\gamma p(\gamma^2 t)$.

b) The Eigenvalue Spectrum to Order One: We will develop the results on the assumption that $c_{\lambda\lambda} = 0$, so that $\gamma p(\gamma^2 t)$ has zero mean value in all eigenfunctions, and is thus a truly random noise term. In this case, it is only at order one that we get the basic nontrivial results. From (9.2.32)

$$U_1^\lambda = -(\lambda + L_1)^{-1}\mathcal{Q}_\lambda \hat{L}_2 p U_0^\lambda = \hat{L}_2 \sum_{\mu\neq\lambda}(\mu-\lambda)^{-1}\mathcal{P}_\mu p \mathcal{P}_\lambda\,, \tag{9.2.38}$$

$$= L_2 \sum_{\mu\neq\lambda}(\mu-\lambda)^{-1}c_{\lambda\mu}\mathcal{P}_{\mu\lambda}\,. \tag{9.2.39}$$

Thus, from (9.2.33)

$$A_1^\lambda = \mathcal{P}_\lambda(L_2 U_1^\lambda + L_3 U_0^\lambda)\,, \tag{9.2.40}$$

$$= \left[L_3 + (\hat{L}_2)^2 \sum_{\mu\neq\lambda}\frac{c_{\lambda\mu}c_{\mu\lambda}}{\mu-\lambda}\right]\mathcal{P}_\lambda\,. \tag{9.2.41}$$

Hence, to this order the eigenvalue equation (9.2.17) becomes

$$\left[L_3 + (\hat{L}_2)^2 \sum_{\mu\neq\lambda}\frac{c_{\lambda\mu}c_{\mu\lambda}}{\mu-\lambda}\right]|\lambda,\alpha) = -E_{\lambda\alpha}|\lambda,\alpha)\,. \tag{9.2.42}$$

Putting in the specific forms for $|\lambda,\alpha)$ given in (9.2.5), and writing

$$D_\lambda = \sum_{\mu\neq\lambda}\frac{c_{\lambda\mu}c_{\mu\lambda}}{\mu-\lambda}\,, \tag{9.2.43}$$

we find that (9.2.42) becomes

$$\epsilon^2\left[-\frac{\partial}{\partial x}a(x) + D_\lambda\frac{\partial}{\partial x}b(x)\frac{\partial}{\partial x}b(x)\right]f_{\alpha\lambda}(x) = -E_{\lambda\alpha}f_{\alpha\lambda}(x)\,. \tag{9.2.44}$$

c) Fokker-Planck Equation: The corrections to the eigenvalues, $E_{\lambda\alpha}$, are thus the eigenvalues of the Fokker-Planck operator on the LHS of equation (9.2.44) and are of order ϵ^2. Only the change to the eigenvalue $\lambda = 0$ will be significant—the perturbation of the higher eigenvalues is quite infinitesimal compared to the unperturbed values of the eigenvalues. Fortunately, it is not necessary to actually solve (9.2.44) as an eigenvalue equation to obtain the results we desire.

9.2.3 Construction of the Conditional Probability

a) Eigenfunctions of L: We now construct these from those of A_λ by using the operator U_λ as defined in Sect.9.2.1. As shown in (9.2.18), the eigenfunctions of A_λ are

$$|\lambda\alpha) \equiv P_\lambda(p) f_{\alpha\lambda}(x) , \tag{9.2.45}$$

and we know that, to order one ,

$$U_\lambda = U_0^\lambda + \epsilon U_1^\lambda + \cdots = \mathscr{P}_\lambda + \epsilon \sum_{\mu\neq\lambda} (\mu - \lambda)^{-1} \mathscr{P}_{\mu\lambda} c_{\lambda\mu} \hat{L}_2 . \tag{9.2.46}$$

Thus, the eigenfunctions of L can be constructed by putting these together thus:

$$\begin{aligned} |\lambda\alpha\rangle &= U_\lambda|\lambda\alpha) = P_\lambda(p) f_{\alpha\lambda}(x) + \epsilon \sum_{\mu,\lambda} \frac{c_{\lambda\mu}}{\lambda - \mu} P_\mu(p) \hat{L}_2 f_{\alpha\lambda}(x) , \\ &\equiv G_{\lambda\alpha}(p, x) . \end{aligned} \tag{9.2.47}$$

b) Conditional Probability: This can be expressed as a linear combination of these eigenfunctions multiplied by the time dependence appropriate to the eigenvalue thus:

$$P(x, p, \tau \,|\, x_0, p_0, 0) = \sum_{\lambda\neq\alpha} e^{-(\lambda+\epsilon^2 E_{\lambda\alpha})\tau} G_{\lambda\alpha}(p, x) R_{\lambda\alpha}(p_0, x_0) , \tag{9.2.48}$$

where the $R_{\lambda\alpha}(p_0, x_0)$ are coefficients to be determined from initial conditions. For convenience, we define the quantity

$$K_\lambda(x, t \,|\, x_0, p_0) \equiv \sum_\alpha e^{-E_{\lambda\alpha} t} R_{\lambda\alpha}(p_0, x_0) f_{\alpha\lambda}(x) , \tag{9.2.49}$$

which is a solution of the equation of the Fokker-Planck form

$$\left[-\frac{\partial}{\partial x} a(x) + D_\lambda \frac{\partial}{\partial x} b(x) \frac{\partial}{\partial x} b(x) \right] K_\lambda = \frac{\partial K_\lambda}{\partial t} , \tag{9.2.50}$$

since the $f_{\alpha\lambda}(x)$ are solutions of (9.2.44). We can thus write the conditional probability as

$$\begin{aligned} &P(x, p, t \,|\, x_0, p_0, 0) = \\ &\quad \sum_\lambda P_\lambda(p) \left[e^{-\lambda\tau} K_\lambda(x, t \,|\, x_0, p_0) + \epsilon \sum_{\lambda\neq\mu} \frac{c_{\mu\lambda}}{\lambda - \mu} \hat{L}_2 K_\mu(x, t \,|\, x_0, p_0) e^{-\mu\tau} \right] . \end{aligned} \tag{9.2.51}$$

Using the initial condition of $P(x, p, t \,|\, x_0, p_0, 0)$ in the form

$$\begin{aligned} P(x, p, 0 \,|\, x_0, p_0, 0) &= \delta(x - x_0)\delta(p - p_0) \\ &= \delta(x - x_0) \sum_\lambda Q_\lambda(p_0) P_\lambda(p) , \end{aligned} \tag{9.2.52}$$

we find the initial condition for K_λ is

$$\delta(x - x_0) Q_\lambda(p_0) = K_\lambda(x, 0 \,|\, x_0, p_0) + \epsilon \sum_{\mu\neq\lambda} \frac{c_{\mu\lambda}}{\lambda - \mu} \hat{L}_2 K_\mu(x, 0 \,|\, x_0, p_0) . \tag{9.2.53}$$

This can be solved to first order in ϵ as

$$K_\lambda(x, 0 \,|\, x_0, p_0) = \delta(x - x_0) Q_\lambda(p_0) - \epsilon \sum_{\mu\neq\lambda} \frac{c_{\mu\lambda}}{\lambda - \mu} \hat{L}_2 \delta(x - x_0) . \tag{9.2.54}$$

We now define the quantity $H_\lambda(x, 0 \,|\, x_0)$ as the solution of (9.2.50) subject to the initial condition

$$H_\lambda(x, 0 \,|\, x_0) = \delta(x - x_0) , \tag{9.2.55}$$

(which is essentially, a conditional probability for the Markov process represented by (9.2.50)). Notice that the backward operator L_2^* satisfies

$$L_2^*(x_0)\delta(x-x_0) = \hat{L}_2(x)\delta(x-x_0)\,, \tag{9.2.56}$$

so that

$$K_\lambda(x,t\,|\,x_0,p_0) = Q_\lambda(p_0)H_\lambda(x,t\,|\,x_0) - \epsilon \sum_{\mu\neq\lambda} \frac{c_{\mu\lambda}Q_\mu(p_0)}{\lambda-\mu}\hat{L}_2^*(x_0)H_\lambda(x,t\,|\,x_0)\,. \tag{9.2.57}$$

The solution for the conditional probability, to first order in ϵ, is therefore

$$\begin{aligned} P(x,p,t\,|\,x_0,p_0,0) = \sum_\lambda P_\lambda(p)\bigg\{ & e^{-\lambda\tau}Q_\lambda(p_0)H_\lambda(x,t\,|\,x_0) \\ & + \epsilon \sum_{\mu\neq\lambda} \frac{c_{\mu\lambda}Q_\mu(p_0)}{\lambda-\mu}\Big[e^{-\mu\tau}\hat{L}_2(x)H_\mu(x,t\,|\,x_0) - e^{-\lambda\tau}\hat{L}_2^*(x_0)H_\lambda(x,t\,|\,x_0)\Big]\bigg\}\,, \end{aligned} \tag{9.2.58}$$

in which $\tau = t/\epsilon^2$.

c) Two-time scale structure: The solution shows a two-time scale structure, in which we see that in a time of order $\gg \epsilon^2$ all exponential terms with non zero argument will be negligible. After that time, the solution will take the form

$$\begin{aligned} P(x,p,t\,|\,x_0,p_0,0) \approx{}& P_0(p)H_0(x,t\,|\,x_0) \\ & + \epsilon \sum_{\lambda\neq 0} \frac{1}{\lambda}\Big\{P_\lambda(p)c_{0\lambda}L_2(x) + P_0(p)Q_\lambda(p_0)c_{\lambda 0}\hat{L}_2^*(x_0)\Big\}H_0(x,t\,|\,x_0)\,. \qquad (t \gg \epsilon^2) \end{aligned} \tag{9.2.59}$$

The zero order term in this expression is the solution given in Chap. 8, since $H_0(x,t\,|\,x_0)$ is a solution to the adiabatic elimination Fokker-Planck equation. However, we now see that there are corrections, which depend on the eigenfunctions of L_1, as well as the solution to lowest order.

d) Connection with Adiabatic Elimination Results: The expression (9.2.43) for D_0 can be written

$$D_0 = \sum_{\mu\neq 0} \frac{c_{\mu 0}c_{0\mu}}{\mu}\,. \tag{9.2.60}$$

From (9.1.15), it can be seen that

$$c_{0\mu} = c_{\mu 0} = \int dp\, p P_\mu(p)\,, \tag{9.2.61}$$

so that, using the result for the stationary correlation function (5.4.22),

$$D_0 = \sum_\mu \int_0^\infty dt\,\Big\{\int dp\, pP_\mu(p)\int dp'\, p'P_\mu(p')e^{-\mu t}\Big\} = \int_0^\infty dt\,\langle p(t)p(0)\rangle_s\,. \tag{9.2.62}$$

The formula for D_0 is exactly that found by the methods of Chap. 8.

9.2.4 Stationary Solution $P_s(x, p)$

This is given by the long time limit of (9.2.58) and hence of (9.2.59). We define

$$H_s(x) = \lim_{t\to\infty} H_0(x, t \mid x_0), \tag{9.2.63}$$

and can consequently write

$$P_s(x, p) = P_0(p)H_s(x) + \epsilon \sum_{\lambda\neq 0} P_\lambda(p)\frac{c_{0\lambda}}{\lambda} L_2(x)H_s(x), \tag{9.2.64}$$

the other term vanishing, since $H_s(x)$ is independent of x_0.

9.2.5 Examples

a) Example—An Ornstein Uhlenbeck Process for L_1: In this case, we take

$$L_1 = \frac{\partial}{\partial p}p + \frac{1}{2}\frac{\partial^2}{\partial p^2}, \tag{9.2.65}$$

for which the eigenvalues and eigenfunctions are, as given in Sect. 5.4.3c,

$$\lambda_n = n = 0, 1, 2, \ldots, \tag{9.2.66}$$
$$Q_n(p) = (2^n n!)^{-1/2} H_n(p), \tag{9.2.67}$$
$$P_n(p) = \sqrt{\pi}\exp(-p^2)Q_n(p), \tag{9.2.68}$$

where $H_n(p)$ are the Hermite polynomials. From the recursion relations for Hermite polynomials, we have

$$c_{\lambda'\lambda} = c_{\lambda\lambda'} = \delta_{\lambda,\lambda'-1}\sqrt{\tfrac{1}{2}(\lambda+1)} + \delta_{\lambda,\lambda'+1}\sqrt{\tfrac{1}{2}\lambda}, \tag{9.2.69}$$

and we can compute that

$$D_\lambda = \tfrac{1}{2}. \tag{9.2.70}$$

In this case, then, all the evolution equations (9.2.50) are the same, and hence all the $H_\lambda(x, t \mid x_0)$ are the same. The expansion (9.2.58) is consequently greatly simplified.

b) Random Telegraph Process for L_1: We use the formulation of Sect. 3.8.5. The variable p takes on the values $(1, -1)$, and the equation for $P(p, t \mid p_0, 0)$ is

$$\begin{aligned}\partial_t P(1, t \mid p_0, 0) &= -\frac{1}{2}P(1, t \mid p_0, 0) + \frac{1}{2}P(-1, t \mid p_0, 0),\\ \partial_t P(-1, t \mid p_0, 0) &= \frac{1}{2}P(1, t \mid p_0, 0) - \frac{1}{2}P(-1, t \mid p_0, 0).\end{aligned} \tag{9.2.71}$$

The eigenvalues are 0 and -1, with eigenfunctions

$$P_0(p) = \begin{pmatrix}1/2\\1/2\end{pmatrix}, \qquad P_1(p) = \begin{pmatrix}1/2\\-1/2\end{pmatrix}, \tag{9.2.72}$$

from which we find

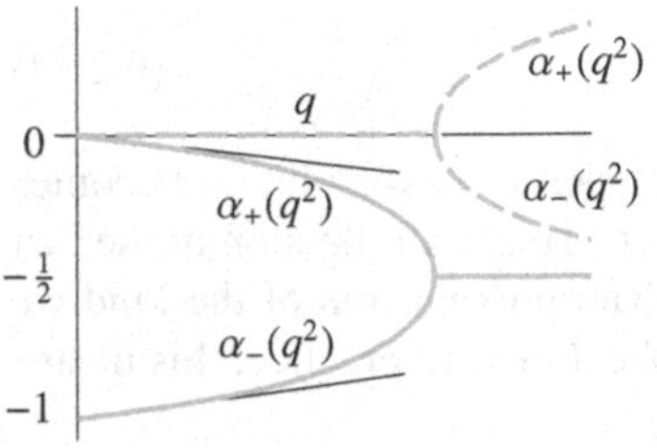

Fig. 9.1. Behaviour of the eigenvalue spectrum (9.2.84). The real part is the solid line, and the imaginary part is the dashed line. The perturbative solution follows the tangents (drawn as fine black lines) and becomes inaccurate for larger q.

$$c_{00} = c_{11} = 0\,, \qquad c_{10} = c_{01} = 1\,, \tag{9.2.73}$$

and hence

$$D_0 = 1\,, \qquad D_1 = -1\,. \tag{9.2.74}$$

The fact that D_1 is negative looks disturbing, since it means that the Fokker-Planck equation for $H_1(x, t\,|\,x_0)$ is unstable. At first glance this might appear to be an artifact of the approximation, but we can show that it is a completely accurate representation of what really happens.

Let us set $L_3 = 0$ for simplicity, so that in this case the equation (9.1.18) takes the explicit form

$$\partial_\tau P(x,1,\tau) = \epsilon\frac{\partial}{\partial x}b(x)P(x,1,\tau) - \frac{1}{2}P(x,1,\tau) + \frac{1}{2}P(x,-1,\tau)\,, \tag{9.2.75}$$

$$\partial_\tau P(x,-1,\tau) = -\epsilon\frac{\partial}{\partial x}b(x)P(x,-1,\tau) - \frac{1}{2}P(x,-1,\tau) + \frac{1}{2}P(x,1,\tau)\,. \tag{9.2.76}$$

Defining

$$\phi(x,\tau) = P(x,1,\tau) + P(x,-1,\tau)\,, \tag{9.2.77}$$

$$\psi(x,\tau) = P(x,1,\tau) - P(x,-1,\tau)\,, \tag{9.2.78}$$

we readily deduce the equations

$$\partial_\tau\phi(x,\tau) = \epsilon\frac{\partial}{\partial x}b(x)\psi(x,\tau)\,, \tag{9.2.79}$$

$$\partial_\tau\psi(x,\tau) = \epsilon\frac{\partial}{\partial x}b(x)\phi(x,\tau) - \psi(x,\tau)\,. \tag{9.2.80}$$

Exact solutions are possible if $b(x) = b$, a constant. Writing

$$\phi(x,\tau) = \textstyle\int dq\, \exp(iqx)\tilde{\phi}(q,\tau)\,, \tag{9.2.81}$$

we find that $\tilde{\phi}(q,\tau)$ and $\tilde{\psi}(q,\tau)$ both obey the equation.

$$\partial_\tau^2\tilde{\phi}(q,\tau) + \partial_\tau\tilde{\phi}(q,\tau) = \epsilon^2 b^2 q^2\tilde{\phi}(q,\tau)\,. \tag{9.2.82}$$

The solutions of (9.2.72) are of the form

$$\tilde{\phi}(q,\tau) \propto \exp(\alpha_\pm(q^2)\tau)\,, \tag{9.2.83}$$

where

$$\alpha_{\pm}(q^2) = -\frac{1}{2} \pm \sqrt{\frac{1}{4} - \epsilon^2 b^2 q^2}\,. \tag{9.2.84}$$

The two branches are plotted in Fig. 9.1. Clearly, the perturbative solution becomes inaccurate for sufficiently high q, whatever ϵ is chosen. This is a reflection of the fact that $\hat{L}_2$ is an unbounded operator, and that a perturbation expansion of the kind we use requires bounded operators for its universal applicability. In practice, this means that there must be some kind of cutoff in q

9.2.6 Generalisation to a system driven by several Markov Processes

The results developed so far depend on the Markov nature of the process being eliminated; the variable p. Since the sum of variables, all of which are Markovian, is not necessarily Markovian, we can obtain much more general results by developing the same formalism in the case that we are dealing with many variables p_r. The results generalise rather obviously. We assume that each variable p_r has a corresponding development operator L_1^r, and hence

$$L_1 = \sum_r L_1^r\,, \tag{9.2.85}$$

and the eigenvalues of L_1^r are λ_r. The p_r are therefore assumed independent. We then write

$$L_2 = \sum_r p_r \hat{L}_2^r\,, \tag{9.2.86}$$

so that the driving occurs by all the p_r, with a different coupling L_2^r.

We use the notation $P_{\lambda_r}(p_r)$ for the eigenfunctions of L_1^r, and assume as before that

$$p_r P_{\lambda_r}(p_r) = \sum_{\lambda_r'} c_{\lambda_r \lambda_r'} P_{\lambda_r'}(p_r). \tag{9.2.87}$$

We use the notation

$$\mathscr{P}_{\boldsymbol{\lambda}}\,, \qquad \boldsymbol{\lambda} = \{\lambda_1, \lambda_2, \ldots\}\,, \tag{9.2.88}$$

for the projector onto the eigenspace $\boldsymbol{\lambda}$. Following through the perturbation theory, we find that at order zero

$$U_0^{\boldsymbol{\lambda}} = \mathscr{P}_{\boldsymbol{\lambda}}\,, \tag{9.2.89}$$

$$A_0^{\boldsymbol{\lambda}} = \mathscr{P}_{\boldsymbol{\lambda}}\left(\sum_r p_r L_2^r\right)\mathscr{P}_{\boldsymbol{\lambda}} = \sum_r \mathscr{P}_{\boldsymbol{\lambda}} c_{\lambda_r \lambda_r} = 0. \tag{9.2.90}$$

where we again assume

$$c_{\lambda_r \lambda_r} = 0\,. \tag{9.2.91}$$

At first order

$$U_1^\lambda = -\left[\sum_r (\lambda_r + L_1^r)\right]^{-1} \mathcal{Q}_\lambda L_2 U_0^\lambda , \tag{9.2.92}$$

$$= -\sum_{\lambda \neq \lambda'} \left[\sum_r (\lambda_r - \lambda'_r)\right]^{-1} \mathcal{P}_{\lambda'\lambda} \sum_r L_2^r c_{\lambda'_r \lambda_r} \delta(\tilde{\lambda}, \tilde{\lambda}') , \tag{9.2.93}$$

where $\tilde{\lambda}$ is the vector of all components of λ except λ_r. Hence

$$U_1^\lambda = \sum_r \sum_{\lambda \neq \lambda'} (\lambda_r - \lambda'_r)^{-1} \mathcal{P}_{\lambda'_r \lambda_r} \mathcal{P}_{\tilde{\lambda}} L_2^r c_{\lambda_r \lambda'_r} . \tag{9.2.94}$$

We can then compute

$$A_1^\lambda = \mathcal{P}_\lambda \left[L_3 + \sum_r D_r(\lambda_r)(L_2^r)^2\right] , \tag{9.2.95}$$

where

$$D_r(\lambda_r) = \sum_{\lambda_r \neq \lambda'_r} \frac{c_{\lambda_r \lambda'_r} c_{\lambda'_r \lambda_r}}{\lambda'_r - \lambda_r} . \tag{9.2.96}$$

The conditional probability can now be constructed as before. It can be seen from (9.2.94) that we simply add results for each r: thus

$$P(x, p, \tau \mid x_0, p_0, 0) = \sum_\lambda P_\lambda(p) \Bigg[e^{-\sum_r \lambda_r \tau} Q_\lambda(p_0) H(x, t \mid x_0)$$

$$+\epsilon \sum_r \sum_{\mu_r \neq \lambda_r} \frac{c_{\mu_r \lambda_r}}{\lambda_r - \mu_r} Q_{\tilde{\lambda}}(\hat{p}_0) Q_{\mu_r}(p_{r,0}) e^{-\sum_{s\neq r} \lambda_s \tau} \times$$

$$\left[e^{-\mu_r \tau} L_2^r(x) H_{\tilde{\lambda},\mu_r}(x, \tau \mid x_0) - e^{-\lambda_r \tau} L_2^{r*}(x_0) H_\lambda(x, \tau \mid x_0)\right]\Bigg] . \tag{9.2.97}$$

The stationary distribution becomes

$$P_s(x, p) \approx P_0(p) H_s(x, t) + \epsilon \sum_r \sum_{\lambda_r \neq 0} \frac{c_{0\lambda_r}}{\lambda_r} L_2^r(x) H_s(x, t \mid x_0) P_{\tilde{\lambda}}(\hat{p}) P_{\lambda_r}(p_r) . \tag{9.2.98}$$

In these, $H(x, t \mid x_0)$ is the solution of

$$\left[L_3 + \sum_r D_r(\lambda_r)(L_2^r)^2\right] F = \frac{\partial F}{\partial t} , \tag{9.2.99}$$

with the initial condition $\delta(x - x_0)$, while $H_s(x)$ is the stationary solution of (9.2.99) for $\lambda = 0$.

Significance of the Result: The basic method works only for systems coupled to Markov Processes. However, by choosing all the L_2^r to be the same, we obtain results valid for a system driven by the sum of an arbitrary number of Markov variables, which is not in general a Markov Process. This enables significant generalisations to be made.

9.3 Computation of Correlation Functions

Using the formula for conditional probabilities, we are now in a position to compute the correlation functions between system variables x and the driving or noise variables p. The calculations are straightforward and we merely summarise them here.

a) Computation of $\langle x(t)\,|\,[x_0, p_0, 0]\rangle$

$$\langle x(t)\,|\,[x_0, p_0, 0]\rangle = \int\int dp\, dx\, xP(x, p, t\,|\,x_0, p_0, 0)\,, \tag{9.3.1}$$

$$\begin{aligned} &= \int dx\, xH_0(x, t\,|\,x_0) \\ &\quad -\epsilon \sum_{\mu\neq 0} \frac{Q_\mu(p_0)c_{\mu 0}}{\mu}\left[\int dx\, xL_2(x)H_\mu(x, t\,|\,x_0)e^{-\mu\tau} - L_2^*(x_0)\int dx\, xH_0(x, t\,|\,x_0)\right]. \end{aligned} \tag{9.3.2}$$

b) Computation of $\langle x(t)p(0)\rangle_s$ for $t > 0$

When $t > 0$,

$$\langle x(t)p(0)\rangle_s = \int dx_0\, dp_0\, p_0\langle x(t)\,|\,[x_0, p_0, 0]\rangle P_s(x, p_0)\,. \tag{9.3.3}$$

We now use the expression (9.3.2), and the expression (9.2.64) for $P_s(x, p)$, and retain only terms up to order ϵ^1. Noting $c_{\mu\lambda} = c_{\lambda\mu}$, using the definition (9.2.43) for D_0, we find that this can be reduced to two terms thus:

$$\begin{aligned} \langle x(t)p(0)\rangle_s &= 2\epsilon D_0\langle x(t)L_2(x(0))\rangle_s^r \\ &\quad -\epsilon \sum_{\mu\neq 0} \frac{c_{\mu 0}c_{0\mu}}{\mu}e^{-\mu\tau}\left[\int\int dx\, dx_0 xL_2(x)H_\mu(x, t\,|\,x_0)H_s(x_0)\right]. \end{aligned} \tag{9.3.4}$$

The notation $\langle\ \ \rangle^r$ is taken to mean the average in the reduced process obtained by adiabatic elimination, namely the process in the subspace $\lambda = 0$, whose evolution equation is given by (9.2.50) with $\lambda = 0$. Thus

$$\langle x(t)L_2(x(0))\rangle_s^r s = \int\int dx\, dx_0 xH_0(x, t\,|\,x_0)\,[L_2(x_0)H_s(x_0)]\,. \tag{9.3.5}$$

The first term of (9.3.4) is very simple and elegant, while the second is not. Notice that the second term is a function of $\tau = t/\epsilon^2$, i.e., it contains effects of fast eigenvalues. For $t \gg \epsilon^2$, we find that this term vanishes, and we are left with a much simpler result. By integrating by parts we find, in this case

$$\begin{aligned} \langle x(t)p(0)\rangle_s &\approx 2\epsilon D_0 \int dx_0 H_s(x_0)b(x_0)\frac{\partial}{\partial x_0}\int dx\, xH(x, t\,|\,x_0)\,, \\ &= 2\epsilon D_0\langle b(x_0)\frac{\partial}{\partial x_0}\langle x(t)\,|\,[x_0, 0]\rangle\,\rangle_s \qquad (t \gg 1/\epsilon^2)\,. \end{aligned} \tag{9.3.6}$$

In the case that $a(x)$ is linear and $b(x)$ is constant the right hand side reduces to the response of the system to a delta function input $x_0 b(x_0)\delta(t - t_0)$. The expression (9.3.6) is thus the mean response of the system with noise to an additional delta function input.

c) Computation of $\langle x(0)p(0)\rangle_s$

This can of course be done directly from the stationary distribution function, or we can set $t = \tau = 0$ in (5.5). It is not difficult to see that the second term becomes half the first term; we get

$$\langle x(0)p(0)\rangle_s = \epsilon D_0 \langle x(0) L_2(x(0))\rangle_s^r . \tag{9.3.7}$$

d) Computation of $\langle x(0)p(t)\rangle_s$ for $t > 0$

We first compute

$$\langle p(t) | [x_0, p_0, 0]\rangle = \int\int dx\, dp\, p P(x, p, t | x_0, p_0, 0) , \tag{9.3.8}$$

$$= \sum_\lambda \int dp\, p P_\lambda(p) \Bigg[Q_\lambda(p_0) e^{-\lambda\tau} + \epsilon \sum_{\mu\neq\lambda} \frac{Q_\mu(p_0) c_{\mu\lambda}}{\lambda-\mu} \left\{ \int dx\, L_2(x) H_\mu(x, t | x_0) e^{-\mu\tau} - \int dx\, L_2^*(x_0) H_\lambda(x, t | x_0) e^{-\lambda\tau} \right\} \Bigg] . \tag{9.3.9}$$

Note that

$$\int dx\, L_2(x)\phi(x) = 0 , \tag{9.3.10}$$

for any function $\phi(x)$, since $L_2(x) = \frac{\partial}{\partial x} b(x)$, and that

$$\int dx\, H_\lambda(x, t | x_0) = 1 , \tag{9.3.11}$$

so that

$$\int dx\, L_2^*(x_0) H_\lambda(x, t | x_0) = 0 . \tag{9.3.12}$$

Hence

$$\langle p(\tau) | [x_0, p_0, 0]\rangle = \sum_\lambda c_{\lambda 0} Q_\lambda(p_0) e^{-\lambda\tau} . \tag{9.3.13}$$

Now

$$\langle x(0)p(\tau)\rangle = \int\int dx_0 dp_0 \langle p(\tau) | [x_0, p_0, 0]\rangle x_0 P_s(x_0, p_0) , \tag{9.3.14}$$

$$= \epsilon \sum_{\lambda\neq 0} \frac{c_{\lambda 0} c_{0\lambda}}{\lambda} \langle x L_2(x)\rangle_s^r . \tag{9.3.15}$$

after a small amount of manipulation to get to the second line.

Hence

$$\langle x(0)p(\tau)\rangle = \epsilon \langle x L_2(x)\rangle_s^r \int_\tau^\infty \langle p(\tau')p(0)\rangle_s d\tau' . \tag{9.3.16}$$

e) Summary

For $t \geq 0$

$$\langle x(t)p(0)\rangle_s = 2\epsilon D_0\langle x(t)L_2(x(0))\rangle_s^r ,$$
$$-\epsilon \sum_{\mu\neq 0} \frac{c_{\mu 0}c_{0\mu}}{\mu} e^{-\mu\tau}\left(\int\int dx\, dx_0 x L_2(x) H_\mu(x,t\,|\,x_0)H_s(x_0)\right) , \qquad (9.3.17)$$

$$\langle x(0)p(0)\rangle_s = \epsilon D_0\langle x L_2(x)\rangle_s^r , \qquad (9.3.18)$$

$$\langle x(0)p(t)\rangle_s = \epsilon\langle x L_2(x)\rangle_s^r \int_\tau^\infty \langle p(\tau')p(0)\rangle_s d\tau' . \qquad (9.3.19)$$

Notice that although the functional forms (9.3.17–9.3.19) are different, the function so defined is continuous as t passes through zero. These results can be further simplified if the higher order terms in an expansion in powers of ϵ are not of interest. The second term in (9.3.17) contains the terms $\exp(-\mu\tau) = \exp(-\mu t/\epsilon^2)$, which become vanishingly small if τ is much larger than one, that is if t is much larger than ϵ^2. The coefficients are functions of t, which will therefore only change to order ϵ^2 from their value at t = 0. This means that we can simplify (9.3.17) to

$$\langle x(t)p(0)\rangle = \epsilon D_0\langle x(t)L_2(x(0))\rangle_s^r - \epsilon\langle x L_2(x)\rangle_s^r \int_\tau^\infty \langle p(\tau')p(0)\rangle_s d\tau' . \qquad (9.3.20)$$

9.3.1 Special Results for Ornstein-Uhlenbeck $p(t)$

As noted in Sect.9.2.5a, in the case that $p(t)$ is an Ornstein Uhlenbeck process, $D_\lambda = 1/2$. This means that all the $H_\mu(x,t\,|\,x_0)$ are identical, and in (9.3.17) we can make the simplification in the last bracket, by replacing H_μ with H_0,

$$\int\int dx\, dx_0 x L_2(x) H_0(x,t\,|\,x_0)H_s(x_0) = \int dx\, x L_2(x)H_s(x) = \langle x L_2(x)\rangle_s^r , \qquad (9.3.21)$$

by using the Markov property. Thus, for the Ornstein Uhlenbeck process we find, using (9.2.69), that (9.3.20) is in fact valid even without the argument used previously.

9.3.2 Generalisation to Arbitrary Gaussian Inputs

If we have

$$L_2 = \hat{L}_2(x)\sum_r p^r , \qquad (9.3.22)$$

where the p^r are different Ornstein-Uhlenbeck processes then $\sum_r p^r$ is Gaussian, but no longer Markov. It seems reasonable that any Gaussian process could be approximated by a limit of sums of such variables. It is clear from (9.2.80) that the results (9.3.18–9.3.20) will still apply, that is, these results are valid for a very large class of systems driven by Gaussian $p(t)$. We thus deduce that the results, valid to order ϵ in an expansion in powers of ϵ, are

$$\langle x(t)p(0)\rangle = \epsilon D\langle x(t)L_2(x(0))\rangle_s^r - \epsilon\langle xL_2(x)\rangle_s^r \int_\tau^\infty \langle p(\tau')p(0)\rangle_s d\tau' , \tag{9.3.23}$$

$$\langle x(0)p(0)\rangle = -\epsilon D\langle xL_2(x)\rangle_s^r , \tag{9.3.24}$$

$$\langle x(0)p(t)\rangle = -\epsilon\langle xL_2(x)\rangle_s^r \int_\tau^\infty \langle p(\tau')p(0)\rangle_s d\tau' . \tag{9.3.25}$$

9.4 The White Noise Limit

The original stochastic differential equation is (9.1.1), in which we see that (noting $\epsilon = 1/\gamma$) the white noise limit is

$$\frac{1}{\epsilon}p(t/\epsilon^2) \to \xi(t) \qquad \text{as } \epsilon \to 0 , \tag{9.4.1}$$

in which $\xi(t)$ is Gaussian (Stratonovich) white noise. The limit of all our general results is

$$\langle x(t)\xi(0)\rangle = \begin{cases} D_0\langle x(t)L_2(x(0))\rangle_s^r , & t > 0, \\ \frac{1}{2}D_0\langle x(0)L_2(x(0))\rangle_s^r , & t = 0, \\ 0 , & t < 0. \end{cases} \tag{9.4.2}$$

Thus the white noise limit is unique, and independent of whatever form of Markov process or sum of Markov processes is used to model the white noise. Only the details of the behaviour of the correlation function in the region $t \approx \pm 1/\gamma^2$ depend on the nature of the eliminated variables.

9.4.1 Relation of the White Noise Limit of $\langle x(t)\xi(0)\rangle$ to the Impulse Response Function

The impulse response function requires careful definition in the case of a nonlinear system. I shall define the impulse response function $I(t)$ of a system obeying a white noise (Stratonovich) stochastic differential equation

$$\dot{x} = a(x) + b(x)\xi(t) , \tag{9.4.3}$$

as the mean response of the system to an additional delta function driving term, obtained by the replacement

$$\xi(t) \to \xi(t) + \lambda\delta(t) , \tag{9.4.4}$$

under the condition that the system is stationary for $t < 0$. That is, if $x(\lambda, t)$ is the solution of (9.4.3) with the replacement (9.4.4), then

$$I(t) = \frac{\partial}{\partial\lambda}\langle x(\lambda, t)\rangle\bigg|_{\lambda\to 0} . \tag{9.4.5}$$

Now we can make a more explicit formula for (9.4.5). Firstly, suppose that the value of $x(t)$ just before $t = 0$ is x_0 and the value infinitesimally after $t = 0$ is $\bar{x}(\lambda, x_0)$. Then we can find this value by integrating (9.4.3) (with the replacement (9.4.4)) over an infinitesimal time interval round $t = 0$, and by noting that only the delta function is significant, we readily find that

$$\int_{x_0}^{\bar{x}(\lambda,x_0)} \frac{dx}{b(x)} = \lambda\,, \tag{9.4.6}$$

so that

$$\frac{\partial \bar{x}(\lambda, x_0)}{\partial \lambda} = b(\bar{x}(\lambda, x_0))\,. \tag{9.4.7}$$

The system evolves to time t with the equation (9.4.3); the mean of $x(\lambda, t)$ given the initial condition x_0 before $t = 0$ is

$$\langle x(\lambda, t) \,|\, [x_0, 0] \rangle = \int dx\, x P(x, t \,|\, \bar{x}(\lambda, x_0), 0\rangle\,, \tag{9.4.8}$$

and using (9.4.7), the response function is obtained by the $\lambda = 0$ value of

$$\int dx_0 P_s(x_0) \frac{\partial}{\partial \lambda} \langle x(\lambda, t) \,|\, [x_0, 0] \rangle = \int dx_0 \left[\int dx \frac{\partial}{\partial \lambda} P(x, t \,|\, \bar{x}(\lambda, x_0), 0\rangle \right] P_s(x_0) \tag{9.4.9}$$

$$= \int dx_0 P_s(x_0) \int dx\, b(\bar{x}(\lambda, 0)) \left[\frac{\partial}{\partial x} P(x, t \,|\, \bar{x}, 0) \right]_{\bar{x} = \bar{x}(\lambda, 0)}\,, \tag{9.4.10}$$

and setting $\lambda \to 0$,

$$= -\int dx_0 \int dx\, x P(x, t \,|\, x_0, 0) \frac{\partial}{\partial x_0} b(x_0) P_s(x_0)\,, \tag{9.4.11}$$

so that

$$I(t) = \langle x(t) \hat{L}_2(x_0) \rangle\,, \tag{9.4.12}$$

$$= \langle x(t) \xi(0) \rangle / D_0 \qquad (t > 0)\,, \tag{9.4.13}$$

from the result (9.4.2). The result (9.4.12) depends on the definition (9.4.5) of the impulse response function *with the noise term present.* In the case of a linear system (i.e., one with linear $a(x)$ and constant $b(x)$), the impulse response function in the presence of noise is the same as that in the absence of noise, but it is quite clear that this is otherwise not the case. The response of the equation $\dot{x} = a(x)$ to an additional $\lambda b(x) \delta(t)$ is obviously a quite complicated function.

10. Lévy Processes and Financial Applications

Outside the microscopic world exemplified by physics, chemistry and similar sciences, there is a range of phenomena whose behaviour it seems reasonable to describe by stochastic processes using similar tools, such as stochastic differential equations and master equations. In fact the very first formulation of the mathematics behind the theory of Brownian motion was that of *Bachelier* [10.1], who is therefore the originator of the idea that human behaviour could possibly be modelled as having an underlying dynamics described in terms of stochastic processes. Bachelier's formulations were based on rather limited data, and did not claim to be anything other than a basic conceptual description of the stock market. In re-introducing Bachelier's ideas to finance, but modified to use *geometric* or (to use Samuelson's terminology) *economic* Brownian motion, *Samuelson* [10.2] acknowledged the priority of several others in publishing the idea, and possibly even of conceiving it. The work of *Osborne* [10.3] very nicely demonstrated that the behaviour of observed stock prices was very much better described by geometric Brownian motion than by simple Brownian motion, and he later [10.4] gave a history of the idea, which he traced as far back in time as 1738 to a paper of *Daniel Bernoulli* [10.5, 10.6], who is indeed the true founder of the theory of relative value.

10.1 Stochastic Description of Stock Prices

As noted in Sect. 1.3.1, a model based on *geometric Brownian motion* has been found empirically to be a convenient description of the fluctuating values of stocks, or of the prices of any commodity such as wheat,coffee or cotton which traded on a regular basis in a market situation. In this case, if the value of one item of stock as traded on the stock market is $S(t)$, then the appropriate equation for the time dependence of this value is written as the stochastic differential equation

$$dS(t) = \mu(t)S(t)\,dt + \sigma(t)S(t)\,dW(t)\,. \tag{10.1.1}$$

Here the parameter $\mu(t)$ is conventionally referred to as the *drift*, and the parameter $\sigma(t)$ is called the *volatility*.

The solution for the stock price is easily obtained using Ito calculus, and is

$$S(t) = \exp\left\{\int_{t_0}^{t}\left(\mu(t) - \tfrac{1}{2}\sigma(t)^2\right)dt + \int_{t_0}^{t}\sigma(t)\,dW(t)\right\}. \tag{10.1.2}$$

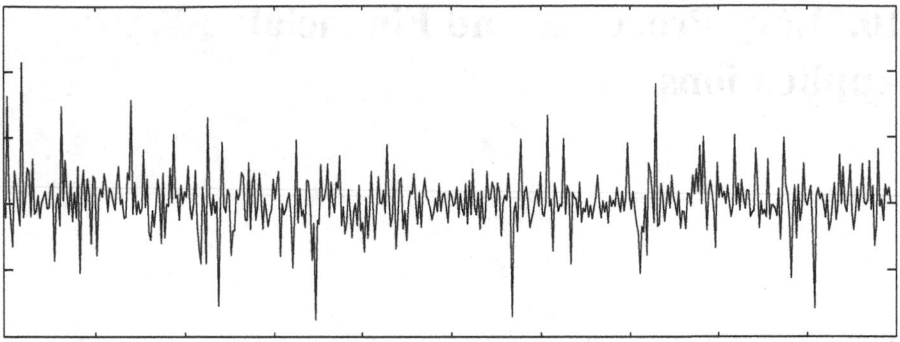

Fig. 10.1. Stock returns, showing Brownian-like behaviour as well as large jumps.

In the simplest case, we take the drift and volatility to be constant, and the solution shows that the $\log S(t)$ has a Gaussian distribution. The information is most commonly expressed in terms of the *return* over a specific period τ (very often one day, although with electronic markets the relevant period can be very much less)

$$R(t,\tau) \equiv \log S(t+\tau) - \log S(t), \tag{10.1.3}$$

$$\approx \frac{\Delta_\tau S(t)}{S(t)} \equiv \frac{S(t+\tau) - S(t)}{S(t)}. \tag{10.1.4}$$

However, the data for stock prices and similar prices when collected more extensively and analysed carefully showed that this simple picture, which implies that the returns are Gaussian, cannot be exact, or even realistic. The famous paper of *Mandelbrot* [10.7] showed that the probability distribution of cotton price returns was very poorly described by Gaussian models. The tails of the observed distribution tended to zero very slowly for large deviations from the mean—certainly much more slowly than the rapidly dropping $\exp(-x^2/2\sigma^2)$ of the Gaussian. Such distributions are now known as *heavy tailed* distributions. Since the only continuous Markov stochastic processes are Gaussian, this also necessarily means that jumps are an essential feature of financial markets.

Mandelbrot suggested that such behaviour could be described by a class of non-Gaussian probability laws, which he called *stable Paretian*, which were first introduced by in 1922 by *Lévy*. These processes are treated in some detail in Sect. 10.3, and are illustrated in Fig. 10.5–Fig. 10.7. For Paretian processes it is quite possible to have distributions for a variable x which look very similar to the Gaussian for moderate fluctuations, but which fall off slowly, according to a power law of the form $x^{-\alpha}$, for $-2 < \alpha < 0$, as opposed to the very rapid $\exp(-x^2/2\sigma^2)$ behaviour of the Gaussian. Other models can give a behaviour like $\exp(-k|x|)$, which is also consistent with the existence of jumps.

The heavy tails are necessary to describe a feature of all markets, that the returns do not change continuously, as required by any Brownian motion description, but are a mixture of apparently continuous motion and not infrequent large jumps. There does not yet seem to be any agreement currently on the "correct" model of financial markets; however, there is a body of well established information on their behaviour known as *stylised facts*, which is summarised in Sect. 10.5.1. In the short term a Brownian description can work, since in practice large jumps are not frequent. Although they do appear to be more realistic, descriptions in terms of Lévy processes do not seem to correspond exactly with reality; neither do they provide the relatively simple analytic tools which can be derived out of the Brownian models.

10.2 The Brownian Motion Description of Financial Markets

The description in terms Brownian motion is equivalent to the use of the stochastic differential equations of Chap. 4, and is so attractive, so fruitful, and so profitable—in spite of its manifest defects—that we will start our exposition by describing the techniques it provides and the results it produces, before moving on to the issue of more realistic descriptions.

10.2.1 Financial Assets

In financial markets, we can distinguish three broad classes of asset, as follows:

i) *Bonds* : These are essentially cash in the bank, and earn interest at an appropriate rate r. For simplicity, it is assumed that all bonds earn the same rate of interest. Thus, if the value of a bond at time t is $B(t)$, then this obeys the differential equation

$$dB(t) = r\,B(t)\,dt\,. \tag{10.2.1}$$

ii) *Stocks* : These are securities which are traded on the stock market, and have a value which fluctuates with time depending on market conditions. If the value of one item of stock as traded on the stock market is $S(t)$, then the appropriate equation for the time dependence of this value is written as the stochastic differential equation

$$dS(t) = \mu(t)S(t)\,dt + \sigma(t)S(t)\,dW(t)\,. \tag{10.2.2}$$

Here the parameter $\mu(t)$ is conventionally referred to as the *drift*, and the parameter $\sigma(t)$ is called the *volatility*.

iii) *Derivatives* : The general concept of a derivative security is a right or an obligation to effect a sale or purchase of some other security, as discussed in Sect. 1.3.2. In particular, we will be considering the most relevant derivatives for the stock market, namely *options*, which convey the right to purchase (a "call" option) or sell (a "put" option) a certain amount of stock at a definite price K (the "strike" price) at some future time T.

10.2.2 "Long" and "Short" Positions

In finance one can either possess a quantity of certain assets, or acquire a debt of value equivalent to a certain quantity of assets. The latter is done by selling the quantity of assets without actually possessing any to sell, but with a view to acquiring the assets in time to deliver them when required. This became known as "selling the assets short", and is conventionally known as *taking a short position on the asset*. The opposite, owning a quantity of assets, is then referred to as *taking a long position on the asset*. A variant terminology is to say assets are *held long* or *held short* as the case may be.

This means that in finance, we can reasonably consider both positive and negative quantities of assets.

10.2.3 Perfect Liquidity

The assumption of *perfect liquidity* is often made in finance, and by this it is meant that one can acquire any amount positive or negative of any asset, and that assets of all kinds can always be freely traded in the market place. This is obviously an idealisation, but the behaviour of such ideal systems yields valuable insights, in much the same way as the ideal gas or ideal frictionless motion are valuable concepts in physics and chemistry, which yield very powerful theoretical structures.

10.2.4 The Black-Scholes Formula

The fundamental question when buying an option is what price to pay for it. Within the geometric Brownian motion description of the stock market, there is a precise answer, developed by *Black* and *Scholes* [10.8] and re-derived by *Merton* [10.9]. The argument is most simply presented as follows.

a) The Value of the Option: We suppose that an option to buy one unit of stock has well defined value $F(S(t),t)$, which depends only on the *current* value of the stock, and not on its history. It is this function that we want to determine. Ito's formula says that this value will change with time according to the stochastic differential equation

$$dF(S(t),t) = \frac{\partial F(S(t),t)}{\partial t}\,dt + \frac{\partial F(S(t),t))}{\partial s}\,dS(t) + \frac{1}{2}\frac{\partial^2 F(S(t),t))}{\partial s^2}\,dS(t)^2, \tag{10.2.3}$$

$$= \left\{\frac{\partial F(S(t),t))}{\partial t} + \mu(t)S(t)\frac{\partial F(S(t),t))}{\partial s} + \tfrac{1}{2}\sigma(t)^2 S(t)^2 \frac{\partial^2 F(S(t),t))}{\partial s^2}\right\}dt + \sigma(t)S(t)\frac{\partial F(S(t),t))}{\partial s}\,dW(t). \tag{10.2.4}$$

The option is seen to have a fluctuating term proportional to $dW(t)$, the noise source in the stock market. Is it possible to construct a *portfolio* of stocks, options and bonds so that all the fluctuations cancel?

Let us consider taking a short position on one option, that is, one acquires a debt of size $F(S(t), t)$. This is balanced with a quantity $\Delta(t)$ of stock, which is an asset, not a debt. We want to choose $\Delta(t)$ so that the fluctuation in the stock (held long) exactly balances the fluctuation in the option (held short).

b) The Portfolio: To put this all into practice, it is necessary to consider a *portfolio* consisting of;

i) *The option (held short)* :	of value	$-F(S(t), t)$,
ii) *The amount* $\Delta(t)$ *of stock (held long)* :	of value	$\Delta(t)S(t)$,
iii) *A quantity* $\beta(t)$ *of bonds stock (held long)* :	of value	$\beta(t)B(t)$.

The total value of the portfolio is

$$P(t) = -F(S(t), t) + \Delta(t)S(t) + \beta(t)B(t), \tag{10.2.5}$$

and the equation for the change of this as a function of time is the stochastic differential equation

$$dP(t) = -\left\{\frac{\partial F(S(t), t)}{\partial t}dt + \frac{\partial F(S(t), t))}{\partial s}dS(t) + \frac{1}{2}\frac{\partial^2 F(S(t), t))}{\partial s^2}dS(t)^2\right\}$$
$$+ d\left\{\Delta(t)S(t) + \beta(t)B(t)\right\}. \tag{10.2.6}$$

c) The Self-Financing Condition: We want to try and vary the quantity of stock by trading bonds for stock, so that the change in the value, $\Delta(t)S(t) + \beta(t)B(t)$, of stocks and bonds arises only from the changes in values of the stocks and bonds themselves, not by any net inflow or outflow of capital. In the presence of white noise fluctuations, this requires some careful specification. To do this, let us use a discretised description of the time development of the portfolio of the kind

$$\left.\begin{aligned} S(t) &\to S_n, & S(t) + dS(t) &\to S_{n+1}, \\ B(t) &\to B_n, & B(t) + dB(t) &\to B_{n+1}, \\ \Delta(t) &\to \Delta_n, & \Delta(t) + d\Delta(t) &\to \Delta_{n+1}, \\ \beta(t) &\to \beta_n, & \beta(t) + d\beta(t) &\to \beta_{n+1}. \end{aligned}\right\} \tag{10.2.7}$$

This means that when we advance from time n to time $n + 1$, we rebalance the portfolio by changing the quantity of stocks by $\Delta_{n+1} - \Delta_n$ and the quantity of bonds by $\beta_{n+1} - \beta_n$. The total change in value *as a result of this rebalancing* is given by

$$Z_{n\to n+1} = (\Delta_{n+1} - \Delta_n) S_{n+1} + (\beta_{n+1} - \beta_n) B_{n+1}. \tag{10.2.8}$$

The stock and bond values are those for the time step $n + 1$, since the changed values of Δ and β apply at time $n + 1$.

The *self financing condition requires this change in value to be zero*, reflecting the fact that we can only purchase shares by exchanging them for an appropriate amount of bonds. Expressing (10.2.8) in differentials, we get the condition

$$\left\{S(t) + dS(t)\right\} d\Delta(t) + \left\{B(t) + dB(t)\right\} d\beta(t) = 0. \tag{10.2.9}$$

This condition can now be substituted into the second line of (10.2.6), which leads to the form

$$d\left\{\Delta(t)S(t) + \beta(t)B(t)\right\} = \Delta(t)\,dS(t) + \beta(t)\,dB(t). \tag{10.2.10}$$

In fact the term $dB(t)\,d\beta(t)$ in (10.2.9) is zero, since by (10.2.1), $dB(t)$ has no noise term—thus Ito rules will require any such term to vanish. The resulting condition is

$$d\beta(t) = -\big(S(t) + dS(t)\big)\frac{d\Delta(t)}{B(t)}, \tag{10.2.11}$$

amounts to a stochastic differential equation for $\beta(t)$, which fixes $\beta(t)$ if $S(t)$, $B(t)$ and $\Delta(t)$ are known.

d) The No Arbitrage Condition: If we now use the self-financing condition in the form (10.2.10), and make the choice

$$\Delta(t) = \frac{\partial F(S(t), t))}{\partial s}, \tag{10.2.12}$$

the equation for $P(t)$ becomes

$$dP(t) = -\left\{\frac{\partial F(S(t), t)}{\partial t} + \tfrac{1}{2}\sigma(t)^2 S(t)^2\,\frac{\partial^2 F(S(t), t))}{\partial s^2} - r\beta(t)B(t)\right\} dt. \tag{10.2.13}$$

Here we have used the stochastic differential equations (10.2.1, 10.2.2), and Ito rules.

By making this particular choice for $\Delta(t)$, we are left with a time development equation with no noise term. Thus, in the short term this does not fluctuate. Since the self-financing condition means that we are not putting any new investment into it either, we deduce that $P(t)$ behaves like a bond, which is an investment to which no further investment is being added, and whose rate of change is given by a simple differential equation. This equation therefore must be equivalent to the equation of the bond, i.e., to the equation

$$dP(t) = rP(t)\,dt. \tag{10.2.14}$$

If this were not so, *arbitrage* could occur, in which bonds could be borrowed at the rate r and invested in the portfolio, thus making a risk free gain (or loss, depending on which is more profitable.) Putting together these three equations, (10.2.5, 10.2.13, 10.2.14), we deduce the *Black-Scholes equation*

$$\boxed{\frac{\partial F(s,t)}{\partial t} = rF(s,t) - rs\frac{\partial F(s,t)}{\partial s} - \tfrac{1}{2}\sigma(t)^2 s^2\frac{\partial^2 F(s,t)}{\partial s^2}.} \tag{10.2.15}$$

10.2.5 Explicit Solution for the Option Price

An explicit solution for $F(S, t)$ was given by Black and Scholes for the case where the volatility $\sigma(t)$ has the constant value σ. We will show how to get their solution using stochastic methods.

The equation (10.2.15) for the option price is very similar to a backward Fokker-Planck equation, so let us define a conditional probability $P(x, T \mid s, t)$ which, in the case of constant volatility, satisfies the backward Fokker-Planck equation and final condition

$$\left.\begin{aligned} \frac{\partial P(x,T\mid s,t)}{\partial t} &= -rs\frac{\partial P(x,T\mid s,t)}{\partial s} - \tfrac{1}{2}\sigma^2 s^2 \frac{\partial^2 P(x,T\mid s,t)}{\partial s^2}, \\ P(x,T\mid s,T) &= \delta(x-s). \end{aligned}\right\} \tag{10.2.16}$$

Then the solution of the (10.2.15) can be written as

$$F(s,t) = e^{r(t-T)} \int dx\, P(x,T\mid s,t)\, F(x,T). \tag{10.2.17}$$

The stochastic differential equation corresponding to the backward Fokker-Planck equation (10.2.16) is

$$dx(t) = x(t)[r\,dt + \sigma\, dW(t)], \tag{10.2.18}$$

and this can be solved in the same way as that given for geometric Brownian motion in Sect. 4.5.2. We define $y = \log x$, and using Ito calculus find the equation of motion for y is

$$dy = \left(r - \tfrac{1}{2}\sigma^2\right) dt + \sigma\, dW(t), \tag{10.2.19}$$

whose solution at time τ is

$$y(\tau) = y(t) + \left[r - \tfrac{1}{2}\sigma^2\right](\tau - t) + \sigma\,[W(\tau) - W(t)]. \tag{10.2.20}$$

The corresponding conditional probability for the variable $y(t)$ to have the value $\bar{y}$ is then

$$p(\bar{y},\tau\mid y,t) = \frac{1}{\sqrt{2\pi(\tau-t)}} \exp\left\{-\frac{\left[\bar{y} - y - \left(r - \tfrac{1}{2}\sigma^2\right)(\tau - t)\right]^2}{2(\tau-t)\sigma^2}\right\}. \tag{10.2.21}$$

Then from (10.2.17)

$$F(s,t) = e^{r(t-T)} \int d\bar{y}\, p(\bar{y},T\mid y,t) F\left(e^{\bar{y}}, T\right). \tag{10.2.22}$$

a) The Final Condition: The value of the option at time T will depend on whether the strike price K is greater than or less than the value x of the stock at that time.

If $K > x$, the option has no value, since it is cheaper to buy the stock on the market than to exercise the option. If $K < x$, the value is $x - K$, the profit one could make by buying the stock at the strike price K and selling it on the open market at the current value x. Thus

$$F(x,T) = \begin{cases} 0, & x < K, \\ x - K, & x > K. \end{cases} \tag{10.2.23}$$

b) The Option Pricing Formula: For convenience define a quantity M by $K = e^M$; then the formula (10.2.22) becomes

$$F(s,t) = e^{r(t-T)} \int_M^\infty d\bar{y}\, \left(e^{\bar{y}} - e^M\right) p(\bar{y},T\mid y,t). \tag{10.2.24}$$

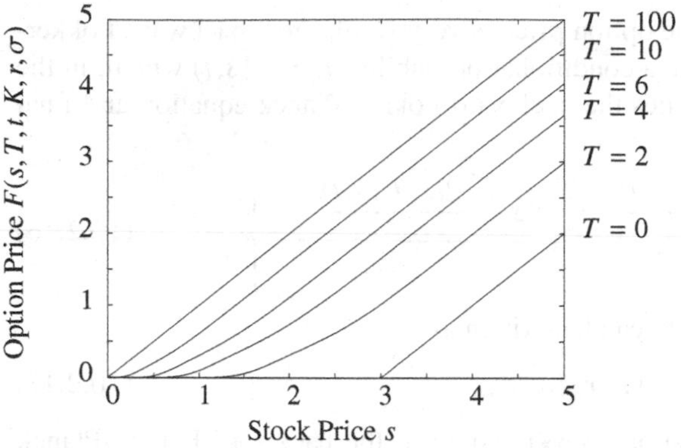

Fig. 10.2. The Black-Scholes option price formula (10.2.26) plotted for various maturity times T, and for initial time $t = 0$, strike price $K = 3$, interest rate $r = 0..2$, volatility $\sigma = 0.2$.

Using the Gaussian nature of the result (10.2.21), this integral can be evaluated in terms of the cumulative Gaussian function

$$\mathrm{N}(z) \equiv \int_{-\infty}^{z} \exp\left(-\tfrac{1}{2}x^2\right) dx, \tag{10.2.25}$$

as

$$F(s,t) = s\,\mathrm{N}(d_1) - Ke^{r(t-T)}\,\mathrm{N}(d_2), \tag{10.2.26}$$

$$d_1 = \frac{\log(s/K) + \left(r + \frac{1}{2}\sigma^2\right)(T-t)}{\sigma\sqrt{T-t}}, \tag{10.2.27}$$

$$d_2 = \frac{\log(s/K) + \left(r - \frac{1}{2}\sigma^2\right)(T-t)}{\sigma\sqrt{T-t}}. \tag{10.2.28}$$

This is the celebrated *Black-Scholes option pricing formula.*

10.2.6 Analysis of the Formula

The option value formula is plotted in Fig. 10.2 as a function of T and s for representative values of the parameters. The behaviour is a rather unsurprising smooth transition from the final condition at $T = 0$, to the value being equal to that of the stock when T is very large. The merit of the formula is not its appearance, but its quantitative behaviour.

In Fig. 10.3 two scenarios are plotted for the evolution of the Black-Scholes portfolio P given by (10.2.5). It should be borne in mind that the portfolio formula is used only to prove the Black-Scholes formula, and is not a realistic or sensible investment choice. The first scenario shows the stock price rising quite rapidly, exceeding the current interest rate r, and the option price increasing more rapidly than that. To keep the growth of the portfolio at r, the investor increases the amount of stock purchased, until at the time when the option price reaches it maximal value $s - K$, we

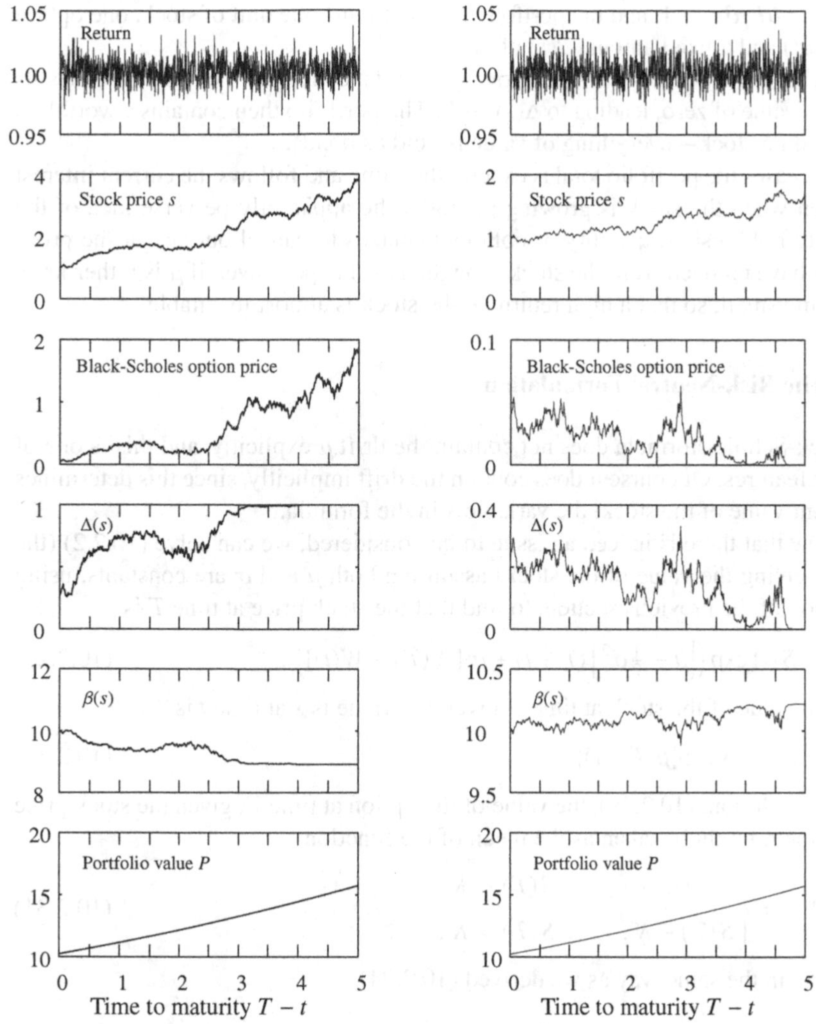

Fig. 10.3. Two scenarios for the evolution of the Black-Scholes portfolio and its components; The initial conditions are the same for both, with $s(0) = 1$, $B(0) = 1$, $\beta(0) = 10$, $K = 2$, $\sigma = 0.15$, $r = 0.0862$, $\mu = 0.1823$—the only difference is the particular realisation of Brownian motion. (However, note that the vertical axis scales differ.)
Left: A relatively large rise in stock price over the time period leads to $\Delta(s) \to 1$ before the end of the time period, and a portfolio with one unit of stock for each unit of options held short;
Right: A poor performance of the stock leads to the value of the option becoming zero, and the portfolio consists of a valueless option and no stocks, and thus with all assets held in bonds.

find $\Delta(s) \equiv \partial F/\partial s \to 1$, and the portfolio now contains one unit of stock, one option (held short) and a fixed amount $\beta(s)$ of bonds.

The other scenario shows a stock price growing slowly, and the option reaches its minimum value of zero, leading to $\Delta(s) \to 0$. The portfolio then contains a worthless option and no stock—everything of value is held as bonds.

In both cases the portfolio total is exactly the same and follows the current interest rate. Even when the stock is growing strongly, the apparently perverse idea of the investor to hold a short quantity of options manages to cancel out any of the profit he might have enjoyed from the stock growth. This happens even if μ is rather large and σ quite small, so that a high return on the stock is almost inevitable!

10.2.7 The Risk-Neutral Formulation

The Black-Scholes formula does not contain the drift μ explicitly, and this is one of its major features. Of course it does contain the drift implicitly, since this determines the current value of the stock, the variable s in the formula.

To show that there is indeed an issue to be considered, we can solve (10.2.2) (the equation giving the value of the stock) assuming both μ and σ are constants, using the method of the previous section, to find that the stock price at time T is

$$S(T) = S(t)\exp\left(\left[\mu - \tfrac{1}{2}\sigma^2\right](T-t) + \sigma\left[W(T) - W(t)\right]\right) . \tag{10.2.29}$$

The mean value of the stock at time T given the value is s at time t is

$$\langle S(T) \,|\, s, t\rangle = s\exp[\mu(T-t)] . \tag{10.2.30}$$

Using this solution, (10.2.29), the value of the option at time T, given the stock price $S(t)$ at time t, might be taken as the mean of the function

$$H(S(T)) = \begin{cases} 0, & S(T) < K, \\ S(T) - K, & S(T) > K, \end{cases} \tag{10.2.31}$$

and this is, in the same way as we derived (10.2.24),

$$\langle H(S(T)) \,|\, s, t\rangle = \int_M^\infty d\bar{y}\left(e^{\bar{y}} - e^M\right) p_\mu(\bar{y}, T \,|\, y, t), \tag{10.2.32}$$

$$= \int_K^\infty dS\,(S-K)\,P_\mu(S, T \,|\, s, t) . \tag{10.2.33}$$

Here, the subscript μ means that the probability densities are to be calculated using μ, not r as in the derivation of the Black-Scholes formula.

If I wish to sell my option at time t, the price I could reasonably demand is $\langle H(S(T)) \,|\, s, t\rangle$, discounted appropriately by the interest rate factor $e^{r(t-T)}$, or possibly by $e^{\mu(t-T)}$—but in neither case do I arrive at the same value $F(s,t)$ as given by (10.2.24), unless $\mu = r$. How shall we compare the two estimates of value? The Black-Scholes argument is compelling, and gives a definite value, with no fluctuation. The value given by this argument is only a mean value—it is therefore risky, in the same way as the estimate of the value of the stock in the future is risky.

The Black-Scholes result can be derived by using this methodology, after imposing a *risk-neutral behaviour* on the owner of the option. This behaviour is characterised by the investor's decision to take no account of any information on *individual* growth rates (determined in this case by μ) and his attribution of only the current *risk-free* interest rate r to all estimates of value; thus the risk-neutral investor will calculate values by assuming the stock behaviour is given by the equation

$$S(T) = S(t)\exp\left(\left[r - \tfrac{1}{2}\sigma^2\right](T-t) + \sigma\left[W(T) - W(t)\right]\right), \tag{10.2.34}$$

thus suppressing all knowledge of μ by replacing it everywhere with r, but at the same time continuing to accept the volatility value σ. In that case, the *risk neutral valuation of the option* is identical to the Black-Scholes valuation.

10.2.8 Change of Measure and Girsanov's Theorem

The argument that leads to the *risk neutral* formulation requires no construction of a risk-free portfolio, but yields exactly the same answer—possibly it is better viewed as a rationalisation for the correctness of the Black-Scholes formula than a derivation. The risk neutral valuation argument is in fact more widely applicable than the original Black-Black-Scholes argument, since we do not require anything more than a conditional probability for the stock price at time T given its price at time t—thus the evolution of the stock price can be given by any Markov process, allowing for a much wider range of stock return models than geometric Brownian motion. However:

i) The correct way to replace the drift μ by the interest rate r becomes one of the main tasks in applying models;

ii) In non-Brownian models the Black-Scholes argument is no longer available to provide a solid rationale for the procedure.

a) Equivalent Probability Measures: The central issue is to ask under what conditions is it possible to have different views on the correct stochastic differential equation for the stock price, and the response to those who raise this issue is to introduce the idea of *equivalent probability measures* into the theory of stochastic differential equations. In probability theory, two probability measures P and Q are said to be *equivalent* if for any set A in the probability space

$$P(A) > 0 \Longleftrightarrow Q(A) > 0. \tag{10.2.35}$$

This means that all events which are possible under the measure P are possible under the measure Q. If we suppose that the equivalence (10.2.35) is not true—for example for some set X it may be found that $P(X) = 0$ while $Q(X) > 0$ —then according to the measure P the event X is impossible, while according to the measure Q the event is possible. The two measures are then inequivalent—the worlds they describe are different.

b) Application to Stochastic Differential Equations: We can show that two stochastic differential equations can be considered *equivalent* if their noise terms are the same even if their drift terms are different. Let us show this with a simple example. Suppose a process $x(t)$ has the stochastic differential equation on the interval $[0, T]$

$$dx(t) = f(t)\,dt + dW(t). \tag{10.2.36}$$

In a discretised form this is

$$\Delta x_i = f_i\,\Delta t_i + \Delta W_i. \tag{10.2.37}$$

The measure used for this equation is the Wiener measure

$$\mathscr{P}(W) = \prod_i \frac{1}{\sqrt{2\pi\Delta t_i}} \exp\left(-\frac{\Delta W_i^2}{2\Delta t_i}\right). \tag{10.2.38}$$

If we define

$$\Delta V_i = \Delta W_i + f_i\,\Delta t_i, \tag{10.2.39}$$

then the measure on V is

$$\mathscr{Q}(V) = \prod_i \frac{1}{\sqrt{2\pi\Delta t_i}} \exp\left(-\frac{(\Delta V_i - f_i\,\Delta t_i)^2}{2\Delta t_i}\right), \tag{10.2.40}$$

$$= \exp\left(\sum_i \{f_i\,\Delta V_i - \tfrac{1}{2} f_i^2\,\Delta t_i\}\right), \tag{10.2.41}$$

$$= \exp\left(\int_0^T \left\{f(t)\,dV(t) - \tfrac{1}{2} f(t)^2\,dt\right\}\right) \mathscr{P}(V). \tag{10.2.42}$$

Since the factor multiplying $\mathscr{P}(V)$ is always positive, we can conclude that $\mathscr{P}(V)$ and $\mathscr{Q}(V)$ are equivalent; that is, any set of sample paths which is possible under $\mathscr{P}(V)$ is possible under $\mathscr{Q}(V)$ and conversely.

The stochastic differential equation (10.2.36) can be written

$$dx(t) = dV(t). \tag{10.2.43}$$

If we assign the measure $\mathscr{P}(V)$ to $V(t)$ then $x(t)$ is the Wiener process, whereas if we assign the measure $\mathscr{Q}(V)$ to $V(t)$, this says that $x(t)$ follows the stochastic differential equation (10.2.36). In other words, the possible sample paths from the two equations are identical, but depending on the choice of measure for the underlying driving process $V(t)$ the relative frequency of the paths is different.

c) Girsanov's Theorem: More generally, the same procedure can be used to show that the stochastic differential equations

$$dx(t) = a(t)\,dt + b(t)\,dW(t), \tag{10.2.44}$$

$$dy(t) = f(t)\,dt + g(t)\,dW(t), \tag{10.2.45}$$

are equivalent if $b(t) = g(t)$.

This result is Girsanov's theorem. It means that we can write

$$dy(t) = a(t)\,dt + b(t)\,dV(t), \tag{10.2.46}$$

where the measure for V is given in terms of the Wiener measure (10.2.38) in the form

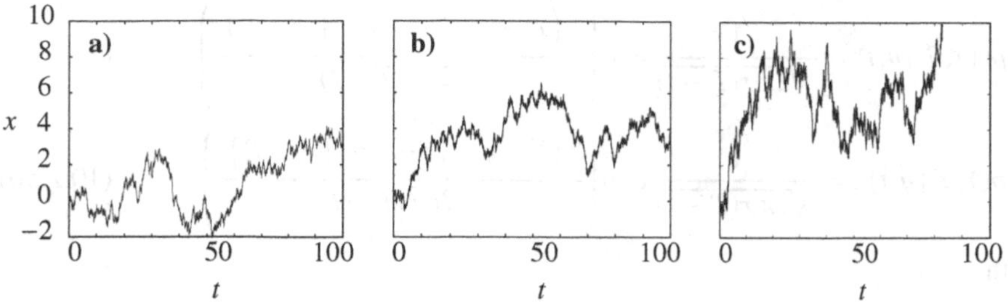

Fig. 10.4. Plots of simulations of the stochastic differential equation $dx = a\,dt + b\,dW(t)$ for: **a)** $a = 0, b = 0.5$; **b)** $a = 0.1, b = 0.5$; **c)** $a = 0, b = 1.0$. While a) and b) look qualitatively indistinguishable, the more intense noise in c) is immediately evident.

$$\mathcal{Q}(V) = \exp\left(\int_0^T \left\{\phi(t)\,dV(t) - \tfrac{1}{2}\phi(t)^2\,dt\right\}\right)\mathcal{P}(V), \tag{10.2.47}$$

and where

$$\phi(t) = \frac{f(t) - a(t)}{b(t)}. \tag{10.2.48}$$

When $b(t) \neq g(t)$, the procedure breaks down, and in fact it can be shown that in this case the processes are never equivalent [10.10, 10.11]. An illustration of what this means intuitively is given in Fig. 10.4. The two plots on the left are simulations of stochastic differential equations with different drifts, but the same noise, and it is quite credible that either could be a simulation of the other equation. On the other hand, the right hand plot is a simulation of a process with the same drift as the left hand plot, but with a different noise coefficient, and the increased noise intensity is a very obvious characteristic.

d) Financial Interpretation: Girsanov's theorem is now the justification for use of the drift rate r instead of μ in the valuation of options using the risk-neutral procedure. The noise term is identical for both cases, and in this case we can say that the two processes can be seen as arising from the choice of a different probability measure to the same set of sample paths. In some sense it can be shown that this is a rigorously justifiable procedure [10.11], although not everyone would accept that. However, the use of change of measure is now an accepted part of the procedure for valuing options and other derivatives when one goes beyond the simple geometric Brownian motion picture.

e) Conditional Probabilities and Change of Measure: An alternative way of viewing the change of measure is to note that the conditional probabilities for the processes (10.2.29, 10.2.23) are

$$p_\mu(Y,T\,|\,y,t) = \frac{1}{\sqrt{2\pi\sigma(T-t)}}\exp\left(-\frac{\left(Y-y-[\mu-\frac{1}{2}\sigma^2](T-t)\right)^2}{2\sigma^2(T-t)}\right), \tag{10.2.49}$$

$$p(Y,T\,|\,y,t) = \frac{1}{\sqrt{2\pi\sigma(T-t)}}\exp\left(-\frac{\left(Y-y-[r-\frac{1}{2}\sigma^2](T-t)\right)^2}{2\sigma^2(T-t)}\right), \tag{10.2.50}$$

with

$$Y = \log S\,, \qquad y = \log s\,. \tag{10.2.51}$$

The two probability densities can be related by the formula

$$p(Y,T\,|\,y,t) = \mathcal{N}^{-1}\mathrm{e}^{\frac{(r-\mu)(Y-y)}{\sigma^2}}\,p_\mu(Y,T\,|\,y,t)\,, \tag{10.2.52}$$

$$\text{with } \mathcal{N} = \int_{-\infty}^{\infty}\mathrm{e}^{\frac{(r-\mu)(Y-y)}{\sigma^2}}\,p_\mu(Y,T\,|\,y,t)\,dY\,, \tag{10.2.53}$$

$$= \exp\left(\frac{(\mu-r)(\mu+r+\sigma^2)(T-t)}{2\sigma^2}\right). \tag{10.2.54}$$

The two densities are related by a positive factor with no zeroes, and an appropriate normalisation, and are therefore equivalent probability densities, as noted above. This is of course exactly the same measure change as would be obtained using the Girsanov theorem procedure of b) and c) above.

This particular methodology is one which can be generalised to a variety of situations, and one which we will develop more fully in Sect. 10.5.4.

10.3 Heavy Tails and Lévy Processes

Samuelson [10.2] recognised that the most important thing in the description of the stock market was the concept of the *return* as the increment in the logarithm of the price; his work does not require in any essential sense that the return be described by a Gaussian white-noise stochastic differential equation. His pricing formula was overtaken by the Back-Scholes description, which on the other hand does rely on a Gaussian white-noise stochastic differential equation. Mandelbrot was the first, however, to attempt a serious and explicit description of financial markets in terms of specific non-Gaussian models. His, and many other models can be formulated in terms of *Lévy Processes*, which we will now proceed to formulate and apply.

10.3.1 Lévy Processes

Lévy processes arise from the differential Chapman-Kolmogorov equation by requiring the process be homogeneous in time and in the probability space variables. The Wiener process has this property, but there is a more extensive class of such processes, which can all be described by a master equation or by a limit of a master equation.

In one dimension, the kind of process we want to study can be described by a differential Chapman-Kolmogorov equation in which the parameters of (3.4.22) take the form

$$A(x,t) \to a\,, \tag{10.3.1}$$

$$B(x,t) \to \tfrac{1}{2}\sigma^2\,, \tag{10.3.2}$$

$$W(z\,|\,x,t) \to w(z-x)\,. \tag{10.3.3}$$

In this case the differential Chapman-Kolmogorov equation takes the form

$$\partial_t p(z,t) = -a\partial_z p(z,t) + \tfrac{1}{2}\sigma^2\partial_z^2 p(z,t) + \fint du\, w(u)\,\{p(z-u,t) - p(z,t)\}\,, \tag{10.3.4}$$

where $p(z,t)$ is shorthand for $p(z\,|\,y,t)$.

If we write $p(z,t)$ in terms of the characteristic function thus

$$\phi(s,t) = \int_{\infty}^{\infty} dz\, e^{isz} p(z,t)\,, \tag{10.3.5}$$

we can rewrite (10.3.4) as

$$\partial_t \phi(s,t) = \left(ias - \tfrac{1}{2}\sigma^2 s^2 + \fint_{-\infty}^{\infty} du\,\left(e^{isu} - 1\right) w(u)\right)\phi(s,t)\,. \tag{10.3.6}$$

Therefore the characteristic function of a Lévy process which starts at position $z = 0$ can be written as

$$\phi(s,t) = \int_{-\infty}^{\infty} dz\, e^{isz} p(z\,|\,0,t) = \exp\left\{\left(ias - \tfrac{1}{2}\sigma^2 s^2 + \fint_{-\infty}^{\infty} du\,\left(e^{isu} - 1\right) w(u)\right) t\right\}. \tag{10.3.7}$$

10.3.2 Infinite Divisibility

The property of *infinite divisibility* arises in probability theory—a probability density $p(x)$ is infinitely divisible if the random variable X whose distribution is $p(x)$ is such that for every positive integer n there exist n independent identically distributed random variables $X_1, X_2, \ldots, X_n$ whose sum has the probability density $p(x)$. The probability distribution of the X_i is in general different from $p(x)$. Not every probability density is infinitely divisible, and the proof that any particular probability density is infinitely divisible is not straightforward.

Lévy processes have the property of infinite divisibility, which is automatic from their definition in terms of homogeneous Markov processes. In a homogeneous Markov process with conditional probability $p(x', t+\tau\,|\,x,t) \equiv p(x', \tau\,|\,x, 0)$, this property is a direct consequence of the Chapman-Kolmogorov equation, and conversely. Thus, the representation in the form (10.3.7)—or more accurately, the more refined version of (10.3.24)—is valid for any infinitely divisible probability density.

If the characteristic function of an infinitely divisible distribution $P(x)$ is known to be $\Phi(s)$, we can define the characteristic function $\phi(s,t)$ of a conditional probability in terms of some specific time scale τ (usually taken to be 1) by

$$\phi(s,t) \equiv \{\Phi(s)\}^{t/\tau}. \tag{10.3.8}$$

Because the distribution $P(x)$ is infinitely divisible, the right hand side is a valid characteristic function for any t. However, one must be very careful to choose the correct Riemann sheet if $\Phi(s)$ is not real and positive—see [10.12].

10.3.3 The Poisson Process

The simplest Lévy process is the Poisson process, described by $a = \sigma = 0$, and

$$w(u) = d\,\delta(u-1). \tag{10.3.9}$$

This characteristic function is that of the Poisson process $N(t)$, given in (3.8.49), that is

$$\langle \exp(isN(t))\rangle = \exp\left(\lambda t[e^{is}-1]\right) \tag{10.3.10}$$

The *compensated Poisson process* $\bar{N}(t)$ is obtained by subtracting the mean value from the Poisson process; thus

$$\bar{N}(t) = N(t) - \langle N(t)\rangle = N(t) - \lambda t. \tag{10.3.11}$$

The characteristic function of the compensated Poisson process is

$$\left\langle \exp(is\bar{N}(t))\right\rangle = \exp\left(\lambda t[e^{is}-1-is]\right). \tag{10.3.12}$$

In the description of shot noise in Sect. 1.5.1, the differential noise term $d\eta(t)$ in (1.5.19) is the differential of a compensated Poisson process. The concept of the compensated Poisson process is essentially that of the fluctuations about a mean drift λt, picture which makes sense if this value is very much larger than 1.

10.3.4 The Compound Poisson Process

More generally, the *compound Poisson process*is obtained by setting $a = \sigma = 0$ and requiring a normalisable $w(u)$, i.e., such that

$$\int_{-\infty}^{\infty} w(u)\,du \equiv \lambda < \infty. \tag{10.3.13}$$

This quantity λ will be called the *intensity* of the process, and of course is equal to the inverse of the mean time between jumps.

Following the methodology of Sect. 3.5.1b, the sample paths of the system so described are given by a sequence of jumps U_n occurring at time intervals t_n which are exponentially distributed and have the probability density

$$\text{Prob}(t < t_n < t+dt) = \exp(-\lambda t)\,dt, \tag{10.3.14}$$

We can introduce a Poisson process variable (as described in Sect. 3.5.1) $N(t)$, which takes on the values n at the time t_n. The jumps U_n have a probability density

$$\text{Prob}(u < U_n < u+du) = \frac{w(u)}{\lambda}du. \tag{10.3.15}$$

The position $Z(t)$ after a time t can then be written

$$Z(t) = \sum_{n}^{N} U_n , \tag{10.3.16}$$

where N is that integer such that

$$t_N < t \le t_{N+1} . \tag{10.3.17}$$

10.3.5 Lévy Processes with Infinite Intensity

a) The Meaning of Infinite Intensity: If $w(u) \to \infty$ as $u \to 0$, intuitively this corresponds to a process with jumps of an infinitesimal size occurring at an infinite rate in such a way that the net result is a well-defined limiting process. In this limit, the concept of a jump process becomes very similar to that of a diffusion process. The precise formulation of this concept was first made by *Lévy* [10.13]. In our formulation, the characterisation of the resulting process depends on the particular behaviour of $w(u)$ at $u = 0$.

b) The Definition of the Principal Value Integral: The principal value integral in (10.3.7) can be defined so as to admit its existence when $w(u)$ is quite singular near $u = 0$, in fact it can be defined when

$$w(u) \sim |u|^{-\alpha-1} \text{ as } |u| \to 0 , \text{ provided } \alpha < 2 . \tag{10.3.18}$$

When $\alpha \le 1$, the process has a finite intensity, and there is no need for a principal value integral, since $e^{isu} - 1 \sim isu$ near $u = 0$.

However for $1 < \alpha < 2$, the intensity is infinite, and a rather unusual specification of the principal value integral has to be made as follows.

i) Let us assume that $w(u)$ is asymmetric, in such a way that near $u = 0$

$$w(u) \approx \begin{cases} A|u|^{-\alpha-1} , & u < 0 , \\ Bu^{-\alpha-1} , & u > 0 . \end{cases} \tag{10.3.19}$$

Then we can define the principal value integral as

$$⨍ du\, w(u)\left(e^{isu} - 1\right) \equiv \lim_{\epsilon\to 0} \left\{ \int_{-\infty}^{-\delta(\epsilon)} du\, w(u)\left(e^{isu} - 1\right) + \int_{\epsilon}^{\infty} du\, w(u)\left(e^{isu} - 1\right) \right\} , \tag{10.3.20}$$

where the function $\delta(\epsilon)$ is defined by

$$A\delta(\epsilon)^{-\alpha+1} = B\epsilon^{-\alpha+1} + \kappa . \tag{10.3.21}$$

(Here we exclude the case $\alpha = 1$, which is treated in Sect. 10.4b.) Using this prescription, for any value of κ, the divergence at the upper limit of the first integral cancels that in the lower limit of the second integral provided $\alpha < 2$.

The arbitrary constant κ is an expression of the ambiguity in definition of the principal value integral, which is not that which appears in the derivation of the Chapman-Kolmogorov equation given in Sect. 3.4. The symmetric definition given there arises from the imposition of the conditions i) to ii), which are more restrictive than absolutely necessary for a well-defined stochastic process.

ii) If $\alpha > 2$, then using a higher order expansion $e^{isu} - 1 \sim isu - \frac{1}{2}s^2u^2$, the integral arising from second term is divergent, and since the integrand is positive, this cannot be evaded by any choice of δ or ϵ.

iii) In fact it is clear that for any reasonable behaviour of $w(u)$ near $u = 0$, provided $\int_{-\delta}^{\epsilon} u^2 w(u)\, du$ exists for positive ϵ, δ, we can choose a $\delta(\epsilon)$ such that the principal value integral (10.3.21) is defined.

iv) Looking back at the differential Chapman-Kolmogorov equation in the form (10.3.4), it is clear that this choice of definition of the principal value integral is also that required to ensure its convergence in that equation too.

10.3.6 The Lévy-Khinchin Formula

The Lévy-Khinchin formula evades this rather precise definition of the principal value integral by noting that we can say

$$\fint_{-1}^{1} du\, isu\, w(u) \equiv \lim_{\epsilon\to 0}\left\{ \int_{-1}^{-\delta(\epsilon)} du\, isu\, w(u) + \int_{\epsilon}^{1} du\, isu\, w(u) \right\}, \tag{10.3.22}$$

$$\equiv iA_L s, \tag{10.3.23}$$

where A_L can be evaluated in any particular case, including if necessary the arbitrary constant κ. This enables the characteristic function to be rewritten as

$$\phi(s,t) = \exp\left\{\left(ia's - \tfrac{1}{2}\sigma^2 s^2 + \int_{-\infty}^{\infty} du\left(e^{isu} - 1 - isu\chi(|u|<1)\right) w(u)\right)t\right\}, \tag{10.3.24}$$

$$a' \equiv a + A_L, \tag{10.3.25}$$

$$\chi(|u|<1) = \begin{cases} 1, & |u| < 1, \\ 0, & |u| \ge 1. \end{cases} \tag{10.3.26}$$

This the *Lévy-Khinchin formula* for the characteristic function of a Lévy process. The formula does not require the curious and particular choice of the principal value integral (10.3.20), but is consistent with it.

The formula also makes a connection with the original concept of a Lévy process. The integral in (10.3.24) represents a kind of compound Poisson process, in which however,

i) Both positive and negative jumps can occur;

ii) Jumps of magnitude $|u| < 1$ are represented by a compensated Poisson process.

10.4 The Paretian Processes

The particular choice

$$a = \sigma = 0, \tag{10.4.1}$$

$$w(u) = \begin{cases} A|u|^{-\alpha-1}, & -\infty < u < 0, \\ Bu^{-\alpha-1}, & 0 < u < \infty, \end{cases} \tag{10.4.2}$$

with

$$0 < \alpha < 2, \tag{10.4.3}$$

yields the class of *Paretian processes*. Specifying that the arbitrary constant of (10.3.21) is given by $\kappa = 0$, the characteristic function can be evaluated from (10.3.7) as

$$\phi(s,t) = \exp\left\{|s|^\alpha t\,\Gamma(-\alpha)\left((A+B)\cos\frac{\pi\alpha}{2} + \frac{\mathrm{i}s}{|s|}(A-B)\sin\frac{\pi\alpha}{2}\right)\right\}. \tag{10.4.4}$$

Choosing another value of κ adds a term $\mathrm{i}\kappa s$ in the exponential in (10.4.4), adding a constant drift term. This is best regarded as arising from an appropriately modified coefficient a in the term $-a\partial_z p(z,t)$ when the principal value integral definition is chosen with $\kappa = 0$.

The characteristic function is then normally parametrised using the notation

$$\gamma = -(A+B)\,\Gamma(-\alpha)\cos\frac{\pi\alpha}{2}, \tag{10.4.5}$$

$$\beta = \frac{A-B}{A+B}, \tag{10.4.6}$$

which gives the form

$$\phi(s,t) = \exp\left\{-|s|^\alpha\,t\gamma\left(1 + \mathrm{i}\beta\frac{s}{|s|}\tan\frac{\pi\alpha}{2}\right)\right\}, \tag{10.4.7}$$

$$\equiv \int du\, e^{isu}\,\mathrm{Par}\,(\alpha,\beta,\gamma t;u). \tag{10.4.8}$$

This is the characteristic function of *Paretian process*; the last equation is the definition of its conditional probability Par $(\alpha, \beta, \gamma t; u)$.

a) The Wiener Process: Although the formula was derived for $\alpha < 2$, setting $\alpha = 2$ and $\gamma = \frac{1}{2}$ in the formula (10.4.7) yields the characteristic function of the Wiener process.

b) The Cauchy Process: If we set $\alpha \to 1$ in the original formula (10.4.4), we note that

$$\Gamma(-\alpha) \to \infty, \tag{10.4.9}$$

$$\Gamma(-\alpha)\cos\frac{\pi\alpha}{2} \to -\frac{\pi}{2}, \tag{10.4.10}$$

$$\Gamma(-\alpha)\sin\frac{\pi\alpha}{2} \to \infty. \tag{10.4.11}$$

i) If $A = B$, this yields the characteristic function of the *Cauchy process* of Sect. 3.3.1, namely

$$\phi_{\mathrm{Cauchy}} = \exp(-\gamma t|s|). \tag{10.4.12}$$

ii) If $A \neq B$, the condition (10.3.21) required to define the principal value integral takes a logarithmic form

$$A \log(\delta(\epsilon)) = B \log(\epsilon) + \kappa . \tag{10.4.13}$$

Here, κ is an arbitrary constant, for any value of which the principal value integral is well defined. The resulting characteristic function takes the form

$$\phi(s,t) = \exp\left\{-|s|\, t\left(\tfrac{1}{2}\pi(A+B) + \frac{is}{|s|}\left[\kappa + (A-B)(\gamma_{\text{Euler}} + \log|s|)\right]\right)\right\} , \tag{10.4.14}$$

$$\gamma_{\text{Euler}} = 0.57721\,56649\ldots \text{ is Euler's constant.} \tag{10.4.15}$$

The notation to be used in this case is

$$\gamma = \tfrac{1}{2}\pi(A+B) , \tag{10.4.16}$$

$$\beta = \frac{A-B}{A+B} . \tag{10.4.17}$$

Instead of the choice $\kappa = 0$, in this case it is most convenient to standardise on

$$\kappa = (B-A)\gamma_{\text{Euler}} , \tag{10.4.18}$$

giving the standard notation for the characteristic function

$$\phi(s,t) = \exp\left\{-|s|\, t\gamma\left(1 + \beta\frac{is}{|s|}\frac{2\log|s|}{\pi}\right)\right\} . \tag{10.4.19}$$

$$\equiv \int du\, e^{isu}\, \mathrm{ParI}(\beta, \gamma t; u) \tag{10.4.20}$$

This is the characteristic function of a *Paretian process* for $\alpha = 1$; the last equation is the definition of its conditional probability $\mathrm{ParI}(\beta, \gamma t; u)$. The term proportional to μ is of the same form as a displacement from the origin, and does not turn up in the case of other Paretian processes.

c) Divergent Moments: The characteristic function is nonanalytic in s except for $\alpha = 2$. This means that all moments higher that the first diverge if $\alpha \leq 2$. The first moment only exists for $1 < \alpha < 2$, and is then zero, even if $\beta \neq 0$, meaning the distribution is not symmetric.

For $0 < \alpha \leq 1$, even the first moment is divergent.

d) The Case $\beta = \pm 1$: The formulae (10.4.7) and (10.4.19) give quite sensible results even when either A or B vanishes, corresponding to $\beta = \pm 1$. However, the particular cancellation process chosen with the arbitrary constant κ of (10.3.21) set equal to zero does not work. The formulae for these cases arise by choosing a drift term ias in (10.3.7) which cancels with the corresponding term arising from any finite value of κ chosen. Thus, although the characteristic function formula makes sense and looks very like the case for $|\beta| < 1$, the underlying process is slightly different.

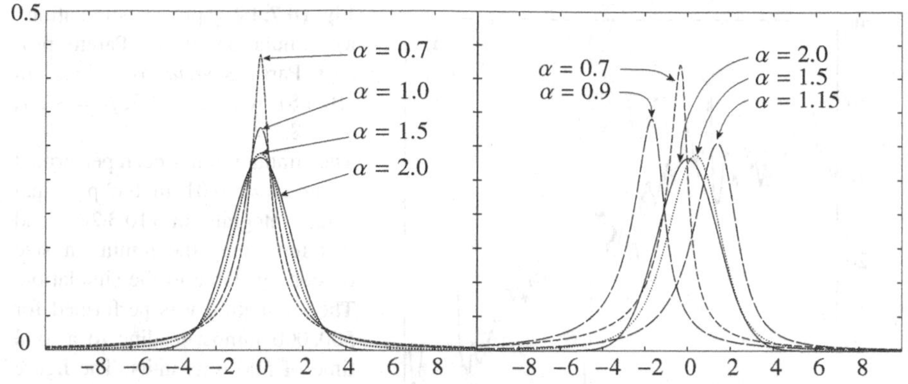

Fig. 10.5. Pareto stable distributions Par($\alpha, \beta, \gamma t; x$) as defined in (10.4.8); Left: for various α and $\beta = 0$, $\gamma t = 1$; Right: for various α and $\beta = 0.3$, $\gamma t = 1$.

10.4.1 Shapes of the Paretian Distributions

a) The Stable Paretian Distributions: This is the case when α and β have any allowable values with the exception of the case for which $\alpha = 1$ and $\beta \neq 0$ simultaneously. The distributions are plotted for various values of the parameters in Fig. 10.5. As can be seen, even for $\alpha = 1.5$, the central features are very like the Gaussian form visible for $\alpha = 2$, but the very much slower decay of the tail of the former is very evident. On the right of the figure, the asymmetric cases with $\beta = 0.3$ and $\alpha < 1$ are qualitatively different from those with the same value of β and $\alpha > 1$—as α approaches 1 from above the peak moves further to the right, and eventually recedes to infinity, reappearing from negative infinity when δ becomes less than 1.

b) Shapes of Paretian Distributions for $\alpha = 1$: We plot the distributions for $\alpha = 1$ in Fig. 10.6—these are not qualitatively very different from the distributions for α near 1, and similar β.

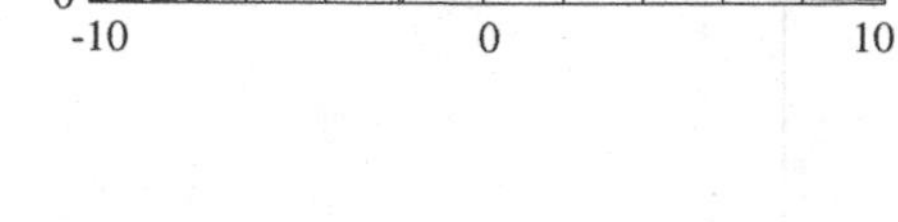

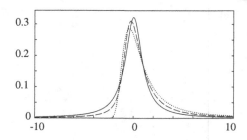

Fig. 10.6. Shapes of Paretian distributions ParI ($\beta, \gamma t; x$) for $\beta = 0$ (solid line); $\beta = 0.5$ (dashed line) and $\beta = 1$ (dotted line).

10.4.2 The Events of a Paretian Process

To simulate a Paretian process one uses the algorithm described in Sect. 3.5.1a, but since $w(u)$ is given by (10.4.2), the simulation must be carried out by omitting an

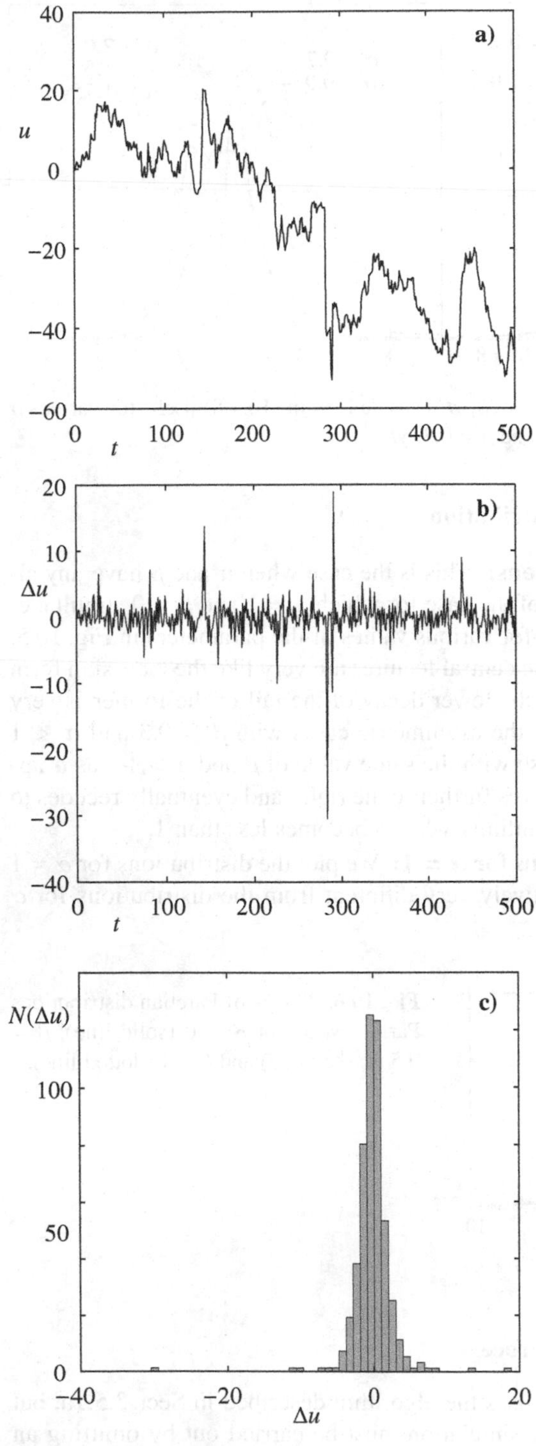

Fig. 10.7. Lévy process simulation:
a) Simulation of the Pareto process $\mathrm{Par}(\alpha, \beta, \gamma t; u)$ as defined in (10.4.8), with $\alpha = 1.7, \beta = -0.1, \gamma = 88.3$.
The simulation has been performed using $\epsilon = 0.01$ in the principal value integrals in (10.3.20), and this represents the minimum size of step available to the simulation. The simulation was performed for 500,000 jumps, leading to a total time of 500 time units. The figure plots the value of u at times which are multiples of $\Delta t = 0.5$ time units.

b) The jumps Δu between successive values of u at times which are multiples of $\Delta t = 0.5$ obtained from the simulation in a).

c) Simulation of the distribution of $\mathrm{Par}(\alpha, \beta, \gamma\Delta t; x)$ at time $\Delta t = 0.5$ obtained from the data of b).

interval $-\delta < u < \epsilon$, exactly as is necessary in calculating the characteristic function. This means that jumps of very small sizes, down to the span of the interval occur very rapidly, and if the interval is made smaller, the jumps occur more rapidly, and are even smaller. The behaviour approaches that of the Wiener process as $\alpha \to 2$ and as the omitted interval becomes smaller. The behaviour is illustrated in Fig. 10.7, which also illustrates that occasional large jumps also occur.

The probability distribution of these jumps can be extracted from the simulation, and is shown in Fig. 10.7, where the occasional large jumps are also evident.

10.4.3 Stable Processes

The Paretian processes are known as *stable processes*, because they possess a property of stability, which is a generalisation of the property possessed by Gaussian variables—that a linear combination of two Gaussian variables is also Gaussian.

a) Strictly Stable Processes: The case of $\alpha = 1$ and $\beta \neq 0$ is special; for all other cases it is clear from the expressions (10.4.7, 10.4.8) that

$$\mathrm{Par}\,(\alpha, \beta, \gamma t; \lambda u) = \mathrm{Par}\,(\alpha, \beta, \gamma t/\lambda^{\alpha}; u)\,. \tag{10.4.21}$$

From this it follows that the distribution has a width which is proportional to $(\gamma t)^{1/\alpha}$. This, in the case of a Gaussian the width is proportional to $\sqrt{\gamma t}$, in the case of a Cauchy distribution, it is proportional to γt.

This generalises to arbitrary linear combinations; thus in the case of two variables U_1, U_2, distributed according to the Pareto law, that is, such that

$$U_1 \text{ has the distribution } \mathrm{Par}\,(\alpha, \beta, \gamma t_1; u)\,, \tag{10.4.22}$$

$$U_2 \text{ has the distribution } \mathrm{Par}\,(\alpha, \beta, \gamma t_2; u)\,, \tag{10.4.23}$$

the linear combination $aU_1 + bU_2$ has the distribution $\mathrm{Par}\,(\alpha, \beta, \gamma(a^{\alpha}t_1 + b^{\alpha}t_2); u)$.

Thus the *shape* of the distribution of the linear combination is the same as that of the components, and the scale factor of the resultant is given by $(\gamma t)^{1/\alpha}$, where

$$t = (a^{\alpha}t_1 + b^{\alpha}t_2)\,. \tag{10.4.24}$$

This property expresses a kind of stability or invariance under addition which has come to be referred to as *strict stability*, and distributions with this property are called *strictly stable distributions*—the shape and location of the distribution is *stable* under arbitrary scaling and linear combinations. In the case of Gaussians with variances σ_1^2 and σ_2^2 this means that the distribution of a linear combination is also Gaussian with variance $\sigma^2 = a^2\sigma_1^2 + b^2\sigma_2^2$. It is only for the Gaussian case that the stability property can be expressed in terms of the variances, since for all other cases the variance is divergent. However, the concept of stability is of the preservation of the *shape* of the resulting distribution, which, although its variance is infinite, nevertheless has a definite width, given for example by the *full width at half maximum*.

b) The Special Case $\alpha = 1, \beta \neq 0$: The scaling law is different in this case from that given by (10.4.21), and takes the form

$$\mathrm{ParI}\,(\beta, \gamma t; \lambda u) = \mathrm{ParI}\,(\beta, \gamma t/\lambda; u + 2\gamma\beta \log \lambda/\pi)\,. \tag{10.4.25}$$

Thus the distribution changes to another of the same kind under scaling, but there is an additional displacement of the distribution as a whole. These distributions are therefore not *strictly stable*. The class of *stable processes* comprises strictly stable processes, and those which also are shifted by scaling in this way.

10.4.4 Other Lévy processes

According to Sect. 10.3.2, any infinitely divisible distribution gives rise to a corresponding Lévy process, and this has been used by [10.12, 10.14, 10.15]

a) The Generalised Inverse Gaussian Distribution: It can be shown that the density [10.16],

$$d_{IG}(x) = \frac{\sqrt{\psi/\gamma}}{2K_1(\sqrt{\psi\gamma})} \exp\left\{-\tfrac{1}{2}\left(\gamma x^{-1} + \psi x\right)\right\}, \qquad x > 0, \tag{10.4.26}$$

(where ψ and γ are positive parameters) is infinitely divisible.

b) The Hyperbolic Distribution: This infinitely divisible distribution is given by

$$\mathrm{hyp}(x) = \frac{\sqrt{\alpha^2 - \beta^2}}{2\alpha\delta K_1\left(\delta\sqrt{\alpha^2 - \beta^2}\right)} \exp\left(-\alpha\sqrt{\delta^2 + (x-\mu)^2} + \beta(x-\mu)\right). \tag{10.4.27}$$

This can be related to the inverse Gaussian [10.16].

c) The Related Lévy Processes: These distributions are not stable distributions in any sense, and therefore the relationship to the relevant Lévy process is not direct, and must proceed as in Sect. 10.3.2. The related Lévy processes have been applied with significant success to financial markets, as will be seen in Sect. 10.5.5.

10.5 Modelling the Empirical Behaviour of Financial Markets

The "correct" stochastic description of financial markets has been under discussion for over 40 years, and has not reached a definitive resolution. However, there is a substantial amount of well classified empirical data common to a wide set of financial assets. This has been carefully collated and analysed by *Rama Cont* [10.17, 10.11], which he lists as the following.

10.5.1 Stylised Statistical Facts on Asset Returns

1. *Absence of autocorrelations* : (Linear) autocorrelations of asset returns are often insignificant, except for very small intraday time scales ($\approx$ 20 minutes) for which microstructure effects come into play.
2. *Heavy tails* : The (unconditional) distribution of returns seems to display a power-law or Pareto-like tail, with a tail index which is finite, higher than two and less than five for most data sets studied. In particular this excludes stable laws with infinite variance and the normal distribution. However the precise form of the tails is difficult to determine.

3. *Gain-loss asymmetry* : One observes large drawdowns in stock prices and stock index values but not equally large upward movements.
4. *Aggregational Gaussianity* : As one increases the time scale Δt over which returns are calculated, their distribution looks more and more like a normal distribution. In particular, the shape of the distribution is not the same at different time scales.
5. *Intermittency* : Returns display, at any time scale, a high degree of variability. This is quantified by the presence of irregular bursts in time series of a wide variety of volatility estimators.
6. *Volatility clustering* : Different measures of volatility display a positive autocorrelation over several days, which quantifies the fact that high-volatility events tend to cluster in time.
7. *Conditional heavy tails* : Even after correcting returns for volatility clustering, the residual time series still exhibit heavy tails. However, the tails are less heavy than in the unconditional distribution of returns.
8. *Slow decay of autocorrelation in absolute returns* : The autocorrelation function of absolute returns decays slowly as a function of the time lag, roughly as a power law with an exponent ≈ 0.2–0.4. This is sometimes interpreted as a sign of long-range dependence.
9. *Leverage effect* : Most measures of volatility of an asset are negatively correlated with the returns of that asset.
10. *Volume-volatility correlation* : Trading volume is correlated with all measures of volatility.
11. *Asymmetry in time scales* : Coarse-grained measures of volatility predict fine-scale volatility better than the other way round.

10.5.2 The Paretian Process Description

The introduction of Paretian processes by Mandelbrot [10.7, 10.18] demonstrated convincingly that many features of the actual data (cotton prices in the USA) were present in Paretian models. Nevertheless, one of the principal features he observed—that the returns appeared to have infinite variance—does not seem to be present in financial markets, since this contradicts Sect. 10.5.1 No. 2.

10.5.3 Implications for Realistic Models

The main features of the facts listed above can be modelled by a simple description of the same form as (10.1.1), but with a different kind of driving noise.

a) Doléans-Dade Exponential: The most obvious thing to do is to write a stochastic differential equation of the kind

$$dS(t) = \mu(t)S(t)\,dt + \sigma(t)S(t)\,dX(t)\,, \tag{10.5.1}$$

in which

i) The noise term $X(t)$ is a Lévy process

ii) The drift $\mu(t)$ and the volatility $\sigma(t)$ are determined empirically, and the volatility may itself be stochastic quantity.

However, the solution of this kind of equation is not quite as simple as in the Gaussian case. In the case of constant μ and σ, the solution is given by the *Doléans-Dade exponential* (also known as the *stochastic exponential*)

$$S(t) = S(0)\exp(\mu t + \sigma X(t)) \prod_{s \le t} (1 + \sigma \Delta X(s)) e^{-\sigma \Delta X(s)} . \tag{10.5.2}$$

Here $\Delta X(s)$ denotes the jump at time s, if there is one, so that the product is over the discrete set of points s at which there are jumps.

The simple stochastic differential equation (10.5.1) thus has a rather complicated solution, and one which has the unacceptable property of being possibly negative, since negative jumps such that $\sigma \Delta X(s) < -1$ cannot be excluded.

b) Exponential Lévy Process: The more acceptable generalisation of the geometric Brownian motion description is to match the solutions; thus one can choose

$$S(t) = S(0)\exp\left\{ \int_0^t \mu(s)\, ds + \int_0^t \sigma(s)\, dX(s) \right\} , \tag{10.5.3}$$

and this corresponds to the rather complex stochastic differential equation

$$dS(t) = \mu(t) S(t)\, dt + \sigma(t) S(t)\, dX(t) + S(t)\left(e^{\sigma(t)\Delta X(t)} - 1 - \sigma(t)\Delta X(t)\right) . \tag{10.5.4}$$

The definition of the stochastic integrals is most easily understood as being via a Riemann-Stieltjes integral of each sample path.

The second description is known as an *exponential Lévy process*. The description is clearly different from that based on a stochastic exponential, since the exponential Lévy process is always positive, unlike the stochastic exponential. However, it can be shown [10.11] that every exponential Lévy process can also be written as a stochastic exponential, based on another Lévy process—the converse does not of course hold. We shall therefore concentrate on the exponential Lévy process descriptions.

10.5.4 Equivalent Martingale Measure

The Black-Scholes argument on options pricing does not work for the kinds of stochastic differential equation such as (10.5.1, 10.5.4), but the *risk-neutral* formulation given in Sect. 10.2.7 can be generalised in an acceptable, but possibly non-unique way. Let us take the exponential Lévy process equation (10.5.3), and ask if there is a way of producing a process related to it in the same way as the process (10.2.34) is related to the process (10.2.29). In finance terminology this is viewed as a change of probability measure for the driving process $X(t)$ of the representation (10.5.3), that is the same paths are weighted according to the risk-neutral judgment of the investor, rather than the more objective, but individualised judgement of an unbiased observer.

In writing a stochastic differential equation such as (10.5.1) it is implicitly understood that the stochastic increment $dX(t)$ has zero mean and is independent of $S(t)$, so that we can write

$$\langle dS(t)\rangle = \mu(t)\langle S(t)\rangle\, dt\,, \tag{10.5.5}$$

and $\mu(t)$ does correspond to the average growth rate. However, if $X(t)$ is a general Lévy process, this will not necessarily be true. The concept of an *equivalent Martingale measure* is to adjust the measure on the underlying Lévy process $X(t)$ so as to achieve the desired average growth rate, which in finance is the interest rate r. Put precisely, this means that we will assume that the stock price $S(t)$ is given in the form of an exponential Lévy process which can be written in terms of an underlying Lévy process $Z(t)$ as

$$S(t) = S(0)\exp(Z(t))\,. \tag{10.5.6}$$

where $Z(t)$ can be written in the form like the exponent in (10.5.3). We want to choose and equivalent measure $\mathcal{Q}$ such that

$$\langle e^{-rt}S(t)\rangle_{\mathcal{Q}} = S(0)\,. \tag{10.5.7}$$

Thus, $e^{-rt}S(t)$ is a martingale under the measure $\mathcal{Q}$, and the measure $\mathcal{Q}$ is called the *equivalent martingale measure*.

One straightforward way of achieving the equivalent martingale measure is by the use of the *Esscher transform*, which we shall describe below. This is by no means the only way, and for a range of methods the reader is referred to [10.11].

a) Moment Generating Function: It is convenient to use the *moment generating function* for the Lévy process $S(t)$, which can be defined in terms of the characteristic function $\phi(u,t)$ by

$$\Psi(p,t) = \langle S(t)^p\rangle = \left\langle e^{pZ(t)}\right\rangle = \phi(-ip,t)\,. \tag{10.5.8}$$

then the moments of $S(t)$ are given by

$$\langle S(t)^n\rangle = S(0)\langle\exp(nZ(t))\rangle\,, \tag{10.5.9}$$

$$= \Psi(n,t)\,. \tag{10.5.10}$$

Since $Z(t)$ is a Lévy process, we can use (10.3.24) to write the moment generating function in the form

$$\Psi(p,t) = \exp(g(p)t)\,, \tag{10.5.11}$$

where

$$g(p) = a'p + \tfrac{1}{2}\sigma^2p^2 + \int_{-\infty}^{\infty} du\,(e^{pu} - 1 - pu\chi(|u|<1))\,w(u)\,. \tag{10.5.12}$$

b) Options Pricing: A method which reproduces the results of Sect. 10.2.7 using the concept of the Esscher transform proceeds as follows.

i) Suppose the probability density of $Z(t)$ is $f(z,t)$; define a new density by

$$f(z,t;\theta) \equiv \frac{e^{\theta z}f(z,t)}{\int_{-\infty}^{\infty} e^{\theta y}f(y,t)\,dy} = \frac{e^{\theta z}f(z,t)}{\Psi(\theta,t)} \tag{10.5.13}$$

Here θ is a quantity to be determined, and $f(z,t;\theta)$ is the *Esscher transform* of $f(z,t)$. The new density is still the density of a Lévy process, provided that it is normalisable. This excludes all Paretian processes, but for any density which decays faster than an exponential as $|z| \to \infty$ there will be a range of values of θ for which $f(z, t; \theta)$ is normalisable. This includes the hyperbolic process, and the *tempered Paretian processes*, obtained by multiplying the Paretian density by $\exp(-\epsilon|z|)$ for some positive ϵ.

ii) Now choose θ so that

$$S(0) = e^{-rt}\langle S(t)\rangle_\theta , \tag{10.5.14}$$

$$= S(0)e^{-rt}\langle e^{Z(t)}\rangle_\theta , \tag{10.5.15}$$

$$= S(0)e^{-rt}\frac{\langle e^{(1+\theta)Z(t)}\rangle}{\Psi(\theta, t)} , \tag{10.5.16}$$

$$= S(0)e^{-rt}\frac{\Psi(1+\theta, t)}{\Psi(\theta, t)} , \tag{10.5.17}$$

$$= S(0)\exp\left\{t\left[g(1+\theta) - g(\theta) - r\right]\right\} . \tag{10.5.18}$$

The solution for θ is obtained by requiring

$$r = g(1+\theta) - g(\theta) . \tag{10.5.19}$$

iii) This then defines a new stochastic process; for example, for the stock growth process defined by (10.2.29), we find that

$$\theta = \frac{r-\mu}{\sigma^2} , \tag{10.5.20}$$

and the transformed process is described by the risk-neutral version (10.2.34).

10.5.5 Hyperbolic Models

This kind of formulation was first introduced by *Eberlein, Keller* and *Prause* [10.19]. They showed that the choice of an underlying Lévy process $Z(t)$ given by the hyperbolic density (10.4.27), yielded a good fit to a large quantity of data on financial markets.

10.5.6 Choice of Models

An accessible and comprehensive treatment of the application of jump-processes to finance can be found in the book by *Cont and Tankov* [10.11], who give a thorough review of almost every model which has been tried. Unlike the situation in physics or chemistry, there is no real theoretical foundation upon which to build; rather, one attempts to fit the observed facts in as simple and reliable way as possible in order to exploit the predictive power of the model so determined.

A relatively recent piece of work by *Barndorff-Nielsen* and *Shephard* [10.15] introduces a number of rather complex but realistic models based on generalised Ornstein-Uhlenbeck processes. These obey the standard Ornstein-Uhlenbeck stochastic differential equation, with the Wiener process increment $dW(t)$ replaced by the increment of an appropriate Lévy process. This paper is most notable because of the extensive discussion section attached, in which about 40 experts in the field comment (often in great detail) on the paper, and the authors respond. The comments which give a vivid picture of the state of the field, range from uncritical praise to Mandelbrot's downright condemnation, which receives a tactful but pointed response from the authors.

Mathematical finance will always be controversial, since there is no reason to believe that there is any "correct" mathematical description of financial markets.

10.6 Epilogue—the Crash of 2008

As this book goes to press, the world is experiencing a global collapse of financial markets, which many blame on the creation and uncritical trading in derivatives, some of them of a far more exotic nature than those described in this chapter. The confidence engendered by the mathematical description of financial markets has been seen to be ill-founded, and many of Samuelson's "high-paid consultants to Wall Street" (Sect. 1.3.3) have found themselves jobless. The connection of the theory of financial markets with reality has naturally come to be questioned. While some may take that point of view, others would point out that a massive set of changes is to be expected occasionally in any system governed by probability laws with heavy tails, as is undoubtedly the case in all the more careful and accurate models.

Even though the disastrous events of October 2008 came unforseen by financial experts, this does not mean the insights given by mathematical finance are specious, but rather, that they are still incomplete. The future of mathematical finance will depend on its ability to adapt to the new financial world order which may be about to happen, and to what extent it can actually assist in developing a financial system with greater stability.

11. Master Equations and Jump Processes

It is very often the case that in systems involving numbers of particles, or individual objects (animals, bacteria, etc) that a description in terms of a jump process can be very plausibly made. In such cases we find, as first mentioned in Sect. 1.1, that in an appropriate limit *macroscopic deterministic* laws of motion arise, about which the random nature of the process generates a fluctuating part. However the deterministic motion and the fluctuations arise directly out of the same description in terms of individual jumps, or transitions. In this respect, a description in terms of a jump process (and its corresponding master equation) is very satisfactory.

In contrast, we could model such a system approximately in terms of stochastic differential equations, in which the deterministic motion and the fluctuations have a completely independent origin. In such a model this independent description of fluctuations and deterministic motion is an embarrassment, and fluctuation dissipation arguments are necessary to obtain some information about the fluctuations. In this respect the master equation approach is a much more complete description.

However the existence of the macroscopic deterministic laws is a very significant result, and we will show in this chapter that there is often a limit in which the solution of a master equation can be approximated asymptotically (in terms of a large parameter Ω describing the system size) by a deterministic part (which is the solution of a deterministic differential equation), plus a fluctuating part, describable by a stochastic differential equation, whose coefficients are given by the original master equation. Such asymptotic expansions have already been noted in Sect. 3.8.3, when we dealt with the Poisson process, a very simple jump process, and are dealt with in detail in Sect. 11.2.

The result of these expansions is the development of rather simple rules for writing Fokker-Planck equations equivalent (in an asymptotic approximation) to master equations, and in fact it is often in practice quite simple to write down the appropriate approximate Fokker-Planck equation without ever formulating the master equation itself. There are several different ways of formulating the first-order approximate Fokker-Planck equation, all of which are equivalent. However, there is as yet only one way of systematically expanding in powers of Ω^{-1}, and that is the system size expansion of van Kampen.

11.1 Birth-Death Master Equations—One Variable

The one dimensional prototype of all birth-death systems consists of a population of individuals X in which the number that can occur is called x, which is a non-negative integer. We are led to consider the conditional probability $P(x,t\,|\,x',t')$ and its corresponding master equation. The concept of birth and death is usually that *only a finite number of X are created (born) or destroyed (die) in a given event.* The simplest case is when the X are born or die one at a time, with a time independent probability so that the transition probabilities $W(x\,|\,x',t)$ can be written

$$W(x\,|\,x',t) = t^+(x')\delta_{x,x'+1} + t^-(x')\delta_{x,x'-1}\,. \tag{11.1.1}$$

Thus there are two processes,

$$x \to x+1: \quad t^+(x) = \text{transition probability per unit time.} \tag{11.1.2}$$

$$x \to x-1: \quad t^-(x) = \text{transition probability per unit time.} \tag{11.1.3}$$

The general master equation (3.5.10) then takes the form

$$\begin{aligned}\partial_t P(x,t\,|\,x',t') = {}& t^+(x-1)P(x-1,t\,|\,x',t') + t^-(x+1)P(x+1\,|\,x',t') \\ & -[t^+(x)+t^-(x)]P(x,t\,|\,x',t')\,.\end{aligned} \tag{11.1.4}$$

There are no general methods of solving this equation, except in the time-independent situation.

11.1.1 Stationary Solutions

We can write the equation for the stationary solution $P_s(x)$ as

$$0 = J(x+1) - J(x)\,, \tag{11.1.5}$$

with

$$J(x) = t^-(x)P_s(x) - t^+(x-1)P_s(x-1)\,, \tag{11.1.6}$$

and defines the probability current for a jump process. We now take note of the fact that x is a non-negative integer; we cannot have a negative number of individuals. This requires

i) Since there is no probability of an individual dying if there are none present, we require

$$t^-(0) = 0\,. \tag{11.1.7}$$

ii) In addition, we must also require

$$P(x,t\,|\,x',t') = 0 \quad \text{for } x<0 \text{ or } x'<0\,. \tag{11.1.8}$$

This means that

$$J(0) = t^-(0)P_s(0) - t^+(-1)P_s(-1) = 0\,. \tag{11.1.9}$$

We now sum (11.1.5) so

$$0 = \sum_{z=0}^{x-1}[J(z+1) - J(z)] = J(x) - J(0)\,. \tag{11.1.10}$$

Hence,

$$J(x) = 0\,, \tag{11.1.11}$$

and thus

$$P_s(x) = \frac{t^+(x-1)}{t^-(x)} P_s(x-1)\,, \tag{11.1.12}$$

so that

$$P_s(x) = P_s(0) \prod_{z=1}^{x} \frac{t^+(z-1)}{t^-(z)}\,. \tag{11.1.13}$$

a) Detailed Balance Interpretation: The condition $J(x) = 0$ can be viewed as a *detailed balance requirement*, in which x is an even variable. For, it is clear that it is a form of the detailed balance condition (6.3.42), which takes the form here of

$$P(x, \tau \,|\, x', 0) P_s(x') = P(x', \tau \,|\, x, 0) P_s(x)\,. \tag{11.1.14}$$

Setting $x' = x \pm 1$ and taking the limit $\tau \to 0$, and noting that by definition (3.4.1),

$$W(x \,|\, x', t) = \lim_{\tau \to 0} P(x, t+\tau \,|\, x', t)/\tau\,, \tag{11.1.15}$$

the necessity of this condition is easily proved.

b) Rate Equations: We notice that the mean of x satisfies

$$d_t \langle x(t) \rangle = \partial_t \sum_{x=0}^{\infty} x P(x, t \,|\, x', t')\,, \tag{11.1.16}$$

$$= \sum_{x=0}^{\infty} x[t^+(x-1)P(x-1, t \,|\, x', t') - t^+(x)P(x, t \,|\, x', t')]$$

$$+ \sum_{x=0}^{\infty} x[t^-(x+1)P(x+1, t \,|\, x', t') - t^-(x)P(x, t \,|\, x', t')]\,, \tag{11.1.17}$$

$$= \sum_{x=0}^{\infty} [(x+1)t^+(x) - xt^+(x) + (x-1)t^-(x) - xt^-(x)] P(x, t \,|\, x', t')\,, \tag{11.1.18}$$

i.e.,

$$\frac{d}{dt} \langle x(t) \rangle = \langle t^+[x(t)] \rangle - \langle t^-[x(t)] \rangle\,. \tag{11.1.19}$$

The corresponding deterministic equation is that which would be obtained by neglecting fluctuations, i.e.,

$$\frac{dx}{dt} = t^+(x) - t^-(x)\,. \tag{11.1.20}$$

Notice that a stationary state occurs deterministically when

$$t^+(x) = t^-(x)\,. \tag{11.1.21}$$

Corresponding to this, notice that the maximum value of $P_s(x)$ occurs when

$$P_s(x)/P_s(x-1) \simeq 1\,, \tag{11.1.22}$$

which from (11.1.12) corresponds to

$$t^+(x-1) = t^-(x)\,. \tag{11.1.23}$$

Since the variable x takes on only integral values, for sufficiently large x (11.1.21) and (11.1.23) are essentially the same.

Thus, the *modal value of* x, which corresponds to (11.1.23), is the stationary stochastic analogue of the deterministic steady state which corresponds to (11.1.21).

11.1.2 Example: Chemical Reaction $X \rightleftharpoons A$

We treat the case of a reaction $X \underset{k_2}{\overset{k_1}{\rightleftharpoons}} A$, in which it is assumed that A is a fixed concentration. Thus, we assume

$$t^+(x) = k_2 a\,, \qquad t^-(x) = k_1 x\,, \tag{11.1.24}$$

so that the Master equation takes the simple form [where we abbreviate $P(x,t\,|\,x',t')$ to $P(x,t)$]

$$\partial_t P(x,t) = k_2 a P(x-1,t) + k_1(x+1)P(x+1,t) - (k_1 x + k_2 x)P(x,t)\,. \tag{11.1.25}$$

a) Generating Function: To solve the equation, we introduce the generating function (as in Sects. 1.5.1, 2.8.3, 3.8.2)

$$G(s,t) = \sum_{x=0}^{\infty} s^x P(x,t)\,, \tag{11.1.26}$$

so that

$$\partial_t G(s,t) = k_2 a(s-1)G(s,t) - k_1(s-1)\partial_s G(s,t)\,. \tag{11.1.27}$$

If we substitute

$$\phi(s,t) = G(s,t)\exp(-k_2 as/k_1)\,, \tag{11.1.28}$$

(11.1.27) becomes

$$\partial_t \phi(s,t) = -k_1(s-1)\partial_s \phi(s,t)\,. \tag{11.1.29}$$

The further substitution,

$$s - 1 = \mathrm{e}^z\,, \qquad \phi(s,t) = \psi(z,t)\,, \tag{11.1.30}$$

gives

$$\partial_t \psi(z,t) + k_1 \partial_z \psi(z,t) = 0\,, \tag{11.1.31}$$

whose solution is an arbitrary function of $k_1 t - z$. For convenience, write this as

$$\psi(z,t) = F[\exp(-k_1 t + z)]\mathrm{e}^{-k_2 a/k_1} = F[(s-1)\mathrm{e}^{-k_1 t}]\mathrm{e}^{-k_2 a/k_1}\,, \tag{11.1.32}$$

so

$$G(s,t) = F[(s-1)\mathrm{e}^{-k_1 t}]\exp[(s-1)k_2 a/k_1]\,. \tag{11.1.33}$$

Normalisation requires $G(1,t) = 1$, and hence

$$F(0) = 1\,. \tag{11.1.34}$$

b) Conditional Probability: The initial condition determines F. We will choose the initial condition to correspond to exactly N particles at an initial time $t' = 0$; thus

$$P(x,0\,|\,N,0) = \delta_{x,N}\,, \tag{11.1.35}$$

which means

$$G(s,0) = s^N = F(s-1)\exp\left[(s-1)k_2a/k_1\right], \tag{11.1.36}$$

so that

$$G(s,t) = \exp\left[\frac{k_2a}{k_1}(s-1)(1-\mathrm{e}^{-k_1t})\right]\left[1+(s-1)\mathrm{e}^{-k_1t}\right]^N. \tag{11.1.37}$$

This can now be expanded in a power series in s giving

$$P(x,t\,|\,N,0) = \sum_{r=0}^{x}\frac{N!}{(N-r)!r!(x-r)!}\left(\frac{k_2a}{k_1}\right)^{x-r}(1-\mathrm{e}^{-k_1t})^{N+x-2r}\mathrm{e}^{-k_1tr}$$
$$\times\exp\left[-\frac{k_2a}{k_1}(1-\mathrm{e}^{-k_1t})\right]. \tag{11.1.38}$$

This very complicated answer is a complete solution to the problem but is of very little practical use. It is better to work either directly from (11.1.37), the generating function, or from the equations for mean values.

From the generating function we can compute

$$\langle x(t)\rangle = \partial_s G(s=1,t) = \frac{k_2a}{k_1}(1-\mathrm{e}^{-k_1t}) + N\mathrm{e}^{-k_1t}\,, \tag{11.1.39}$$

$$\langle x(t)[x(t)-1]\rangle = \partial_s^2 G(s=1,t) = \langle x(t)\rangle^2 - N\mathrm{e}^{-2k_1t}\,, \tag{11.1.40}$$

$$\mathrm{var}[x(t)] = \left(N\mathrm{e}^{-k_1t} + \frac{k_2a}{k_1}\right)(1-\mathrm{e}^{-k_1t})\,. \tag{11.1.41}$$

c) Moment Equations: From the differential equation (11.1.27) we have

$$\partial_t\left[\partial_s^n G(s,t)\right] = \left\{n\left[k_2a\partial_s^{n-1} - k_1\partial_s^n\right] + (s-1)\left[k_2a\partial_s^n - k_1\partial_s^{n+1}\right]\right\}G(s,t)\,. \tag{11.1.42}$$

We set $s = 1$ and use

$$\partial_s^n G(s,t)\,|_{s=1} = \langle x(t)^n\rangle_f\,, \tag{11.1.43}$$

where the notation $\langle x^n\rangle_\mathrm{f}$ is the *factorial moment* defined in (2.8.15). We then find

$$\frac{d}{dt}\langle x(t)^n\rangle_f = n[k_2a\langle x(t)^{n-1}\rangle_f - k_1\langle x(t)^n\rangle_f]\,, \tag{11.1.44}$$

and these equations form a closed hierarchy. Naturally, the mean and variance solutions correspond to (11.1.39, 11.1.41).

d) Autocorrelation Function and Stationary Distribution: As $t \to \infty$ for any F, we find from (11.1.33, 11.1.34)

$$G(s,t\to\infty) = \exp\left[(s-1)k_2a/k_1\right], \tag{11.1.45}$$

corresponding to the *Poissonian solution:*

$$P_s(x) = \frac{\exp(-k_2a/k_1)(k_2a/k_1)^x}{x!}. \tag{11.1.46}$$

Since the equation of time evolution for $\langle x(t)\rangle$ is *linear*, we can apply the methods of Sect. 3.7.4, namely, the regression theorem, which states that the stationary autocorrelation function has the same time dependence as the mean, and its value at $t = 0$ is the stationary variance. Hence,

$$\langle x(t)\rangle_s = k_2a/k_1, \tag{11.1.47}$$

$$\mathrm{var}[x(t)]_s = k_2a/k_1, \tag{11.1.48}$$

$$\langle x(t), x(0)\rangle_s = e^{-k_1t}k_2a/k_1. \tag{11.1.49}$$

The Poissonian stationary solution also follows from (11.1.13) by direct substitution.

e) Poissonian Time-Dependent Solutions: A very interesting property of this equation is the existence of *Poissonian time-dependent solutions*. For if we choose

$$P(x,0) = \frac{e^{-\alpha_0}\alpha_0^x}{x!}, \tag{11.1.50}$$

then

$$G(s,0) = \exp[(s-1)\alpha_0], \tag{11.1.51}$$

and from (11.1.33) we find

$$G(s,t) = \exp\{(s-1)[\alpha_0 e^{-k_1t} + (k_2a/k_1)(1-e^{-k_1t})]\}, \tag{11.1.52}$$

corresponding to

$$P(x,t) = \frac{e^{-\alpha(t)}\alpha(t)^x}{x!}, \tag{11.1.53}$$

with

$$\alpha(t) = \alpha_0 e^{-k_1t} + (k_2a/k_1)(1-e^{-k_1t}). \tag{11.1.54}$$

Here $\alpha(t)$ is seen to be the solution of the deterministic equation

$$\frac{dx}{dt} = k_2a - k_1x, \tag{11.1.55}$$

with the initial condition $x(0) = \alpha_0$. This result can be generalised to many variables and forms the rationale for the *Poisson representation* which will be developed in Sect. 12.1. The existence of Poissonian propagating solutions is a consequence of the linearity of the system.

11.1.3 A Chemical Bistable System

We consider the system

$$A + 2X \underset{k_2}{\overset{k_1}{\rightleftharpoons}} 3X, \tag{11.1.56}$$

$$A \underset{k_4}{\overset{k_3}{\rightleftharpoons}} X, \tag{11.1.57}$$

which has been studied by many authors [11.1]. The concentration of A is held fixed so that we have

$$\left.\begin{aligned} t^+(x) &= k_1 A x(x-1) + k_3 A\,, \\ t^-(x) &= k_2 x(x-1)(x-2) + k_4 x\,. \end{aligned}\right\} \tag{11.1.58}$$

The corresponding deterministic equation is, of course,

$$\frac{dx}{dt} = t^+(x) - t^-(x) \simeq -k_2 x^3 + k_1 A x^2 - k_4 x + k_3 A\,, \tag{11.1.59}$$

where it is assumed that $x \gg 1$ so that we set $x(x-1)(x-2) \simeq x^3$, etc. The solution of this equation, with the initial condition $x(0) = x_0$, is given by

$$\begin{aligned} &\left(\frac{x - x_1}{x_0 - x_1}\right)^{x_3 - x_2} \left(\frac{x - x_2}{x_0 - x_2}\right)^{x_1 - x_3} \left(\frac{x - x_3}{x_0 - x_3}\right)^{x_2 - x_1} \\ &\qquad = \exp[-k_2(x_1 - x_2)(x_2 - x_3)(x_3 - x_1)t]\,. \end{aligned} \tag{11.1.60}$$

Here, x_1, x_2, x_3 are roots of

$$k_2 x^3 - k_1 A x^2 + k_4 x - x_3 A = 0\,, \tag{11.1.61}$$

with $x_3 \geqslant x_2 \geqslant x_1$.

Clearly these roots are the stationary values of the solutions $x(t)$ of (11.1.59). From (11.1.59) we see that

$$\begin{aligned} x < x_1 &\Rightarrow \frac{dx}{dt} > 0\,, \\ x_2 > x > x_1 &\Rightarrow \frac{dx}{dt} < 0\,, \\ x_3 > x > x_2 &\Rightarrow \frac{dx}{dt} > 0\,, \\ x > x_3 &\Rightarrow \frac{dx}{dt} < 0\,. \end{aligned} \tag{11.1.62}$$

Thus, in the region $x < x_2$, $x(t)$ will be attracted to x_1 and in the region $x > x_2$, $x(t)$ will be attracted to x_3. The solution $x(t) = x_2$ will be unstable to small perturbations. This yields a system with two deterministically stable stationary states.

a) Stochastic Stationary Solution: From (11.1.13)

$$P_s(x) = P_s(0) \prod_{z=1}^{x} \left\{ \frac{B[(z-1)(z-2) + P]}{z[(z-1)(z-2) + R]} \right\}, \tag{11.1.63}$$

where

$$B = k_1 A / k_2\,, \qquad R = k_4 / k_2\,, \qquad P = k_3 / k_1\,. \tag{11.1.64}$$

Notice that if $P = R$, the solution (11.1.63) is Poissonian with mean B. In this case, we have a stationary state in which reactions (11.1.56, 11.1.57) are simultaneously in balance. This is *chemical equilibrium*, in which, as we will show later, there is always a Poissonian solution (Sects. 11.5.1 and 12.1b) and The maxima of (11.1.63) occur, according to (11.1.21), when

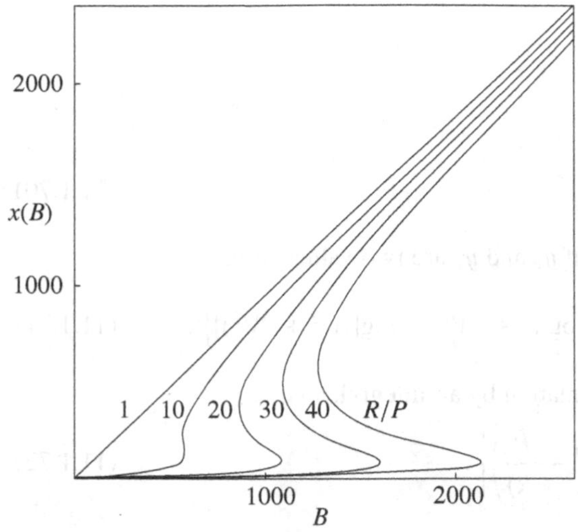

Fig. 11.1. Plot of $x(B)$ against B, as given by the solution of (11.1.65) for various values of R/P, and $P = 10,000$.

$$B = \frac{x(x-1)(x-2) + Rx}{P + x(x-1)} \,. \tag{11.1.65}$$

The function $x = x(B)$, found by inverting (11.1.65), gives the maxima (or minima) corresponding to that value of B for a given P and R.

There are the two asymptotic forms:

$$\begin{aligned} x(B) &\sim B\,, && \text{large } B\,, \\ x(b) &\sim PB/R\,, && \text{small } B\,. \end{aligned} \tag{11.1.66}$$

If $R > 9P$, we can show that the slope of $x(B)$ becomes negative for some range of $x > 0$ and thus we get three solutions for a given B, as shown in Fig. 11.1. The transition from one straight line to the other gives the kink that can be seen.

Notice also that for the choice of parameters shown, the bimodal shape is significant over a very small range of B. This range is very much narrower than the range over which $P(x)$ is two peaked, since the ratio of the heights of the peaks can be very high.

b) Asymptotic Behaviour for Large V: A more precise result can be given. Suppose the volume V of the system becomes very large and the concentration y of X, given by

$$y = x/V\,, \tag{11.1.67}$$

is constant. Clearly the transition probabilities must scale like V, since the rate of production of X will scale like $x = yV$.

Hence,

$$k_1 A \sim 1/V\,, \quad k_3 A \sim V\,, \quad k_2 \sim 1/V^2\,, \quad k_4 \sim 1\,, \tag{11.1.68}$$

which means that

$$B \sim V\,, \quad R \sim V^2\,, \quad P \sim V^2\,. \tag{11.1.69}$$

We then write

$$B = \tilde{B}V\,, \quad R = \tilde{R}V^2\,, \quad P = \tilde{P}V^2$$

so that (11.1.65) becomes

$$\tilde{B} \approx \frac{y(y^2 + \tilde{R})}{y^2 + \tilde{P}}\,. \tag{11.1.70}$$

Using the explict form (11.1.63), if y_1 and y_2 are two values of y,

$$\log\left(\frac{P_s(y_2)}{P_s(y_1)}\right) \approx \sum_{z=y_1V}^{y_2V} \left\{\log \tilde{B}V + \log(z^2 + \tilde{P}V^2) - \log[z(z^2 + \tilde{R}V^2)]\right\}\,, \tag{11.1.71}$$

and we now approximate the summation by an integral, so that

$$\log\left(\frac{P_s(y_2)}{P_s(y_1)}\right) \approx V\int_{y_1}^{y_2} dy \left[\log\left(\frac{\tilde{B}(y^2 + \tilde{P})}{y(y^2 + \tilde{R})}\right)\right]\,. \tag{11.1.72}$$

Hence,

$$\frac{P_s(y_2)}{P_s(y_1)} \approx \exp\left[V\int_{y_1}^{y_2} \log\left(\frac{\tilde{B}(y^2 + \tilde{P})}{y(y^2 + \tilde{R})}\right)\right]\,, \tag{11.1.73}$$

and as $V \to \infty$, depending on the sign of the integral, this ratio becomes either zero or infinity. Thus, in a large volume limit, the two peaks, unless precisely equal in height, become increasingly unequal and only one survives.

c) Variance: The variance of the distribution can be obtained by a simple trick. Notice from (11.1.63) that $P_s(x)$ can be written

$$P_s(x) = B^x G(x)\,, \tag{11.1.74}$$

where $G(x)$ is a function defined through (11.1.63). Then,

$$\langle x^k \rangle = \left[\sum_{x=0}^{\infty} x^k B^x G(x)\right]\left[\sum_{x=0}^{\infty} B^x G(x)\right]^{-1}\,, \tag{11.1.75}$$

and therefore

$$B\frac{\partial}{\partial B}\langle x^k \rangle = \langle x^{k+1} \rangle - \langle x \rangle\langle x^k \rangle\,, \tag{11.1.76}$$

so that

$$\mathrm{var}[x] = B\frac{\partial\langle x \rangle}{\partial B}\,. \tag{11.1.77}$$

From this we note that as $V \to \infty$,

$$\mathrm{var}[y] \sim \frac{1}{V} \to 0\,. \tag{11.1.78}$$

So a deterministic limit is approached. Further, notice that if $\langle x \rangle$ is proportional to B, the variance is equal to the mean, in fact, we find on the two branches (11.1.66),

$$\langle x(B) \rangle = \mathrm{var}[x(B)] = \begin{cases} B\,, & \text{large } B\,, \\ PB/R\,, & \text{small } B\,, \end{cases} \tag{11.1.79}$$

which means that the distributions are roughly Poissonian on these limiting branches.

The stochastic mean is not, in fact, given exactly by the peak values but approximates it very well. Of course, for any B there is one well defined $\langle x(B)\rangle$, not three values. Numerical computations show that the mean closely follows the lower branch and then suddenly makes a transition at B_c to the upper branch. This will be the value at which the two maxima have equal height and can, in principle, be determined from (11.1.73).

d) Time-Dependent Behaviour: This is impossible to deduce exactly. Almost all approximation methods depend on the large volume limit, whose properties in a stationary situation have just been noted and which will be dealt with systematically in the next section.

11.2 Approximation of Master Equations by Fokker-Planck Equations

The existence of the parameter V in terms of which well-defined scaling laws are valid leads to the concept of *system size expansions*, first put on a systematic basis by *van Kampen* [11.2]. There is a confused history attached to this, which arises out of repeated attempts to find a limiting form of the Master equation in which a Fokker-Planck equation arises. However, the fundamental result is that a diffusion process can always be approximated by a jump process, not the reverse.

11.2.1 Jump Process Approximation of a Diffusion Process

The prototype result is that found for the random walk in Sect. 3.8.2, that in the limit of infinitely small jump size, the Master equation becomes a Fokker-Planck equation. Clearly the jumps must become more probable and smaller, and this can be summarised by a *scaling assumption:* that there is a parameter δ, such that the average step size and the variance of the step size are proportional to δ, and such that the jump probabilities increase as δ becomes small.

We assume that the jump probabilities can be written

$$W_\delta(x' \mid x) = \Phi\left(\frac{x' - x - A(x)\delta}{\sqrt{\delta}}, x\right)\delta^{-3/2}, \tag{11.2.1}$$

where

$$\int dy\, \Phi(y, x) = Q, \tag{11.2.2}$$

and

$$\int dy\, y\, \Phi(y, x) = 0. \tag{11.2.3}$$

This means that

$$\begin{aligned}
\alpha_0(x) &\equiv \int dx' W_\delta(x' \mid x) = Q/\delta, \\
\alpha_1(x) &\equiv \int dx' (x' - x) W_\delta(x' \mid x) = A(x) Q, \\
\alpha_2(x) &\equiv \int dx' (x' - x)^2 W_\delta(x' \mid x) = \int dy\, y^2 \Phi(y, x).
\end{aligned} \tag{11.2.4}$$

We further assume that $\Phi(y, x)$ vanishes sufficiently rapidly as $y \to \infty$, so that

$$\lim_{\delta\to 0} W_\delta(x' \mid x) = \lim_{y\to\infty}\left[\left(\frac{y}{x'-x}\right)^3 \Phi(y,x)\right] = 0\,, \quad \text{for } x' \neq x\,. \tag{11.2.5}$$

The conditions (11.2.4, 11.2.5) are very similar to those in Sect. 3.4, namely, (3.4.1), (3.4.4) and (3.4.5) and by taking a twice differentiable function $f(z)$, one can carry out much the same procedure as that used in Sect. 3.4 to show that

$$\lim_{\delta\to 0}\left\langle \frac{\partial f(z)}{\partial t}\right\rangle = \left\langle \alpha_1(z)\frac{\partial f}{\partial z} + \tfrac{1}{2}\alpha_2(z)\frac{\partial^2 f}{\partial z^2}\right\rangle, \tag{11.2.6}$$

implying that in the limit $\delta \to 0$, the Master equation

$$\frac{\partial P(x)}{\partial t} = \int dx'[W(x\,|\,x')P(x') - W(x'\,|\,x)P(x)]\,, \tag{11.2.7}$$

becomes the FPE

$$\frac{\partial P(x)}{\partial t} = -\frac{\partial}{\partial x}\alpha_1(x)P(x) + \tfrac{1}{2}\frac{\partial^2}{\partial x^2}\alpha_2(x)P(x)\,. \tag{11.2.8}$$

Thus, given (11.2.8), one can always construct a Master equation depending on a parameter δ which approximates it as closely as desired. Such a Master equation will have transition probabilities which satisfy the criteria of (11.2.4). If they do *not* satisfy these criteria, then this approximation is not possible. Some examples are appropriate.

a) Random Walk: Using the notation of Sect. 3.8.2, let $x = nl$, then

$$W(x\,|\,x') = d(\delta_{x',x-l} + \delta_{x',x+l})\,, \tag{11.2.9}$$

and

$$\alpha_0(x) = 2d\,, \quad \alpha_1(x) = 0\,, \quad \alpha_2(x) = 2l^2 d\,. \tag{11.2.10}$$

Let

$$\delta = l^2\,, \tag{11.2.11}$$

$$D = l^2 d\,. \tag{11.2.12}$$

Then all requirements are met, so the limiting equation is

$$\frac{\partial P}{\partial t} = D\frac{\partial^2 P}{\partial x^2}\,, \tag{11.2.13}$$

as found in Sect. 3.8.2.

b) Poisson Process: Using the notation of Sect. 3.8.3, and letting $x = nl$,

$$W(x\,|\,x') = d\delta_{x,x'+l}\,, \tag{11.2.14}$$

and

$$\alpha_0(x) = d\,, \alpha_1(x) = ld\,, \alpha_2(x) = l^2 d\,. \tag{11.2.15}$$

There is no way of parametrising l and d in terms of δ such that $l \to 0$ and both $\alpha_1(x)$ and $\alpha_2(x)$ are finite. In this case, there is no Fokker-Planck limit.

c) General Approximation of Diffusion Process by a Birth-Death Master Equation: Suppose we have a Master equation such that

$$W_\delta(x' \mid x) = \left(\frac{A(x)}{2\delta} + \frac{B(x)}{2\delta^2}\right)\delta_{x', x+\delta} + \left(-\frac{A(x)}{2\delta} + \frac{B(x)}{2\delta^2}\right)\delta_{x', x-\delta}, \tag{11.2.16}$$

so that for sufficiently small δ, $W_\delta(x' \mid x)$ is positive and we assume that this is uniformly possible over the range of x of interest. The process then takes place on a range of x composed of integral multiples of δ. This is *not* of the form of (11.2.1) but, nevertheless, in the limit $\delta \to 0$ gives a Fokker-Planck equation. For

$$\alpha_0(x) = B(x)/\delta^2, \quad \alpha_1(x) = A(x), \quad \alpha_2(x) = B(x), \tag{11.2.17}$$

and

$$\lim_{\delta \to 0} W_\delta(x' \mid x) = 0, \quad \text{for } x' \neq x. \tag{11.2.18}$$

Here, however, $\alpha_0(x)$ diverges like $1/\delta^2$, rather than like $1/\delta$ as in (11.2.4) and the picture of a jump taking place according to a smooth distribution is no longer valid. The proof carries through, however, since the behaviour of $\alpha_0(x)$ is irrelevant and the limiting Fokker-Planck equation is

$$\frac{\partial P(x)}{\partial t} = -\frac{\partial}{\partial x}A(x)P(x) + \frac{1}{2}\frac{\partial^2}{\partial x^2}B(x)P(x). \tag{11.2.19}$$

In this form, we see that we have a possible tool for simulating a diffusion process by an approximating birth-death process. The method fails if $B(x) = 0$ anywhere in the range of x, since this leads to negative $W_\delta(x' \mid x)$.

Notice that the stationary solution of the Master equation in this case is

$$\begin{aligned} P_s(x) &= P_s(0) \prod_{z=\delta}^{x} \left[\frac{\delta A(z-\delta) + B(z-\delta)}{-\delta A(z) + B(z)}\right], \\ &= P_s(0)\left[\frac{-\delta A(0) + B(0)}{\delta A(x) + B(x)}\right] \prod_{z=0}^{x} \left[\frac{1 + \delta A(z)/B(z)}{1 - \delta A(z)/B(z)}\right], \end{aligned} \tag{11.2.20}$$

so that, for small enough δ

$$\log P_s(x) \to -\log B(x) + \sum_{z=0}^{x} 2\delta A(z)/B(z) + \text{constant}, \tag{11.2.21}$$

i.e.,

$$P_s(x) \to \frac{\mathcal{N}}{B(x)} \exp\left[2\int_0^x dz\, A(z)/B(z)\right], \tag{11.2.22}$$

as required. The limit is clearly uniform in any finite interval of x provided $A(x)/B(x)$ is bounded there.

11.2.2 The Kramers-Moyal Expansion

A simple but nonrigorous derivation was given by *Kramers* [11.3] and considerably improved by *Moyal* [11.4]. It was implicitly used by *Einstein* [11.5] as explained in Sect. 1.2.1.

In the Master equation (11.2.7), we substitute x' by defining

$$y = x - x' \quad \text{in the first term, and},$$
$$y = x' - x \quad \text{in the second term}.$$

Defining

$$t(y, x) = W(x + y \,|\, x), \tag{11.2.23}$$

the master equation becomes

$$\frac{\partial P(x)}{\partial t} = \int dy\, [t(y, x - y)P(x - y) - t(y, x)P(x)]. \tag{11.2.24}$$

We now expand in power series, so that

$$\frac{\partial P(x)}{\partial t} = \int dy \sum_{n=1}^{\infty} \frac{(-y)^n}{n!} \frac{\partial^n}{\partial x^n}[t(y, x)P(x)], \tag{11.2.25}$$

$$= \sum_{n=1}^{\infty} \frac{(-1)^n}{n!} \frac{\partial^n}{\partial x^n}[\alpha_n(x)P(x)], \tag{11.2.26}$$

where the n th *derivate moment* is defined by

$$\alpha_n(x) = \int dx'(x' - x)^n\, W(x' \,|\, x) = \int dy\, y^n\, t(y, x). \tag{11.2.27}$$

This expansion is known as the Kramers-Moyal expansion. By terminating the series (11.2.26) at the second term, we obtain the Fokker-Planck equation (11.2.8).

In introducing the system size expansion, van Kampen criticised this "proof", because there is no consideration of what small parameter is being considered. Nevertheless, this procedure enjoyed wide popularity—mainly because of the convenience and simplicity of the result. However, the demonstration in Sect. 11.2.1 shows that there are limits to its validity. Indeed, if we assume that $W(x' \,|\, x)$ has the form (11.2.1), we find that

$$\alpha_n(x) = \delta^{n/2-1} \int dy\, y^n \Phi(y, x). \tag{11.2.28}$$

So that as $\delta \to 0$, terms higher than the second in the expansion (11.2.26) do vanish. And indeed in his presentation, *Moyal* [11.4] did require conditions equivalent to (11.2.4) and (11.2.5).

11.2.3 Van Kampen's System Size Expansion

Birth-death Master equations provide good examples of cases where the Kramers-Moyal expansion fails. The simplest of these is the Poisson process mentioned in Sect. 11.2.1. In all of these, the size of the jump is ± 1 or some small integer, whereas typical sizes of the variable may be large, e.g., the number of molecules or the position of the random walker on a long lattice.

For such cases, *van Kampen* [11.2] introduced a system size parameter Ω such that the transition probabilities can be written in terms of the intensive variables x/Ω etc. For example, in the reaction of Sect. 11.1.3, Ω was the volume V and x/Ω the concentration. Let us use van Kampen's notation:

$a \propto \Omega \quad$ = extensive variable (number of molecules, etc.),
$x = a/\Omega =$ intensive variable (concentration of molecules etc.).

The limit of interest is large Ω at fixed x, corresponding to the approach to a macroscopic system. We can rewrite the transition probability as

$$W(a\,|\,a') = W(a';\Delta a)\,, \quad \text{where } \Delta a = a - a'\,. \tag{11.2.29}$$

The essential point is that the size of the jump is expressed in terms of the extensive quantity Δa, but the dependence on a' is better expressed in terms of the intensive variable x.

Thus, we make a *scaling assumption*

$$W(a';\Delta a) = \Omega\psi\left(\frac{a'}{\Omega}\,,\,\Delta a\right). \tag{11.2.30}$$

If this is the case, we can now make an expansion. We choose a new variable z so that

$$a = \Omega\phi(t) + \Omega^{1/2}z\,, \tag{11.2.31}$$

where $\phi(t)$ is a function to be determined. It will now be the case that the $\alpha_n(a)$ are proportional to Ω: we will write

$$\alpha_n(a) = \Omega\tilde{\alpha}_n(x)\,. \tag{11.2.32}$$

We now take the Kramers-Moyal expansion (11.2.26) and change the variable to get

$$\frac{\partial P(z,t)}{\partial t} - \Omega^{1/2}\phi'(t)\frac{\partial P(z,t)}{\partial z} = \sum_{n=1}^{\infty}\frac{\Omega^{1-n/2}}{n!}\left(-\frac{\partial}{\partial z}\right)^n \tilde{\alpha}_n[\phi(t) + \Omega^{-1/2}z]P(z,t)\,. \tag{11.2.33}$$

The terms of order $\Omega^{1/2}$ on either side will cancel if $\phi(t)$ obeys

$$\phi'(t) = \tilde{\alpha}_1[\phi(t)]\,, \tag{11.2.34}$$

which is the *deterministic equation* expected. We expand $\tilde{\alpha}_n[\phi(t) + \Omega^{-1/2}z]$ in powers of $\Omega^{-1/2}$, rearrange and find

$$\frac{\partial P(z,t)}{\partial t} = \sum_{m=2}^{\infty}\frac{\Omega^{-(m-2)/2}}{m!}\sum_{n=1}^{m}\frac{m!}{n!(m-n)!}\tilde{\alpha}_n^{(m-n)}[\phi(t)]\left(-\frac{\partial}{\partial z}\right)^n z^{m-n}P(z,t)\,. \tag{11.2.35}$$

Taking the large Ω limit, only the $m = 2$ term survives giving

$$\frac{\partial P(z,t)}{\partial t} = -\tilde{\alpha}_1^1[\phi(t)]\frac{\partial}{\partial z}z\,P(z,t) + \frac{1}{2}\tilde{\alpha}_2[\phi(t)]\frac{\partial^2}{\partial z^2}P(z,t)\,. \tag{11.2.36}$$

a) Comparison with Kramers-Moyal Result: The Kramers-Moyal Fokker-Planck equation, obtained by terminating (11.2.26) after two terms, is

$$\frac{\partial P(a)}{\partial t} = -\frac{\partial}{\partial a}[\alpha_1(a)P(a)] + \frac{1}{2}\frac{\partial^2}{\partial a^2}[\alpha_2(a)P(a)]\,, \tag{11.2.37}$$

and changing variables to $x = a/\Omega$, we get

$$\frac{\partial P(x)}{\partial t} = -\frac{\partial}{\partial x}[\tilde{\alpha}_1(x)P(x)] + \frac{1}{2\Omega}\frac{\partial^2}{\partial x^2}[\tilde{\alpha}_2(x)P(x)]. \tag{11.2.38}$$

We can now use the small noise theory of Sect. 7.3, with

$$\varepsilon^2 = \frac{1}{\Omega}, \tag{11.2.39}$$

and we find that substituting

$$z = \Omega^{1/2}[x - \phi(t)], \tag{11.2.40}$$

the lowest-order Fokker-Planck equation for z is exactly the same as the lowest-order term in van Kampen's method (11.2.36). This means that if we are only interested in the lowest order, we may use the Kramers-Moyal Fokker-Planck equation which may be easier to handle than van Kampen's method. The results will differ, but to lowest order in $\Omega^{-1/2}$ will agree, and each will only be valid to this order.

Thus, if a Fokker-Planck equation has been obtained from a Master equation, its validity depends on the kind of limiting process used to derive it. If it has been derived in a limit $\delta \to 0$ of the kind used in Sect. 11.2.1, then it can be taken seriously and the full nonlinear dependence of $\alpha_1(a)$ and $\alpha_2(a)$ on a can be exploited.

On the other hand, if it arises as the result of an Ω expansion like that in Sect. 11.2.3, only the small noise approximation has any validity. There is no point in considering anything more than the linearisation, (11.2.36), about the deterministic solution. The solution of this equation is given in terms of the corresponding stochastic differential equation

$$dz = \tilde{\alpha}_1^1[\Phi(t)]z\, dt + \sqrt{\tilde{\alpha}_2[\phi(t)]}\, dW(t), \tag{11.2.41}$$

by the results of Sect. 4.5.7 (4.5.91), or Sect. 4.5.9 (4.5.109).

b) Example: Chemical Reaction $X \rightleftharpoons A$: From Sect. 11.1.2, we have

$$W(x\,|\,x') = \delta_{x,x'+1}k_2 a + \delta_{x,x'-1}k_1 x'. \tag{11.2.42}$$

The assumption is

$$a = a_0 V, \quad x = x_0 V, \tag{11.2.43}$$

where V is the volume of the system. This means that we assume the total amounts of A and X to be proportional to V (a reasonable assumption) and that the rates of production and decay of X are proportional to a and x, respectively.

Thus,

$$W(x_0'; \Delta x) = V\left(k_2\, a_0\, \delta_{\Delta x,1} + k_1\, x_0'\, \delta_{\Delta x,-1}\right), \tag{11.2.44}$$

which is in the form of (11.2.30), with $\Omega \to V, a \to x$, etc.

Thus,

$$\begin{aligned} \alpha_1(x) &= \textstyle\sum(x' - x)W(x'\,|\,x) = k_2 a - k_1 x = V(k_2 a_0 - k_1 x_0), \\ \alpha_2(x) &= \textstyle\sum(x' - x)^2 W(x'\,|\,x) = k_2 a + k_1 x = V(k_2 a_0 + k_1 x_0). \end{aligned} \tag{11.2.45}$$

The deterministic equation is

$$\phi'(t) = [k_2 a_0 - k_1 \phi(t)], \tag{11.2.46}$$

whose solutions is

$$\phi(t) = \phi(0)\mathrm{e}^{-k_1 t} + \frac{k_2 a_0}{k_1}(1 - \mathrm{e}^{-k_1 t}). \tag{11.2.47}$$

The Fokker-Planck equation is

$$\frac{\partial P(z)}{\partial t} = k_1 \frac{\partial}{\partial z} z P(z) + \frac{1}{2}\frac{\partial^2}{\partial z^2}[k_2 a_0 + k_1 \phi(t)]P(z). \tag{11.2.48}$$

From (4.5.110), (4.5.111) we can compute that

$$\langle z(t) \rangle = z(0)\mathrm{e}^{-k_1 t}. \tag{11.2.49}$$

Usually, one would assume $z(0) = 0$, since the initial condition can be fully dealt with by the initial condition on ϕ. Assuming $z(0)$ is zero, we find

$$\mathrm{var}[z(t)] = \left[\frac{k_2 a_0}{k_1} + \phi(0)\mathrm{e}^{-k_1 t}\right](1 - \mathrm{e}^{-k_1 t}), \tag{11.2.50}$$

so that

$$\langle x(t) \rangle = V\phi(t) = V\phi(0)\mathrm{e}^{-k_1 t} + \frac{k_2 a}{k_1}(1 - \mathrm{e}^{-k_1 t}), \tag{11.2.51}$$

$$\mathrm{var}[x(t)] = V\,\mathrm{var}[z(t)] = \left[\frac{k_2 a}{k_1} + V\phi(0)\mathrm{e}^{-k_1 t}\right](1 - \mathrm{e}^{-k_1 t}). \tag{11.2.52}$$

With the identification $V\phi(0) = N$, these are exactly the same as the exact solutions (11.1.39), (11.1.41). The stationary solution of (11.2.48) is

$$P_s(z) = \mathcal{N} \exp\left(-\frac{k_1 z^2}{2 k_2 a_0}\right), \tag{11.2.53}$$

which is Gaussian approximation to the exact Poissonian.

The stationary solution of the Kramers-Moyal equation is

$$P_s(x) = \frac{\mathcal{N}}{\alpha_2(x)} \exp\left[\int_0^x \frac{2\alpha_1(x')}{\alpha_2(x')} dx'\right] = \mathcal{N}(k_2 a + k_1 x)^{-1+4k_2 a/k_1} \mathrm{e}^{-2x}. \tag{11.2.54}$$

In fact, one can explicitly check the limit by setting

$$x = V(k_2 a_0 / k_1) + \delta, \tag{11.2.55}$$

so that

$$(11.2.54) \approx \mathcal{N}(2V k_2 a_0 + k_1 \delta)^{-1+4V k_2 a_0/k_1} \mathrm{e}^{-2V k_2 a_0/k_1 - 2\delta}, \tag{11.2.56}$$

so that

$$\log P_s(x) \approx -\frac{k_2}{2 k_2 a_0 V}(\delta - \delta^2) + \text{constant}. \tag{11.2.57}$$

Using the exact Poissonian solution, making the same substitution and using Stirling's formula

$$\log x! \sim \left(x + \frac{1}{2}\right)\log x - x + \text{constant}, \tag{11.2.58}$$

one finds the same result as (11.2.57). However, the exact results are different, in the sense that even the ratio of the logarithms is different. The term linear in δ is, in fact, of lower order in V: because using (11.2.40), we find $\delta = z\sqrt{V}$ and

$$\log P_s(z) \sim -\frac{k_1}{2k_2 a_0}\left(\frac{z}{\sqrt{V}} - z^2\right) + \text{constant}, \qquad (11.2.59)$$

so that in the large V limit, we have a simple Gaussian with zero mean.

c) Moment Hierarchy: From the expansion (11.2.35), we can develop equations for the moments

$$\langle z^k \rangle = \int dz\, P(z,t) z^k\,, \qquad (11.2.60)$$

by direct substitution and integration by parts:

$$\frac{d}{dt}\langle z^k \rangle = \sum_{m=2}^{\infty} \frac{\Omega^{-(m-2)/2}}{m!} \sum_{n=1}^{\min(m,k)} \frac{m!k!}{n!(m-n)!(k-n)!} \tilde{\alpha}_n^{(m-n)}[\phi(t)] \langle z^{m+k-2n} \rangle\,. \qquad (11.2.61)$$

One can develop a hierarchy by expanding $\langle z^k \rangle$ in inverse powers of $\Omega^{1/2}$:

$$\langle z^k \rangle = \sum_{r=0}^{\infty} M_r^k \Omega^{-r/2}\,. \qquad (11.2.62)$$

From such a hierarchy one can compute stationary moments and autocorrelation functions using the same techniques as those used in handling the moment hierarchy for the small noise expansion of the Fokker-Planck equation in Sect. 7.3.1. *Van Kampen* [11.2] has carried this out.

11.2.4 Kurtz's Theorem

Kurtz [11.6] has demonstrated that in a certain sense, the Kramers-Moyal expansion can give rise to a slightly stronger result than van Kampen's expansion. For the restricted class of birth-death processes with polynomial transition probabilities, he has shown the following. We consider the stochastic process obeying a birth-death master equation

$$\partial_t P(a,t) = \sum_{a'} W(a\,|\,a')P(a',t) - \sum_{a'} W(a'\,|\,a)P(a,t)\,, \qquad (11.2.63)$$

in which the scaling condition (11.2.30) is satisfied.

Then the process $b(t)$, satisfying the stochastic differential equation

$$db(t) = \alpha_1(b)\,dt + \sqrt{\alpha_2(b)}\,dW(t)\,, \qquad (11.2.64)$$

exists, and to each sample path $a(t)$ of (11.2.63) a sample path of $b(t)$ of (11.2.64) exists such that

$$|\,b(t) - a(t)\,| \sim \log V\,, \qquad (11.2.65)$$

for all *finite* t.

This result implies the lowest order result of van Kampen. For, we make the substitution of the form (11.2.31)

$$a(t) = V\phi(t) + V^{1/2}z(t)\,, \tag{11.2.66}$$
$$b(t) = V\phi(t) + V^{1/2}y(t)\,. \tag{11.2.67}$$

Then the characteristic function of $z(t)$ is

$$\begin{aligned}\langle \exp[\mathrm{i}sz(t)]\rangle &= \left\langle \exp\left[\mathrm{i}sV^{-1/2}a(t) - \mathrm{i}sV^{1/2}\phi(t)\right]\right\rangle\,,\\ &= \exp[-\mathrm{i}sV^{1/2}\phi(t)]\left\langle \exp\left[\mathrm{i}sV^{-1/2}b(t)\right]\right\rangle + O(V^{-1/2}\log V)\,,\\ &= \langle \exp[\mathrm{i}sy(t)]\rangle + O(V^{-1/2}\log V)\,.\end{aligned} \tag{11.2.68}$$

Using now the asymptotic expansion for the Fokker-Planck equation we know the distribution function of $y(t)$ approaches that of the Fokker-Planck equation (11.2.36) to $O(V^{-1/2})$ and the result follows with, however, a slightly weaker convergence because of the $\log V$ term involved. Thus, in terms of quantities which can be calculated and measured, means, variances, etc, Kurtz's apparently stronger result is equivalent to van Kampen's system size expansion.

11.2.5 Critical Fluctuations

The existence of a system size expansion as outlined in Sect. 11.2.3 depends on the fact that $\tilde{\alpha}_1'(a)$ does not vanish. It is possible, however, for situations to arise where

$$\tilde{\alpha}_1'(\phi_s) = 0\,, \tag{11.2.69}$$

where ϕ_s is a stationary solution of the deterministic equation. This occurs, for example, when we consider the reaction of Sect. 11.1.3 for which (using the notation of that section)

$$\tilde{\alpha}_1(y) = (\tilde{B}y^2 + \tilde{p} - y^3 - y\tilde{R})\tilde{k}_2\,, \tag{11.2.70}$$

where $\tilde{k}_2 = V^2 k_2$.

Two situations can occur, corresponding to A and B in Fig. 11.2. The situation A corresponds to an unstable stationary state—any perturbation to the left will eventually lead to C, but B is stable. Clearly the deterministic equation takes the form

$$\dot{y} = -\tilde{k}_2(y - \phi_s)^3\,, \tag{11.2.71}$$

and we have a Master equation analogue of the cubic process of Sect. 7.2.4a.

Van Kampen [11.7] has shown that in this case we should write

$$a = \Omega\,\phi(t) + \Omega^{\mu}u\,, \tag{11.2.72}$$

in which case (11.2.33) becomes

$$\frac{\partial P(z,t)}{\partial t} - \Omega^{1-\mu}\phi'(t)\frac{\partial P(z,t)}{\partial z} = \sum_{n=1}^{\infty}\frac{\Omega^{1-\mu n}}{n!}\left(-\frac{\partial}{\partial z}\right)^n \tilde{\alpha}_n[\phi(t) + \Omega^{\mu-1}z]P(z,t)\,. \tag{11.2.73}$$

Suppose now that the first $q-1$ derivatives of $\tilde{\alpha}_1(\phi_s)$ vanish. Then if we choose ϕ_s for $\phi(t)$, (11.2.73) becomes to lowest order

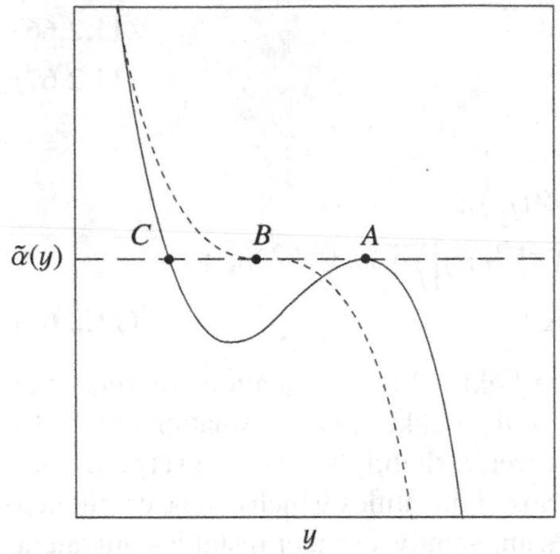

Fig. 11.2. Graph showing different kinds of behaviour of $\alpha_1(y)$ which lead to $\alpha_1'(y) = 0$.

$$\frac{\partial P(z,t)}{\partial t} = -\frac{1}{q!}\tilde{\alpha}_1^{(q)}(\phi_s)\Omega^{(1-q)(1-\mu)}\frac{\partial}{\partial z}(z^q p) + \frac{1}{2}\alpha_2(\phi_s)\Omega^{1-2\mu}\frac{\partial^2 p}{\partial z^2} + \text{higher terms.} \tag{11.2.74}$$

To make sure z remains of order unity, set

$$(1-q)(1-\mu) = (1-2\mu), \text{ i.e., } \mu = \frac{q}{q+1}, \tag{11.2.75}$$

so the result is

$$\frac{\partial P}{\partial t} = \Omega^{-(q-1)/(q+1)}\left(-\frac{1}{q!}\tilde{\alpha}_1^{(q)}\frac{\partial}{\partial z}z^q p + \frac{1}{2}\tilde{\alpha}_2\frac{\partial^2 P}{\partial z^2}\right), \tag{11.2.76}$$

(where $\tilde{\alpha}_1^{(q)}$ and $\tilde{\alpha}_2$ are evaluated at ϕ_s.) The fluctuations now vary on a slower time scale τ given by

$$\tau = t\Omega^{-(q-1)/(q+1)}, \tag{11.2.77}$$

and the equation for the average is

$$\frac{d\langle x\rangle}{d\tau} = \frac{1}{q!}\tilde{\alpha}_1^{(q)}\langle x^q\rangle, \tag{11.2.78}$$

which is no longer that associated with the linearised deterministic equation. Of course, stability depends on the sign of $\tilde{\alpha}_1^{(q)}$ and whether q is odd or even. The simplest stable case occurs for $q = 3$ which occurs at the critical point B of Fig. 11.2, and in this case we have the cubic process of Sect. 7.2.4a. The long-time scale is

$$\tau = \Omega^{-1/2}t. \tag{11.2.79}$$

We see that for large Ω, the system's time dependence is given as a function of $\tau = \Omega^{-1/2}t$. Only for times $t \gtrsim \Omega^{1/2}$ does τ become a significant size, and thus it is only for very long times t that any significant time development of the system takes place. Thus, the motion of the system becomes *very slow* at large Ω.

The condition (11.2.69) is normally controllable by some external parameter, (say, for example, the temperature), and the point in the parameter space where (11.2.69) is satisfied is called a *critical point*. This property of very slow time development at a critical point is known as *critical slowing down*.

11.3 Boundary Conditions for Birth-Death Processes

For birth-death processes, we have a rather simple way of implementing boundary conditions. For a process confined within an interval $[a, b]$, it is clear that reflecting and absorbing boundary conditions are obtained by forbidding the exit from the interval or the return to it, respectively. Namely,

	Reflecting	Absorbing,	
Boundary at a	$t^-(a) = 0$	$t^+(a-1) = 0$,	(11.3.1)
Boundary at b	$t^+(b) = 0$	$t^-(b+1) = 0$.	

It is sometimes useful, however, rather than to set certain transition probabilities equal to zero, to impose boundary conditions similar to those used for Fokker-Planck equations Sect. 5.1, so that the resulting solution in the interval $[a, b]$ is a solution of the Master equation with the appropriate vanishing transition probabilities. This may be desired in order to preserve the particular analytic form of the transition probabilities, which may have a certain convenience.

a) Forward Master Equation: We can write the forward Master equation as

$$\partial_t P(x,t\,|\,x',t') = t^+(x-1)P(x-1,t\,|\,x',t') + t^-(x+1)P(x+1,t\,|\,x',t') \\ -[t^+(x) + t^-(x)]P(x,t\,|\,x',t'). \tag{11.3.2}$$

Suppose we want a reflecting barrier at $x = a$. Then this could be obtained by requiring

$$t^-(a) = 0, \tag{11.3.3}$$

and

$$P(a-1,t\,|\,x',t) = 0. \tag{11.3.4}$$

The only equation affected by this requirement is that for $\partial_t P(a,t\,|\,x',t')$ for which the same equation can be obtained by not setting $t^-(a) = 0$ but instead introducing a *fictitious* $P(a-1,t\,|\,x',t')$ such that

$$t^+(a-1)P(a-1,t\,|\,x',t') = t^-(a)P(a,t\,|\,x',t'). \tag{11.3.5}$$

This can be viewed as the analogue of the zero current requirement for a reflecting barrier in a Fokker-Planck equation.

If we want an absorbing barrier at $x = a$, we can set

$$t^+(a-1) = 0. \tag{11.3.6}$$

After reaching the point $a-1$, the process never returns and its behaviour is now of no interest. The only equation affected by this is that for $\partial_t P(a,t\,|\,x',t')$ and the same equation can be again obtained by introducing a *fictitious* $P(a-1,t\,|\,x',t')$ such that

$$P(a-1,t\,|\,x',t') = 0\,. \tag{11.3.7}$$

Summarising, we have the alternative formulation of imposed boundary conditions which yield the same effect in $[a,b]$ as (11.3.1):

(11.3.8)

Forward Master Equation on interval $[a,b]$	*Reflecting*	*Absorbing*
Boundary at a	$t^-(a)P(a) = t^+(a-1)P(a-1)$	$P(a-1)=0$
Boundary at b	$t^+(b)P(b) = t^-(b+1)P(b+1)$	$P(b+1)=0$

b) Backward Master Equation: The backward Master equation is (see Sect. 3.6)

$$\begin{aligned}\partial_{t'} P(x,t|x',t') = {} & t^+(x')[P(x,t\,|\,x'+1,t') - P(x,t\,|\,x',t')] \\ & + t^-(x')[P(x,t\,|\,x'-1,t') - P(x,t\,|\,x',t')]\,.\end{aligned} \tag{11.3.9}$$

In the case of a reflecting barrier, at $x=a$, it is clear that $t^-(a)=0$ is equivalent to constructing a fictitious $P(x,t\,|\,a-1,t')$ such that

$$P(x,t\,|\,a-1,t') = P(x,t\,|\,a,t')\,. \tag{11.3.10}$$

In the absorbing barrier case, none of the equations for $P(x,t\,|\,x',t')$ with $x,x' \in [a,b]$ involve $t^+(a-1)$. However, because $t^+(a-1)=0$, the equations in which $x' < a-1$ will clearly preserve the condition

$$P(x,t\,|\,x',t') = 0, \quad x\in[a,b], \quad x' \leqslant a-1\,, \tag{11.3.11}$$

and the effect of this on the equation with $x'=a$ will be to impose

$$P(x,t\,|\,a-1,t') = 0\,, \tag{11.3.12}$$

which is therefore the required boundary condition. Summarising:

(11.3.13)

Backward Master Equation on interval $[a,b]$	*Reflecting*	*Absorbing*
Boundary at a	$P(\cdot\,|\,a-1) = P(\cdot\,|\,a)$	$P(\cdot\,|\,a-1)=0$
Boundary at b	$P(\cdot\,|\,b+1) = P(\cdot\,|\,b)$	$P(\cdot\,|\,b+1)=0$

11.4 Mean First Passage Times

The method for calculating these in the simple one-step case parallels that of the Fokker-Planck equation (Sect. 5.5) very closely. We assume the system is confined to the range

$$a \leqslant x \leqslant b\,, \tag{11.4.1}$$

and is absorbed or reflected at either end, as the case may be. For definiteness we take a system with

Reflecting barrier at $x = a$,
Absorbing state at $x = b + 1$.

The argument is essentially the same as that in Sect. 5.5 and we find that $T(x)$, the mean time for a particle initially at x to be absorbed, satisfies the equation related to the *backward Master equation* (11.3.9):

$$t^+(x)[T(x+1) - T(x)] + t^-(x)[T(x-1) - T(x)] = -1\,, \tag{11.4.2}$$

with the boundary conditions corresponding to (5.5.22a, 5.5.22b) and arising from (11.3.13):

$$T(a-1) = T(a)\,, \tag{11.4.3a}$$
$$T(b+1) = 0\,. \tag{11.4.3b}$$

Define

$$U(x) = T(x+1) - T(x)\,, \tag{11.4.4}$$

so (11.4.2) becomes

$$t^+(x)U(x) - t^-(x)U(x-1) = 1\,. \tag{11.4.5}$$

Define

$$\phi(x) = \prod_{z=a+1}^{x} \frac{t^-(z)}{t^+(z)}\,, \tag{11.4.6}$$
$$S(x) = U(x)/\phi(x)\,, \tag{11.4.7}$$

then (11.4.5) is equivalent to

$$t^+(x)\phi(x)[S(x) - S(x-1)] = -1\,, \tag{11.4.8}$$

with a solution

$$S(x) = -\sum_{z=a}^{x} 1/[t^+(z)\phi(z)]\,. \tag{11.4.9}$$

This satisfies the boundary condition (11.4.3a) which implies that

$$U(a-1) = S(a-1) = 0\,. \tag{11.4.10}$$

Hence,

$$T(x+1) - T(x) = -\phi(x)\sum_{z=a}^{x} 1/[t^+(z)\phi(z)]\,, \tag{11.4.11}$$

and

$$T(x) = \sum_{y=x}^{b} \phi(y) \sum_{z=a}^{y} 1/[t^+(z)\phi(z)] \qquad \begin{array}{l} a \text{ reflecting} \\ b \text{ absorbing} \\ b > a \end{array} \tag{11.4.12}$$

which also satisfies the boundary condition $T(b+1) = 0$, (11.4.3b).

Similarly, if a is absorbing and b reflecting

$$T(x) = \sum_{y=a}^{x} \phi(y) \sum_{z=y}^{b} 1/[t^{+}(z)\phi(z)] \qquad \begin{array}{l} a \text{ absorbing} \\ b \text{ reflecting} \\ b > a \end{array} \tag{11.4.13}$$

and a formula corresponding to (5.5.21) for both a and b absorbing can be similarly deduced.

11.4.1 Probability of Absorption

The mean time to absorption is always finite when a and b are finite. If, however, b is at infinity and is reflecting, the mean time may diverge. This does not itself mean that there is a finite probability of not being absorbed. The precise result (see [11.8] Sect. 4.7) is the following.

If the process takes place on the interval (a, ∞) and a is absorbing, then the probability of absorption into state $a-1$ from state x is given as follows. Define the function $M(x)$ by

$$M(x) = \sum_{y=a}^{\infty} \left[\prod_{z=a}^{y} \frac{t^{+}(y)}{t^{-}(y)} \right]. \tag{11.4.14}$$

Then if $M(x) < \infty$, the probability of absorption at $a-1$, from state x, is

$$\frac{M(x)}{1+M(x)}, \tag{11.4.15}$$

and if $M(x) = \infty$, this probability is one. If this probability is 1, then the mean time to absorption is (11.4.13).

11.4.2 Comparison with Fokker-Planck Equation

The formulae (11.4.12, 11.4.13) are really very similar to the corresponding formulae (11.4.1, 11.4.2) for a diffusion process. In fact, using the model of Sect. 11.2.1c it is not difficult to show that in the limit $\delta \to 0$ the two become the same.

If we wish to deal with the kind of problem related to escape over a potential barrier, such as in Sect. 5.5c, which turn up in the context of this kind of master equation, for example, in the bistable reaction discussed in Sect. 11.1.3, very similar approximations can be made. In this example, let us consider the mean first passage time from the stable stationary state x_1 to the other stable stationary state x_3.

Then the point $x = 0$ is a reflecting barrier, so the interval under consideration is $(0, x_3)$ with initial point x_1. Notice that

$$\phi(x) = \prod_{z=1}^{x} \frac{t^{-}(z)}{t^{+}(z)} = \frac{P_s(0)\, t^{+}(0)}{P_s(x)\, t^{+}(x)}, \tag{11.4.16}$$

so that

$$T(x_1 \to x_3) = \sum_{y=x_1}^{x_3} [P_s(y)t^+(y)]^{-1} \sum_{z=0}^{y} P_s(z). \tag{11.4.17}$$

If we assume that $P_s(y)^{-1}$ has a *sharp* maximum at the unstable point x_2, we can set $y = x_2$ in all other factors in (11.4.17) to obtain

$$T(x_1 \to x_3) \sim \frac{n_1}{t^+(x_2)} \sum_{y=x_1}^{x_3} [P_s(y)]^{-1}, \tag{11.4.18}$$

where

$$n_1 = \sum_{z=0}^{x_2} P_s(z), \tag{11.4.19}$$

and is the total probability of being in the lower peak of the stationary distribution. The result is a discrete analogue of those obtained in Sect. 5.2.7c.

11.5 Birth-Death Systems with Many Variables

There is a very wide class of systems whose time development can be considered as the result of individual encounters between members of some population. These include, for example,

i) Chemical reactions, which arise by transformations of molecules on collision;
ii) Population systems, which die, give birth, mate and consume each other;
iii) Systems of epidemics, in which diseases are transmitted from individual to individual by contact.

All of these can usually be modelled by what I call *combinatorial kinetics*, in which the transition probability for a certain transformation consequent on that encounter is proportional to the number of possible encounters of that type.

For example, in a chemical reaction $X \rightleftharpoons 2Y$, the reaction $X \to 2Y$ occurs by spontaneous decay, a degenerate kind of encounter, involving only one individual. The number of encounters of this kind is the number of X; hence, we say

$$t(x \to x-1, y \to y+2) = k_1 x. \tag{11.5.1}$$

For the reverse reaction, one can assemble pairs of molecules of Y in $y(y-1)/2$ different ways. Hence

$$t(x \to x+1, y \to y-2) = k_2 y(y-1). \tag{11.5.2}$$

In general, we can consider encounters of many kinds between molecules, species, etc., of many kinds. Using the language of chemical reactions, we have the general formulation as follows.

Consider an n-component reacting system involving s different reactions:

$$\sum_a N_a^A X_a \underset{k_A^-}{\overset{k_A^+}{\rightleftharpoons}} \sum_a M_a^A X_a, \qquad (A = 1, 2 \dots s). \tag{11.5.3}$$

The coefficient N_a^A of X_a is the number of molecules of X_a involved on the left and M_a^A is the number involved on the right. We introduce a vector notation so that if x_a is the number of molecules of X_a, then

$$\left.\begin{aligned} \boldsymbol{x} &= (x_1, x_2 \dots x_n)\,, \\ \boldsymbol{N}^A &= (N_1^A, N_2^A \dots N_n^A)\,, \\ \boldsymbol{M}^A &= (M_1^A, M_2^A, \dots, M_n^A)\,, \end{aligned}\right\} \tag{11.5.4}$$

and we also define

$$\boldsymbol{r}^A = \boldsymbol{M}^A - \boldsymbol{N}^A\,. \tag{11.5.5}$$

Clearly, as reaction A proceeds one step in the forward direction,

$$\boldsymbol{x} \to \boldsymbol{x} + \boldsymbol{r}^A\,, \tag{11.5.6}$$

and in the backward direction,

$$\boldsymbol{x} \to \boldsymbol{x} - \boldsymbol{r}^A\,. \tag{11.5.7}$$

The rate constants are defined by

$$\begin{aligned} t_A^+(\boldsymbol{x}) &= k_A^+ \prod_a \frac{x_a!}{(x_a - N_a^A)!}\,, \\ t_A^-(\boldsymbol{x}) &= k_A^- \prod_a \frac{x_a!}{(x_a - M_a^A)!}\,, \end{aligned} \tag{11.5.8}$$

which are proportional, respectively, to the number of ways of choosing the combination $\boldsymbol{N}^A$ or $\boldsymbol{M}^A$ from $\boldsymbol{x}$ molecules. The Master equation is thus

$$\begin{aligned} \partial_t P(\boldsymbol{x}, t) = \sum_A \{ & [t_A^-(\boldsymbol{x} + \boldsymbol{r}^A) P(\boldsymbol{x} + \boldsymbol{r}^A, t) - t_A^+(\boldsymbol{x}) P(\boldsymbol{x}, t)] \\ & + [t_A^+(\boldsymbol{x} - \boldsymbol{r}^A) P(\boldsymbol{x} - \boldsymbol{r}^A, t) - t_A^-(\boldsymbol{x}) P(\boldsymbol{x}, t)]\}\,. \end{aligned} \tag{11.5.9}$$

This form is, of course, a completely general way of writing a time-homogeneous Master equation for an integer variable $\boldsymbol{x}$ in which steps of size $\boldsymbol{r}^A$ can occur. It is only by making the special choice (11.5.8) for the transition probabilities per unit time that the general *combinatorial Master equation* arises. Another name is the *chemical Master equation*, since such equations are particularly adapted to chemical reactions.

11.5.1 Stationary Solutions when Detailed Balance Holds

In general, there is no explicit way of writing the stationary solution in a practical form. However, if detailed balance is satisfied, the stationary solution is easily derived. The variable $\boldsymbol{x}$, being simply a vector of numbers, can only be an even variable, hence, detailed balance must take the form (from Sect. 6.3.5)

$$t_A^-(\boldsymbol{x} + \boldsymbol{r}^A) P_s(\boldsymbol{x} + \boldsymbol{r}^A) = t_A^+(\boldsymbol{x}) P_s(\boldsymbol{x}) \qquad \text{for all } A. \tag{11.5.10}$$

The requirement that this holds for all A puts quite stringent requirements on the $t_A^\pm$. This arises from the fact that (11.5.10) provides a way of calculating $P_s(\boldsymbol{x}_0 + n\boldsymbol{r}^A)$ for any n and any initial $\boldsymbol{x}_0$. Using this method for all available A, we can generate $P_s(\boldsymbol{x})$ on the space of all $\boldsymbol{x}$ which can be written

$$\boldsymbol{x} = \boldsymbol{x}_0 + \sum n_A \boldsymbol{r}^A ; \qquad (n_A \text{ integral}), \tag{11.5.11}$$

but the solutions obtained may be ambiguous since, for example, from (11.5.10) we may write

$$\left.\begin{aligned} P_s[(\boldsymbol{x}+\boldsymbol{r}^A)+\boldsymbol{r}^B] &= \frac{P_s(\boldsymbol{x})t_A^+(\boldsymbol{x})t_B^+(\boldsymbol{x}+\boldsymbol{r}^A)}{t_A^-(\boldsymbol{x}+\boldsymbol{r}^A)t_B^-(\boldsymbol{x}+\boldsymbol{r}^A+\boldsymbol{r}^B)}\,, \\ \text{but}\,, & \\ P_s[(\boldsymbol{x}+\boldsymbol{r}^B)+\boldsymbol{r}^A] &= \frac{P_s(\boldsymbol{x})t_B^+(\boldsymbol{x})t_A^+(\boldsymbol{x}+\boldsymbol{r}^B)}{t_B^-(\boldsymbol{x}+\boldsymbol{r}^B)t_A^-(\boldsymbol{x}+\boldsymbol{r}^A+\boldsymbol{r}^B)} \end{aligned}\right\}, \tag{11.5.12}$$

Using the combinatorial forms (11.5.8) and substituting in (11.5.12), we find that this condition is automatically satisfied.

The condition becomes nontrivial when the same two points can be connected to each other in two essentially different ways, i.e., if, for example,

$$\boldsymbol{N}^A \neq \boldsymbol{N}^B\,, \qquad \boldsymbol{M}^A \neq \boldsymbol{M}^B\,, \tag{11.5.13}$$

but $\boldsymbol{r}^A = \boldsymbol{r}^B \equiv \boldsymbol{r}$.

In this case, uniqueness of $P_s(\boldsymbol{x}+\boldsymbol{r}^A)$ in (11.5.10) requires

$$\frac{t_A^+(\boldsymbol{x})}{t_A^-(\boldsymbol{x}+\boldsymbol{r})} = \frac{t_B^+(\boldsymbol{x})}{t_B^-(\boldsymbol{x}+\boldsymbol{r})}\,, \tag{11.5.14}$$

and this means

$$\frac{k_A^+}{k_A^-} = \frac{k_B^+}{k_B^-}\,. \tag{11.5.15}$$

If there are two chains $A, B, C, \ldots$, and $A', B', C', \ldots$, of reactions such that

$$\boldsymbol{r}^A + \boldsymbol{r}^B + \boldsymbol{r}^C + \ldots = \boldsymbol{r}^{A'} + \boldsymbol{r}^{B'} + \boldsymbol{r}^{C'} + \ldots\,. \tag{11.5.16}$$

Direct substitution shows that

$$P_s(\boldsymbol{x} + \boldsymbol{r}^A + \boldsymbol{r}^B + \boldsymbol{r}^C + \ldots) = P_s(\boldsymbol{x} + \boldsymbol{r}^{A'} + \boldsymbol{r}^{B'} + \boldsymbol{r}^{C'} + \ldots)\,, \tag{11.5.17}$$

only if

$$\frac{k_A^+}{k_A^-}\frac{k_B^+}{k_B^-}\frac{k_C^+}{k_C^-}\cdots = \frac{k_{A'}^+}{k_{A'}^-}\frac{k_{B'}^+}{k_{B'}^-}\frac{k_{C'}^+}{k_{C'}^-}\cdots \tag{11.5.18}$$

which is, therefore, the condition for detailed balance in a Master equation with combinatorial kinetics.

A solution for $P_s(\boldsymbol{x})$ in this case is a multivariate Poisson

$$P_s(\boldsymbol{x}) = \prod_a \frac{\alpha_a^{x_a}\,\mathrm{e}^{-\alpha_a}}{x_a!}\,, \tag{11.5.19}$$

which we check by substituting into (11.5.10) which gives

$$\prod_a \frac{\alpha_a^{(x_a+r_a^A)} e^{-\alpha_a}}{(x_a+r_a^A)!} \frac{k_A^-(x_a+r_a^A)!}{(x_a+r_a^A-M_a^A)!} = \prod_a \frac{\alpha_a^{x_a} e^{-\alpha_a}}{x_a!} \frac{k_A^+ x_a!}{(x_a-N_a^A)!} . \tag{11.5.20}$$

Using the fact that

$$r_a^A = M_a^A - N_a^A , \tag{11.5.21}$$

we find that

$$k_A^+ \prod_a \alpha_a{}^{N_a^A} = k_A^- \prod_a \alpha_a^{M_a^A} . \tag{11.5.22}$$

However, the most general solution will have this form only subject to conservation laws of various kinds. For example, in the reaction

$$X \rightleftharpoons 2Y , \tag{11.5.23}$$

the quantity $2x + y$ is conserved. Thus, the stationary distribution is

$$\frac{e^{-\alpha_1}\alpha_1^x}{x!} \frac{e^{-\alpha_2}\alpha_2^y}{y!} \phi(2x+y) , \tag{11.5.24}$$

where ϕ is an arbitrary function. Choosing $\phi(2x + y) = 1$ gives the Poissonian solution. Another choice is

$$\phi(2x+y) = \delta(2x+y, N) , \tag{11.5.25}$$

which corresponds to fixing the total of $2x + y$ at N.

As a degenerate form of this, one sometimes considers a reaction written as

$$A \rightleftharpoons 2Y , \tag{11.5.26}$$

in which, however, A is considered a fixed, deterministic number and the possible reactions are

$$\begin{aligned} A \to 2Y : \quad & t^+(y) = k_1 a , \\ 2Y \to A : \quad & t^-(y) = k_2 y(y-1) . \end{aligned} \tag{11.5.27}$$

In this case, the conservation law is now simply that y is always even, or always odd. The stationary solution is of the form

$$P_s(y) = \frac{\alpha^y}{y!}\psi(y,\alpha) , \tag{11.5.28}$$

where $\psi(y, \alpha)$ is a function which depends on y only through the evenness or oddness of y.

11.5.2 Stationary Solutions Without Detailed Balance (Kirchoff's Solution)

There is a method which, in principle, determines stationary solutions in general, though it does not seem to have found great practical use. The interested reader is referred to *Haken* (see [11.9], Sect. 4.8) and *Schnakenberg* [11.10] for a detailed treatment. In general, however, approximation methods have more to offer.

11.5.3 System Size Expansion and Related Expansions

In general we find that in chemical Master equations a system size expansion does exist. The rate of production or absorption is expected to be proportional to Ω, the size of the system. This means that as $\Omega \to \infty$, we expect

$$x \sim \Omega \rho\,, \tag{11.5.29}$$

where ρ is the set of chemical concentrations. Thus, we must have $t_A^{\pm}(x)$ proportional to Ω as $\Omega \to \infty$, so that this requires

$$\left.\begin{aligned} k_A^+ &\sim \kappa_A^+ \Omega^{-\sum_a N_a^A + 1}\,, \\ k_A^- &\sim \kappa_A^- \Omega^{-\sum_a M_a^A + 1}\,. \end{aligned}\right\} \tag{11.5.30}$$

Under these circumstances, a multivariate form of van Kampen's system size expansion can be developed. This is so complicated that it will not be explicitly derived here, but as in the single variable case, we have a Kramers-Moyal expansion whose first two terms give a diffusion process whose asymptotic form is the same as that arising from a system size expansion.

The Kramers-Moyal expansion from (11.5.9) can be derived in exactly the same way as in Sect. 11.2.2, in fact, rather more easily, since (11.5.9) is already in the appropriate form. Thus we have

$$\partial_t P(x,t) = \sum_{A,n} \left\{ \frac{(r^A \cdot \nabla)^n}{n!} [t_A^-(x) P(x,t)] + \frac{(-r^A \cdot \nabla)^n}{n!} [t_A^+(x) P(x,t)] \right\}, \tag{11.5.31}$$

and we now truncate this to second order to obtain

$$\partial_t P(x,t) = -\sum_a \partial_a [A_a(x) P(x,t)] + \tfrac{1}{2} \sum_{a,b} \partial_a \partial_b [B_{ab}(x) P(x,t)]\,, \tag{11.5.32}$$

$$A_a(x) = \sum_A r_a^A [t_A^+(x) - t_A^-(x)]\,, \tag{11.5.33}$$

$$B_{ab}(x) = \sum_A r_a^A r_b^A [t_A^+(x) + t_A^-(x)]\,. \tag{11.5.34}$$

In this form we have the *chemical Fokker-Planck equation* corresponding to the Master equation. However, we note that this is really only valid as an approximation, whose large volume asymptotic expansion is identical to order $1/\Omega$ with that of the corresponding Master equation.

If one is satisfied with this degree of approximation, it is often simpler to use the Fokker-Planck equation than the Master equation.

11.6 Some Examples

11.6.1 $X + A \rightleftharpoons 2X$

Here,

$$t^+(x) = k_1 ax, \qquad t^-(x) = k_2 x(x-1).$$

Hence,

$$A(x) = k_1 ax - k_2 x(x-1) \sim k_1 ax - k_2 x^2 \quad \text{to order } 1/\Omega, \tag{11.6.1}$$

$$B(x) = k_1 ax + k_2 x(x-1) \sim k_1 ax + k_2 x^2 \quad \text{to order } 1/\Omega. \tag{11.6.2}$$

11.6.2 $X \underset{k}{\overset{\gamma}{\rightleftharpoons}} Y \underset{\gamma}{\overset{k}{\rightleftharpoons}} A$

Here we have

$$\left.\begin{aligned} t_1^+(\boldsymbol{x}) &= \gamma y, \\ t_1^-(\boldsymbol{x}) &= kx, \end{aligned}\right\} \qquad \boldsymbol{r}^1 = (1,-1), \tag{11.6.3}$$

$$\left.\begin{aligned} t_2^+(\boldsymbol{x}) &= ka, \\ t_2^-(\boldsymbol{x}) &= \gamma y. \end{aligned}\right\} \qquad \boldsymbol{r}^2 = (0,1). \tag{11.6.4}$$

Hence,

$$\boldsymbol{A}(\boldsymbol{x}) = \begin{bmatrix} 1 \\ -1 \end{bmatrix}(\gamma y - kx) + \begin{bmatrix} 0 \\ 1 \end{bmatrix}(ka - \gamma y), \tag{11.6.5}$$

$$= \begin{bmatrix} \gamma y - kx, \\ kx + ka - 2\gamma y \end{bmatrix}, \tag{11.6.6}$$

$$\mathsf{B}(\boldsymbol{x}) = \left\{\begin{bmatrix} 1 \\ -1 \end{bmatrix}(1,-1)\right\}(\gamma y + kx) + \left\{\begin{bmatrix} 0 \\ 1 \end{bmatrix}(0,1)\right\}(ka + \gamma y), \tag{11.6.7}$$

$$= \begin{bmatrix} \gamma y + kx & -\gamma y - kx \\ -\gamma y - kx & 2\gamma y + kx + ka \end{bmatrix}. \tag{11.6.8}$$

If we now use the linearised form about the stationary state,

$$\gamma y = kx = ka, \tag{11.6.9}$$

$$\mathsf{B}_s = \begin{bmatrix} 2ka & -2ka \\ -2ka & 4ka \end{bmatrix}. \tag{11.6.10}$$

11.6.3 Prey-Predator System

The prey-predator system of Sect. 1.4 provides a good example of the kind of system in which we are interested. As a chemical reaction we can write it as

$$\left.\begin{array}{lll} \text{i)} & X + A \to 2X & \boldsymbol{r}^1 = (1,0), \\ \text{ii)} & X + Y \to 2Y & \boldsymbol{r}^2 = (-1,1), \\ \text{iii)} & \quad Y \to B & \boldsymbol{r}^3 = (0,-1). \end{array}\right\} \tag{11.6.11}$$

The reactions are all irreversible (though reversibility may be introduced) so we have

$$t_A^-(\boldsymbol{x}) = 0, \qquad (A = 1,2,3) \tag{11.6.12}$$

but

$$t_1^+(\boldsymbol{x}) = k_1 a \frac{x!}{(x-1)!}\frac{y!}{y!} = k_1 a x\,,$$
$$t_2^+(\boldsymbol{x}) = k_2 \frac{x!}{(x-1)!}\frac{y!}{(y-1)!} = k_2 x y\,, \qquad (11.6.13)$$
$$t_3^+(\boldsymbol{x}) = k_3 \frac{x!}{x!}\frac{y!}{(y-1)!} = k_3 y\,.$$

The Master equation can now be explicitly written out using (11.5.9): one obtains

$$\partial_t P(x,y) = k_1 a(x-1)P(x-1,y) + k_2(x+1)(y-1)P(x+1,y-1)\,,$$
$$+ k_3(y+1)P(x,y+1) - (k_1 a x + k_2 x y + k_3 y)P(x,y)\,. \qquad (11.6.14)$$

There are no exact solutions of this equation, so approximation methods must be used.

a) Use of the Kramers-Moyal Approximation: From (11.5.32–11.5.34)

$$\boldsymbol{A}(\boldsymbol{x}) = \begin{bmatrix}1\\0\end{bmatrix} k_1 a x + \begin{bmatrix}-1\\1\end{bmatrix} k_2 x y + \begin{bmatrix}0\\-1\end{bmatrix} k_3 y\,, \qquad (11.6.15)$$
$$= \begin{bmatrix}k_1 a x - k_2 x y\\ k_2 x y - k_3 y\end{bmatrix}, \qquad (11.6.16)$$
$$\mathsf{B}(\boldsymbol{x}) = \begin{bmatrix}1\\0\end{bmatrix}(1,0)\,k_1 a x + \begin{bmatrix}-1\\1\end{bmatrix}(-1,1)\,k_2 x y + \begin{bmatrix}0\\-1\end{bmatrix}(0,-1)\,k_3 y\,, \qquad (11.6.17)$$
$$= \begin{bmatrix}k_1 a x + k_2 x y & -k_2 x y\\ -k_2 x y & k_2 x y + k_3 y\end{bmatrix}. \qquad (11.6.18)$$

The deterministic equations are

$$\frac{d}{dt}\begin{bmatrix}x\\y\end{bmatrix} = \begin{bmatrix}k_1 a x - k_2 x y\\ k_2 x y - k_3 y\end{bmatrix}. \qquad (11.6.19)$$

b) Stability of the Stationary State: This occurs for the values

$$\begin{bmatrix}x_s\\y_s\end{bmatrix} = \begin{bmatrix}k_3/k_2\\ k_1 a/k_2\end{bmatrix}. \qquad (11.6.20)$$

To determine the stability of this state, we check the stability of the linearised deterministic equation

$$\frac{d}{dt}\begin{bmatrix}\delta x\\ \delta y\end{bmatrix} = \frac{\partial \boldsymbol{A}(\boldsymbol{x}_s)}{\partial x_s}\delta x + \frac{\partial \boldsymbol{A}(\boldsymbol{x}_s)}{\partial y_s}\delta y\,,$$
$$= \begin{bmatrix}k_1 a - k_2 y_s\\ k_2 y_s\end{bmatrix}\delta x + \begin{bmatrix}-k_2 x_s\\ k_2 x_s - k_3\end{bmatrix}\delta y\,, \qquad (11.6.21)$$
$$= \begin{bmatrix}0 & -k_3\,,\\ k_1 a & 0\end{bmatrix}\begin{bmatrix}\delta x\\ \delta y\end{bmatrix}. \qquad (11.6.22)$$

The eigenvalues of the matrix are

$$\lambda = \pm\, \mathrm{i}(k_1 k_3 a)^{1/2}\,, \qquad (11.6.23)$$

which indicates a periodic motion of any small deviation from the stationary state. We thus have neutral stability, since the disturbance neither grows nor decays.

This is related to the existence of a conserved quantity

$$V = k_2(x+y) - k_3 \log x - k_1 a \log y\,, \tag{11.6.24}$$

which can readily be checked to satisfy $dV/dt = 0$. Thus, the system conserves V and this means that there are different circular trajectories of constant V.

Writing again

$$\begin{aligned} x &= x_s + \delta x\,, \\ y &= y_s + \delta y\,, \end{aligned} \tag{11.6.25}$$

and expanding to second order, we see that

$$V = \frac{k_2^2}{2}\left(\frac{\delta x^2}{k_3^2} + \frac{\delta y^2}{(k_1 a)^2}\right), \tag{11.6.26}$$

so that the orbits are initially elliptical (this can also be deduced from the linearised analysis).

As the orbits become larger, they become less elliptic and eventually either x or y may become zero.

If x is the first to become zero (all the prey have been eaten), one sees that y inevitably proceeds to zero as well. If y becomes zero (all predators have starved to death), the prey grow unchecked with exponential growth.

c) Stochastic Behaviour: Because of the conservation of the quantity V, the orbits have neutral stability which means that when the fluctuations are included, the system will tend to change the size of the orbit with time. We can see this directly from the equivalent stochastic differential equations

$$\begin{bmatrix} dx \\ dy \end{bmatrix} = \begin{bmatrix} k_1 ax - k_2 xy \\ k_2 xy - k_3 y \end{bmatrix} dt + \mathsf{C}(x,y) \begin{bmatrix} dW_1(t)\,, \\ dW_2(t) \end{bmatrix}, \tag{11.6.27}$$

where

$$\mathsf{C}(x,y)\mathsf{C}(x,y)^{\mathrm{T}} = \mathsf{B}(\boldsymbol{x})\,. \tag{11.6.28}$$

Then using Ito's formula

$$dV(x,y) = \frac{\partial V}{\partial x}dx + \frac{\partial V}{\partial y}dy + \frac{1}{2}\left(\frac{\partial^2 V}{\partial x^2}dx^2 + 2\frac{\partial^2 V}{\partial x \partial y}dx\,dy + \frac{\partial^2 V}{\partial y^2}dy\right), \tag{11.6.29}$$

so that

$$\langle dV(x,y)\rangle = \left\langle \frac{\partial V}{\partial x}(k_1 ax - k_2 xy) + \frac{\partial V}{\partial y}(k_2 xy - k_3 y)\right\rangle dt + \left\langle B_{11}\frac{k_3}{2x^2} + B_{22}\frac{k_1}{2y^2}\right\rangle dt\,. \tag{11.6.30}$$

The first average vanishes since V is deterministically conserved and we find

$$\langle dV(x,y)\rangle = \frac{1}{2}\left\langle \frac{k_3 k_1 a}{x} + \frac{k_3 k_2 y}{x} + \frac{k_1 k_2 a x}{y} + \frac{k_1 k_3 a}{y}\right\rangle. \tag{11.6.31}$$

All of these terms are of order Ω^{-1} and are positive when x and y are positive. Thus, in the mean, $V(x,y)$ increases steadily. Of course, eventually one or other of the axes is hit and similar effects occur to the deterministic case. We see that when x or y vanish, $V = \infty$.

d) Equations for Moments: Direct implementation of the system size expansion is very cumbersome in this case, and moment equations prove more useful. These can be derived directly from the Master equation or from the Fokker-Planck equation. The results differ slightly from each other, by terms of order inverse volume. For simplicity, we use the Fokker-Planck equation so that

$$\frac{d}{dt}\begin{bmatrix}\langle x\rangle\\ \langle y\rangle\end{bmatrix} = \begin{bmatrix}k_1a\langle x\rangle - k_2\langle xy\rangle\\ k_2\langle xy\rangle - k_3\langle y\rangle\end{bmatrix}, \tag{11.6.32}$$

$$\frac{d}{dt}\begin{bmatrix}\langle x^2\rangle\\ \langle xy\rangle\\ \langle y^2\rangle\end{bmatrix} = \frac{1}{dt}\begin{bmatrix}\langle 2x\,dx + dx^2\rangle\\ \langle x\,dy + y\,dx + dx\,dy\rangle\\ \langle 2y\,dy + dy^2\rangle\end{bmatrix}, \tag{11.6.33}$$

$$= \begin{bmatrix}2k_1a\langle x^2\rangle - 2k_2\langle x^2y\rangle + k_1a\langle x\rangle + k_2\langle xy\rangle\\ k_2\langle x^2y - y^2x\rangle + (k_1a - k_3 - k_2)\langle xy\rangle\\ 2k_2\langle xy^2\rangle - 2k_3\langle y^2\rangle + k_2\langle xy\rangle + k_3\langle y\rangle\end{bmatrix}. \tag{11.6.34}$$

Knowing a system size expansion is valid means that we know all correlations and variances are of order $1/\Omega$ compared with the means. We therefore write

$$\begin{aligned} x &= \langle x\rangle + \delta x\,,\\ y &= \langle y\rangle + \delta y\,, \end{aligned} \tag{11.6.35}$$

and keep terms only of lowest order. Noting that terms arising from $\langle dx^2\rangle$, $\langle dxdy\rangle$ and $\langle dy^2\rangle$ are one order in Ω smaller than the others, we get

$$\frac{d}{dt}\begin{bmatrix}\langle x\rangle\\ \langle y\rangle\end{bmatrix} = \begin{bmatrix}k_1a\langle x\rangle - k_2\langle x\rangle\langle y\rangle\\ k_2\langle x\rangle\langle y\rangle - k_3\langle x\rangle\langle y\rangle\end{bmatrix} + \begin{bmatrix}-k_2\langle \delta x\delta y\rangle\\ k_2\langle \delta x\delta y\rangle\end{bmatrix}, \tag{11.6.36}$$

$$\frac{d}{dt}\begin{bmatrix}\langle \delta x^2\rangle\\ \langle \delta x\delta y\rangle\\ \langle \delta y^2\rangle\end{bmatrix} = \begin{bmatrix}k_1a\langle x\rangle + k_2\langle x\rangle\langle y\rangle\\ -k_2\langle x\rangle\langle y\rangle\\ k_2\langle x\rangle\langle y\rangle + k_3\langle y\rangle\end{bmatrix}$$
$$+\begin{bmatrix}2k_1a - 2k_2\langle y\rangle & -2k_2\langle x\rangle & 0\\ k_2\langle y\rangle & k_1a - k_3 + k_2(\langle x\rangle - \langle y\rangle) & -k_2\langle x\rangle\\ 0 & 2k_2\langle y\rangle & 2k_2\langle x\rangle - 2k_3\end{bmatrix}\begin{bmatrix}\langle \delta x^2\rangle\\ \langle \delta x\delta y\rangle\\ \langle \delta y^2\rangle\end{bmatrix} \tag{11.6.37}$$

We note that the means, to lowest order, obey the deterministic equations, but to next order, the term containing $\langle \delta x\delta y\rangle$ will contribute. Thus, let us choose a simplified case in which

$$k_1a = k_3 = 1,\; k_2 = \alpha\,, \tag{11.6.38}$$

which can always be done by rescaling variables. Also abbreviating

$$\langle x\rangle \to x,\; \langle y\rangle \to y,\; \langle \delta x^2\rangle \to f,\; \langle \delta x\delta y\rangle \to g,\; \langle \delta y^2\rangle \to h\,, \tag{11.6.39}$$

we obtain

$$\frac{d}{dt}\begin{bmatrix} x \\ y \end{bmatrix} = \begin{bmatrix} x - \alpha xy \\ \alpha xy - y \end{bmatrix} + \begin{bmatrix} -\alpha g \\ \alpha g \end{bmatrix},$$

$$\frac{d}{dt}\begin{bmatrix} f \\ g \\ h \end{bmatrix} = \begin{bmatrix} x + \alpha xy \\ -\alpha xy \\ \alpha xy + y \end{bmatrix} + \begin{bmatrix} 2 - 2\alpha y & -2\alpha x & 0 \\ \alpha y & \alpha(x-y) & -\alpha x \\ 0 & 2\alpha y & 2\alpha x - 2 \end{bmatrix}\begin{bmatrix} f \\ g \\ h \end{bmatrix}. \tag{11.6.40}$$

We can attempt to solve these equations in a stationary state. Bearing in mind that f, g, h, are a factor Ω^{-1} smaller than x and y, this requires α to be of order Ω^{-1} [this also follows from the scaling requirements (11.5.30)]. Hence α is small. To lowest order one has

$$x_s = y_s = 1/\alpha. \tag{11.6.41}$$

But the equations for f, g, h in the stationary state then become

$$\begin{bmatrix} 2g \\ h - f \\ -2g \end{bmatrix} = \begin{bmatrix} 2/\alpha \\ -1/\alpha \\ 2/\alpha \end{bmatrix}, \tag{11.6.42}$$

which are inconsistent. Thus this method does not yield a stationary state. Alternatively one can solve all of (11.6.40) in a stationary state.

After some manipulation one finds

$$\begin{aligned} x_s &= y_s, \\ g_s &= \alpha^{-1}(x_s - \alpha x_s^2), \end{aligned} \tag{11.6.43}$$

so that

$$\begin{aligned} f_s &= x_s(-2\alpha x_s^2 + x_s(2-\alpha) - 1)/(2 - 2\alpha x_s), \\ h_s &= x_s(-2\alpha x_s^2 + x_s(2+\alpha) + 1)/(2 - 2\alpha x_s), \end{aligned} \tag{11.6.44}$$

and the equation for g_s gives

$$-\alpha x_s^2 + \alpha x_s(f_s - h_s) = 0, \tag{11.6.45}$$

giving a solution for x_s, y_s, etc.

$$\left.\begin{aligned} x_s &= y_s = \tfrac{1}{2}, \\ f_s &= \tfrac{1}{2}\alpha/(\alpha - 2), \\ g_s &= \tfrac{1}{4}(2-\alpha)/\alpha, \\ h_s &= -1/(\alpha - 2), \end{aligned}\right\}, \tag{11.6.46}$$

and for small α in which the method is valid, this leads to a negative value for f_s, which is by definition, postive. Thus there is no stationary solution.

By again approximating $x_s = y_s = 1/\alpha$, the differential equations for f, g, and h can easily be solved. We find, on assuming that initially the system has zero variances and correlations,

$$\left.\begin{aligned} f(t) &= \frac{1}{2\alpha}(\cos 2t - 1) + \frac{2t}{\alpha}, \\ g(t) &= -\frac{1}{2\alpha}\sin 2t, \\ h(t) &= -\frac{1}{2\alpha}(\cos 2t - 1) + \frac{2t}{\alpha}. \end{aligned}\right\} \tag{11.6.47}$$

Notice that $f(t)$ and $h(t)$ are, in fact, always positive and increase steadily. The solution is valid only for a short time since the increasing value of $g(t)$ will eventually generate a time-dependent mean.

11.6.4 Generating Function Equations

In the case of combinatorial kinetics, a relatively simple differential equation can be derived for the generating function:

$$G(\boldsymbol{s},t) = \sum_{\boldsymbol{x}}\left(\prod_a s_a^{x_a}\right)P(\boldsymbol{x},t). \tag{11.6.48}$$

For we note that

$$\partial_t G(\boldsymbol{s},t) = \partial_t^+ G(\boldsymbol{s},t) + \partial_t^- G(\boldsymbol{s},t), \tag{11.6.49}$$

where the two terms correspond to the t^+ and t^- parts of the master equation. Thus

$$\begin{aligned} \partial_t^+ G(\boldsymbol{s},t) = \sum_{A,x} k_A^+ \Bigg\{ &\prod_a \left[\frac{(x_a - r_a^A)!}{(x_a - r_a^A - N_a^A)!} s_a^{x_a}\right] P(\boldsymbol{x} - \boldsymbol{r}^A, t) \\ &- \prod_a \left[\frac{x_a!}{(x_a - N_a^A)!} s_a^{x_a}\right] P(\boldsymbol{x},t) \Bigg\}. \end{aligned} \tag{11.6.50}$$

Changing the summation variable to $\boldsymbol{x} - \boldsymbol{r}^A$ and renaming this as $\boldsymbol{x}$, in the first term we find

$$\partial_t^+ G(\boldsymbol{s},t) = \sum_{A,x} k_A^+ \left[\prod_a \frac{x_a!}{(x_a - N_a^A)!} s_a^{x_a + r_a^A} - \prod_a \frac{x_a!}{(x_a - N_a^A)!} s_a^{x_a}\right] P(\boldsymbol{x},t). \tag{11.6.51}$$

Note that

$$\prod_a \frac{s_a^{x_a} x_a!}{(x_a - N_a^A)!} = \prod_a \left(\partial_a^{N_a^A} s_a^{x_a}\right) s_a^{N_a^A}, \tag{11.6.52}$$

and that

$$\prod_a \frac{s_a^{x_a + r_a^A} x_a!}{(x_a - N_a^A)!} = \prod_a \left(\partial_a^{N_a^A} s_a^{x_a}\right) s_a^{M_a^A}, \tag{11.6.53}$$

so that

$$\partial_t^+ G(\boldsymbol{s},t) = \sum_A k_A^+ \left(\prod_a s_a^{M_a^A} - \prod_a s_a^{N_a^A}\right) \partial_a^{N_a^A} G(\boldsymbol{s},t). \tag{11.6.54}$$

Similarly, we derive a formula for $\partial_t^- G(\boldsymbol{s},t)$ and put these together to get

$$\partial_t G(\boldsymbol{s},t) = \sum_A \left(\prod_a s_a^{M_a^A} - \prod_a s_a^{N_a^A}\right)\left(k_A^+ \prod_a \partial_a^{N_a^A} - k_A^- \prod_a \partial_a^{M_a^A}\right) G(\boldsymbol{s},t), \tag{11.6.55}$$

which is the general formula for a generating function differential equation. We now give a few examples.

a) An Exactly Soluble Model: Reactions: (A, B, C held fixed)

$$A + X \xrightarrow{k_2} 2X + D \qquad N^1 = 1, M^1 = 2; \quad r^1 = 1,$$
$$k_1^+ = k_2 A, \tag{11.6.56}$$
$$k_1^- = 0,$$

$$B + X \underset{k_3}{\overset{k_1}{\rightleftharpoons}} C \qquad N^2 = 1, M^2 = 0; \quad r^2 = -1,$$
$$k_2^+ = k_1 B, \tag{11.6.57}$$
$$k_2^- = k_3 C.$$

Hence, from (11.6.55), the generating function equation is

$$\partial_t G = (s^2 - s)(k_2 A \partial_s G) + (1 - s)(k_1 B \partial_s G - k_3 CG). \tag{11.6.58}$$

Solve by characteristics.

Set

$$k_1 B = \beta, \quad k_2 A = \alpha, \quad k_3 C = \gamma. \tag{11.6.59}$$

The characteristics are

$$\frac{dt}{1} = -\frac{ds}{(1-s)(\beta - \alpha s)} = \frac{dG}{\gamma(1-s)G}, \tag{11.6.60}$$

which have the solutions

$$\left(\frac{1-s}{\beta - \alpha s}\right) e^{(\alpha-\beta)t} = u, \tag{11.6.61}$$

$$(\beta - \alpha s)^{\gamma/\alpha} G = v. \tag{11.6.62}$$

The general solution can be written $v = F(u)$, i.e.,

$$G = (\beta - \alpha s)^{-\gamma/\alpha} F\left\{ e^{(\alpha-\beta)t} \left(\frac{1-s}{\beta - \alpha s}\right)\right\}. \tag{11.6.63}$$

From this we can find various time-dependent solutions. The conditional probability $P(x, t \,|\, y, 0)$ comes from the initial condition

$$G_y(s, 0) = s^y, \tag{11.6.64}$$

$$\Rightarrow F(z) = (1 - \beta z)^y (1 - \alpha z)^{-y/\alpha - y} (\beta - \alpha)^{\gamma/\alpha}, \tag{11.6.65}$$

$$\Rightarrow G_y(s,t) = \lambda^{\gamma/\alpha} [\beta(1 - e^{-\lambda t}) - s(\alpha - \beta e^{-\lambda t})]^y [(\beta - \alpha e^{-\lambda t}) - \alpha s(1 - e^{-\lambda t})]^{-\gamma/\alpha - y}, \tag{11.6.66}$$

with $\lambda = \beta - \alpha$.

As $t \to \infty$, a stationary state exists only if $\beta > \alpha$ and is

$$G_y(s, \infty) = (\beta - \alpha s)^{-\gamma/\alpha} (\beta - \alpha)^{\gamma/\alpha}, \tag{11.6.67}$$

$$\Rightarrow P_s(x) = \frac{\Gamma(x + \gamma/\alpha)(\alpha/\beta)^x}{\Gamma(\gamma/\alpha) x!} (\beta - \alpha)^{\gamma/\alpha}. \tag{11.6.68}$$

We can also derive moment equations from the generating function equations by noting

$$\left.\begin{aligned} \partial_s G(s,t)|_{s=1} &= \langle x(t)\rangle\,, \\ \partial_s^2 G(s,t)|_{s=1} &= \langle x(t)[x(t)-1]\rangle\,. \end{aligned}\right\} \tag{11.6.69}$$

Proceeding this way we have

$$\frac{d}{dt}\langle x(t)\rangle = (k_2A - k_1B)\langle x(t)\rangle + k_3C\,, \tag{11.6.70}$$

and

$$\frac{d}{dt}\langle x(t)[x(t)-1]\rangle = 2(k_2A - k_1B)\langle x(t)[x(t)-1]\rangle + 2k_2A\langle x(t)\rangle + 2k_3C\langle x(t)\rangle\,., \tag{11.6.71}$$

These equations have a stable stationary solution provided

$$k_2A < k_1B, \quad \text{i.e.,} \quad \alpha < \beta\,. \tag{11.6.72}$$

In this case, the stationary mean and variance are

$$\langle x\rangle_s = k_3C/(k_1B - k_2A)\,, \tag{11.6.73}$$

$$\text{var}[x]_s = k_1k_3BC/(k_2A - k_1B)^2\,. \tag{11.6.74}$$

This model is a simple representation of the processes taking place in a nuclear reactor. Here X is a neutron. The first reaction represents the fission by absorption of a neutron by A to produce residue(s) D plus two neutrons. The second represents absorption of neutrons and production by means other than fission.

As k_2A approaches k_1B, we approach a critical situation where neutrons are absorbed and created in almost equal numbers. For $k_2A > k_1B$, an explosive chain reaction occurs. Notice that $\langle x_s\rangle$ and $\text{var}[x]_s$ both become very large as a critical point is approached and, in fact,

$$\frac{\text{var}[x]_s}{\langle x\rangle_s} = \frac{k_1B}{k_1B - k_2A} \to \infty\,. \tag{11.6.75}$$

Thus, there are *very* large fluctuations in $\langle x_s\rangle$ near the critical point.

Note also that the system has linear equations for the mean and is Markovian, so the methods of Sect. 3.7.4 (the regression theorem) show that

$$\langle x(t), x(0)\rangle_s = \exp[(k_2A - k_1\,B)t]\text{var}[x]_s\,, \tag{11.6.76}$$

so that the fluctuations become vanishingly slow as the critical point is approached, i.e., the time correlation function decays very slowly with time.

b) Chemical Reaction $X_1 \underset{k_2}{\overset{k_1}{\rightleftharpoons}} X_2$: One reaction

$$\begin{aligned} &N = \begin{bmatrix} 1 \\ 0 \end{bmatrix}, \quad M = \begin{bmatrix} 0 \\ 1 \end{bmatrix}, \quad r = \begin{bmatrix} -1 \\ 1 \end{bmatrix}, \\ &k_1^+ = k_1, \qquad k_1^- = k_2\,, \\ &\partial_t G(s_1,s_2,t) = (s_2 - s_1)(k_1\partial_{s_1} - k_2\partial_{s_2})G(s_1,s_2,t)\,. \end{aligned} \tag{11.6.77}$$

This can be solved by characteristics. The generating function is an arbitrary function of solutions of

$$\frac{dt}{1} = -\frac{ds_1}{k_1(s_2 - s_1)} = \frac{ds_2}{k_2(s_2 - s_1)} . \tag{11.6.78}$$

Two integrals are solutions of

$$k_2 ds_1 + k_1 ds_2 = 0 \Rightarrow k_2 s_1 + k_1 s_2 = v, , \tag{11.6.79}$$

$$(k_1 + k_2)dt = \frac{d(s_2 - s_1)}{s_2 - s_1}, \tag{11.6.80}$$

$$\Rightarrow (s_2 - s_1)\mathrm{e}^{-(k_1+k_2)t} = u ,$$

$$\therefore G(s_1, s_2, t) = F[k_2 s_1 + k_1 s_2, (s_2 - s_1)\mathrm{e}^{-(k_1+k_2)t}] . \tag{11.6.81}$$

The initial condition (Poissonian)

$$G(s_1, s_2, 0) = \exp[\alpha(s_1 - 1) + \beta(s_2 - 1)] , \tag{11.6.82}$$

gives the Poissonian solution:

$$G(s_1, s_2, t) = \exp\left\{\frac{k_2\beta - k_1\alpha}{k_1 + k_2}(s_2 - s_1)\mathrm{e}^{-(k_1+k_2)t} + \frac{\alpha + \beta}{k_1 + k_2}[k_1(s_2 - 1) + k_2(s_1 - 1)]\right\} . \tag{11.6.83}$$

In this case, the stationary solution is not unique because $x+y$ is a conserved quantity. From (11.6.81) we see that the general stationary solution is of the form

$$G(s_1, s_2, \infty) = F(k_2 s_1 + k_1 s_2, 0) . \tag{11.6.84}$$

Thus,

$$k_1^n \frac{\partial^n G}{\partial s_1^n} = k_2^n \frac{\partial^n G}{\partial s_2^n} , \tag{11.6.85}$$

which implies that, setting $s_1 = s_2 = 1$,

$$k_1^n \langle x_1^n \rangle_f = k_2^n \langle x_2^n \rangle_f . \tag{11.6.86}$$

12. The Poisson Representation

For master equations governed by combinatorial kinetics (as defined in Sect. 11.5), an expansion of the probability distribution in Poisson distributions yields a very useful technique which can be used to set up a Fokker-Planck equation *exactly* equivalent to master equations of the form (11.5.9). This method, devised by the author [12.1] and co-workers, is called the *Poisson representation*, and in it the system size expansion arises as a small noise expansion of the Poisson representation Fokker-Planck equation.

The Poisson representation at its simplest can give stochastic differential equations which can be simulated without difficulty, but in there are cases where it is only possible to formulate the relevant stochastic differential equations in the complex plane, and in these there are frequently instability issues which have until recently made practical simulations impracticable. The introduction by *Drummond* of the *gauge Poisson representation* [12.2] has to a large extent solved or at least ameliorated the problems arising from these instabilities, and recently the Poisson representation has been used in the population biology [12.3], where it can provide a natural way of integrating environmental noise, expressed in terms of noisy time-dependent parameters, with the fluctuations arising form a birth-death process.

12.1 Formulation of the Poisson Representation

We *assume* that we can expand $P(\boldsymbol{x},t)$ as a superposition of multivariate uncorrelated Poissons:

$$P(\boldsymbol{x},t) = \int d\boldsymbol{\alpha} \prod_a \frac{e^{-\alpha_a}\alpha_a^{x_a}}{x_a!} f(\boldsymbol{\alpha},t) . \tag{12.1.1}$$

This means that the generating function $G(\boldsymbol{s},t)$ can be written

$$G(\boldsymbol{s},t) = \int d\boldsymbol{\alpha} \exp\left[\sum_a (s_a - 1)\alpha_a\right] f(\boldsymbol{\alpha},t) . \tag{12.1.2}$$

We substitute this in the generating function equation (11.6.55) to get

$$\begin{aligned}\partial_t G(\boldsymbol{s},t) = \sum_A \int d\boldsymbol{\alpha} \Bigg\{ & \left[\prod_a \left(\frac{\partial}{\partial\alpha_a} + 1\right)^{M_a^A} - \prod_a \left(\frac{\partial}{\partial\alpha_a} + 1\right)^{N_a^A}\right] \\ & \times \left(k_A^+ \prod_a \alpha_a^{N_a^A} - k_A^- \prod_a \alpha_a^{M_a^A}\right) \exp\left[\sum_a (s_a - 1)\alpha_a\right]\Bigg\} f(\boldsymbol{\alpha},t) . \end{aligned} \tag{12.1.3}$$

We now integrate by parts, drop surface terms and finally equate coefficients of the exponential to obtain

$$\frac{\partial f(\alpha,t)}{\partial t} = \sum_A \left[\prod_a \left(1 - \frac{\partial}{\partial \alpha_a}\right)^{M_a^A} - \prod_a \left(1 - \frac{\partial}{\partial \alpha_a}\right)^{N_a^A} \right] \left[k_A^+ \prod_a \alpha_a^{N_a^A} - k_A^- \prod_a \alpha_a^{M_a^A} \right] f(\alpha,t) . \tag{12.1.4}$$

a) Fokker-Planck Equations for Bimolecular Reaction Systems: This equation is of the Fokker-Planck form if we have, as is usual in real chemical reactions,

$$\sum_a M_a^A \leqslant 2 , \quad \sum_a N_a^A \leqslant 2 , \tag{12.1.5}$$

which indicates that only pairs of molecules at the most participate in reactions. The Fokker-Planck equation can then be simplified as follows. Define the currents

$$J_A(\alpha) = k_A^+ \prod_a \alpha_a^{N_a^A} - k_A^- \prod_a \alpha_a^{M_a^A} , \tag{12.1.6}$$

the drifts

$$A_a[J(\alpha)] = \sum_A r_a^A J_A(\alpha) , \tag{12.1.7}$$

and the diffusion matrix elements by

$$B_{ab}[J(\alpha)] = \sum_A J_A(\alpha)(M_a^A M_b^A - N_a^A N_b^A - \delta_{a,b} r_a^A) . \tag{12.1.8}$$

Then the Poisson representation Fokker-Planck equation is

$$\frac{\partial f(\alpha,t)}{\partial t} = -\sum_a \frac{\partial}{\partial \alpha_a} \{A_a[J(\alpha)] f(\alpha,t)\} + \frac{1}{2} \sum_{a,b} \frac{\partial^2}{\partial \alpha_a \partial \alpha_b} \{B_{ab}[J(\alpha)] f(\alpha,t)\} . \tag{12.1.9}$$

Notice also that if we use the explicit volume dependence of the parameters given in equation (11.5.30) of Sect. 11.5.3 and define

$$\eta_a = \alpha_a / V , \tag{12.1.10}$$

$$\varepsilon = V^{-1/2} , \tag{12.1.11}$$

and $F(\eta,t)$ is the quasiprobability in the η variable, then the Fokker-Planck equation for the η variable takes the form of

$$\frac{\partial F(\eta,t)}{\partial t} = -\sum_a \frac{\partial}{\partial \eta_a} [\hat{A}_a(\eta) F(\eta,t)] + \frac{\varepsilon^2}{2} \sum_{a,b} \frac{\partial^2}{\partial \eta_a \partial \eta_b} [\hat{B}_{a,b}(\eta) F(\eta,t)] , \tag{12.1.12}$$

with

$$\hat{A}_a(\eta) = \sum_A r_a^A \hat{J}_A(\eta) , \tag{12.1.13}$$

$$\hat{J}_A(\eta) = \kappa_A^+ \prod_a \eta_a^{N_a^A} - \kappa_A^- \prod_a \eta_a^{M_a^A} , \tag{12.1.14}$$

$$\hat{B}_{ab}(\eta) = \sum_A \hat{J}_A(\eta)(M_a^A M_b^A - N_a^A N_b^A - \delta_{a,b} r_a^A) . \tag{12.1.15}$$

In this form we see how the system size expansion in $V^{-1/2}$ corresponds exactly to a small noise expansion in η of the Fokker-Planck equation (12.1.12). For such birth-death Master equations, this method is technically much simpler than a direct system size expansion.

b) Unimolecular Reactions: If for all A,

$$\sum_a M_a^A \leqslant 1 \quad \text{and} \quad \sum_a N_a^A \leqslant 1\,, \tag{12.1.16}$$

then it is easily checked that the diffusion coefficient $\hat{B}_{ab}(\eta)$ in (12.1.15) vanishes, and we have a Liouville equation. An initially Poissonian $P(\boldsymbol{x}, t_0)$, corresponds to a delta function $F(\boldsymbol{\eta}, t_0)$, and the time evolution generated by this Liouville equation will generate a delta function solution, $\delta(\boldsymbol{\eta} - \bar{\boldsymbol{\eta}}(t))$, where $\bar{\boldsymbol{\eta}}(t)$ is the solution of

$$d\boldsymbol{\eta}/dt = \boldsymbol{A}(\boldsymbol{\eta})\,. \tag{12.1.17}$$

This means that $P(\boldsymbol{x}, t)$ will preserve a Poissonian form, with mean equal to $\bar{\boldsymbol{\eta}}(t)$. Thus we derive the general result, that there exist propagating multipoissonian solutions for any unimolecular reaction system. Non Poissonian solutions also exist—these correspond to initial $F(\boldsymbol{\eta}, t_0)$ which are not delta functions.

c) Example: As an example, consider the reaction pair

$$\left.\begin{aligned} &\text{i)}\ A + X \underset{k_4}{\overset{k_2}{\rightleftharpoons}} 2X\,, \\ &\text{ii)}\ B + X \underset{k_3}{\overset{k_1}{\rightleftharpoons}} C\,, \end{aligned}\right\} \tag{12.1.18}$$

for which the parameters are

$$\left.\begin{aligned} N^1 &= 1, & M^1 &= 2, & k_1^+ &= k_2 A, & k_1^- &= k_4, \\ N^2 &= 1, & M^2 &= 0, & k_2^+ &= k_1 B, & k_2^- &= k_3 C, \end{aligned}\right\} \tag{12.1.19}$$

so that (12.1.4) takes the form

$$\frac{\partial f}{\partial t} = \left[\left(1 - \frac{\partial}{\partial \alpha}\right)^2 - \left(1 - \frac{\partial}{\partial \alpha}\right)\right](k_2 A\alpha - k_4\alpha^2) f + \left[1 - \left(1 - \frac{\partial}{\partial \alpha}\right)\right](k_1 B\alpha - k_3 C) f\,, \tag{12.1.20}$$

$$\frac{\partial f}{\partial t} = \left\{-\frac{\partial}{\partial \alpha}[k_3 C + (k_2 A - k_1 B)\alpha - k_4\alpha^2] + \frac{\partial^2}{\partial \alpha^2}[k_2 A\alpha - k_4\alpha^2]\right\} f\,. \tag{12.1.21}$$

This is of the Fokker-Planck form, provided $k_2 A\alpha - k_4\alpha^2 > 0$. Furthermore, there is the simple relationship between moments, which takes the form (in the case of one variable)

$$\langle x^r \rangle_f \equiv \sum_x \int d\alpha [x(x-1)\ldots(x-r+1)\frac{\mathrm{e}^{-\alpha}\alpha^x}{x!} f(\alpha) = \int d\alpha\, \alpha^r f(\alpha) \equiv \langle \alpha^r \rangle\,. \tag{12.1.22}$$

This follows from the factorial moments of the Poisson distribution, as given in Sect. 2.8.3. However, $f(\alpha)$ is not a probability or at least, is not guaranteed to be a probability in the simple minded way it is defined. This is clear, since any positive superposition of Poisson distributions must have a variance at least as wide as the Poisson distribution. Hence any $P(x)$ for which the variance is less than that of the Poisson distribution cannot be represented by a positive $f(\alpha)$.

A representation in terms of *distributions* is always possible, at least formally. For if we define

$$f_y(\alpha) = (-1)^y \delta^y(\alpha) e^\alpha , \tag{12.1.23}$$

then integrating by parts

$$\int d\alpha \, f_y(\alpha) e^{-\alpha} \frac{\alpha^x}{x!} = \int d\alpha \, \frac{\alpha^x}{x!} \left(-\frac{d}{d\alpha}\right)^y \delta(\alpha) = \delta_{x,y} . \tag{12.1.24}$$

This means that we can write

$$P(x) = \int d\alpha \, \frac{e^{-\alpha}\alpha^x}{x!} \left[\sum_y (-1)^y P(y) \delta^y(\alpha) e^\alpha \right] , \tag{12.1.25}$$

so that in a *formal* sense, an $f(\alpha)$ can always be found for any $P(x)$.

The rather singular form just given does not, in fact, normally arise since, for example, we can find the stationary solution of the Fokker-Planck equation (12.1.21) as the potential solution (up to a normalisation)

$$f_s(\alpha) = e^\alpha (k_2 A - k_4 \alpha)^{(k_1 B/k_4 - k_3 C/k_2 A - 1)} \alpha^{k_3 C/k_2 A - 1} , \tag{12.1.26}$$

which is a relatively smooth function. However, an interpretation as a probability is only possible if $f_s(\alpha)$ is positive or zero and is normalisable.

If we define

$$\delta = (k_1 B/k_4 - k_3 C/k_2 A) , \tag{12.1.27}$$

then $f_s(\alpha)$ is normalisable on the interval $(0, k_2 A/k_4)$ provided that

$$\delta > 0 , \quad k_3 > 0 . \tag{12.1.28}$$

Clearly, by definition, k_3 must be positive.

It must further be checked that the integrations by parts used to derive the Fokker-Planck equation (12.1.4) are such that under these conditions, surface terms vanish. For an interval (a, b) the surface terms which would arise in the case of the reaction (12.1.18) can be written

$$\left[e^{(s-1)\alpha} \left\{ (k_2 A\alpha - k_4\alpha^2 - k_1 B\alpha + k_3 C) f - \frac{\partial}{\partial\alpha}\left[(k_2\alpha - k_4\alpha^2) f \right] \right\} \right]_b^a$$
$$+ \left[(s-1) e^{(s-1)\alpha} \left\{ (k_2\alpha - k_4\alpha^2) f \right\} \right]_b^a . \tag{12.1.29}$$

Because of the extra factor $(s-1)$ on the second line, each line must vanish separately. It is easily checked that on the interval $(0, k_2 A/k_4)$, each term vanishes at each end of the interval for the choice (12.1.26) of f, provided δ and k_3 are both greater than zero.

In the case where k_3 and δ are both positive, we have a genuine Fokker-Planck equation equivalent to the stochastic differential equation

$$d\alpha = \left[k_3C + (k_2A - k_1B)\alpha - k_4\alpha^2\right] dt + \sqrt{2(k_2A\alpha - k_4\alpha^2)}\, dW(t)\,. \tag{12.1.30}$$

The motion takes place on the range $(0, k_2A/k_4)$ and both boundaries satisfy the criteria for *entrance boundaries*, which means that it is not possible to leave the range $(0, k_2A/k_4)$—see Sect. 5.1.

If either of the conditions (12.1.28) is violated, it is found that the drift vector is such as to take the point outside the interval $(0, k_2A/k_4)$. For example, near $\alpha = 0$ we have

$$d\alpha \approx k_3C\, dt\,, \tag{12.1.31}$$

and if k_3C is negative, α will proceed to negative values. In this case, the coefficient of $dW(t)$ in (12.1.30) becomes imaginary and interpretation is no longer possible without further explanation.

Of course, viewed as a stochastic differential equation in the *complex* variable

$$\alpha = \alpha_x + \mathrm{i}\alpha_y\,, \tag{12.1.32}$$

the stochastic differential equation is perfectly sensible and is really a pair of stochastic differential equations for the two variables α_x and α_y. However, the corresponding Fokker-Planck equation is no longer the one variable equation (12.1.21) but a two-variable Fokker-Planck equation. We can derive such a Fokker-Planck equation in terms of variations of the Poisson representation, which we now treat.

12.2 Kinds of Poisson Representations

Let us consider the case of one variable and write

$$P(x) = \int_{\mathscr{D}} d\mu(\alpha) \frac{\mathrm{e}^{-\alpha}\alpha^x}{x!} f(\alpha)\,. \tag{12.2.1}$$

Then $\mu(\alpha)$ is a measure which we will show may be chosen in three ways which all lead to useful representations, and $\mathscr{D}$ is the domain of integration, which can take on various forms, depending on the choice of measure.

12.2.1 Real Poisson Representations

Here we choose

$$d\mu(\alpha) = d\alpha\,, \tag{12.2.2}$$

and $\mathscr{D}$ is a section of the real line. As noted in the preceding example, this representation does not always exist, but where it does, a simple interpretation in terms of Fokker-Planck equations is possible.

12.2.2 Complex Poisson Representations

Here,

$$d\mu(\alpha) = d\alpha\,, \tag{12.2.3}$$

and $\mathscr{D}$ is a contour C in the complex plane. We can show that this exists under certain restrictive conditions. For, instead of the form (12.1.23), we can choose

$$f_y(\alpha) = \frac{y!}{2\pi\mathrm{i}}\alpha^{-y-1}\mathrm{e}^{\alpha} \tag{12.2.4}$$

and C to be a contour surrounding the origin. This means that

$$P_y(x) = \frac{1}{2\pi\mathrm{i}}\frac{y!}{x!}\oint_C d\alpha\,\alpha^{x-y-1} = \delta_{x,y}\,. \tag{12.2.5}$$

By appropriate summation, we may express a given $P(x)$ in terms of an $f(\alpha)$ given by

$$f(\alpha) = \frac{1}{2\pi\mathrm{i}}\sum_y P(y)\mathrm{e}^{\alpha}\alpha^{-y-1}y!\,. \tag{12.2.6}$$

If the $P(y)$ are such that $y!P(y)$ is bounded for all y, the series has a finite radius of convergence outside which $f(\alpha)$ is analytic. By choosing C to be outside this circle of convergence, we can take the integration inside the summation to find that $P(x)$ is given by

$$P(x) = \oint_C d\alpha\,\frac{\mathrm{e}^{-\alpha}\alpha^x}{x!}f(\alpha)\,. \tag{12.2.7}$$

a) Example: We take a chemical reaction model given by the reactions

$$(1) \qquad A + X \leftrightharpoons 2X\,, \tag{12.2.8}$$

$$(2) \qquad B + X \leftrightharpoons C\,. \tag{12.2.9}$$

We use the notation of Sect. 12.1 and distinguish three cases, depending on the magnitude of δ, as defined in (12.1.27). The quantity δ gives a measure of the direction in which the reaction system (12.1.18) is proceeding when a steady state exists. If $\delta > 0$, we find that when x has its steady state value, reaction (1) is producing X while reaction (2) consumes X. When $\delta = 0$, both reactions balance separately—thus we have chemical equilibrium. When $\delta < 0$, reaction (1) consumes X while reaction (2) produces X.

i) $\boldsymbol{\delta > 0}$: According to (12.1.28), this is the condition for $f_s(\alpha)$ to be a valid quasiprobability on the real interval $(0, k_2A/k_4)$. In this range, the diffusion coefficient $(k_2A\alpha - k_4\alpha_2)$ is positive. The deterministic mean of α, given by

$$\alpha = \frac{k_2A - k_1B + [(k_2A - k_1B)^2 + 4k_3k_4C]^{1/2}}{2k_4}\,, \tag{12.2.10}$$

lies *within* the interval $(0, k_2A/k_4)$. We are therefore dealing with the case of a genuine Fokker-Planck equation and $f_s(\alpha)$ is a function vanishing at both ends of the interval and peaked near the deterministic steady state.

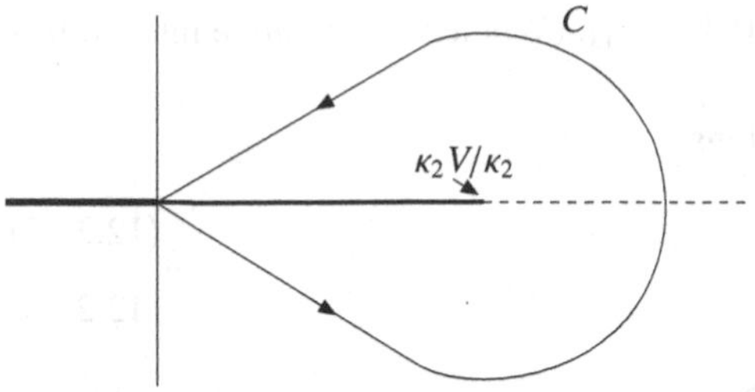

Fig. 12.1. Contour C in the complex plane for the evaluation of (12.2.13)

ii) $\boldsymbol{\delta = 0}$: Since both reactions now balance separately, we expect a Poissonian steady state. We note that $f_s(\alpha)$ in this case has a pole at $\alpha = k_2A/k_4$ and we choose the range of α to be a contour in the complex plane enclosing this pole. Since this is a closed contour, there are no boundary terms arising from partial integration and $P_s(x)$ given by choosing this type of Poisson representation clearly satisfies the steady state master equation. Now using the calculus of residues, we see that

$$P_s(x) = \frac{e^{-\alpha_0}\alpha_0{}^x}{x!}, \tag{12.2.11}$$

with

$$\alpha_0 = k_2A/k_4\,. \tag{12.2.12}$$

iii) $\boldsymbol{\delta < 0}$: When $\delta < 0$ we meet some very interesting features. The steady state solution (12.1.26) now no longer satisfies the condition $\delta > 0$. However, if the range of α is chosen to be a contour C in the complex plane as in Fig. 12.1, and we employ the complex Poisson representation, then $P_s(x)$ constructed as

$$P_s(x) = \int\limits_{C} d\alpha f_s(\alpha)\frac{e^{-\alpha}\alpha^x}{x!}, \tag{12.2.13}$$

is a solution of the Master equation. The deterministic steady state now occurs at a point on the real axis to the right of the singularity at $\alpha = k_2A/k_4$, and asymptotic evaluations of means, moments, etc., may be obtained by choosing C to pass through the saddle point that occurs there. In doing so, one finds that the variance of α, defined as

$$\mathrm{var}[\alpha] = \langle\alpha^2\rangle - \langle\alpha\rangle^2, \tag{12.2.14}$$

is negative, so that

$$\mathrm{var}[x] = \langle x^2\rangle - \langle x\rangle^2 = \langle\alpha^2\rangle - \langle\alpha^2\rangle + \langle\alpha\rangle, \tag{12.2.15}$$

$$< \langle x\rangle. \tag{12.2.16}$$

This means that the steady state is narrower than the Poissonian. Finally, it should be noted that all three cases can be obtained from the contour C. In the case where $\delta = 0$, the cut from the singularity at $\alpha = k_2A/k_4$ to $-\infty$ vanishes and C may be distorted to a simple contour round the pole, while if $\delta > 0$, the singularity at $\alpha = k_2A/k_4$ is now integrable so the contour may be collapsed onto the cut and the integral evaluated as a discontinuity integral over the range $[0, k_2A/k_4]$. (When δ is a positive integer, this argument requires modification).

b) Example: We treat the pair of reactions

$$(3) \qquad B \xrightarrow{k_1} X\,, \tag{12.2.17}$$

$$(4) \qquad 2X \xrightarrow{k_2} A\,, \tag{12.2.18}$$

for which the Fokker-Planck equation is

$$\frac{\partial f(\alpha,t)}{\partial t} = -\frac{\partial}{\partial\alpha}[(\kappa_1 V - 2\kappa_2 V^{-1}\alpha^2)f(\alpha,t)] - \frac{\partial^2}{\partial\alpha^2}[\kappa_2 V^{-1}\alpha^2 f(\alpha,t)]\,, \tag{12.2.19}$$

where and V is the system volume, $\kappa_1 V = k_1 B$ and $\kappa_2 V^{-1} = k_2$. Note that the diffusion coefficient in the above Fokker-Planck equation is negative on all the real lines.

The potential solution of (12.2.19) is (up to a normalisation factor)

$$f(\alpha) = \alpha^{-2}\exp(2\alpha + aV^2/\alpha)\,, \tag{12.2.20}$$

with $a = 2\kappa_2/\kappa_1$ and the α integration is to be performed along a closed contour encircling the origin. Of course, in principle, there is another solution obtained by solving the stationary Fokker-Planck equation in full. However, only the potential solution is single valued and allows us to choose an acceptable contour on which partial integration is permitted.

Thus, by putting $\alpha = \eta V$, we get

$$\langle x^r\rangle_f = \frac{V^r \oint d\eta\, e^{V(2\eta + a/\eta)}\eta^{r-2}}{\oint d\eta\, e^{V(2\eta+a/\eta)}\eta^{-2}}\,. \tag{12.2.21}$$

On the real line the function $(2\eta + a/\eta)$ does not have a maximum at the deterministic steady state—in fact, it has a minimum at the deterministic steady state $\eta = +(a/2)^{1/2}$. However, in the complex η plane this point is a saddle point and provides the dominant contribution to the integral.

Thus, the negative diffusion coefficient in (12.2.19) reflects itself by giving rise to a saddle point at the deterministic steady state, which results in the variance in X being less than $\langle x\rangle$.

From (12.2.21) all the steady state moments can be calculated exactly. The results are

$$\langle x^r\rangle_f = \left(V\sqrt{a/2}\right)^r \frac{I_{r-1}\left(2(2a)^{1/2}V\right)}{I_1\left(2(2a)^{1/2}V\right)}\,, \tag{12.2.22}$$

where $I_r\left(2(2a)^{1/2}V\right)$ are the modified Bessel functions. Using the large-argument expansion for $I_r\left(2(2a)^{1/2}V\right)$, we get

$$\langle x\rangle = V(a/2)^{1/2} + \frac{1}{8} + O(1/V)\,, \tag{12.2.23}$$

$$\mathrm{var}[x] = \frac{3}{4}V(a/2)^{1/2} - \frac{1}{16} + O(1/V)\,.$$

These asymptotic results can also be obtained by directly applying the method of steepest descents to (12.2.21). In general, this kind of expansion will always be possible after explicitly exhibiting the volume dependence of the parameters.

c) Summary of Advantages: The complex Poisson representation yields stationary solutions in analytic form to which asymptotic or exact methods are easily applicable. It is not so useful in the case of time-dependent solutions. The greatest advantages, however, occur in quantum mechanical systems where similar techniques can be used for complex P representations which can give information that is not otherwise extractable. These are treated in [12.4] and [12.5].

12.2.3 The Positive Poisson Representation

Here we choose α to be a complex variable $\alpha_x + i\alpha_y$,

$$d\mu(\alpha) = d^2\alpha = d\alpha_x d\alpha_y\,, \tag{12.2.24}$$

and $\mathscr{D}$ is the whole complex plane. We show below that *for any $P(x)$, a positive $f(\alpha)$ exists such that*

$$P(x) = \int d^2\alpha \frac{e^{-\alpha}\alpha^x}{x!} f(\alpha)\,; \tag{12.2.25}$$

thus, the positive P representation always exists. It is not unique, however. For example, choose

$$f_P(\alpha) = \frac{1}{2\pi\sigma^2}\exp\left(-\frac{|\alpha-\alpha_0|^2}{2\sigma^2}\right) \tag{12.2.26}$$

and note that if $g(\alpha)$ is any analytic function of α, we can write

$$g(\alpha) = g(\alpha_0) + \sum_{n=1}^{\infty} g^{(n)}(\alpha_0)(\alpha-\alpha_0)^n/n! \tag{12.2.27}$$

so that

$$\int d^2\alpha \frac{1}{2\pi\sigma^2}\exp\left(-\frac{|\alpha-\alpha_0|^2}{2\sigma^2}\right) g(\alpha) = g(\alpha_0)\,, \tag{12.2.28}$$

since the terms with $n \geqslant 1$ vanish when integrated in (12.2.28). Noting that the Poisson form $e^{-\alpha}\alpha^x/x!$ is itself analytic in α, we obtain for any positive value of σ^2

$$P(x) = \int d^2\alpha f_P(\alpha)\frac{e^{-\alpha}\alpha^x}{x!} = \frac{e^{-\alpha_0}\alpha_0^x}{x!}\,. \tag{12.2.29}$$

In practice, this nonuniqueness is an advantage rather than a problem.

a) Existence of the Positive Poisson Representation: A proof of the existence of the Positive Poisson representation is not straightforward, and surprisingly, proceeds most naturally from its quantum mechanical analogue, the Positive P-representation [12.4], which is described in detail in Chap. 6 of the book *Quantum Noise* [12.5] The proof we give here requires the techniques developed in that book, and uses the notation developed there.

Given a probability distribution $p(x)$ over the integers, we can always define a corresponding positive density matrix by

$$\rho = \sum_n |n\rangle\langle n|\, p(n)\,, \tag{12.2.30}$$

and a P-representation for ρ gives the corresponding Poisson representation for $p(x)$. Thus

$$p(x) = \langle x|\rho|x\rangle = \left\langle x \left| \int d\mu(\alpha,\beta) P(\alpha,\beta) \frac{|\alpha\rangle\langle\beta^*|}{\langle\beta^*\alpha\rangle} \right| x \right\rangle, \tag{12.2.31}$$

$$= \int d\mu(\alpha,\beta) \frac{\mathrm{e}^{-\alpha\beta}(\alpha\beta)^x}{x!} P(\alpha,\beta)\,. \tag{12.2.32}$$

Hence on can write

$$p(x) = \int d\mu(\alpha_1) f(\alpha_1) \frac{\mathrm{e}^{-\alpha_1}\alpha_1{}^x}{x!}\,, \tag{12.2.33}$$

with

$$d\mu(\alpha_1) f(\alpha_1) = \int d\mu(\alpha,\beta)\delta_\mu(\alpha\beta - \alpha_1) P(\alpha,\beta)\,, \tag{12.2.34}$$

and $\delta_\mu(\alpha_1 - \alpha_2)$ is a Dirac delta function with respect to the measure $\mu(\alpha_1)$, that is

$$\int d\mu(\alpha_1)\,\delta_\mu(\alpha_1 - \alpha_2)\,\phi(\alpha_1) = \phi(\alpha_2)\,. \tag{12.2.35}$$

Thus we can assert from Theorem 3 of Sect. 6.4.2 of *Quantum Noise* that a positive Poisson representation always exists.

a) Fokker-Planck Equations: We make use of the analyticity of the Poisson and its generating function to produce Fokker-Planck equations with positive diffusion matrices. A Fokker-Planck equation of the form of (12.1.9) arises from a generating function equation

$$\partial_t G(\boldsymbol{s},t) = \int d^2\boldsymbol{\alpha} f(\boldsymbol{\alpha},t)\left(\sum_a A_a \frac{\partial}{\partial\alpha_a} + \sum_{a,b}\frac{1}{2}B_{ab}\frac{\partial^2}{\partial\alpha_a\partial\alpha_b}\right)\exp\left[\sum_a (s_a - 1)\alpha_a\right]. \tag{12.2.36}$$

We now take explicit account of the fact that $\boldsymbol{\alpha}$ is a complex variable

$$\boldsymbol{\alpha} = \boldsymbol{\alpha}_x + \mathrm{i}\boldsymbol{\alpha}_y\,, \tag{12.2.37}$$

and also write

$$\boldsymbol{A}(\boldsymbol{\alpha}) = \boldsymbol{A}_x(\boldsymbol{\alpha}) + \mathrm{i}\boldsymbol{A}_y(\boldsymbol{\alpha})\,. \tag{12.2.38}$$

We further write

$$\mathsf{B}(\boldsymbol{\alpha}) = \mathsf{C}(\boldsymbol{\alpha})\mathsf{C}^{\mathrm{T}}(\boldsymbol{\alpha})\,, \tag{12.2.39}$$

and

$$\mathsf{C}(\boldsymbol{\alpha}) = \mathsf{C}_x(\boldsymbol{\alpha}) + \mathrm{i}\mathsf{C}_y(\boldsymbol{\alpha})\,, \tag{12.2.40}$$

For brevity we use

$$\partial_a = \frac{\partial}{\partial \alpha_a}, \qquad \partial_a^x = \frac{\partial}{\partial \alpha_{x,a}}, \qquad \partial_a^y = \frac{\partial}{\partial \alpha_{y,a}}. \tag{12.2.41}$$

Because of the analyticity of $\exp[\sum_a (s_a - 1)\alpha_a]$ in the generating function equation (12.2.36) we can always make the interchangeable choice

$$\partial_a \longleftrightarrow \partial_a^x \longleftrightarrow -\mathrm{i}\partial_a^y. \tag{12.2.42}$$

We then substitute the form (12.2.38) for B_{ab}, and replace ∂_a by either ∂_a^x or $-\mathrm{i}\partial_a^y$ according to whether the corresponding index on A or C is x or y respectively. We then derive

$$\begin{aligned}\partial_t G(s,t) = \int d^2\alpha f(\alpha,t)\Bigg\{\Bigg[&\sum_a (A_{a;x}\partial_a^x + A_{a;y}\partial_a^y)\\ &+ \tfrac{1}{2}\sum_{a,b,c}(C_{a,c;x}C_{c,b;x}\partial_a^x\partial_b^x + C_{a,c;y}C_{c,b;y}\partial_a^y\partial_b^y)\\ &+ 2C_{a,c;x}C_{c,b;y}\partial_a^x\partial_b^y\Bigg]\exp\Big[\sum_a (s_a-1)\alpha_a\Big]\Bigg\}.\end{aligned} \tag{12.2.43}$$

Integrating by parts and discarding the surface terms to get a Fokker-Planck equation in the variables $(\boldsymbol{\alpha}_x, \boldsymbol{\alpha}_y)$,

$$\begin{aligned}\partial_t f(\alpha,t) = \Bigg[&-\sum_a (\partial_a^x A_{a;x} + \partial_a^y A_{a;y}) + \tfrac{1}{2}\sum_{a,b,c}(\partial_a^x\partial_b^x C_{a,c;x}C_{c,b;x}\\ &+ \partial_a^y\partial_b^y C_{a,c;y}C_{c,b;y} + 2\partial_a^x\partial_b^y C_{a,c;x}C_{c,b;y})\Bigg] f(\alpha,t).\end{aligned} \tag{12.2.44}$$

In the space of doubled dimensions, this is a Fokker-Planck equation with positive semidefinite diffusion. For, we have for the variable $(\boldsymbol{\alpha}_x, \boldsymbol{\alpha}_y)$ the drift vector

$$\mathscr{A}(\alpha) = [\boldsymbol{A}_x(\alpha), \boldsymbol{A}_y(\alpha)], \tag{12.2.45}$$

and the diffusion matrix

$$\mathscr{B}(\alpha) = \begin{bmatrix} \mathsf{C}_x\mathsf{C}_x^{\mathrm{T}} & \mathsf{C}_x\mathsf{C}_y^{\mathrm{T}} \\ \mathsf{C}_y\mathsf{C}_x^{\mathrm{T}} & \mathsf{C}_y\mathsf{C}_y^{\mathrm{T}} \end{bmatrix} = \mathsf{C}(\alpha)\mathsf{C}(\alpha)^{\mathrm{T}}, \tag{12.2.46}$$

where

$$\mathscr{C}(\alpha) = \begin{bmatrix} \mathsf{C}_x & 0 \\ \mathsf{C}_y & 0 \end{bmatrix}, \tag{12.2.47}$$

so that $\mathscr{B}(\alpha)$ is explicitly positive semidefinite.

c) Stochastic Differential Equations: Corresponding to the drift and diffusion (12.2.45, 12.2.46) we have a stochastic differential equation

$$\begin{bmatrix} d\alpha_x \\ d\alpha_y \end{bmatrix} = \begin{bmatrix} \boldsymbol{A}_x(\alpha) \\ \boldsymbol{A}_y(\alpha) \end{bmatrix} dt + \begin{bmatrix} \mathsf{C}_x\, d\boldsymbol{W}(t) \\ \mathsf{C}_y\, d\boldsymbol{W}(t) \end{bmatrix}, \tag{12.2.48}$$

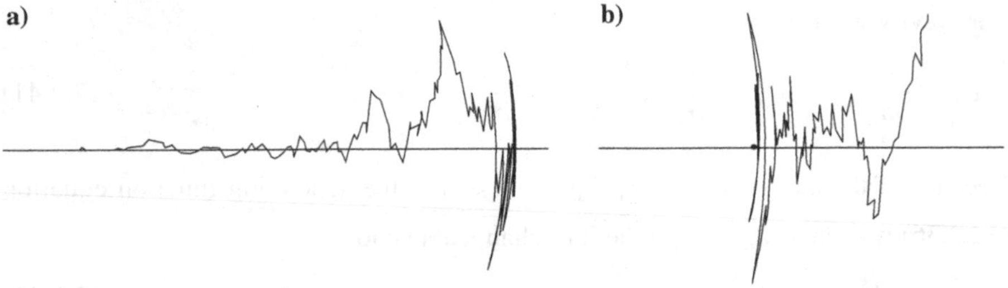

Fig. 12.2. **a)** Path followed by a point obeying the stochastic differential equation (12.2.51); **b)** Simulation of the path of a point obeying the stochastic differential equation (12.2.53).

where $\boldsymbol{W}(t)$ is a Wiener process of the same dimension as $\boldsymbol{\alpha}_x$. Note that the same Wiener process occurs in both lines because of the two zero entries $\mathscr{C}(\boldsymbol{\alpha})$ as written in (12.2.47).

Recombining real and imaginary parts, we find the stochastic differential equation for the complex variable $\boldsymbol{\alpha}$:

$$d\boldsymbol{\alpha} = \boldsymbol{A}(\boldsymbol{\alpha})dt + \mathsf{C}(\boldsymbol{\alpha})\,d\boldsymbol{W}(t)\,. \tag{12.2.49}$$

This is of course, exactly the same stochastic differential equation which would arise if we used the usual rules for converting Fokker-Planck equations to stochastic differential equations directly on the Poisson representation Fokker-Planck equation (12.1.9), and ignored the fact that $\mathsf{C}(\boldsymbol{\alpha})$ so defined would have complex elements if B was not a positive semidefinite diffusion matrix.

d) Examples of Stochastic Differential Equations in the Complex Plane: We again consider the reactions (Sect. 12.1b)

$$A + X \underset{k_4}{\overset{k_2}{\rightleftharpoons}} 2X\,, \qquad B + X \underset{k_3}{\overset{k_1}{\rightleftharpoons}} C\,. \tag{12.2.50}$$

The use of the positive Poisson representation applied to this system yields the stochastic differential equation, arising from the Fokker-Planck equation (12.1.21):

$$d\alpha = \left[k_3 C + (k_2 A - k_1 B)\alpha - k_4\alpha^2\right] dt + \left[2(k_2 A\alpha - k_4\alpha^2)\right]^{1/2} dW(t)\,. \tag{12.2.51}$$

Again, the sign of the quantity δ, defined in (12.1.27) plays a determining role. In the case $\delta > 0$, we note that the noise term vanishes at $\alpha = 0$ and at $\alpha = k_2A/k_4$, is positive between these points and the drift term is such as to return α to the range $[0, k_2A/k_4]$ whenever it approaches the end points. Thus, for $\delta > 0$, (12.2.51) represents a real stochastic differential equation on the real interval $[0, k_2A/k_4]$.

In the case $\delta < 0$, the stationary point lies outside the interval $[0, k_2A/k_4]$, and a point initially in this interval will move along this interval governed by (12.2.51) until it meets the right-hand end, where the noise vanishes and the drift continues to drive it towards the right. One leaving the interval, the noise becomes imaginary and the point will follow a path like that shown in Fig. 12.2a until it eventually reaches the interval $[0, k_2A/k_4]$ again.

The case of $\delta = 0$ is not very dissimilar, except that once the point reaches the right-hand end of the interval $[0, k_2 A/k_4]$, both drift and diffusion vanish so it remains there from then on.

In the case of the system

$$B \to X, \qquad 2X \to A, \tag{12.2.52}$$

we can compute that the stochastic differential equation coming from the Fokker-Planck equation (12.2.19) is

$$d\eta = \left[\kappa_1 - 2\kappa_2\eta^2\right] dt + \mathrm{i}\varepsilon(2\kappa_2)^{1/2}\eta\, dW(t), \tag{12.2.53}$$

where $\alpha = \eta V$ and $\varepsilon = V^{-1/2}$.

The stochastic differential equation (12.2.53) can be computer simulated and a plot of motion in the complex η plane generated. Fig. 12.2b illustrates the behaviour. The point is seen to remain in the vicinity of $\mathrm{Re}(\alpha) = (a/2)^{1/2}$ but to fluctuate mainly in the imaginary direction on either side, thus giving rise to a negative variance in α.

12.3 Time Correlation Functions

The time correlation function of a Poisson variable α is not the same as that for the variable x. This can be seen, for example, in the case of a reaction $X \leftrightharpoons Y$ which gives a Poisson Representation Fokker-Planck equation with no diffusion term. Hence, the Poisson variable does not fluctuate, even though the master equation variable does fluctuate. We now show what the relationship is. For clarity, the demonstration is carried out for one variable only.

We define

$$\langle\alpha(t)\alpha(s)\rangle = \int d\mu(\alpha)d\mu(\alpha')\alpha\alpha' f(\alpha, t\,|\,\alpha', s)f(\alpha', s). \tag{12.3.1}$$

We note that

$$f(\alpha, s\,|\,\alpha', s) = \delta_\mu(\alpha - \alpha'), \tag{12.3.2}$$

which means that

$$\int d\mu(\alpha)\frac{\mathrm{e}^{-a}\alpha^x}{x!} f(\alpha, s\,|\,\alpha', s) = \frac{\mathrm{e}^{-\alpha'}\alpha'^x}{x!}, \tag{12.3.3}$$

so that

$$\int d\mu(\alpha)\,\alpha f(\alpha, t\,|\,\alpha', s) = \sum_{x,x'} xP(x, t\,|\,x', s)\frac{\mathrm{e}^{-\alpha'}\alpha'^{x'}}{x'!}. \tag{12.3.4}$$

Hence,

$$\langle\alpha(t)\alpha(s)\rangle = \sum_{x,x'} xP(x,t\,|\,x',s)\int d\mu(\alpha')\frac{\alpha'^{x'+1}\mathrm{e}^{-\alpha'}}{x'!}f(\alpha',s),$$

$$= \sum_{x,x'} xP(x,t\,|\,x',s)\int d\mu(\alpha')\left[\left(-\alpha'\frac{\partial}{\partial\alpha'}+x'\right)\frac{\alpha'^{x'}\mathrm{e}^{-\alpha'}}{x!}\right]f(\alpha',s), \tag{12.3.5}$$

$$= \sum_{x,x'} xx'P(x,t\,|\,x',s)P(x',s)$$

$$-\int d\mu(\alpha')f(\alpha',s)\alpha'\frac{\partial}{\partial\alpha'}\sum_{x,x'} xP(x,t\,|\,x',s)\frac{\alpha'^{x'}\mathrm{e}^{-\alpha'}}{x!}. \tag{12.3.6}$$

We define

$$\langle\alpha(t)\,|\,[\alpha',s]\rangle \equiv \int d\alpha\,\alpha f(\alpha,t\,|\,\alpha',s), \tag{12.3.7}$$

as the mean of $\alpha(t)$ given the initial condition α' at s. Then the second term can be written

$$-\int d\mu(\alpha')\alpha'\frac{\partial}{\partial\alpha'}\langle\alpha(t)\,|\,[\alpha',s]\rangle f(\alpha',s) \equiv \left\langle\alpha'\frac{\partial}{\partial\alpha'}\langle\alpha(t)\,|\,[\alpha,s]\rangle\right\rangle, \tag{12.3.8}$$

so we have

$$\langle x(t)x(s)\rangle = \langle\alpha(t)\alpha(s)\rangle + \left\langle\alpha'\frac{\partial}{\partial\alpha'}\langle\alpha(t)\,|\,[\alpha,s]\rangle\right\rangle. \tag{12.3.9}$$

Taking into account a many-variable situation and noting that

$$\langle \boldsymbol{x}(t)\rangle = \langle\boldsymbol{\alpha}(t)\rangle \qquad \text{always}, \tag{12.3.10}$$

we have

$$\langle x_a(t), x_b(s)\rangle = \langle\alpha_a(t),\alpha_b(s) + \left\langle\alpha'_b\frac{\partial}{\partial\alpha'_b}\langle\alpha_a(t)\,|\,[\alpha',s]\rangle\right\rangle. \tag{12.3.11}$$

This formula explicitly shows the fact that the Poisson representation gives a process which is closely related to the Birth-Death Master equation, but not isomorphic to it. The stochastic quantities of interest, such as time correlation functions, can all be calculated but are not given directly by those of the Poisson variable.

12.3.1 Interpretation in Terms of Statistical Mechanics

We assume for the moment that the reader is acquainted with the statistical mechanics of chemical systems. If we consider a system composed of chemically reacting components $A, B, C, \ldots$, the distribution function in the grand canonical ensemble is given by

$$P(I) = \exp\left\{\beta\left[\Omega + \sum_i \mu_i x_i(I) - E(I)\right]\right\}, \tag{12.3.12}$$

where I is an index describing the microscopic state of the system, $x_i(I)$ is the number of molecules of X_i in the state I, $E(I)$ is the energy of the state, μ_i, is the chemical potential of component X_i, Ω is a normalisation factor, and

$$\beta = 1/kT\,. \tag{12.3.13}$$

The fact that the components can react requires certain relationships between the chemical potentials to be satisfied, since a state I can be transformed into a state J only if

$$\sum_i \nu_i^A x_i(I) = \sum_i \nu_i^A x_i(J), \qquad A = 1,2,3,\ldots \tag{12.3.14}$$

where ν_a^A are certain integers. The relations (12.3.14) are the stoichiometric constraints.

The canonical ensemble for a reacting system is defined by requiring

$$\sum_i \nu_i^A x_i(I) = \tau^A\,, \tag{12.3.15}$$

for some τ^A, whereas the grand canonical ensemble is defined by requiring

$$\sum_i P(I) \sum_i \nu_i^A x_i(I) \equiv \sum_i \nu_i^A \langle x_i \rangle = \tau^A\,, \tag{12.3.16}$$

Maximisation of entropy subject to the constraint (12.3.16) (and the usual constraints of fixed total probability and mean energy) gives the grand canonical form (12.3.12) in which the chemical potentials also satisfy the relation

$$\mu_i = \sum_A k_A \nu_i^A\,. \tag{12.3.17}$$

When one takes the ideal solution or ideal gas limit, in which interaction energies (but not kinetic or internal energies) are neglected, there is no difference between the distribution function for an ideal reacting system and an ideal nonreacting system, apart from the requirement that the chemical potentials be expressible in the form of (12.3.17).

The canonical ensemble is not so simple, since the constraints must appear explicitly as a factor of the form

$$\prod_A \delta\Big(\sum_i \nu_i^A x_i(I), \tau^A\Big)\,, \tag{12.3.18}$$

and the distribution function is qualitatively different for every kind of reacting system (including a nonreacting system as a special case).

The distribution in total numbers $\boldsymbol{x}$ of molecules of reacting components in the grand canonical ensemble of an ideal reacting system is easily evaluated, namely,

$$P(\boldsymbol{x}) = \exp\Big[\beta\Big(\Omega + \sum_i \mu_i x_i\Big)\Big] \sum_i \prod_i \delta\big(x_i(I), x_i\big) \exp\big(-\beta E(I)\big)\,. \tag{12.3.19}$$

The sum over states is the same as that for the canonical ensemble of an ideal nonreacting mixture so that

$$P(\boldsymbol{x}) = \exp\Big[\beta\Big(\Omega + \sum_i \mu_i x_i\Big)\Big] \prod_i \frac{1}{x_i!}\Big\{\sum_k \exp[-\beta E_k(i)]\Big\}^{x_i}\,, \tag{12.3.20}$$

where $E_k(i)$ are the energy eigenstates of a single molecule of the substance A. This result is a multivariate Poisson with mean numbers given by

$$\log\langle x_i\rangle = \beta\mu_i - \log\Big[\sum_k e^{-\beta E_k(i)}\Big], \tag{12.3.21}$$

which, as is well known, when combined with the requirement (12.3.16) gives the law of mass action.

The canonical ensemble is obtained by maximizing entropy subject to the stronger constraint (12.3.15), which implies the weak constraint (12.3.16). Thus, the distribution function in total numbers for the canonical ensemble will simply be given by

$$P(\boldsymbol{x}) \propto \left\{\prod_i \frac{1}{x_i!}\Big(\sum_k e^{-\beta E_k(i)}\Big)^{x_i}\right\} \sum_A \delta\Big(\sum_i \nu_i^A x_i, \tau^A\Big). \tag{12.3.22}$$

In terms of the Poisson representation, we have just shown that in equilibrium situations, the quasiprobability (in a grand canonical ensemble) is

$$f(\boldsymbol{\alpha})_{eq} = \delta[\boldsymbol{\alpha} - \boldsymbol{\alpha}(eq)], \tag{12.3.23}$$

since the $\boldsymbol{x}$ space distribution is Poissonian. For the time correlation functions there are two results of this.

i) The variables $\boldsymbol{\alpha}(t)$ and $\boldsymbol{\alpha}(s)$ are nonfluctuating quantities with values $\boldsymbol{\alpha}(eq)$. Thus,

$$\langle \alpha_r(t), \alpha_s(s)\rangle_{eq} = 0\,. \tag{12.3.24}$$

ii) The equilibrium mean in the second term is trivial. Thus,

$$\langle x_a(t), x_b(s)\rangle = \Big[\alpha'_b \frac{\partial}{\partial \alpha'_b}\langle \alpha_a(t)\,|\,[\boldsymbol{\alpha}', s]\rangle\Big]_{\boldsymbol{\alpha}'=\boldsymbol{\alpha}(eq)}. \tag{12.3.25}$$

This result is, in fact, exactly that of *Bernard* and *Callen* [12.6] which relates a two-time correlation function to a derivative of the mean of a quantity with respect to a thermodynamically conjugate variable.

Let us consider a system in which the numbers of molecules of chemical species $X_1, X_2, \dots$ corresponding to a configuration I of the system are $x_1(I), x_2(I), \dots$ and it is understood that these chemical species may react with each other. Then in a grand canonical ensemble, as demonstrated above, the equilibrium distribution function is

$$Z^{-1}(\boldsymbol{\mu}) \exp\Big[\beta\Big(\sum_i \mu_i x_i(I) - E(I)\Big)\Big], \tag{12.3.26}$$

with

$$Z(\boldsymbol{\mu}) = \exp(-\Omega\beta), \tag{12.3.27}$$

where $Z(\boldsymbol{\mu})$ is the grand canonical partition function. As pointed out above, the chemical potentials μ_i for a reacting system cannot be chosen arbitrarily but must be related by the stoichiometric constraints (12.3.16) of the allowable reactions.

Now we further define the quantities

$$\langle x_i, t\,|\,[I, s]\rangle, \tag{12.3.28}$$

to be the mean values of the quantities x_i at time t under the condition that the system was in a configuration I at time s. Then a quantity of interest is the mean value of (12.3.28) over the distribution (12.3.26) of initial conditions, namely,

$$\langle x_i, t \,|\, [\boldsymbol{\mu}, s]\rangle = \sum_j \langle x_i, t \,|\, [J, s]\rangle Z^{-1}(\boldsymbol{\mu}) \exp\Big\{\beta\Big[\sum_j \mu_j x_j(J) - E(J)\Big]\Big\}. \tag{12.3.29}$$

When the chemical potentials satisfy the equilibrium constraints, this quantity will be time independent and equal to the mean of x_i in equilibrium, but otherwise it will have a time dependence. Then, with a little manipulation one finds that

$$\left[kT\frac{\partial}{\partial \mu_j}\langle x_i, t \,|\, [\boldsymbol{\mu}, s]\rangle\right]_{\mu=\mu(eq)} = \langle x_i(t), x_j(s)\rangle_{\mathrm{eq}}\,. \tag{12.3.30}$$

The left-hand side is a reponse function of the mean value to the change in the chemical potentials around equilibrium and is thus a measure of dissipation, while the right-hand side, the two-time correlation function in equilibrium, is a measure of fluctuations.

To make contact with the Poisson representation result (12.3.25) we note that the chemical potentials μ_j in ideal solution theory are given by

$$\mu_i(\langle \boldsymbol{x}\rangle) = kT \log\langle x_i\rangle + \text{const}\,. \tag{12.3.31}$$

Using (12.3.31), we find that (12.3.30) becomes

$$\langle x_i(t), x_j(s)\rangle = \left[\langle x_j\rangle \frac{\partial}{\partial \langle x_j\rangle}\langle x_i, t \,|\, [\boldsymbol{\mu}(\langle \boldsymbol{x}\rangle), s]\rangle\right]_{\langle \boldsymbol{x}\rangle=\langle \boldsymbol{x}\rangle_{eq}}\,. \tag{12.3.32}$$

Since the ideal solution theory gives rise to a distribution in x_i that is Poissonian, it follows that in that limit

$$\langle x_i, t \,|\, [\boldsymbol{\mu}(\langle \boldsymbol{x}\rangle), s]\rangle = \langle \alpha_i, t \,|\, [\boldsymbol{\alpha}', s]\rangle\,, \tag{12.3.33}$$

with $\boldsymbol{\alpha}' = \langle \boldsymbol{x}\rangle$. Thus, (12.3.32) becomes

$$\langle x_i(t), x_j(s)\rangle = \left[\alpha'_j \frac{\partial}{\partial \alpha_j}\langle \alpha_i, t \,|\, [\boldsymbol{\alpha}', s]\rangle\right]_{\alpha'=\alpha(eq)}\,. \tag{12.3.34}$$

Thus, (12.3.25) is the ideal solution limit of the general result (12.3.32).

The general formula (12.3.11) can be considered as a generalisation of the Bernard-Callen result to systems that are not in thermodynamic equilibrium.

However, it is considerably different from the equilibrium result and the two terms are directly interpretable. The second term is the equilibrium contribution, a response function, but since the system is not in a well-defined equilibrium state, we take the average of the equilibrium result over the various contributing α space states. The first term is the contribution from the α-space fluctuations themselves and is not directly related to a response function. It represents the fluctuations in excess of equilibrium.

12.3.2 Linearised Results

The general differential equation, arising from the use of the positive Poisson representation, and corresponding to the Fokker-Planck equation (12.1.12), is

$$d\boldsymbol{\eta} = \hat{\boldsymbol{A}}(\boldsymbol{\eta})\,dt + \varepsilon\hat{\mathsf{C}}(\eta)\,d\boldsymbol{W}(t)\,, \tag{12.3.35}$$

where

$$\hat{\mathsf{C}}\hat{\mathsf{C}}^{\mathrm{T}} = \hat{\mathsf{B}}\,. \tag{12.3.36}$$

We may now make a first-order small noise expansion about the stationary state $\boldsymbol{\eta}$ by following the procedure of Sect. 7.3. Thus, writing

$$\boldsymbol{\eta}(t) = \bar{\boldsymbol{\eta}} + \varepsilon\boldsymbol{\eta}_1(t)\,, \quad (\varepsilon = V^{1/2})\,, \tag{12.3.37}$$

to lowest order we have

$$\begin{aligned} \hat{\boldsymbol{A}}(\bar{\boldsymbol{\eta}}) &= 0\,, \\ d\boldsymbol{\eta}_1 &= -\boldsymbol{F}\boldsymbol{\eta}_1\,dt + \mathsf{G}\,d\boldsymbol{W}(t)\,, \end{aligned} \tag{12.3.38}$$

where

$$F_{rs} = -\frac{\partial}{\partial\eta_s}\hat{A}_r(\bar{\boldsymbol{\eta}})\,, \tag{12.3.39}$$

$$\mathsf{G} = \hat{\mathsf{C}}(\bar{\eta})\,. \tag{12.3.40}$$

Then,

$$\langle\alpha_r(t), \alpha_s(0)\rangle_s = V\sum_{r'}[\exp(-\boldsymbol{F}t)]_{rr'}\langle\eta_{r',1}, \eta_{s,1}\rangle_s\,, \tag{12.3.41}$$

and

$$\frac{\partial}{\partial\alpha'_s}\langle\alpha_r(t)\,|\,[\boldsymbol{\alpha}',0]\rangle = \frac{\partial}{\partial\eta'_{s,l}}\langle\eta_{r,1}(t)\,|\,[\boldsymbol{\eta}'_1,0]\rangle = [\exp(-\boldsymbol{F}t)]_{r,s}\,. \tag{12.3.42}$$

Hence,

$$\begin{aligned} \langle x_r(t), x_s(0)\rangle_s &= V\sum_{r'}[\exp(-\boldsymbol{F}t)]_{rr'}\left(\langle\eta_{r',1}, \eta_{s,1}\rangle_s + \delta_{r',s}\bar{\eta}_S\right) &(12.3.43)\\ &= \sum_{r'}\exp(-\boldsymbol{F}t)_{r,r'}\langle x_r, x_s\rangle_s\,. &(12.3.44) \end{aligned}$$

Thus the linearised result is in agreement with the regression theorem of Sect. 3.7.4 correlation functions for a variety of systems have been computed in [12.1].

12.4 Trimolecular Reaction

In Sect. 11.1.3 we considered a reaction which included a part

$$A + 2X \rightleftharpoons 3X\,, \tag{12.4.1}$$

and set up an appropriate birth-death Master equation for this. However, it is well known in chemistry that such trimolecular steps are of vanishingly small probability and proceed in stages via a short-lived intermediate. Thus, the reaction (12.4.1) presumably occurs as a two-state system

$$\text{i)}\quad A + Y \underset{1}{\overset{1}{\rightleftharpoons}} X + Y\,, \tag{12.4.2a}$$

$$\text{ii)}\quad Y \underset{1}{\overset{\gamma}{\rightleftharpoons}} 2X\,, \tag{12.4.2b}$$

both of which are merely bimolecular, and we have set rate constants equal to one, except for γ (the decay constant of Y) which is assumed as being very large. Thus, Y is indeed a short-lived intermediate. The deterministic rate equations are

$$\left.\begin{aligned}\frac{dx}{dt} &= ay - xy + 2(\gamma y - x^2)\,,\\ \frac{dy}{dt} &= x^2 - \gamma y\,,\end{aligned}\right\} \tag{12.4.3}$$

and the usual deterministic adiabatic elimination procedure sets $y = x^2/\gamma$ and gives

$$\frac{dx}{dt} = (ax^2 - x^3)/\gamma\,. \tag{12.4.4}$$

Although this procedure is straightforward deterministically, it is not clear that the stochastic Master equation of the kind used in Sect. 11.1.3 is a valid adiabatic elimination limit. The adiabatic elimination techniques used in Chap. 8 are not easily adapted to direct use on a Master equation but can be straightforwardly adapted to the case of the Poisson representation Fokker-Planck equation.

12.4.1 Fokker-Planck Equation for Trimolecular Reaction

For the reaction (12.4.1) with forward and backward rate constants equal to $1/\gamma$ to correspond to (12.4.4), the Poisson representation Fokker-Planck equation becomes, from (12.1.4),

$$\frac{\partial f}{\partial t} = \frac{1}{\gamma}\left\{\left(-\frac{\partial}{\partial\alpha} + 2\frac{\partial^2}{\partial\alpha^2} - \frac{\partial^3}{\partial\alpha^3}\right)[\alpha^2(a-\alpha)]f\right\}, \tag{12.4.5}$$

and contains *third-order derivatives*. There is no truly probabilistic interpretation in terms of any real stochastic process in α space, no matter what kind of Poisson representation is chosen. The concept of *third-order noise* will be explained in the next section, which will show how probabilistic methods and stochastic differential equations can still be used.

a) Adiabatic Elimination: Using the rules developed in (11.4.9), the Fokker-Planck equation for the system (12.4.2) with the correspondence

$$\begin{bmatrix} x \\ y \end{bmatrix} \longleftrightarrow \begin{bmatrix} \alpha \\ \beta \end{bmatrix} \tag{12.4.6}$$

is

$$\frac{\partial f}{\partial \alpha} = -\frac{\partial}{\partial \alpha}[(a-\alpha)\beta + 2(\gamma\beta - \alpha^2)] + \frac{\partial}{\partial \beta}(\gamma\beta - \alpha^2)$$
$$+ \frac{\partial^2}{\partial \alpha^2}(\gamma\beta - \alpha^2) + \frac{\partial^2}{\partial \alpha \partial \beta}[(a-\alpha)\beta] . \tag{12.4.7}$$

Adiabatic elimination now proceeds as in Sect. 8.3.1. We define new variables

$$x = \alpha , \qquad y = \gamma\beta - \alpha^2 , \tag{12.4.8}$$

and consequently, changing variables with

$$\frac{\partial}{\partial \alpha} = \frac{\partial}{\partial x} - 2x\frac{\partial}{\partial y} , \qquad \frac{\partial}{\partial \beta} = \gamma\frac{\partial}{\partial y} , \tag{12.4.9}$$

the Fokker-Planck equation becomes

$$\frac{\partial f(x,y)}{\partial t} = \left\{ -\left(\frac{\partial}{\partial x} - 2x\frac{\partial}{\partial y}\right)\left[\frac{(a-x)(y+x^2)}{\gamma} + 2y\right] + \gamma\frac{\partial}{\partial y}y \right.$$
$$+ \left(\frac{\partial}{\partial x} - 2x\frac{\partial}{\partial y}\right)\left(\frac{\partial}{\partial x} - 2x\frac{\partial}{\partial y}\right)y$$
$$\left. + \left(\frac{\partial}{\partial x} - 2x\frac{\partial}{\partial y}\right)\left(\frac{\partial}{\partial y}\right)[(y+x^2)(a-x)] \right\} f . \tag{12.4.10}$$

Since y is to be eliminated, there should be a well-defined limit of the L_1 operator which governs its motion at fixed x. However, this operator is

$$\gamma\frac{\partial}{\partial y}y + \frac{\partial^2}{\partial y^2}[4x^2 y - 2x(y+x^2)(a-x)] , \tag{12.4.11}$$

and the large γ limit turns this into deterministic motion. Setting

$$y = v\gamma^{-1/2} , \tag{12.4.12}$$

transforms (12.4.11) to

$$\gamma L_1(\gamma) = \gamma\left\{\frac{\partial}{\partial v}v + \frac{\partial^2}{\partial v^2}[2x^3(x-a) + (4x^2 - 2x)v\gamma^{-1/2}]\right\} ,$$
$$\underset{\gamma\to\infty}{\longrightarrow} \gamma\left\{\frac{\partial}{\partial v}v + \frac{\partial^2}{\partial v^2}[2x^3(x-a)]\right\} ,$$
$$\equiv \gamma L_1 . \tag{12.4.13}$$

With this substitution, we finally identify

$$\gamma^{-1}L_3 = -\gamma^{-1}\frac{\partial}{\partial x}[x^2(a-x)] , \tag{12.4.14}$$

$$L_2(\gamma) = -\frac{\partial}{\partial x}[(a-x)v\gamma^{-3/2} + 2v\gamma^{-1/2}] - \frac{\partial}{\partial x}2x\frac{\partial}{\partial v}v - 2x\frac{\partial}{\partial v}\frac{\partial}{\partial x}v$$
$$+ \gamma^{-1/2}\frac{\partial^2}{\partial x^2}v + \gamma^{1/2}\frac{\partial}{\partial x}\frac{\partial}{\partial v}[(a-x)(x^2 + v\gamma^{-1/2})] , \tag{12.4.15}$$

and

$$\frac{\partial f}{\partial t} = [\gamma^{-1}L_3 + L_2(\gamma) + \gamma L_1(\gamma)]f \,. \tag{12.4.16}$$

The projection operator P will be onto the null space of L_1 and because L_1 depends on x, we have

$$L_3 P \neq P L_3 \,. \tag{12.4.17}$$

This means that the equation of motion for $Pf \equiv g$ is found by similar algebra to that used in Sect. 8.1.2b. We find

$$\begin{aligned} s\tilde{g}(s) = \gamma^{-1}PL_3\tilde{g}(s) \\ &+ P[L_2(\gamma) + \gamma^{-1}L_3][s - \gamma L_1 - (1-P)L_2(\gamma) - \gamma^{-1}(1-P)L_3]^{-1} \\ &\times [L_2(\gamma) + \gamma^{-1}(1-P)L_3]\tilde{g}(s) + g(0) \,. \end{aligned} \tag{12.4.18}$$

Notice, however, since for any function of v

$$P\phi(v) = p_x(v) \int dv\, \phi(v) \,, \tag{12.4.19}$$

where $p_x(v)$ satisfies

$$L_1 p_x(v) = 0 \,, \tag{12.4.20}$$

that in $PL_2(\gamma)$, all terms with $\partial/\partial v$ in them vanish. Thus, to highest order in γ,

$$PL_2(\gamma) = \gamma^{-1/2}\left(-2v\frac{\partial}{\partial x} + v\frac{\partial^2}{\partial x^2}\right) . \tag{12.4.21}$$

The term $[\;]^{-1}$ in (12.4.18) is asymptotic to $-\gamma^{-1}L_1^{-1}$ and the only term in the remaining bracket which can make the whole expression of order γ^{-1}, like the L_3 term, is the term of order $\gamma^{1/2}$ in $L_2(\gamma)$, i.e.,

$$\gamma^{1/2}\frac{\partial}{\partial x}[(a-x)x^2]\frac{\partial}{\partial v} \,. \tag{12.4.22}$$

Thus, the large γ limit of (12.4.18) is

$$s\tilde{g}(s) = \gamma^{-1}\left\{PL_3\tilde{g} - P\left[-2\frac{\partial}{\partial x} + \frac{\partial^2}{\partial x^2}\right]vL_1^{-1}\,\frac{\partial}{\partial x}\left[(a-x)x^2\frac{\partial}{\partial v}\right]p_x(v)\tilde{p}\right\} + g(0) \,, \tag{12.4.23}$$

where we have written

$$g = p_x(v)p \,, \qquad \tilde{g} = p_x(v)\tilde{p} \,. \tag{12.4.24}$$

We are now lead to the central problem of the evaluation of

$$\int dv' v' L_1^{-1}\frac{\partial}{\partial x}(a-x)x^2\frac{\partial}{\partial v'}p_x(v') \,, \tag{12.4.25}$$

which arises in the evaluation of the second part in the braces in (12.4.23). We wish to bring the $\partial/\partial x$ to the left outside the integral, but since $\partial/\partial x$ and L_1 do not commute, this requires care. Now

$$\left[L_1^{-1}, \frac{\partial}{\partial x}\right] = L_1^{-1}\left[\frac{\partial}{\partial x}, L_1\right]L_1^{-1} \tag{12.4.26}$$

and from (12.4.13),

$$= L_1^{-1}\left\{\frac{\partial^2}{\partial v^2}[8x^3 - 6ax^2]\right\}L_1^{-1}, \tag{12.4.27}$$

$$(12.4.25) = \frac{\partial}{\partial x}\int dv' v' L_1^{-1}(a-x)x^2\frac{\partial}{\partial v'}p_x(v') + \int dv' v' L_1^{-1}\frac{\partial^2}{\partial v'^2}L_1^{-1}$$
$$\times[(8x^3 - 6ax^2)(a-x)x^2]\frac{\partial}{\partial v'}p_x(v'). \tag{12.4.28}$$

The second term vanishes, through the demonstration that this is so is rather specialised. For, we know that L_1 describes an Ornstein-Uhlenbeck process in v and that $p_x(v)$ is its stationary solution. The eigenfunction properties used in Sect. 8.2 show that

$$L_1^{-1}\frac{\partial^2}{\partial v^2}L_1^{-1}\frac{\partial}{\partial v}p_x(v), \tag{12.4.29}$$

is proportional to the third eigenfunction, which is orthogonal to v, the first eigenfunction of the corresponding backward equation. The first term is now easily computed using the fact that L_1 involves the Ornstein-Uhlenbeck process. Using the same techniques as in Sect. 8.3.1, we find that all the x dependence arising from $p_x(v')$ vanishes, and hence

$$(12.4.25) = \frac{\partial}{\partial x}(a-x)x^2. \tag{12.4.30}$$

We similarly find

$$PL_3\tilde{g} = -p_x(v)\int dv'\frac{\partial}{\partial x}[x^2(a-x)]p_x(v')\tilde{p},$$
$$= p_x(v)\frac{\partial}{\partial x}[x^2(a-x)]\tilde{p}, \tag{12.4.31}$$

so in the end

$$\frac{\partial p}{\partial t} = \frac{1}{\gamma}\left[\left(-\frac{\partial}{\partial x} + 2\frac{\partial^2}{\partial x^2} - \frac{\partial^3}{\partial x^3}\right)[(a-x)x^2]p\right], \tag{12.4.32}$$

which is exactly the same as the trimolecular model Fokker-Planck equation (12.4.5). This means the trimolecular Master equation is valid in the same limit.

b) Comments:

i) Notice that this system gives an end result which is *not* in a Stratonovich form but in the Ito form, with all derivatives to the left.

ii) The derivation of (12.4.32) means that techniques for understanding such non-probabilistic Fokker-Planck equations are required. We outline a possible way of doing this in the next section.

12.4.2 Third-Order Noise

To handle the third-order Fokker-Planck equations which arise with trimolecular reactions, we introduce the stochastic variable $V(t)$ whose conditional probability density $p(v,t)$ obeys the third-order partial differential equation

$$\frac{\partial p(v,t)}{\partial t} = -\frac{1}{6}\frac{\partial^3 p(v,t)}{\partial v^3}\,. \tag{12.4.33}$$

Since we have already shown in Sect. 3.4 that no Markov process can possible give a third-order term like this, some fundamental requirement must be violated by $p(v,t)$. It turns out that $p(v,t)$ is not always positive, which is permissible in a quasiprobability. We will see that in spite of this, the formal probabilistic analogy is very useful.

We know that the solution of (12.4.33), subject to the boundary condition

$$p(v,t_0) = \delta(v - v_0)\,, \tag{12.4.34}$$

is given by Fourier transform methods as

$$p(v,t\,|\,v_0,t_0) = \frac{1}{2\pi}\int_{-\infty}^{\infty} dq \exp\left\{\mathrm{i}\left[q(v-v_0) + \frac{1}{6}q^3(t-t_0)\right]\right\}\,. \tag{12.4.35}$$

The moments of V can be calculated, after a partial integration, to be

$$\left.\begin{aligned} \langle [V(t)-V_0]^n\rangle &= 0, \qquad n \text{ not a multiple of } 3,\\ \langle [V(t)-V_0]^{3m}\rangle &= \frac{(t-t_0)^m(3m)!}{6^m m!}\,, \end{aligned}\right\} \tag{12.4.36}$$

Further, we assume the process (12.4.33) is some kind of generalized Markov process, for which the joint probability distribution is given by

$$p(v_2,t_2 : v_1,t_1) = p(v_2,t_2\,|\,v_1,t_1)p(v_1,t_1)\,, \tag{12.4.37}$$

and from (12.4.34) we see that the first factor is a function of only $v_2 - v_1$ and $t_2 - t_1$, so that the variable $V(t_2) - V(t_1)$ is statistically independent of $V(t_1)$ and that this process is a *process with independent increments*. Thus, $dV(t)$ will be independent of $V(t)$.

The rigorous definition of stochastic integration with respect to $V(t)$ is a task that we shall not attempt at this stage. However, it is clear that it will not be too dissimilar to Ito integration and, in fact, *Hochberg* [12.7] has rigorously defined higher-order noises of even degree and carried out stochastic integration with respect to them. We can show, however, that a stochastic differential equation of the form

$$dy(t) = a(y)\,dt + b(y)\,dW(t) + c(y)\,dV(t)\,, \tag{12.4.38}$$

[with $W(t)$ and $V(t)$ independent processes] is equivalent to a third-order Fokker-Planck equation. It is clear that because $W(t)$ and $V(t)$ are processes with independent increments, $y(t)$ is a Markov process. We then calculate

$$\lim_{t\to t_0}\frac{\langle [y(t)-y(t_0)]^n\rangle}{t-t_0} = \lim_{dt_0\to 0}\frac{\langle [dy(t_0)]^n\rangle}{dt_0}\,, \tag{12.4.39}$$

where $y(t_0)$ is a numerical initial value, not a stochastic variable. From (12.4.38), $y(t)$ depends on $W(t')$ and $V(t')$ for only $t' \leqslant t$ and, since $dW(t)$ and $dV(t)$ are independent of $y(t)$, we find

$$\begin{aligned}\langle dy(t_0)\rangle &= \langle a[y(t_0)]\rangle\, dt_0 + \langle b[y(t_0)]\rangle\langle dW(t_0)\rangle + \langle c[y(t_0)]\rangle\langle dV(t_0)\rangle\,, \\ &= \langle a[y(t_0)]\rangle dt_0 = a[y(t_0)]\, dt_0\,,\end{aligned} \tag{12.4.40}$$

because $y(t_0)$ is a numerical initial value. Similarly, to lowest order in dt_0

$$\langle dy(t_0)^2\rangle = b[y(t_0)]^2\langle dW(t_0)^2\rangle = b[y(t_0)]^2 dt_0\,, \tag{12.4.41}$$

$$\langle dy(t_0)^3\rangle = c[y(t_0)]^3\langle dW(t_0)^3\rangle = b[y(t_0)]^3 dt_0\,. \tag{12.4.42}$$

Thus, we find

$$\left.\begin{aligned}\lim_{t\to t_0}\frac{\langle y(t)-y(t_0)\rangle}{(t-t_0)} &= a[y(t_0)]\,, \\ \lim_{t\to t_0}\frac{\langle [y(t)-y(t_0)]^2\rangle}{(t-t_0)} &= b[y(t_0)]^2\,, \\ \lim_{t\to t_0}\frac{\langle [y(t)-y(t_0)]^3\rangle}{(t-t_0)} &= c[y(t_0)]^3\,,\end{aligned}\right\} \tag{12.4.43}$$

and all higher powers give a zero result. By utilising a similar analysis to that of Sect. 3.4, this is sufficient to show that $y(t)$ is a generalized diffusion process whose generalized Fokker-Planck equation is

$$\frac{\partial p(y,t)}{\partial t} = -\frac{\partial}{\partial y}[a(y)p] + \frac{1}{2}\frac{\partial^2}{\partial y^2}[b(y)^2 p] - \frac{1}{6}\frac{\partial^3}{\partial y^3}[c(y)^3 p]\,. \tag{12.4.44}$$

We define a noise source $\zeta(t)$ by

$$dV(t) = \zeta(t)\,dt\,, \tag{12.4.45}$$

where

$$\langle\zeta(t)\rangle = \langle\zeta(t)\zeta(t')\rangle = 0\,, \tag{12.4.46}$$

$$\langle\zeta(t)\zeta(t')\zeta(t'')\rangle = \delta(t-t')\delta(t'-t'')\,, \tag{12.4.47}$$

and higher moments can be readily calculated from the moments of $dV(t)$. The independence of increments means that, as with the Ito integral, integrals that have a delta-function singularity at their upper limit are to be taken as zero.

12.4.3 Example of the Use of Third-Order Noise.

Consider the chemical process

$$A + 2X \underset{k_2}{\overset{k_1}{\rightleftharpoons}} 3X\,, \qquad A \underset{k_4}{\overset{k_3}{\rightleftharpoons}} X\,, \tag{12.4.48}$$

whose Poisson representation Fokker-Planck equation is

$$\begin{aligned}\frac{\partial f(\alpha,t)}{\partial t} &= -\frac{\partial}{\partial \alpha}[\kappa_1 V^{-1}\alpha^2 - \kappa_2 V^{-2}\alpha^3 + \kappa_3 V - \kappa_4\alpha)f(\alpha,t)] \\ &+ \frac{1}{2}\frac{\partial^2}{\partial\alpha^2}[4(\kappa_1 V^{-1}\alpha^2 - \kappa_2 V^{-2}\alpha^3)f(\alpha,t)] \\ &+ \frac{1}{3!}\frac{\partial^3}{\partial\alpha^3}[6(\kappa_1 V^{-1}\alpha^2 - \kappa_2 V^{-2}\alpha^3)f(\alpha,t)]\,, \end{aligned} \tag{12.4.49}$$

where $\kappa_1 V^{-1} = k_1 A, \quad \kappa_2 V^{-2} = k_2, \quad \kappa_3 V = k_3, \quad \kappa_4 = k_4$.

In the steady state, (12.4.49) reduces to a linear second-order differential equation which may be solved in terms of hypergeometric functions, and an asymptotic expansion for the various moments can be obtained using steepest descent methods. This procedure, although possible in principle, is not very practicable. It is in such cases that the method of stochastic differential equations proves to be very useful because of its ease of application.

The stochastic differential equation equivalent to (12.4.49) is

$$\begin{aligned} d\eta(t) &= \{\kappa_1\eta(t)^2 - \kappa_2\eta(t)^3 + \kappa_3 - \kappa_4\eta(t)\}\,dt \\ &+ \mu^3\{4[\kappa_1\eta(t)^2 - \kappa_2\eta(t)^3]\}^{1/2}\,dW(t) \\ &+ \mu^4\{6[\kappa_1\eta(t)^2 - \kappa_2\eta(t)^3]\}^{1/3}\,dV(t)\,, \end{aligned} \tag{12.4.50}$$

where $\alpha = \eta V$ and $\mu = V^{-1/6}$. The noise source $dV(t)$ will be henceforth referred to as the "third-order noise".

Equation (12.4.50) may be solved iteratively by expanding $\eta(t)$:

$$\eta(t) = \eta_0(t) + \mu^3\eta_3(t) + \mu^4\eta_4(t) + \mu^6\eta_6(t) + \mu^8\eta_8(t) + \mu^9\eta_9(t) + \dots \tag{12.4.51}$$

which, when substituted in (12.4.50), yields the deterministic equation in the lowest order and linear stochastic differential equations in the higher orders which may be solved as in Sect. 7.2.

In the stationary state the results are

$$\langle x\rangle = V\eta_0 + \langle\eta_6\rangle + \dots = V\eta_0 + \frac{2ab}{c^2} + \dots \tag{12.4.52a}$$

$$\begin{aligned}\langle x^2\rangle - \langle x^2\rangle &= V\langle\eta_3^2\rangle + [2\langle\eta_9\eta_3\rangle + 2\langle\eta_8\eta_4\rangle + \langle\eta_6^2\rangle - 2\langle\eta_6\rangle^2 + 2\langle\eta_6\rangle] + \dots\,, \\ &= V\left[\frac{2a}{c}\right] + \left[\frac{28}{3}\frac{a^2b^2}{c^4} + \frac{8ab^2\eta_0}{c^3} - \frac{36\kappa_2 a^2}{c^3} + \frac{8ab}{c^2}\right] + \dots \end{aligned} \tag{12.4.52b}$$

$$\begin{aligned}\langle(x-\langle x\rangle)^3\rangle &= V[\langle\eta_4^3\rangle - 3\langle\eta_3^2\rangle\langle\eta_6\rangle + 3\langle\eta_3^2\rangle + \langle\eta_0\rangle + \dots\,, \\ &= V\left[\frac{8a}{c} - \frac{12a^2b}{c^3} + \eta_0\right] + \dots\,, \end{aligned} \tag{12.4.52c}$$

where $a = \kappa_1\eta_0^2 - \kappa_2\eta_0^3, \quad b = 2\kappa_1 - 3\kappa_2\eta_0, \quad c = \kappa_4 - 2\kappa_1\eta_0 + 3\kappa_2\eta_0^2$, and η_0 is the solution of the steady-state deterministic equation

$$\kappa_1\eta_0^2 - \kappa_2\eta_0^3 + \kappa_3 - \kappa_4\eta_0 = 0\,. \tag{12.4.53}$$

Here a few remarks are in order. The third-order noise $\zeta(t)$ contributes to $O(V^{-1})$ to the mean and to $O(1)$ to the variance, but contributes to $O(V)$ to the skewness coefficient. If one is only interested in calculating the mean and the variance to $O(V)$, the third-order noise may be dropped from (12.4.50) and the expansion carried out in powers of $\varepsilon = V^{-1/2}$. Also note that as $c \to 0$, the variance and the higher order corrections become divergent. This, of course, is due to the fact that in this limit, the reaction system exhibits a first-order phase transition type behaviour.

12.5 Simulations Using the Positive Poisson representation

The demonstration that the positive Poisson representation yields genuine stochastic differential equations (albeit in the complex plane) which can be simulated and applied does rely on the neglect of boundary terms arising in the process of integration by parts. While this might appear to be a reasonable procedure, it cannot be guaranteed, and it has to be checked on a case by case basis.

Simulations based on the positive Poisson representation very often yield instabilities which arise after an apparently finite time, and averages calculated using ensembles of such simulations are found to be correct up to this finite time, after which they become quite suddenly inaccurate. This behaviour arises in fact from the nature of the drift term in the Poisson stochastic differential equations,whose stability properties in the complex plane are different from those on the real line. For example, a differential equation of the form (12.2.51) is globally unstable towards rapid growth of α as $\alpha \to -\infty$, a direction which is inaccesible from the positive real line if α is restricted to be real.

12.5.1 Analytic Treatment via the Deterministic Equation

To explain the instability phenomenon, note that the stochastic equation (12.2.51) can be written in form

$$d\alpha = -k_4(\alpha - a)(\alpha - b)\,dt + Q(\alpha)\,dW(t)\,. \tag{12.5.1}$$

Here $b > 0 > a$ are the two roots of the drift function, and the $Q(\alpha)$ is the noise coefficient, which satisfies

$$|Q(\alpha)| \equiv \left|\sqrt{2(k_2A\alpha - k_4\alpha^2)}\right| \sim \sqrt{2k_4}\,|\alpha| \;\text{ as }\; |\alpha| \to \infty. \tag{12.5.2}$$

Since, therefore, for large $|\alpha|$ the drift term dominates in this equation, the behaviour at large $|\alpha|$ is governed by the behaviour of the deterministic part.

The deterministic part of (12.5.1) can be written,

$$\frac{d\alpha}{dt} = -k_4(\alpha - a)(\alpha - b), \tag{12.5.3}$$

and this contains the essence of the problem. There are two stationary points, one at b, which is stable (an attracter) and the other at a, which is unstable (a repeller). The effect of the nonlinearity is to wrap the trajectories around from the repeller to the attracter; see Fig. 12.3. There is a single trajectory from a (the repeller) which

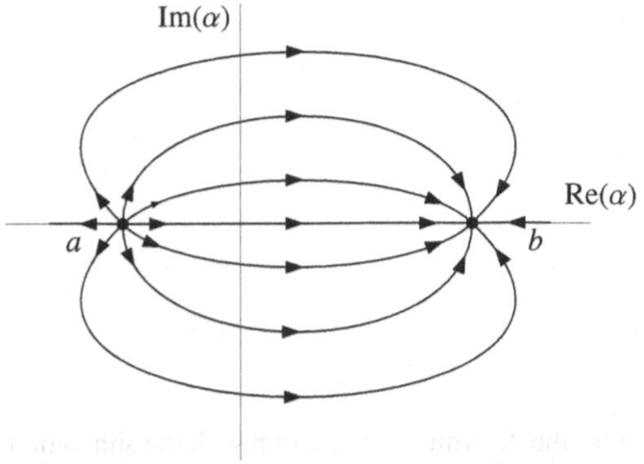

Fig. 12.3. A schematic representation of the phase space motion given by equation (12.5.3). It is of particular importance that there is a single trajectory from a which escapes to $-\infty$, before returning from $+\infty$ to b.

escapes to $-\infty$, before returning from $+\infty$ to b, and which plays the critical role in the validity of the solutions.

a) Solutions of the Deterministic Differential Equation: The solution of (12.5.3) with initial condition $\alpha(t = 0) = n$ is

$$\alpha(t, n) = a + \frac{(b-a)(n-a)}{n(1-e^{\lambda t}) + be^{\lambda t} - a}, \tag{12.5.4}$$

where $\lambda = k_4(b-a)$. The solution clearly has a singularity as a function of n at

$$n = \frac{a - be^{-\lambda t}}{1 - e^{-\lambda t}}, \tag{12.5.5}$$

which starts off at negative infinity when $t = 0$ and moves along the negative real axis reaching the point a at $t = \infty$.

b) Solution with an Initial Gaussian : We choose an initial Gaussian distribution in the complex plane centred on the value $n_0 > 0$, of the form

$$f_G(\alpha) = \frac{1}{2\pi\sigma^2}\exp\left(-\frac{|\alpha - n_0|^2}{2\sigma^2}\right). \tag{12.5.6}$$

This can be thought of as an appropriately weighted sum of radially uniform distributions on concentric rings of radius R, given by $|n - n_0| = R$. The average number at a later time t, averaged over an initial ring of radius R is given by

$$\langle\alpha(t)\rangle_R = \frac{1}{2\pi i}\oint_{C_R}\frac{dn}{n - n_0}\alpha(t, n), \tag{12.5.7}$$

where the contour of the integral C_R is over the circle $|n - n_0| = R$. The average over the Gaussian (12.5.6) is

$$\langle\alpha(t)\rangle_G \equiv \int d^2n\, f_G(n)\alpha(t, n) = \frac{1}{\sigma^2}\int R\, dR\, e^{-R^2/2\sigma^2}\langle\alpha(t)\rangle_R\,. \tag{12.5.8}$$

As long as $\alpha(t, n)$ is analytic in n inside C_R we can use contour integration to get

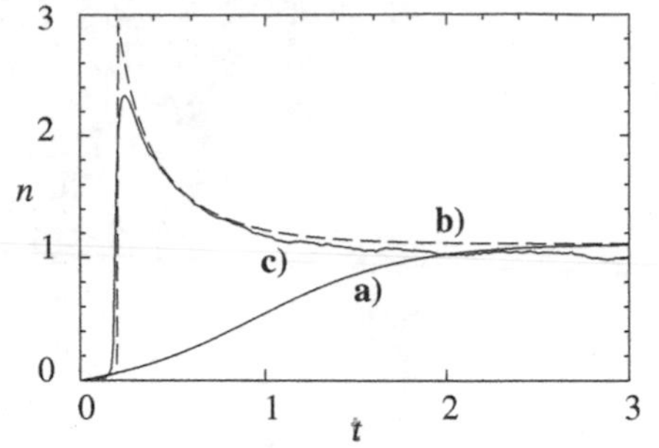

Fig. 12.4. Comparison of analytic treatment for the deterministic equation with the simulation of the full stochastic equation (12.5.1). **a)** Mean number with an initial delta distribution centred at the origin; **b)** (dashed line) analytic curve from (12.5.11); **c)** Simulated mean number over trajectories from an initial 1,000 points distributed on a circle.

$$\langle \alpha(t) \rangle_R = \alpha(t, n_0) . \tag{12.5.9}$$

Thus the mean value will be the solution for the deterministic equation with the initial condition $\alpha = n_0$. However, $\alpha(t, n)$ *does* have a singularity (equation (12.5.5)), and this will hit the circle C_R at the point $n = n_0 - R$, at the time

$$t_e = \begin{cases} \dfrac{1}{\lambda} \ln \left| \dfrac{R + b - n_0}{R + a - n_0} \right| & \text{for } R > n_0 - a , \\ \infty & \text{for } R \leq n_0 - a . \end{cases} \tag{12.5.10}$$

At the same time t_e we find the solution $\alpha(t, -n_0 - R)$ escapes to infinity. Taking into account that the singularity is now within the contour, we can evaluate the contour integral to get the discontinuous solution,

$$\langle \alpha(t) \rangle_R = \alpha(t, n_0) + u(t - t_e) \frac{e^{\lambda t}(a - b)^2}{(1 - e^{\lambda t})(a e^{\lambda t} - b)} . \tag{12.5.11}$$

c) Summary:

i) If the circle $|n - n_0| = R$ does not enclose the repeller at $n = a$, the average over the ring distribution is always equal to the true average.

ii) Otherwise the ring distribution evolves so that at the time t_e it passes through a point at infinity. At this time it is no longer valid to drop the boundary terms. Prior to this time the boundary terms are necessarily negligible since the distribution is bounded.

iii) In this case the solution of the positive Poisson representation is correct up to the time t_e and is incorrect thereafter. Only one trajectory—a set of measure zero—actually escapes to infinity and in practice this trajectory never appears in simulations. However trajectories close to this singular trajectory always appear, and undergo large excursions.

iv) If a simulation of the dynamics is made by averaging trajectories corresponding to $\alpha(t,n)$ for a number of initial values n uniformly distributed on $C_{\bar{R}}$, we obtain the curve like (c) of Fig. 12.5.1. This is very close to the curve (b), which corresponds to (12.5.11), itself an analytic average over all curves, including the singular curve.

v) By using the full Gaussian initial condition, (12.5.13), we obtain a weighted average of behaviours for all $\bar{R}$. If the variance σ^2 of the Gaussian is sufficiently small, the part arising from the spurious term in (12.5.11) will be of order of magnitude $\exp\left(-1/2\sigma^2\right)$ smaller than the other (correct term), and thus negligible.

vi) Problems appear at the *earliest time a deterministic trajectory can escape*. Trajectories near to one which actually escapes give an indication of the time the distribution results break down.

12.5.2 Full Stochastic Case

Consider now the full stochastic equations (12.5.1). The influence of the noise term will distort the circle so that we would expect to see a broadening of the jump in the mean.

a) An Exactly Solvable Model: In the case that $k_3 = k_2 = 0$, the equation (12.2.51)—which is equivalent to (12.5.1)—becomes

$$d\alpha = \left(-k_1 B\alpha - k_4\alpha^2\right) dt + \mathrm{i}\alpha\sqrt{2k_4}\, dW(t)\,. \tag{12.5.12}$$

The transformation $y = -1/\alpha$ turns this into

$$dy = (2k_4 + (k_1 B - 2k_4)\, y)\, dt - \mathrm{i}y\sqrt{2k_4}\, dW(t)\,, \tag{12.5.13}$$

which is a linear equation of the kind treated in Sect. 4.5.7b, where an exact solution is given.

We now implement the exact solution; for this equation, using the notation of that section we find

$$a(t) = 2k_4\,, \tag{12.5.14a}$$

$$b(t) = k_1 B - 2k_4\,, \tag{12.5.14b}$$

$$f(t) = 0\,, \tag{12.5.14c}$$

$$g(t) = -\mathrm{i}\sqrt{2k_4}\,, \tag{12.5.14d}$$

$$\phi(t) = \exp\left\{\int_0^t \left[k_1 B - 2k_4 - \tfrac{1}{2}(-2k_4)\right] dt - \int_0^t \mathrm{i}\sqrt{2k_4}\, dW(t)\right]\,, \tag{12.5.14e}$$

$$= \exp\left\{(k_1 B - k_4)t - \mathrm{i}\sqrt{2k_4}\, W(t)\right\}\,, \tag{12.5.14f}$$

so that the solution for $y(t)$ is

$$y(t) = \phi(t)\left\{y(0) + 2k_4\int_0^t dt'\, \phi(t')^{-1}\right\}\,. \tag{12.5.15}$$

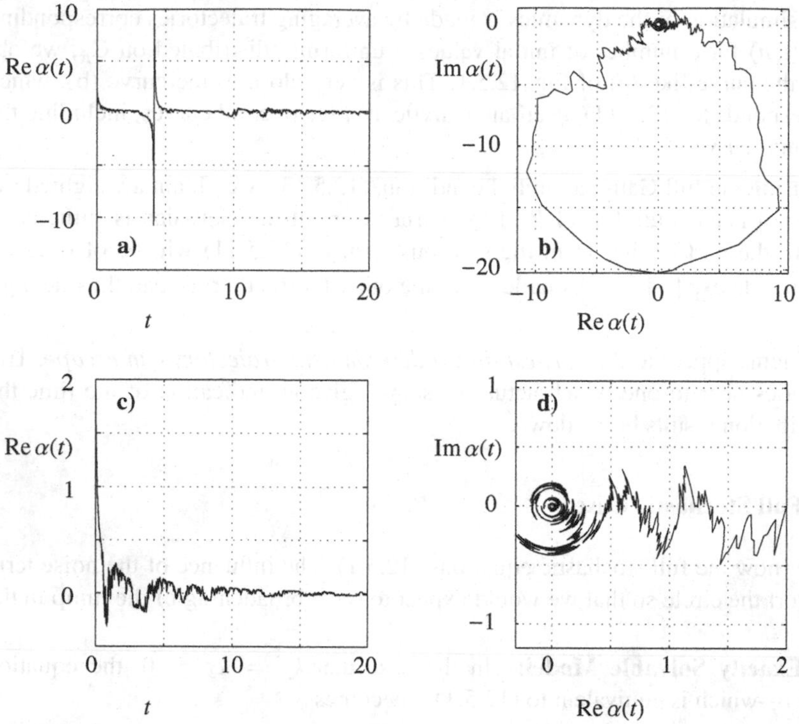

Fig. 12.5. Simulation of (12.5.12) for the case $k_4 = 2$, $\alpha(0) = 2$, and two values of k_1B; **Top:** Here $k_1B = 1.5$ and according to the criterion (12.5.17) an escape to infinity is possible, and the positive Poisson simulation is not valid. In **a)** a "spike" is seen, and in **b)** the corresponding large excursion in the complex plane is visible. **Bottom:** Here $k_1B = 2.1$ and according to the criterion (12.5.17) an escape to infinity is not possible, and the positive Poisson simulation is valid. In **c)** no "spikes" are seen, and in **d)** a steady fluctuating drift towards the stationary state is apparent.

The solution of (12.5.12) is therefore

$$\alpha(t) = \Big[\alpha(0)^{-1} \exp\left\{(k_1B - k_4)t - \mathrm{i}\sqrt{2k_4}\, W(t)\right\}$$

$$+ 2k_4 \int_0^t dt' \exp\left\{(k_1B - k_4)(t-t') - \mathrm{i}\sqrt{2k_4}\,(W(t) - W(t'))\right\}\Big]^{-1} . \tag{12.5.16}$$

This is quite straightforward to compute numerically, and even to understand analytically, as follows.

i) From (12.5.14f), $|\phi(t)| = \exp\{(k_1B - k_4)t\}$, and thus from (12.5.15), we can see that we can write

$$\left|2k_4 \int_0^t dt'\, \phi(t')^{-1}\right| \le 2tk_4 \exp\{(k_4 - k_1B)t\} . \tag{12.5.17}$$

ii) Hence we can see that provided

$$k_4 - k_1 B < 0 \text{ and } y(0) < \frac{2k_4}{k_1 B - k_4}, \tag{12.5.18}$$

it is never possible that $y(t) = 0$, and hence that $\alpha(t) \to \infty$, and thus the positive Poisson simulation can be expected to be valid. This situation is illustrated in Fig. 12.5 c) and d).

iii) More generally, if $y(0)$ is a Gaussian random variable, provided the variance is sufficiently small, there is very little probability of an escape to infinity of $\alpha(t)$, and the simulation will be accurate. More precisely, the simulation will be valid as an asymptotically valid as the variance becomes smaller.

iv) If $k_1 B - k_4 \geq 0$, it is possible that $y(t) \to 0$, and therefore it is possible for $\alpha(t) \to \infty$. Therefore one would expect the simulation to become invalid after the earliest time at which this can occur. This situation is illustrated in Fig. 12.5 a) and b).

12.5.3 Testing the Validity of Positive Poisson Simulations

The behaviour described above is quite generic; thus, in summary, when simulations using the positive Poisson representation are not valid, very definite numerical signatures practice manifested. *If these signatures are not found, the simulation can with high confidence be taken as valid.*

The signatures are:

i) **Presence of Spikes:** The stochastic trajectories occasionally make large excursions into regions of phase space near the unstable manifold, that is "spikes" will occur. The *earliest* time at which one of these spikes occurs is numerically found to be almost exactly the analytically calculated time t_e based purely on the deterministic equation.

ii) **Increase in the Statistical Error:** The large excursions into phase space lead to an increase in the variance of the distribution and the mean photon number and the variance increase dramatically at the time t_e, the time of the discontinuity in the deterministic equation.

iii) **Development of a Power-Law Tail:** The behaviour of the tails of the distribution with time can be explored by binning the trajectories into a set of concentric bins with increasing values of $\bar{R} = |\alpha|$, giving an estimate of the probability of a trajectory reaching a certain radius. This tail of the distribution for begins to fall off as a power-law at the same time t_e that the onset of spiking is observed.

The appearance of a power-law tail at the time t_e can be expected to invalidate the partial integration required to derive the Fokker-Planck equation from a master equation. From this we can conclude that the solution given *after* the time t_e is not a valid solution to the master equation with the given initial condition, even though the power-law tail quickly disappears after this time.

12.6 Application of the Poisson Representation to Population Dynamics

Birth-death master equations with combinatorial kinetics provide a very reasonable description of the population dynamics of a wide variety of biological systems, and one in which there is increasing current interest. Simulations of these equations can be executed using the procedure outlined in Sect. 3.5.1a; a procedure in which the total rate of transitions per unit time determines a Poisson process which governs the time of residence in the starting state, while the particular state to which the system makes a transition is determined by the relative probabilities for the transitions allowed. This procedure becomes less efficient if the transition parameters are time-dependent or are themselves governed by a stochastic law.

12.6.1 The Logistic Model

In such cases the Poisson representation can provide a rather efficient algorithm, and this has been applied by *Drummond* and *Drummond* [12.3] to the *logistic model* of population dynamics, in which the large-scale law of growth of a population $x(t)$ is governed by the differential equation

$$\frac{dx}{dt} = x(g - cx). \tag{12.6.1}$$

Here g is the initial growth rate of the population, and c represents the effect of competition between individuals for the resources available. The system has an equilibrium population $x_c = g/c$, known as its *carrying capacity*.

The logistic model can be implemented by combining three processes

$$\text{Death:} \quad X \xrightarrow{a} 0, \tag{12.6.2a}$$

$$\text{Birth:} \quad X \xrightarrow{b} 2X, \tag{12.6.2b}$$

$$\text{Competition:} \quad 2X \xrightarrow{c} X. \tag{12.6.2c}$$

The model is parametrised by three quantities

$$\left.\begin{array}{lll}\text{The growth rate:} & g & = b - a, \\ \text{The carrying capacity:} & N_c & = g/c, \\ \text{The reproductive ratio:} & R & = b/a. \end{array}\right\} \tag{12.6.3}$$

The model is mathematically identical to a special case of the reaction (12.1.18) with the substitutions

$$k_1 B \to a, \tag{12.6.4a}$$

$$k_3 C \to 0, \tag{12.6.4b}$$

$$k_2 A \to b, \tag{12.6.4c}$$

$$k_4 \quad \to c. \tag{12.6.4d}$$

12.6.2 Poisson Representation Stochastic Differential Equation

Using these substitutions, the appropriate stochastic differential equation corresponding to (12.1.30)is

$$d\alpha = \left[(b-a)\alpha - c\alpha^2\right] dt + \sqrt{2(b\alpha - c\alpha^2)}\, dW(t)\,. \tag{12.6.5}$$

In the notation of (12.5.1), we can write

$$\left.\begin{array}{lll} a \to & 0\,, & \text{attracter if } b < a\,,\ \text{repeller if } b > a\,, \\ b \to & \dfrac{b-a}{c}\,, & \text{repeller if } b < a\,,\ \text{attracter if } b > a\,. \end{array}\right\} \tag{12.6.6}$$

However, even though $\alpha = 0$ can be a repeller, it is always an exit boundary. This can be determined using the criteria of Sect. 5.2.1b, or by noting that the stationary solution of the corresponding Fokker-Planck equation diverges like $1/\alpha$ at the origin.

In the same way as in Sect. 12.2.3, the quantity $\delta \equiv k_1 B/k_4 - k_3 C/k_2 A \to a/c$ determines the kind of Poisson representation required, and since this is clearly positive if the death rate is nonzero, a real Poisson representation is valid on the interval $[0, (b-a)/c]$. The right hand boundary is an entrance boundary, and the vanishing of boundary terms there as specified by (12.1.29) is equivalent to a reflecting boundary condition there.

This there is a rigorously valid Poisson representation on the interval $[0, N_c]$, with an exit boundary on the left and a reflecting boundary on the right. This means that the Poisson representation population α is certain to vanish eventually. This does not include the full range of possible solutions to the problem, for example and initial value of $\alpha = \bar{\alpha} > N_c$ corresponds to an initial Poisson distribution with $\bar{\alpha}$ as mean, which is clearly valid. Initially the stochastic differential equation will have an imaginary noise term, but eventually it will enter the interval $[0, N_c]$. However, the issues of validity treated in the previous section then have to be checked.

For simplicity, we will restrict our investigations to the behaviour of the system on the interval $[0, N_c]$.

12.6.3 Environmental Noise

The technical advantage of the Poisson representation is the simplicity with which one can add environmental noise by allowing the parameters to fluctuate. Models for such fluctuating parameters can be prescribed in a variety of ways, subject to the constraint that the parameters should be positive, and presumably the quantification of the fluctuation should be in terms a value relative to the current value—this is very similar to the situation in financial markets, where *geometric Brownian motion*, as described in Sect. 4.5.2 and Sect. 10.1.

As an example, consider a fluctuation $r(t)$, added to the death rate a, which has a stochastic differential equation description

$$dr(t) = A(r(t))\, dt + B(r(t))\, dW(t)\,, \tag{12.6.7}$$

so that the Poisson representation stochastic differential equation becomes

$$d\alpha = \left[(b - a - r(t))\alpha - c\alpha^2\right] dt + \sqrt{2(b\alpha - c\alpha^2)}\, dW(t) . \tag{12.6.8}$$

If the fluctuations in $r(t)$ are sufficiently rapid we can use the adiabatic elimination methods of Chap. 06-2 to turn this equation into the white noise equation of the form

$$d\alpha = \left[(b - a + \sigma^2)\alpha - c\alpha^2\right] dt + \sqrt{2(b\alpha - c\alpha^2)}\, dW_1(t) + \sqrt{2}\,\sigma\alpha\, dW_2(t) , \tag{12.6.9}$$

$$= \left[(b - a + \sigma^2)\alpha - c\alpha^2\right] dt + \sqrt{2[b\alpha - (c - \sigma^2)\alpha^2]}\, dW(t) . \tag{12.6.10}$$

Here we have introduced σ as a measure of the resultant white noise, and the extra term $\sigma^2\alpha$ in the drift is recognisable as the Stratonovich correction arising from the adiabatic elimination process.

The resulting equation is qualitatively the same as that without this environmental noise. If the adiabatic elimination is not done, the simulation of the environmental noise using its own equation (12.6.7) in conjunction with the Poisson representation equation is quite feasible.

12.6.4 Extinction

The estimation of the time until the ultimate extinction of the species is of great interest in population biology. This is related to the time until $\alpha = 0$, but because the Poisson representation variable is not identical with the actual population x, a correction to this exit time calculation must be made. The source of the correction is quite obvious, if the Poisson variable has the value α, this represents a Poisson distribution in x for which there is a probability $\exp(-\alpha)$ that the population x is already zero.

a) Formulation as a Mean Exit Time Problem: We can modify the argument of Sect. 5.5 as follows:

i) It is difficult, and not usually important, to estimate the time to extinction for a given initial number x_0 of the actual, population, but the Poisson representation is well adapted to a definite initial value α_0 of the Poisson representation variable, corresponding to an initial Poisson distribution of X with mean α_0.

ii) If the Poisson representation conditional probability is $f(\alpha, t \,|\, \alpha_0, 0)$, then the probability that $x \neq 0$ is, as argued above, given by

$$\int d\alpha\, (1 - \mathrm{e}^{-\alpha})\, f(\alpha, t \,|\, \alpha_0, 0) = \mathrm{Prob}(T \geq t) , \tag{12.6.11}$$

where $T(\alpha_0)$ is the time at which $X(T) = 0$, the exit time starting from the Poisson distribution with mean α_0.

iii) The arguments of Sect. 5.5 then show that the mean exit time is given by

$$\langle T(\alpha_0) \rangle = \int_0^\infty dt\, \mathrm{Prob}(T \geq t) , \tag{12.6.12}$$

$$= \int_0^\infty dt \int d\alpha\, (1 - \mathrm{e}^{-\alpha})\, f(\alpha, t \,|\, \alpha_0, 0) , \tag{12.6.13}$$

and that it satisfies the equation

$$[(b - a)\alpha - c\alpha^2]\, \partial_\alpha T(\alpha_0) + (b\alpha - c\alpha^2)\, \partial_\alpha^2 T(\alpha_0) = \mathrm{e}^{-\alpha} - 1 . \tag{12.6.14}$$

b) Application to Simulations: To estimate a mean exit time from a simulation the formula (12.6.13) means that one computes N different sample paths $\alpha_i(t)$ for t less than some maximum time $T_{\max}$ which is sufficiently long for all sampled trajectories to reach the origin. The estimate of the exit time is then

$$\bar{T}(\alpha_0) = \frac{1}{N} \sum_{i=1}^{N} \int_0^{\infty} dt \left(1 - e^{-\alpha_i(t)}\right) . \tag{12.6.15}$$

In practice this is very little altered if one drops the term $\exp(-\alpha_i(t))$, which then corresponds to averaging all of the extinction times for the trajectories, that is, the times at which $\alpha_i(t) = 0$.

13. Spatially Distributed Systems

Reaction diffusion systems are treated in this chapter as a prototype of the host of spatially distributed systems that occur in nature. We introduce the subject heuristically by means of spatially dependent Langevin equations, whose inadequacies are explained. The more satisfactory multivariate master equation description is then introduced, and the spatially dependent Langevin equations formulated as an approximation to this description, based upon a system size expansion. It is also shown how Poisson representation methods can give very similar spatially dependent Langevin equations without requiring any approximation.

We next investigate the consequences of such equations in the spatial and temporal correlation structures which can arise, especially near instability points. The connection between local and global descriptions is then shown. The chapter concludes with a treatment of systems described by a distribution in phase space (i.e. the space of velocity and position). This is done by means of the *Boltzmann Master equation.*

13.1 Background

The concept of space is central to our perception of the world, primarily because well-separated objects do not, in general, have a great deal of influence on each other. This leads to the description of the world, on a macroscopic deterministic level, by *local* quantities such as local density, concentration, temperature, electromagnetic potentials, and so on. Deterministically, these are normally thought of as obeying partial differential equations such as the *Navier-Stokes* equations of hydrodynamics, the *reaction diffusion* equations of chemistry or *Maxwell's* equations of classical electromagnetism.

The simplest cases to consider are reaction diffusion equations, which describe chemical reactions and which form the main topic of this chapter. In order to get some feel of the concept, let us first consider a Langevin equation description for the time evolution of the concentration ρ of a chemical substance. Then the classical reaction-diffusion equation can be derived as follows. A diffusion *current* $\boldsymbol{j}(\boldsymbol{r},t)$ exists such that

$$\boldsymbol{j}(\boldsymbol{r},t) = -D\nabla\rho(\boldsymbol{r},t)\,, \tag{13.1.1}$$

and (13.1.1) is called Fick's law. If there is no chemical reaction, this current obeys a conservation equation. For, considering an arbitrary volume V, the total amount of chemical in this volume can only change because of transport across the boundary $\boldsymbol{S}$, of V. Thus, if N is the total amount in V,

$$\frac{dN}{dt} = \frac{d}{dt}\int_V d^3\boldsymbol{r}\,\rho(\boldsymbol{r},t) = -\int_S d\boldsymbol{S}\cdot \boldsymbol{j}(\boldsymbol{r},t) = -\int_V d^3\boldsymbol{r}\,\nabla\cdot \boldsymbol{j}(\boldsymbol{r},t)\,. \tag{13.1.2}$$

Hence, since V is arbitrary,

$$\partial_t \rho(\boldsymbol{r},t) + \nabla\cdot \boldsymbol{j}(\boldsymbol{r},t) = 0\,. \tag{13.1.3}$$

Substituting Fick's law (13.1.1) into the conservation equation (13.1.3) we get

$$\partial_t\rho(\boldsymbol{r},t) = D\nabla^2\rho(\boldsymbol{r},t)\,, \tag{13.1.4}$$

the diffusion equation. Now how can one add fluctuations? First notice that the conservation equation (13.1.3) is exact; this follows from its derivation. We cannot add a fluctuating term to it. However, Fick's law could well be modified by adding a stochastic source. Thus, we rewrite

$$\boldsymbol{j}(\boldsymbol{r},t) = -D\nabla\rho(\boldsymbol{r},t) + \boldsymbol{f}_{\mathrm{d}}(\boldsymbol{r},t)\,. \tag{13.1.5}$$

Here, $\boldsymbol{f}_{\mathrm{d}}(\boldsymbol{r}, t)$ is a vector Langevin source. The simplest assumption to make concerning its stochastic properties is

$$\langle \boldsymbol{f}_{\mathrm{d}}(\boldsymbol{r},t)\rangle = 0\,, \tag{13.1.6}$$

and

$$\langle f_{\mathrm{d},i}(\boldsymbol{r},t) f_{\mathrm{d},j}(\boldsymbol{r}',t')\rangle = K_{\mathrm{d}}(\boldsymbol{r},t)\delta_{ij}\delta(\boldsymbol{r}-\boldsymbol{r}')\delta(t-t')\,, \tag{13.1.7}$$

that is, the different components are independent of each other at the same time and place, and all fluctuations at different times or places are independent. This is a *locality* assumption. The fluctuating diffusion equation is then

$$\partial_t\rho(\boldsymbol{r},t) = D\nabla^2\rho(\boldsymbol{r},t) - \nabla\cdot \boldsymbol{f}_{\mathrm{d}}(\boldsymbol{r},t)\,. \tag{13.1.8}$$

Notice that

$$\langle \nabla\cdot \boldsymbol{f}_{\mathrm{d}}(\boldsymbol{r},t)\nabla'\cdot \boldsymbol{f}_{\mathrm{d}}(\boldsymbol{r}',t')\rangle = \nabla\cdot\nabla'[K_{\mathrm{d}}(\boldsymbol{r},t)\delta(\boldsymbol{r}-\boldsymbol{r}')]\delta(t-t')\,. \tag{13.1.9}$$

Now consider including a chemical reaction. Fick's law still applies, but instead of the conservation equation we need an equation of the form

$$\frac{dN}{dt} = \frac{d}{dt}\int_V d^3\boldsymbol{r}\,\rho(\boldsymbol{r},t) = -\int_S d\boldsymbol{S}\cdot \boldsymbol{j}(\boldsymbol{r},t) + \int_V d^3\boldsymbol{r}\,F[\rho(\boldsymbol{r},t)]\,, \tag{13.1.10}$$

where $F[p(\boldsymbol{r},t)]$ is a function of the concentration and represents the production of the chemical by a *local* chemical reaction.

Hence we find, before taking fluctuations into account,

$$\partial_t(\boldsymbol{r},t) + \nabla\cdot \boldsymbol{j}(\boldsymbol{r},t) = F[\rho(\boldsymbol{r},t)]\,. \tag{13.1.11}$$

The production of the chemical by a chemical reaction does generate fluctuations, so we can add to (13.1.11) a term $f_{\mathrm{c}}(\boldsymbol{r}, \mathrm{t})$ which satisfies

$$\langle f_{\mathrm{c}}(\boldsymbol{r},t)\rangle = 0\,, \tag{13.1.12}$$

$$\langle f_{\mathrm{c}}(\boldsymbol{r},t) f_{\mathrm{c}}(\boldsymbol{r}',t')\rangle = K_{\mathrm{c}}(\boldsymbol{r},t)\delta(\boldsymbol{r}-\boldsymbol{r}')\delta(t-t')\,, \tag{13.1.13}$$

which expresses the fact that the reaction is *local* (i.e., fluctuations at different points are uncorrelated) and *Markov* (delta correlated in time). The full reaction-diffusion chemical equation now becomes

$$\partial_t \rho(\boldsymbol{r},t) = D\nabla^2 \rho(\boldsymbol{r},t) + F[\rho(\boldsymbol{r},t)] + g(\boldsymbol{r},t)\,, \tag{13.1.14}$$

where

$$g(\boldsymbol{r},t) = -\nabla\cdot \boldsymbol{f}_{\mathrm{d}}(\boldsymbol{r},t) + f_{\mathrm{c}}(\boldsymbol{r},t)\,, \tag{13.1.15}$$

and

$$\langle g(\boldsymbol{r},t) g(\boldsymbol{r}',t')\rangle = \{K_{\mathrm{c}}(\boldsymbol{r}-\boldsymbol{r}',t)\delta(\boldsymbol{r}-\boldsymbol{r}') + \nabla\cdot\nabla'[K_{\mathrm{d}}(\boldsymbol{r},t)\delta(\boldsymbol{r}-\boldsymbol{r}')]\}\delta(t-t')\,. \tag{13.1.16}$$

The simplest procedure for turning a classical reaction diffusion equation into a Langevin equation yields a rather complex expression. Further, we know nothing about $K_{\mathrm{c}}(\boldsymbol{r})$ or $K_{\mathrm{d}}(\boldsymbol{r})$, and this procedure is based on very heuristic models.

Nevertheless, the form derived is essentially correct in that it agrees with the results arising from a more microscopic approach based on Master equations, which, however, specifies all arbitrary constants precisely.

13.1.1 Functional Fokker-Planck Equations

By writing a *stochastic partial differential equation* such as (13.1.14), we immediately raise the question: what does the corresponding Fokker-Planck equation look like? It must be a partial differential equation in a continuously infinite number of variables $\rho(\boldsymbol{r})$, where $\boldsymbol{r}$ is the continuous index which distinguishes the various variables. A simple-minded way of defining functional derivatives is as follows. First, divide space into cubic cells of side l labelled i with position $\boldsymbol{r}_i$, and introduce the variables

$$x_i = l^3 \rho(\boldsymbol{r}_i)\,, \tag{13.1.17}$$

and consider functions of the variables $\boldsymbol{x} = \{x_i\}$.

We now consider calculus of functions $F(\boldsymbol{x})$ of all these cell variables. Partial derivatives are easily defined in the usual way and we formally introduce the functional derivative by

$$\frac{\delta F(\rho)}{\delta \rho(\boldsymbol{r}_i)} = \lim_{l\to 0} l^{-3} \frac{\partial F(\boldsymbol{x})}{\partial x_i}\,. \tag{13.1.18}$$

In what sense this limit exists is, in most applied literature, left completely undefined. Precise definitions can be given and, as is usual in matters dealing with functionals, the precise definition of convergence is important. Further, the "obvious" definition (13.1.18) is not used.

The precise formulation of functional calculus is not within the scope of this book, but an indication of what is normally done by workers who write such equations is appropriate. Effectively, the functional derivative is formally defined by (13.1.18) and a corresponding discretised version of the stochastic differential equation such as (13.1.14) is formulated. Using the same notation, this would be

$$dx_i = \left[\sum_i D_{ij}x_j + \tilde{F}(x_i)\right] dt + \sum_j \tilde{g}_{ij} dW_j(t) \,. \tag{13.1.19}$$

In this equation, D_{ij} are coefficients which yield a discretised approximation to $D\nabla^2$. The coefficients $\tilde{F}$ and $\tilde{g}$ are chosen so that

$$F[\rho(\boldsymbol{r}_i, t)] = \lim_{l\to 0} l^{-3}\tilde{F}(x_i) \,, \tag{13.1.20}$$

$$g(\boldsymbol{r}_i, t) = \lim_{l\to 0} l^{-3} \sum_j \tilde{g}_{ij} dW_j(t) \,. \tag{13.1.21}$$

More precisely, we assume a more general correlation formula than (13.1.16), i.e.,

$$\langle g(\boldsymbol{r}, t) g(\boldsymbol{r}', t)\rangle = G(\boldsymbol{r}, \boldsymbol{r}')\delta(t - t') \,, \tag{13.1.22}$$

and require

$$G(\boldsymbol{r}_i, \boldsymbol{r}_j) = \lim_{l\to 0} l^{-6} \sum_k \tilde{g}_{ik}\tilde{g}_{jk} \,. \tag{13.1.23}$$

In this case, the Fokker-Planck equation for x_i is

$$\partial_t P(\boldsymbol{x}) = -\sum_{ij} \frac{\partial}{\partial x_i}\{[D_{ij}x_j + \delta_{ij}\tilde{F}(x_i)]P(\boldsymbol{x})\} + \tfrac{1}{2}\sum_{ijk} \frac{\partial^2}{\partial x_i \partial x_j}\tilde{g}_{ik}\tilde{g}_{jk}P(\boldsymbol{x}) \,. \tag{13.1.24}$$

Now consider the limit $l^3 \to 0$. Some manipulation gives

$$\begin{aligned} \partial_t P(\rho) = & -\int d^3\boldsymbol{r} \frac{\delta}{\delta\rho(\boldsymbol{r})}\left\{\left[D\nabla^2\rho(\boldsymbol{r}) + F[\rho(\boldsymbol{r})]\right]P(\rho)\right\} \\ & + \tfrac{1}{2}\int\int d^3\boldsymbol{r}\, d^3\boldsymbol{r}' \left[\frac{\delta^2}{\delta\rho(\boldsymbol{r})\delta\rho(\boldsymbol{r}')}G(\boldsymbol{r}, \boldsymbol{r}')P(\rho)\right] . \end{aligned} \tag{13.1.25}$$

$P(\rho)$ is now a kind of functional probability and the definition of its normalisation requires a careful statement of the probability measure on $\rho(\boldsymbol{r})$. This can be done [13.1] but what is normally understood by (13.1.25) is really the discrete version (13.1.24), and almost all calculations implicitly discretise.

The situation is clearly unsatisfactory. The formal mathematical existence of stochastic partial differential equations and their solutions has now been established, but as an everyday computational tool this has not been developed. We refer the reader to [13.1] for more information on the mathematical formulation. Since, however, most work is implicitly discretised, we will mostly formulate matters directly in a discretised form, using continuum notations simply as a convenience in order to give a simpler notation.

13.2 Multivariate Master Equation Description

We assume that the space is divided into cubic *cells* of volume ΔV and side length l. The cells are labelled by an index i and the number of molecules of a chemical X inside cell i is called x_i. Thus we introduce a multivariate probability

$$P(\boldsymbol{x},t) \equiv P(x_1, x_2, \dots, t) \equiv P(x_i, \hat{\boldsymbol{x}}, t)\,. \tag{13.2.1}$$

In the last expression, $\hat{\boldsymbol{x}}$ means the vector of all x's not explicitly written.

We can model diffusion as a Markov process in which a molecule is transferred from cell i to cell j with probability per unit time $d_{ij}x_i$, i.e., the probability of transfer is proportional to the number of molecules in the cell. For a strictly *local* description, we expect that d_{ij} will be nonzero only when i and j are neighbouring cells, but this is not necessary and will not always be assumed in what follows.

In terms of the notation of Sect. 11.5, we can write a birth-death Master equation with parameters given by the replacements:

$$\left.\begin{aligned}
&\qquad\qquad\quad\; i \qquad\qquad\; j\\
\boldsymbol{N}^{i,j} &= (0\dots 0,\ 1,\ 0,\dots 0, 0, 0,\dots)\,,\\
\boldsymbol{M}^{i,j} &= (0\dots 0,\ 0,\ 0,\dots 0, 1, 0,\dots)\,,\\
\boldsymbol{r}^{i,j} &= (0\dots 0, -1, 0,\dots 0, 1, 0,\dots)\,,\\
k^{+}_{(i,j)} &= d_{ij}\,,\\
k^{-}_{(i,j)} &= 0\,.
\end{aligned}\right\} \tag{13.2.2}$$

Hence, the Master equation becomes

$$\partial_t P(\boldsymbol{x},t) = \sum d_{ij}[(x_i+1)P(\hat{\boldsymbol{x}}, x_i+1, x_j-1, t) - x_i P(\mathbf{x},t)]\,. \tag{13.2.3}$$

This equation is a simple linear Master equation and can be solved by various means.

Notice that since

$$\left.\begin{aligned}
\boldsymbol{r}^{(i,j)} &= -\boldsymbol{r}^{(j,i)}\,,\\
\boldsymbol{N}^{(i,j)} &= \boldsymbol{M}^{(j,i)}\,,
\end{aligned}\right\} \tag{13.2.4}$$

we can also restrict to $i > j$ and set $k^{-}_{(i,j)} = d_{ji}$. From (11.5.15, 11.5.18) we see that in this form, *detailed balance* is satisfied provided

$$d_{ij}\langle x_i\rangle_s = d_{ji}\langle x_j\rangle_s\,. \tag{13.2.5}$$

In a system which is diffusing, the stationary solution is homogenous, i.e., $\langle x_t\rangle_s = \langle x_j\rangle_s$, hence, detailed balance requires

$$d_{ij} = d_{ji}\,, \tag{13.2.6}$$

and (13.2.3) possesses a multivariate Poisson stationary solution.

The mean-value equation is

$$\frac{d\langle x_i(t)\rangle}{dt} = \sum_{j,k} r_i^{(j,k)}\left[\langle t^{+}_{jk}(\boldsymbol{x}) - t^{-}_{jk}(\boldsymbol{x})\rangle\right], \tag{13.2.7}$$

$$= \sum_{j,k}(-\delta_{ij} + \delta_{ik})d_{jk}\langle x_j\rangle\,. \tag{13.2.8}$$

Hence,

$$\frac{d\langle x_i(t)\rangle}{dt} = \sum_j \Big(d_{ji} - \delta_{ji}\sum_k d_{jk}\Big)\langle x_j(t)\rangle \tag{13.2.9}$$

$$\equiv \sum_j D_{ji}\langle x_j(t)\rangle\,. \tag{13.2.10}$$

13.2.1 Continuum Form of Diffusion Master Equation

Suppose the centre of cell i is located at $\boldsymbol{r}_i$ and we make the replacement

$$x_i(t) = l^3\rho(\boldsymbol{r}_i, t) \tag{13.2.11}$$

and assume that

$$d_{ij} = \begin{cases} 0, & i, j \text{ not nearest neighbours,} \\ d, & i, j \text{ adjacent.} \end{cases} \tag{13.2.12}$$

Then (13.2.9, 13.2.10) become, in the limit $l \to 0$,

$$\partial_t\langle\rho(\boldsymbol{r}, t)\rangle = D\nabla^2\langle\rho(\boldsymbol{r}, t)\rangle\,, \qquad \text{with } D = l^2 d\,. \tag{13.2.13}$$

Thus, the diffusion equation is recovered. We will generalise this result shortly.

a) Kramers-Moyal or System Size Expansion Equations: We need a parameter in terms of which the numbers and transition probabilities scale appropriately. There are two limits which are possible, both of which correspond to increasing numbers of molecules:

i) Limit of large cells: $l \to \infty$, at fixed concentration;

ii) Limit of high concentration at fixed cell size.

The results are the same for pure diffusion. In either case,

$$t^{+}_{i,j}(\boldsymbol{x}) \to \infty\,, \tag{13.2.14}$$

and a system size expansion is possible. To lowest order, this will be equivalent to a Kramers-Moyal expansion. From (11.5.32, 11.5.33, 11.5.34) we find

$$A_l(\boldsymbol{x}) \;= \sum_j D_{jl}x_j\,, \tag{13.2.15}$$

$$B_{lm}(\boldsymbol{x}) = \delta_{lm}\sum_j(D_{lj}x_l + D_{jl}x_j) - D_{lm}x_l - D_{ml}x_m\,, \tag{13.2.16}$$

where

$$D_{jl} = d_{jl} - \delta_{jl}\sum_k d_{lk}\,, \tag{13.2.17}$$

and thus, in this limit, $P(\boldsymbol{x}, t)$ obeys the Fokker-Planck equation

$$\partial_t P = -\sum_l \partial_l A_l(\boldsymbol{x})P + \tfrac{1}{2}\sum_{l,m}\partial_l\partial_m B_{lm}(\boldsymbol{x})P\,. \tag{13.2.18}$$

b) Continuum Form of Kramers-Moyal Expansion: The continuum form is introduced by associating a point $\boldsymbol{r}$ with a cell i and writing

$$l^3\sum_i \;\to \int d^3\boldsymbol{r}\,, \tag{13.2.19}$$

$$D_{ji} \;\to D(\boldsymbol{r}', \boldsymbol{r}) \equiv \mathscr{D}(\boldsymbol{r}', \boldsymbol{r} - \boldsymbol{r}')\,, \tag{13.2.20}$$

$$l^{-3}\delta_{ij} \to \delta(\boldsymbol{r} - \boldsymbol{r}')\,. \tag{13.2.21}$$

At this stage we make no particular symmetry assumptions on D_{ij}, etc, so that anisotropic inhomogeneous diffusion is included.

However, there are some requirements brought about by the meaning of the concept "diffusion," namely:

i) *Diffusion is observed only when a concentration gradient exists.* This means that the stationary state corresponds to constant concentration and from (13.2.15, 13.2.17) this means that

$$\sum_j D_{jl} = 0\,, \tag{13.2.22}$$

i.e.,

$$\sum_j d_{jl} = \sum_j d_{lj}\,. \tag{13.2.23}$$

Note that detailed balance (13.2.6) implies these.

ii) *Diffusion does not change the total amount of substance in the system,* i.e.,

$$\frac{d}{dt}\sum_l x_l = 0\,, \tag{13.2.24}$$

and this must be true for any value of x_l. From the equation for the mean values, this requires

$$\sum_j D_{lj} = 0\,, \tag{13.2.25}$$

which follows from (13.2.17) and (13.2.23).

iii) *In the continuum notation,* (13.2.22) implies that for any $\boldsymbol{r}$,

$$\int d^3\boldsymbol{\delta}\,\mathscr{D}(\boldsymbol{r}+\boldsymbol{\delta},-\boldsymbol{\delta}) = 0\,, \tag{13.2.26}$$

and from (13.2.25), we also have

$$\int d^3\boldsymbol{\delta}\,\mathscr{D}(\boldsymbol{r},\boldsymbol{\delta}) = 0\,. \tag{13.2.27}$$

iv) *If detailed balance is true,* (13.2.26) is replaced by the equation obtained by substituting (13.2.6) in the definition of D, i.e.,

$$D_{ij} = D_{ji}\,, \tag{13.2.28}$$

which gives in the continuum form

$$\mathscr{D}(\boldsymbol{r}+\boldsymbol{\delta},-\boldsymbol{\delta}) = \mathscr{D}(\boldsymbol{r},\boldsymbol{\delta})\,. \tag{13.2.29}$$

c) The derivation of a continuum form: This now follows in a similar way to that of the Kramers-Moyal expansion.

We define the derivate moments

$$\boldsymbol{M}(\boldsymbol{r}) = \int d^3\boldsymbol{\delta}\,\boldsymbol{\delta}\,\mathscr{D}(\boldsymbol{r},\boldsymbol{\delta})\,, \tag{13.2.30}$$

$$\mathsf{D}(\boldsymbol{r}) = \tfrac{1}{2}\int d^3\boldsymbol{\delta}\,\boldsymbol{\delta}\boldsymbol{\delta}\,\mathscr{D}(\boldsymbol{r},\boldsymbol{\delta})\,, \tag{13.2.31}$$

and it is assumed that derivate moments of higher order vanish in some appropriate limit, similar to those used in the Kramers-Moyal expansion.

The detailed balance requirement (13.2.29) gives

$$M(r) = \int d^3\delta\, \delta\mathscr{D}(r+\delta, -\delta)\,, \tag{13.2.32}$$
$$= \int d^3\delta\, \delta[\mathscr{D}(r,-\delta) + \delta\cdot\nabla\mathscr{D}(r,-\delta) + \ldots]\,, \tag{13.2.33}$$
$$= -M(r) + 2\nabla\cdot \mathsf{D}(r) + \ldots. \tag{13.2.34}$$

Hence, detailed balance requires

$$M(r) = \nabla\cdot\mathsf{D}(r)\,. \tag{13.2.35}$$

The weaker requirement (13.2.26) similarly requires the weaker condition

$$\nabla\cdot[M(r) - \nabla\cdot\mathsf{D}(r)] = 0\,. \tag{13.2.36}$$

We now can make the continuum form of $A_l(x)$:

$$A_l(x) \to \int d^3\delta\, \mathscr{D}(r,\delta)\rho(r+\delta)\,,$$
$$= M(r)\cdot\nabla\rho(r) + \mathsf{D}(r) : \nabla\nabla\rho(\mathbf{r})\,. \tag{13.2.37}$$

If detailed balance is true, we can rewrite, from (13.2.35)

$$A_l(x) \to \nabla\cdot[\mathsf{D}(r)\cdot\nabla\rho(r)]\,. \tag{13.2.38}$$

The *general* form, without detailed balance, can be obtained by defining

$$J(r) = M(r) - \nabla\cdot\mathsf{D}(r)\,. \tag{13.2.39}$$

From (13.2.36)

$$\nabla\cdot J(r) = 0\,, \tag{13.2.40}$$

so that we can write

$$J(r) = \nabla\cdot\mathsf{E}(r)\,, \tag{13.2.41}$$

where $\mathsf{E}(r)$ is an antisymmetric tensor. Substituting, we find that by defining

$$\mathsf{H}(r) = \mathsf{D}(r) + \mathsf{E}(r)\,, \tag{13.2.42}$$

we have defined a nonsymmetric diffusion tensor $\mathsf{H}(r)$ and that

$$A_l(x) \to \nabla\cdot[\mathsf{H}(r)\cdot\nabla\rho(r,t)]\,. \tag{13.2.43}$$

This means that, deterministically,

$$\partial_t\rho(r,t) = \nabla\cdot[\mathsf{H}(r)\cdot\nabla\rho(r,t)]\,, \tag{13.2.44}$$

where $\mathsf{H}(r)$ is symmetric if detailed balance holds.

We now come to the fluctuation term, given by (13.2.16). To compute $B_{lm}(x)$, we first consider the limit of

$$l^3\sum_m B_{lm}\phi_m \to \int dr' B(r,r')\phi(r')\,, \tag{13.2.45}$$

where ϕ_m is an arbitrary function. By similar, but much more tedious computation, we eventually find

$$l^3\sum_m B_{lm}\phi_m \to -2\nabla\cdot[\mathsf{D}(r)\rho(r)\cdot\nabla\phi(r)]\,, \tag{13.2.46}$$

so that

$$B(\boldsymbol{r},\boldsymbol{r}') = 2\nabla'\nabla : [\mathsf{D}(\boldsymbol{r})\rho(\boldsymbol{r})\delta(\boldsymbol{r}-\boldsymbol{r}')] . \tag{13.2.47}$$

The phenomenological theory of Sect. 13.1 now has a rational basis since (13.2.47) is what arises from assuming

$$\boldsymbol{j}(\boldsymbol{r},t) = -\mathsf{H}(\boldsymbol{r})\nabla\rho(\boldsymbol{r},t) - \boldsymbol{\xi}(\boldsymbol{r},t) , \tag{13.2.48}$$

in which

$$\langle \boldsymbol{\xi}(\boldsymbol{r},t)\boldsymbol{\xi}(\boldsymbol{r}',t')\rangle = 2\delta(t-t')\mathsf{D}(\boldsymbol{r})\delta(\boldsymbol{r}-\boldsymbol{r}')\rho(\boldsymbol{r}) , \tag{13.2.49}$$

and hence

$$\partial_t\rho(\boldsymbol{r},t) = \nabla\cdot\mathsf{H}(\boldsymbol{r})\cdot\nabla\rho + \nabla\cdot\boldsymbol{\xi}(\boldsymbol{r},t) . \tag{13.2.50}$$

This corresponds to a theory of inhomogeneous anisotropic diffusion without detailed balance. The more usual case of homogeneous isotropic diffusion with detailed balance corresponds to setting

$$\mathsf{H}(\boldsymbol{r}) = \mathsf{D}(\boldsymbol{r}) = D\mathsf{1} , \tag{13.2.51}$$

and this gives a more familiar equation. Notice that according to (13.2.49), unless D is diagonal, fluctuations in different components of the current are in general correlated.

d) Comparison with Fluctuation-Dissipation Argument: The result (13.2.47) can almost be obtained from a simple fluctuation-dissipation argument in the stationary state, where we know the fluctuations are Poissonian. In that case,

$$\langle x_i, x_j\rangle = \langle x_i\rangle\delta_{ij} , \tag{13.2.52}$$

corresponding to

$$g(\boldsymbol{r},\boldsymbol{r}') = \langle\rho(\boldsymbol{r}),\rho(\boldsymbol{r}')\rangle = \delta(\boldsymbol{r}-\boldsymbol{r}')\langle\rho(\boldsymbol{r})\rangle . \tag{13.2.53}$$

Since the theory is linear, we can apply (4.5.64) of Sect. 4.5.6. Here the matrices A and A^{T} become

$$\left.\begin{aligned} A &\to -\nabla\cdot\mathsf{H}(\boldsymbol{r})\cdot\nabla , \\ A^{\mathrm{T}} &\to -\nabla'\cdot\mathsf{H}(\boldsymbol{r}')\cdot\nabla' . \end{aligned}\right\} \tag{13.2.54}$$

Thus,

$$\begin{aligned} B(\boldsymbol{r},\boldsymbol{r}') &\to BB^{\mathrm{T}} \\ &= A\sigma + \sigma A^{\mathrm{T}} \to [-\nabla\cdot\mathsf{H}(\boldsymbol{r})\cdot\nabla - \nabla'\cdot\mathsf{H}(\boldsymbol{r}')\cdot\nabla']g(\boldsymbol{r},\boldsymbol{r}')] . \end{aligned} \tag{13.2.55}$$

We note that in the stationary state, $\langle\rho(\boldsymbol{r})\rangle = \langle\rho\rangle$, independent of $\boldsymbol{r}$. Thus,

$$\begin{aligned} B(\boldsymbol{r},\boldsymbol{r}') &= [-\nabla\cdot\mathsf{H}(\boldsymbol{r})\cdot\nabla\langle\rho\rangle\delta(\boldsymbol{r}-\boldsymbol{r}') - \nabla'\cdot\mathsf{H}(\boldsymbol{r}')\cdot\nabla'\langle\rho\rangle\delta(\boldsymbol{r}-\boldsymbol{r}')] , \\ &= \nabla\nabla' : \{[\mathsf{H}(\boldsymbol{r}) + \mathsf{H}^{\mathrm{T}}(\boldsymbol{r})]\langle\rho\rangle\delta(\boldsymbol{r}-\boldsymbol{r}')\} , \\ &= 2\nabla\nabla' : [\mathsf{D}(\boldsymbol{r})\langle\rho\rangle\delta(\boldsymbol{r}-\boldsymbol{r}')] . \end{aligned} \tag{13.2.56}$$

However, this result is not as general as (13.2.47) since it is valid in the stationary state only, nor is it the same even in the stationary state because it includes $\langle\rho\rangle$, not $\rho(\boldsymbol{r})$. However, since the Fokker-Planck formalism is valid only as a large cell size limit in which fluctuations are small, to this accuracy (13.2.56) agrees with (13.2.47)

13.2.2 Combining Reactions and Diffusion

We introduce reactions by assuming that molecules within a cell react with each other according to a Master equation like those of Chap. 11. We wish to consider several chemical components so we introduce the notation

$$X = (\boldsymbol{x}_1, \boldsymbol{x}_2, \dots), \tag{13.2.57}$$

where $\boldsymbol{x}_i$ represents a vector whose components $x_{i,a}$ are the numbers of molecules of species X_a in cell i. Thus, we write

$$\begin{aligned}\partial_t P(X,t) &= \partial_t P(\text{diffusion}) \\ &\quad + \sum_{i,A} \Big\{ t_A^+(\boldsymbol{x}_i - \boldsymbol{r}^A) P(\boldsymbol{x}_i - \boldsymbol{r}^A, \hat{X}) + t_A^-(\boldsymbol{x}_i + \boldsymbol{r}^A) P(\boldsymbol{x}_i + \boldsymbol{r}^A, \hat{X}) \\ &\quad - [t_A^+(\boldsymbol{x}_i) + t_A^-(\boldsymbol{x}_i)] P(X) \Big\}\end{aligned} \tag{13.2.58}$$

where the diffusion part has the form of (13.2.3), but is summed over the various components.

This leads via a Kramers-Moyal expansion to a Fokker-Planck equation, with the usual drift and diffusion as given by (11.5.33, 11.5.34).

We can write equivalent stochastic partial differential equations in terms of a spatially dependent Wiener process $\boldsymbol{W}(\boldsymbol{r}, t)$ as follows: we consider an isotropic constant diffusion tensor

$$\mathsf{D}(\boldsymbol{r}) = D\mathbf{1}, \tag{13.2.59}$$

$$d\rho_a = \left[D_a \nabla^2 \rho_a + \sum_A r_a^A \left(\kappa_A^+ \prod_a \rho_a^{N_a^A} - \kappa_A^- \prod_a \rho_a^{M_a^A} \right) \right] dt + dW_a(\boldsymbol{r}, t), \tag{13.2.60}$$

with

$$\begin{aligned} dW_a(\boldsymbol{r},t)\, dW_b(\boldsymbol{r}',t) = \Bigg\{ & 2\delta_{a,b} \nabla' \cdot \nabla [D_a \rho_a(\boldsymbol{r}) \delta(\boldsymbol{r} - \boldsymbol{r}')] \\ & + \delta(\boldsymbol{r} - \boldsymbol{r}') \sum_A r_a^A r_b^A \left(\kappa_A^+ \prod_a \rho_a^{N_a^A} + \kappa_A^- \prod_a \rho_a^{M_a^A} \right) \Bigg\} dt. \end{aligned} \tag{13.2.61}$$

The validity of the Langevin equation depends on the system size expansion. Equations (13.2.60, 13.2.61) depend on the particular scaling of the chemical rate constants with Ω, given in Sect. 11.5.3, equation (11.5.30). The only interpretation of Ω which is valid in this case is

$$\Omega = l^3 = \text{volume of cell}. \tag{13.2.62}$$

Notice, however, that at the same time, the diffusion part scales like l since $l^2 d$ must remain equal to the diffusion coefficient while the terms arising from chemical reactions scale like l^3. This means that as cell volume is increased, we have less and less effect from diffusion, but still more than the correction terms in the chemical part which will be integral powers of l^3 less than the first.

The precise method of comparing diffusion with reaction will be dealt with later.

Example: $X_1 \rightleftharpoons 2X_2$: For this chemical reaction, we find (using the methods of Sect. 11.5)

$$N = \begin{bmatrix} 1 \\ 0 \end{bmatrix}, \quad M = \begin{bmatrix} 0 \\ 2 \end{bmatrix}, \quad r = \begin{bmatrix} -1 \\ 2 \end{bmatrix}, \tag{13.2.63}$$
$$k^+ = k_1 = \kappa_1 \Omega^{-1+1}, \qquad k^- = k_2 = \kappa_1 \Omega^{-2+1}.$$

Thus, substituting

$$\left.\begin{aligned} d\rho_1(\boldsymbol{r}) &= (D_1 \nabla^2 \rho_1 - \kappa_1 \rho_1 + \kappa_2 \rho_2^2) dt + dW_1(\boldsymbol{r}, t) \\ d\rho_2(\boldsymbol{r}) &= (D_2 \nabla^2 \rho_2 + 2\kappa_1 \rho_1 - 2\kappa_2 \rho_2^2) dt + dW_2(\boldsymbol{r}, t) \end{aligned}\right\} \tag{13.2.64}$$

where

$$dW(\boldsymbol{r}, t), dW^{\mathrm{T}}(\boldsymbol{r}', t) =$$
$$\begin{pmatrix} 2\nabla \cdot \nabla'[D_1 \rho_1 \delta(\boldsymbol{r}-\boldsymbol{r}')] + \delta(\boldsymbol{r}-\boldsymbol{r}')[\kappa_1 \rho_1 + \kappa_2 \rho_2^2] & , & -2\delta(\boldsymbol{r}-\boldsymbol{r}')[\kappa_1 \rho_1 + \kappa_2 \rho_2^2] \\ -2\delta(\boldsymbol{r}-\boldsymbol{r}')[\kappa_1 \rho_1 + \kappa_2 \rho_2^2] & , & 2\nabla \cdot \nabla'[D_2 \rho_2 \delta(\boldsymbol{r}-\boldsymbol{r}')] + 4\delta(\boldsymbol{r}-\boldsymbol{r}')[\kappa_1 \rho_1 + \kappa_2 \rho_2] \end{pmatrix} dt. \tag{13.2.65}$$

This equation is valid only as a system size expansion in Ω^{-1}, that is, the cell size, and the continuum limit is to be regarded as a *notation* in which it is understood that we really mean a cell model, and are working on a sufficiently large scale for the cell size to appear small, though the cell itself is big enough to admit many molecules.

Thus, this kind of equation is really only valid as a linearised equation about the deterministic state which is the form in which *Keizer* [13.2] formulated chemical reactions. In this respect, the Poisson representation is better since it gives equations exactly equivalent to the Master equation.

13.2.3 Poisson Representation Methods

For a reaction with no more than bimolecular steps the Poisson representation method [using (12.1.9)] gives a rather simplified Fokker-Planck equation, since for the spatial diffusion [using the formulation (13.2.2)], the *diffusion matrix vanishes*. The generalisation to a spatially dependent system is then carried out in the density variable

$$\eta_a(\boldsymbol{r}) = \alpha_a(\boldsymbol{r})/l^3, \tag{13.2.66}$$

and we find

$$d\eta_a(\boldsymbol{r}) = \left[D_a \nabla^2 \eta_a(\boldsymbol{r}) + \sum_A r_a^A \left(\kappa_A^+ \prod_a \eta_a^{N_a^A} - \kappa_A^- \prod_a \eta_a^{M_a^A} \right) \right] dt + dW_a(\boldsymbol{r}, t), \tag{13.2.67}$$

$$\begin{aligned} &dW_a(\boldsymbol{r}, t)\, dW_b(\boldsymbol{r}', t) \\ &\quad = \delta(\boldsymbol{r} - \boldsymbol{r}') \sum_A \left(\kappa_A^+ \prod_a \eta_a^{N_a^A} - \kappa_A^- \prod_a \eta_a^{N_a^M} \right) (M_a^A M_b^A - N_a^A N_b^A - \delta_{ab} r_a^A)\, dt. \end{aligned} \tag{13.2.68}$$

These equations are very similar to (13.2.60, 13.2.61). When explicitly written out they are simpler. For example, considering again $X_1 \rightleftharpoons 2X_2$, we get

$$d\eta_1(\boldsymbol{r}) = (D_1\nabla^2\eta_1 - \kappa_1\eta_1 + \kappa_2\eta_2^2)\,dt + dW_1(\boldsymbol{r},t)\,, \tag{13.2.69}$$

$$d\eta_2(\boldsymbol{r}) = (D_2\nabla^2\eta_2 + 2\kappa_1\eta_1 - 2\kappa_2\eta_2^2)\,dt + dW_2(\boldsymbol{r},t)\,, \tag{13.2.70}$$

$$d\boldsymbol{W}(\boldsymbol{r},t)\,d\boldsymbol{W}^{\mathrm{T}}(r',t) = \begin{pmatrix} 0 & 0 \\ 0 & 2 \end{pmatrix}(\kappa_1\eta_1 - \kappa_2\eta_2^2)\delta(\boldsymbol{r}-\boldsymbol{r}')\,dt\,. \tag{13.2.71}$$

The simplicity of (13.2.69–13.2.71) when compared to their counterparts (13.2.63, 13.2.64) is quite striking, and it is especially noteworthy that they are exactly equivalent (in a continuum formulation) to the master equation.

13.3 Spatial and Temporal Correlation Structures

We want to consider here various aspects of spatial, temporal and spatio-temporal correlations in linear systems, which are of course all exactly soluble. The correlations that are important are the *factorial correlations* which are defined in terms of factorial moments in the same way as ordinary correlations are defined in terms of moments. The equations which arise are written much more naturally in terms of factorial moments, as we shall see in the next few examples.

13.3.1 Reaction $X \underset{k_2}{\overset{k_1}{\rightleftharpoons}} Y$

We assume homogenous isotropic diffusion with the same diffusion constant for X and Y, and since both the reaction and the diffusion are linear we find Poisson representation Langevin equations for the *concentration* variables η, μ (corresponding, respectively, to X and Y) *with no stochastic source*, i.e.,

$$\left.\begin{aligned} \partial_t\eta(\boldsymbol{r},-t) &= D\nabla^2\eta - k_1\eta + k_2\mu \\ \partial_t\mu(\boldsymbol{r},t) &= D\nabla^2\mu + k_1\eta - k_2\mu\,. \end{aligned}\right\} \tag{13.3.1}$$

a) Spatial Correlations: We now note that

$$\left.\begin{aligned} \langle\eta(\boldsymbol{r},t)\rangle &= \langle\rho_x(\boldsymbol{r},t)\rangle\,, \\ \langle\mu(\boldsymbol{r},t)\rangle &= \langle\rho_y(\boldsymbol{r},t)\rangle\,, \\ \langle\eta(\boldsymbol{r},t),\eta(\boldsymbol{r}',t)\rangle &= \langle\rho_x(\boldsymbol{r},t),\rho_x(\boldsymbol{r}',t)\rangle - \delta(\boldsymbol{r}-\boldsymbol{r}')\langle\rho_x(\boldsymbol{r},t)\rangle \equiv g(\boldsymbol{r},\boldsymbol{r}',t)\,, \\ \langle\eta(\boldsymbol{r},t),\mu(\boldsymbol{r}',t)\rangle &= \langle\rho_x(\boldsymbol{r},t),\rho_y(\boldsymbol{r}',t)\rangle \equiv f(\boldsymbol{r},\boldsymbol{r}',t)\,, \\ \langle\mu(\boldsymbol{r},t),\mu(\boldsymbol{r}',t)\rangle &= \langle\rho_y(\boldsymbol{r},t),\rho_x(\boldsymbol{r}',t)\rangle - \delta(\boldsymbol{r}-\boldsymbol{r}')\langle\rho_y(\boldsymbol{r},t)\rangle = h(\boldsymbol{r},\boldsymbol{r}',t)\,, \end{aligned}\right\} \tag{13.3.2}$$

which are all continuum notation versions of the fact that the Poissonian moments are equal to the factorial moments of the actual numbers.

The equations for the mean concentrations are obviously exactly the same as (13.3.1). Assuming now a homogeneous situation, so that we can assume

$$\left.\begin{aligned}
\langle \rho_x(\boldsymbol{r},t)\rangle &= \langle \rho_x(t)\rangle\,,\\
\langle \rho_y(\boldsymbol{r},t)\rangle &= \langle \rho_y(t)\rangle\,,\\
g(\boldsymbol{r},\boldsymbol{r}',t) &= g(\boldsymbol{r}-\boldsymbol{r}',t)\,,\\
f(\boldsymbol{r},\boldsymbol{r}',t) &= f(\boldsymbol{r}-\boldsymbol{r}',t)\,,\\
h(\boldsymbol{r},\boldsymbol{r}',t) &= h(\boldsymbol{r}-\boldsymbol{r}',t)\,,
\end{aligned}\right\} \tag{13.3.3}$$

and compute equations of motion for $\langle \eta(\boldsymbol{r},t)\eta(\boldsymbol{0},t)\rangle$, etc. We quickly find

$$\left.\begin{aligned}
\frac{\partial g(\boldsymbol{r},t)}{\partial t} &= 2D\nabla^2 g(\boldsymbol{r},t) - 2k_1 g(\boldsymbol{r},t) + 2k_2 f(\boldsymbol{r},t)\,,\\
\frac{\partial f(\boldsymbol{r},t)}{\partial t} &= 2D\nabla^2 f(\boldsymbol{r},t) - (k_1+k_2) f(\boldsymbol{r},t) + k_2 h(\boldsymbol{r},t) + k_1 g(\boldsymbol{r},t)\,,\\
\frac{\partial h(\boldsymbol{r},t)}{\partial t} &= 2D\nabla^2 h(\boldsymbol{r},t) - 2k_2 h(\boldsymbol{r},t) + 2k_1 f(\boldsymbol{r},t)\,,
\end{aligned}\right\} \tag{13.3.4}$$

The stationary solution of these equations has the form

$$g(\boldsymbol{r}) = \xi k_2^2\,,\quad f(\boldsymbol{r}) = \xi k_1 k_2\,,\quad h(\boldsymbol{r}) = \xi k_1^2\,, \tag{13.3.5}$$

where ξ is an arbitrary parameter. The corresponding stationary solutions for the means are

$$\langle x(\boldsymbol{r}_i)\rangle = \lambda k_2\,,\quad \langle y(\boldsymbol{r}_i)\rangle = \lambda k_1\,, \tag{13.3.6}$$

where λ is another arbitrary parameter. If $\xi = 0$, we recover the Poissonian situation where

$$\left.\begin{aligned}
\langle \rho_x(\boldsymbol{r}),\rho_x(\boldsymbol{r}')\rangle &= \langle \rho_x(\boldsymbol{r})\rangle\delta(\boldsymbol{r}-\boldsymbol{r}')\,,\\
\langle \rho_y(\boldsymbol{r}),\rho_y(\boldsymbol{r}')\rangle &= \langle \rho_y(\boldsymbol{r})\rangle\delta(\boldsymbol{r}-\boldsymbol{r}')\,,\\
\langle \rho_x(\boldsymbol{r}),\rho_y(\boldsymbol{r}')\rangle &= 0\,.
\end{aligned}\right\} \tag{13.3.7}$$

By choosing other values of λ, different solutions corresponding to various distributions over the total number of molecules in the system are obtained.

b) Time-Dependent Solutions: These can easily be developed for any initial condition. In the case where the solutions are initially homogeneous, uncorrelated and Poissonian, (13.3.7) is satisfied as an initial condition and thus f, g, and h are initially all zero, and will remain so. Thus, an uncorrelated Poissonian form is preserved in time, as has already been deduced in Sect. 12.1 b.

The problem of relaxation to the Poisson is best dealt with by assuming a specific form for the initial correlation function, for example, an initially uncorrelated but non-Poissonian system represented by

$$g(\boldsymbol{r},0) = \alpha\delta(\boldsymbol{r}),\quad f(\boldsymbol{r},0) = \beta\delta(\boldsymbol{r}),\quad h(\boldsymbol{r},0) = \gamma\delta(\boldsymbol{r})\,. \tag{13.3.8}$$

Time-dependent solutions take the form

$$\begin{bmatrix} g(\boldsymbol{r},t) \\ f(\boldsymbol{r},t) \\ h(\boldsymbol{r},t) \end{bmatrix} = \frac{\exp(-r^2/8Dt)}{(8\pi Dt)^{3/2}} \begin{bmatrix} k_2^2\varepsilon_1 - 2k_2\varepsilon_2 \mathrm{e}^{-(k_1+k_2)t} + \varepsilon_3 \mathrm{e}^{-2(k_1+k_2)t} \\ k_1k_2\varepsilon_1 + (k_2-k_1)\varepsilon_2 \mathrm{e}^{-(k_1+_2)t} - \varepsilon_3 \mathrm{e}^{-2(k_1+k_2)t} \\ k_1^2\varepsilon_1 + 2k_1\varepsilon_2 \mathrm{e}^{-(k_1+k_1)t} + \varepsilon_3 \mathrm{e}^{-2(k_1+k_2)t} \end{bmatrix}, \tag{13.3.9}$$

where

$$\left.\begin{aligned} \varepsilon_1 &= (\alpha + 2\beta + \gamma)/(k_1+k_2)^2 , \\ \varepsilon_2 &= [k_2(\beta+\gamma) - k_1(\alpha+\beta)]/(k_1+k_2)^2 , \\ \varepsilon_3 &= [k_1^2\alpha + k_2^2\gamma - 2k_1k_2\beta]/(k_1+k_2)^2 . \end{aligned}\right\} \tag{13.3.10}$$

Comments :

i) The terms $\varepsilon_1, \varepsilon_2$, and ε_3 correspond, respectively, to deviations from an uncorrelated Poissonian of the quantities $\langle (x_i+y_i),(x_j+y_j)\rangle, \langle (x_i+y_i),(k_1y_j-k_2x_j)\rangle$, and $\langle (k_1y_i-k_2x_i),(k_1y_j-k_2x_j)\rangle$, which are essentially density fluctuations, correlation between density fluctuation and chemical imbalance, and fluctuations in chemical imbalance. We notice a characteristic diffusion form multiplying a chemical time dependence appropriate to the respective terms.

ii) The time taken for the deviation from a Poissonian uncorrelated form given by (13.3.8) to become negligible compared to the Poissonian depends, of course, on the magnitude of the initial deviation. Assuming, however, that $\alpha, \beta, \gamma, \langle x_i \rangle$, and $\langle y_i \rangle$ are all of comparable size, one can make a rough estimate as follows. We consider a small spherical volume of radius R much larger, however, than our basic cells. Then in this small volume V, we find that

$$\mathrm{var}[x(V,0)] = \int_V d^3\boldsymbol{r} \int_V d^3\boldsymbol{r}' \langle \rho_x(\boldsymbol{r},0), \rho_x(\boldsymbol{r}',0)\rangle = \langle x(V,0)\rangle + \alpha V , \tag{13.3.11}$$

and similarly,

$$\mathrm{var}[y(V,0)] = \langle y[V,0]\rangle + \gamma V , \tag{13.3.12}$$

$$\langle x[V,0], y[V,0]\rangle = \beta V , \tag{13.3.13}$$

while after a time $t \gg R^2/4D$, these quantities satisfy approximately

$$\mathrm{var}[x(V,t)] \simeq \langle x(V,t)\rangle + \frac{V^2}{(8\pi Dt)^{3/2}} \left(k_1^2\varepsilon_1 - 2k_2\varepsilon_2 \mathrm{e}^{-(k_1+k_2)t} + \varepsilon_3 \mathrm{e}^{-2(k_1+k_2)t} \right) , \tag{13.3.14}$$

$$\mathrm{var}[y(V,t)] \simeq \langle y[V,t]\rangle + \frac{V^2}{(8\pi Dt)^{3/2}} \left(k_1^2\varepsilon_1 + 2k_1\varepsilon_2 \mathrm{e}^{-(k_1+k_2)t} + \varepsilon_3 \mathrm{e}^{-2(k_1+k_2)t} \right) , \tag{13.3.15}$$

$$\langle x[V,t], y[V,t]\rangle \simeq \frac{V^2}{(8\pi Dt)^{3/2}} \left(k_1k_2\varepsilon_1 + (k_2-k_1)\varepsilon_2 \mathrm{e}^{-(k_1+k_2)t} - \varepsilon_3 \mathrm{e}^{-2(k_1+k_2)t} \right) . \tag{13.3.16}$$

iii) Thus, the diffusion has reduced the overall deviation from Poissonian uncorrelated behaviour by a factor of the order of magnitude of $R^3/(Dt)^{3/2}$. However, notice that in the case of an initial non-Poissonian, but also uncorrelated, situation, corresponding to $\beta = 0$, we find that a transient correlation appears between X and Y, which if the chemical rate constants are sufficiently large, can be quite substantial.

b) Space-Time Correlations: Since the equations of motion here are linear, we may use the linear theory developed in Sect. 3.7.4. Define the stationary two-time correlation matrix $\mathsf{G}(\boldsymbol{r}, t)$ by

$$\mathsf{G}(\boldsymbol{r}, t) = \begin{bmatrix} \langle \rho_x(\boldsymbol{r}, t), \rho_x(\boldsymbol{0}, 0)\rangle_s & \langle \rho_y(\boldsymbol{r}, t), \rho_x(\boldsymbol{0}, 0)\rangle \\ \langle \rho_x(\boldsymbol{r}, t), \rho_y(\boldsymbol{0}, 0)\rangle_s & \langle \rho_y(\boldsymbol{r}, t), \rho_y(\boldsymbol{0}, 0)\rangle \end{bmatrix}. \tag{13.3.17}$$

Then the equation corresponding to (3.7.61) is

$$\partial_t \mathsf{G}(\boldsymbol{r}, t) = \begin{bmatrix} D\nabla^2 - k_1 & k_2 \\ k_1 & D\nabla^2 - k_2 \end{bmatrix} \mathsf{G}(\boldsymbol{r}, t). \tag{13.3.18}$$

The solution can be obtained by Fourier transforming and solving the resultant first order differential matrix equation by standard methods, with boundary conditions at $t = 0$ given by (13.3.7).The result is

$$\mathsf{G}(\boldsymbol{r}, t) = \left(\frac{\langle \rho_x \rangle + \langle \rho_y \rangle}{(k_1 + k_2)^2} \right) \frac{\exp(-r^2/4Dt)}{(4\pi Dt)^{3/2}} \begin{bmatrix} k_2^2 + k_1 k_2 e^{-(k_1+k_2)t} & k_1 k_2 (1 - e^{-(k_1+k_2)t}) \\ k_1 k_2 (1 - e^{-(k_1+k_2)t}) & k_1^2 + k_1 k_2 e^{-(k_1+k_2)t} \end{bmatrix}. \tag{13.3.19}$$

If we define variables

$$n(\boldsymbol{r}, t) = \rho_x(\boldsymbol{r}, t) + \rho_y(\boldsymbol{r}, t), \tag{13.3.20}$$

$$c(\boldsymbol{r}, t) = [k_1 \rho_x(\boldsymbol{r}, t) - k_2 \rho_y(\boldsymbol{r}, t)]/(k_1 + k_2), \tag{13.3.21}$$

the solution (13.3.19) can be written as

$$\langle n(\boldsymbol{r}, t), n(\boldsymbol{0}, 0)\rangle_s = \langle n \rangle_s \frac{\exp(-r^2/4Dt)}{(4\pi Dt)^{3/2}}, \tag{13.3.22}$$

$$\langle n(\boldsymbol{r}, t), c(\boldsymbol{0}, 0)\rangle_s = \langle c(\boldsymbol{r}, t), n(\boldsymbol{0}, 0)\rangle_s = 0, \tag{13.3.23}$$

$$\langle c(\boldsymbol{r}, t), c(\boldsymbol{0}, 0)\rangle_s = \frac{k_1 k_2 \langle n \rangle_s}{(k_1 + k_2)^2} \frac{\exp(-r^2/4Dt)}{(4\pi Dt)^{3/2}} e^{-(k_1+k_2)t}. \tag{13.3.24}$$

The variables n and c correspond to total density and chemical imbalance density, i.e., $\langle c \rangle_s = 0$. Thus we see that (13.3.22) gives the correlation for density fluctuations which is the same as that arising from pure diffusion: (13.3.24) gives the correlation of fluctuations in chemical imbalance, and (13.3.23) shows that these are independent. A characteristic diffusion term multiplies all of these. The simplicity of this result depends on the identity of diffusion constants for the different species.

13.3.2 Reactions $B + X \underset{k_3}{\overset{k_1}{\rightleftharpoons}} C, \quad A + X \underset{k_2}{\longrightarrow} 2X$

This reaction has already been treated in Sect. 11.6.4 a without spatial dependence. We find the Poisson representation equation for the concentration variable $\eta(\boldsymbol{r}, t)$ is [from (13.2.66–13.2.68, 11.6.56, 11.6.58)]

$$d\eta(\boldsymbol{r},t) = \Big[D\nabla^2\eta(\boldsymbol{r},t) + (K_2 - K_1)\eta(\boldsymbol{r},t) + K_3\Big]\,dt + dW(\boldsymbol{r},t)\,, \tag{13.3.25}$$

in which

$$dW(\boldsymbol{r},t)\,dW(\boldsymbol{r}',t) = 2\delta(\boldsymbol{r}-\boldsymbol{r}')K_2\eta(\boldsymbol{r},t)\,dt\,, \tag{13.3.26}$$

$$k_3C = K_3l^3\,, \qquad k_1B = K_1\,, \qquad k_2A = K_2\,. \tag{13.3.27}$$

This system can be solved exactly since it is linear. However, since the second reaction *destroys* X at a rate proportional to X, we do obtain a noise term in the Poisson representation.

In the Kramers-Moyal method, we would find an equation of the same form as (13.3.25) for $p(\boldsymbol{r},t)$, but the $dW(\boldsymbol{r},t)$ would satisfy

$$dW(\boldsymbol{r},t)dW(\boldsymbol{r}',t) = \{2\nabla'\cdot\nabla\,[D\rho(\boldsymbol{r})\delta(\boldsymbol{r}-\boldsymbol{r}')] + \delta(\boldsymbol{r}-\boldsymbol{r}')\,[(K_2+K_1)\rho(\boldsymbol{r},t)+K_3]\}\;dt\,. \tag{13.3.28}$$

a) Spatial Correlations: Define now

$$\begin{aligned} g(\boldsymbol{r},\boldsymbol{r}',t) &= \langle\rho(\boldsymbol{r},t),\rho(\boldsymbol{r}',t)\rangle - \delta(\boldsymbol{r}-\boldsymbol{r}')\langle\rho(\boldsymbol{r},t)\rangle\,,\\ &= \langle\eta(\boldsymbol{r},t),\eta(\boldsymbol{r}',t)\rangle = \langle\eta(\boldsymbol{r},t)\eta(\boldsymbol{r}',t)\rangle - \langle\eta(\boldsymbol{r},t)\rangle\langle\eta(\boldsymbol{r}',t)\rangle\,. \end{aligned} \tag{13.3.29}$$

We consider the stationary homogeneous situation in which clearly

$$\langle\eta(\boldsymbol{r},t)\rangle_{\mathrm{s}} = \langle\rho(\boldsymbol{r},t)\rangle_{\mathrm{s}} = \langle\rho\rangle_{\mathrm{s}}\,. \tag{13.3.30}$$

Then,

$$\begin{aligned} dg(\boldsymbol{r},\boldsymbol{r}',t) &= d\langle\eta(\boldsymbol{r},t)\eta(\boldsymbol{r},t)\rangle\,,\\ &= \langle\eta(\boldsymbol{r},t)d\eta(\boldsymbol{r}',t)\rangle + \langle d\eta(\boldsymbol{r},t)\eta(\boldsymbol{r}',t)\rangle + \langle d\eta(\boldsymbol{r},t)d\eta(\boldsymbol{r}',t)\rangle\,, \end{aligned} \tag{13.3.31}$$

and using the usual Ito rules and (13.3.25, 13.3.26),

$$\begin{aligned} dg(\boldsymbol{r},\boldsymbol{r}',t) &= [D\nabla^2 + D\nabla'^2 + 2(K_2-K_1)]\langle\eta(\boldsymbol{r},t)\eta(\boldsymbol{r}',t)\rangle\\ &\quad +2K_3\langle\rho\rangle_{\mathrm{s}}dt + K_2\delta(\boldsymbol{r}-\boldsymbol{r}')\langle\rho\rangle_{\mathrm{s}}dt\,. \end{aligned} \tag{13.3.32}$$

Note that

$$\langle\rho\rangle_{\mathrm{s}} = K_3/(K_1-K_2)\,, \tag{13.3.33}$$

and that in a spatially homogeneous situation, $g(\boldsymbol{r},\boldsymbol{r}',t)$ is a function of $\boldsymbol{r}-\boldsymbol{r}'$ only, which we call $g(\boldsymbol{r}-\boldsymbol{r}',t)$.

Substitute using (13.3.29, 13.3.30) to obtain

$$\partial_t g(\boldsymbol{r},t) = 2[D\nabla^2 + (K_2-K_1)]g(\boldsymbol{r},t) + 2K_2\langle\rho\rangle_{\mathrm{s}}\delta(\boldsymbol{r}). \tag{13.3.34}$$

The stationary solution $g_{\mathrm{s}}(\boldsymbol{r})$ is best obtained by representing it as a Fourier integral:

$$g_{\mathrm{s}}(\boldsymbol{r}) = \textstyle\int d^3\boldsymbol{q}\,\mathrm{e}^{-\mathrm{i}\boldsymbol{q}\cdot\boldsymbol{r}}\tilde{g}_{\mathrm{s}}(\boldsymbol{q})\,, \tag{13.3.35}$$

which, on using

$$\delta(\boldsymbol{r}) = (2\pi)^{-3}\textstyle\int d^3\boldsymbol{q}\;\mathrm{e}^{-\mathrm{i}\boldsymbol{q}\cdot\boldsymbol{r}}\,, \tag{13.3.36}$$

gives

$$\tilde{g}_s(\boldsymbol{q}) = \frac{K_2\langle\rho\rangle_s}{(2\pi)^3}\frac{1}{Dq^2 + K_1 - K_2}, \tag{13.3.37}$$

whose Fourier inverse is given by

$$g_s(\boldsymbol{r}) = \frac{K_2\langle\rho\rangle_s}{4\pi D r}\exp\left[-r\left(\frac{K_1 - K_2}{D}\right)^{1/2}\right]. \tag{13.3.38}$$

Hence,

$$\langle\rho(\boldsymbol{r},t),\rho(\boldsymbol{r}',t)\rangle_s = \delta(\boldsymbol{r}-\boldsymbol{r}')\langle\rho\rangle_s + \frac{K_2\langle\rho\rangle_s}{4\pi D|\boldsymbol{r}-\boldsymbol{r}'|}\exp\left(-|\boldsymbol{r}-\boldsymbol{r}'|\left[\frac{K_1-K_2}{D}\right]^{1/2}\right). \tag{13.3.39}$$

Comments :

i) We note two distinct parts: a δ correlated part which corresponds to Poissonian fluctuations, independent at different space points, and added to this, a correlation term with a characteristic correlation length

$$l_c = \sqrt{D/(K_1 - K_2)}. \tag{13.3.40}$$

ii) *Approach to equilibrium*: when $K_2 \to 0$, we approach a simple reversible reaction $B{+}X \rightleftharpoons C$; one sees that the correlation in (13.3.38) becomes zero. However, the correlation length itself does *not* vanish.

b) Local and Global Fluctuations: A question raised originally by *Nicolis* [13.3] is the following. Consider the total number of molecules in a volume V:

$$x(V,t) = \int_V d^3\boldsymbol{r}\,\rho(\boldsymbol{r},t). \tag{13.3.41}$$

Then we would like to know what is the variance of this number. We can easily see

$$\mathrm{var}[x(V)]_s = \int_V d^3\boldsymbol{r}\int_V d^3\boldsymbol{r}'\langle\rho(\boldsymbol{r},t),\rho(\boldsymbol{r}',t)\rangle_s, \tag{13.3.42}$$

and using (13.3.39),

$$\mathrm{var}[x(V)]_s = \langle x(V)\rangle_s + \frac{K_2\langle\rho\rangle_s}{4\pi D}\int_V d^3\boldsymbol{r}\int_V d^3\boldsymbol{r}'\frac{\exp(-|\boldsymbol{r}-\boldsymbol{r}'|/l_c)}{|\boldsymbol{r}-\boldsymbol{r}'|}. \tag{13.3.43}$$

i) *Large volume limit* : If V is much larger than l_c^3, then, noting that $g(\boldsymbol{r}) \to 0$ as $\boldsymbol{r} \to \infty$, we integrate the stationary version of (13.3.34) and drop the surface terms arising from integrating the Laplacian by parts:

$$0 = 2(K_2 - K_1)\int_V d^3\boldsymbol{r}\int_V d^3\boldsymbol{r}' g(\boldsymbol{r}-\boldsymbol{r}') + 2K_2\langle\rho\rangle_s\int_V d^3\boldsymbol{r}'. \tag{13.3.44}$$

Thus,

$$\int_V\int_V d^3\boldsymbol{r}d^3\boldsymbol{r}' g(\boldsymbol{r}-\boldsymbol{r}') \sim \frac{K_2\langle x(V)\rangle_s}{K_1 - K_2}, \tag{13.3.45}$$

so that

$$\mathrm{var}[x(V)]_s \approx \frac{K_1\langle x(V)\rangle_s}{K_1 - K_2}. \qquad (V \gg l_c^3). \tag{13.3.46}$$

ii) *Small volume limit* : For a spherical volume of radius $R \ll l_c$, we can neglect the exponential in (13.3.43) and

$$\begin{aligned}\int_V\int_V d^3r d^3r' |r - r'|^{-1} &= \int_V d^3r' \int_0^R 2\pi r^2 dr \int_{-1}^{1} d(\cos\theta)(r^2 + r'^2 - 2rr'\cos\theta)^{-1/2},\\ &= \int_V d^3r' \int_0^R \frac{4\pi r}{r'} dr(|r + r'| - |r - r'|),\\ &= 2R^5(4\pi)^2/15,\end{aligned} \tag{13.3.47}$$

so that

$$\mathrm{var}[x(V)]_s \approx \langle x(V)\rangle_s \left(1 + \frac{2R^2K_2}{5D}\right), \qquad \left(V = \frac{4}{3}\pi R^3 \ll l_c^3\right). \tag{13.3.48}$$

Hence, we see in a small volume that the fluctuations are Poissonian, but in a large volume the fluctuations are non-Poissonian, and in fact the variance is exactly the same value, (11.6.74), as given by the global Master equation for the same reaction.

iii) *Arbitrary spherical volume* : The variance can be evaluated directly (by use of Fourier transform methods) and is given by

$$\mathrm{var}[x(V)] = \langle x(V)\rangle \left\{1 + \frac{3K_2 l_c^5}{2DR^3}\left[1 - \left(\frac{R}{l_c}\right)^2 + \frac{2}{3}\left(\frac{R}{l_c}\right)^3 - e^{-2R/l_c}\left(1 + \frac{R}{l_c}\right)^2\right]\right\}. \tag{13.3.49}$$

This gives the more precise large volume asymptotic form

$$\mathrm{var}[x(V)] \approx \langle x(V)\rangle \left(\frac{K_1}{K_1 - K_2} - \frac{3K_1 l_c}{2R(K_1 - K_2)}\right), \qquad (V \gg l_c^3). \tag{13.3.50}$$

iv) *Interpretation in Fourier space:* The result can also be illustrated by noting that the Fourier representation of $\langle \rho(\boldsymbol{r}, t), \rho(\boldsymbol{r}', t)\rangle$ is given by adding those of $g_s(\boldsymbol{r})$ and $\delta(\boldsymbol{r})\langle\rho\rangle_s$ and is clearly

$$\langle\rho\rangle_s(2\pi)^{-3}\frac{Dq^2 + K_1}{Dq^2 + K_1 - K_2}. \tag{13.3.51}$$

This means that for small $\boldsymbol{q}$, i.e., long wavelength, this is approximately

$$\langle\rho\rangle_s(2\pi)^{-3}\frac{K_1}{K_1 - K_2}, \tag{13.3.52}$$

which is the Fourier transform of

$$\langle\rho\rangle_s\frac{K_1}{K_1 - K_2}\delta(\boldsymbol{r} - \boldsymbol{r}'), \tag{13.3.53}$$

corresponding to the same variance as the global Master equation.

For large $\boldsymbol{q}$, i.e., small wavelengths, we find

$$\langle\rho\rangle_s(2\pi)^{-3}, \tag{13.3.54}$$

corresponding to

$$\langle\rho\rangle_s\delta(\boldsymbol{r}-\boldsymbol{r}')\,, \tag{13.3.55}$$

i.e., Poissonian fluctuations.

v) *Physical interpretation* : Physically, the difference arises from the different scaling of the diffusion and reaction parts of the master equation which was noted in Sect. 13.2.2. Thus, in a small volume the diffusion dominates, since the fluctuations arising from diffusion come about because the molecules are jumping back and forth across the boundary of V. This is a surface area effect which becomes relatively larger than the chemical reaction fluctuations, which arise from the bulk reaction, and are proportional to volume. Conversely, for larger V, we find the surface effect is negligible and only the bulk effect is important.

b) Space-Time Correlations: Since the system is linear, we can, as in Sect. 13.3.1, use the method of Sect. 3.7.4. Define

$$G(\boldsymbol{r},t) = \langle\rho(\boldsymbol{r},t),\rho(\boldsymbol{0},0)\rangle_s\,. \tag{13.3.56}$$

Then the linear equation corresponding to (3.7.60) is

$$\partial_t G(\boldsymbol{r},t) = D\nabla^2 G(\boldsymbol{r},t) - (K_1-K_2)G(\boldsymbol{r},t)\,, \tag{13.3.57}$$

with an initial condition given by

$$G(\boldsymbol{r},0) = \langle\rho(\boldsymbol{r}),\rho(\boldsymbol{0})\rangle_s\,, \tag{13.3.58}$$

which itself is given by (13.3.39). Representing $G(\boldsymbol{r},t)$ as a Fourier integral

$$\tilde{G}(\boldsymbol{r},t) = \int d^3\boldsymbol{q}\,\mathrm{e}^{-i\boldsymbol{q}\cdot\boldsymbol{r}}\tilde{G}(\boldsymbol{q},t)\,, \tag{13.3.59}$$

(13.3.57) becomes

$$\partial_t\tilde{G}(\boldsymbol{q},t) = -(Dq^2+K_1-K_2)\tilde{G}(\boldsymbol{q},t)\,, \tag{13.3.60}$$

whose solution is, utilising the Fourier representation (13.3.51) of the initial condition (13.3.58),

$$\tilde{G}(\boldsymbol{q},t) = (2\pi)^{-3}\frac{Dq^2+K_1}{Dq^2+K_1-K_2}\exp[-(Dq^2+K_1-K_1)t]\,. \tag{13.3.61}$$

If desired, this can be inverted by quite standard means though, in fact, the Fourier transform correlation function is often what is desired in practical cases and being usually easier to compute, it is favoured.

Thus, we have

$$\begin{aligned}G(\boldsymbol{r},t) = {} & \frac{\langle\rho\rangle\exp(-r^2/4Dt-Dt/l_c^2)}{(4\pi Dt)^{3/2}}\\ & +\frac{K_2\langle\rho\rangle}{8\pi Dr}\left\{[\exp(-r/l_c)]\mathrm{erfc}\left[\frac{(Dt)^{1/2}}{l_c}-\frac{r}{2(Dt)^{1/2}}\right]\right.\\ & \left.+\ [\exp(r/l_c)\mathrm{erfc}\left[\frac{(Dt)^{1/2}}{l_c}+\frac{r}{2(Dt)^{1/2}}\right]\right\}.\end{aligned} \tag{13.3.62}$$

For small t we find

$$G(r,t) \to \langle\rho\rangle \exp\left(\frac{-r^2}{4Dt}\right)\left[(4\pi Dt)^{-3/2} - \frac{4K_2(Dt)^{3/2}}{r^3\sqrt{\pi}}\right] + \frac{K_2\langle\rho\rangle e^{-r/l_c}}{4\pi Dr}, \tag{13.3.63}$$

while for large t,

$$G(r,t) \sim \left(\frac{\exp\left(-r^2/4Dt\right)}{(4\pi Dt)^{3/2}} + \frac{K_2 l_c}{4\pi r(\pi D^2 t)^{1/2}}\right)\langle\rho\rangle \exp(-Dt/l_c^2). \tag{13.3.64}$$

c) Behaviour at the Instability Point: As $K_1 \to K_2$, the reaction approaches instability, and when $K_1 = K_2$ there are no longer any stationary solutions. We see that simultaneously,

i) The correlation length $l_c \to \infty$ (13.3.40);

ii) The variance of the fluctuations in a volume $V \gg l_c^3$ will become infinite [(13.3.46)]. However, as $l_c \to \infty$ at finite V, one reaches a point at which $l_c^3 \sim V$ and (13.3.46) is no longer valid. Eventually $l_c^3 \gg V$; the volume now appears small and the fluctuations within it take the Poissonian form;

iii) The correlation time is best taken to be

$$\tau_c = l_c^2/D \tag{13.3.65}$$

being the coefficient of t in the exponent of the long-time behaviour of the space time correlation function (13.3.64) (ignoring the diffusive part). Thus, $\tau_c \to \infty$ also, near the instability. We thus have a picture of long-range correlated slow fluctuations. This behaviour is characteristic of many similar situations.

13.3.3 A Nonlinear Model with a Second-Order Phase Transition

We consider the reactions

$$A + X \underset{k_4}{\overset{k_2}{\rightleftharpoons}} 2X\,, \qquad B + X \underset{k_4}{\overset{k_1}{\rightleftharpoons}} C\,, \tag{13.3.66}$$

considered previously in Sects. 12.1 c, 12.2.2 a 12.2.3 c This model was first introduced by *Schlögl* [13.4] and has since been treated by many others.

The deterministic equation is

$$\partial_t \rho(\boldsymbol{r},t) = D\nabla^2\rho + \kappa_3 + (\kappa_2 - \kappa_1)\rho - \kappa_4\rho^2\,, \tag{13.3.67}$$

whose stationary solution is given by

$$\rho(\boldsymbol{r}) = \rho_s = [\kappa_2 - \kappa_1 + \sqrt{(\kappa_2-\kappa_1)^2 + 4\kappa_4\kappa_3}]/2\kappa_4\,. \tag{13.3.68}$$

[The κ's are as defined by (11.5.30), with $\Omega = l^3$]. For small κ_3, this stationary solution represents a transition behaviour as illustrated in Fig. 13.1.

The appropriate stochastic differential equation in the Poisson representation is

$$\begin{aligned} d\eta(\boldsymbol{r},t) &= \left[(\mathscr{D}\eta)(\boldsymbol{r},t) + \kappa_3 + (\kappa_2-\kappa_1)\eta(\boldsymbol{r},t) - \kappa_4\eta^2(\boldsymbol{r},t)\right]dt \\ &\quad + \sqrt{2}\left[\kappa_2\eta(\boldsymbol{r},t) - \kappa_4\eta^2(\boldsymbol{r},t)\right]^{1/2} dW(\boldsymbol{r},t)\,, \end{aligned} \tag{13.3.69}$$

where

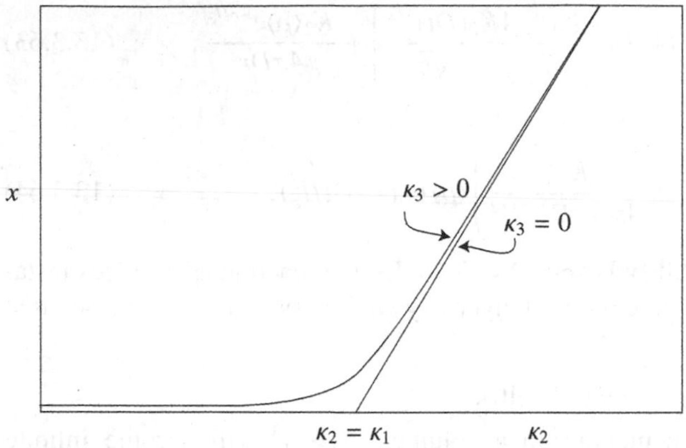

Fig. 13.1. Plot of $\langle x \rangle$ vs κ_2 for the second-order phase transition model

$$dW(\boldsymbol{r},t)dW(\boldsymbol{r}',t) = \delta(\boldsymbol{r}-\boldsymbol{r}')dt\,. \tag{13.3.70}$$

Here we have *not* taken the continuum limit developed in Sect. 13.2.1, but have preserved a possible nonlocality of diffusion, i.e.,

$$(\mathscr{D}\eta)(\boldsymbol{r},t) = -\int \mathscr{D}(|\boldsymbol{r}-\boldsymbol{r}'|)\eta(\boldsymbol{r}',t)d^3\boldsymbol{r}'\,, \tag{13.3.71}$$

but we have considered only homogeneous isotropic diffusion.

In this form, there is no small parameter multiplying the noise term $dW(\boldsymbol{r},t)$ and we are left without any obvious expansion parameter. We formally introduce a parameter λ in (13.3.69) as

$$\begin{aligned} d\eta(\boldsymbol{r},t) &= \left[(\mathscr{D}\eta)(\boldsymbol{r},t) + \kappa_3 + (\kappa_2-\kappa_1)\eta(\boldsymbol{r},t) - \kappa_4\eta^2(\boldsymbol{r},t)\right] dt \\ &+\lambda\sqrt{2}\left[\kappa_2\eta(\boldsymbol{r},t) - \kappa_4\eta^2(\boldsymbol{r},t)\right]^{1/2} dW(\boldsymbol{r},t)\,, \end{aligned} \tag{13.3.72}$$

and expand $\eta(\boldsymbol{r},t)$ in powers of λ to

$$\eta(\boldsymbol{r},t) = \eta_0(\boldsymbol{r},t) + \lambda\eta_1(\boldsymbol{r},t) + \lambda^2\eta_2(\boldsymbol{r},t) + \ldots \tag{13.3.73}$$

and set λ equal to one at the end of the calculation. However, if it is understood that all Fourier variable integrals have a cutoff l^{-1}, this will still be in fact a $(l^3)^{-1}$ expansion.

Substituting (13.3.73) in (13.3.69), we get

$$\frac{d\eta_0(\boldsymbol{r},t)}{dt} = (\mathscr{D}\eta_0)(\boldsymbol{r},t) + \kappa_3 + (\kappa_2 - \kappa_1)\eta_0(\boldsymbol{r},t) - \kappa_4\eta_0^2(\boldsymbol{r},t)\,, \tag{13.3.74}$$

$$d\eta_1(\boldsymbol{r},t) = \{(\mathscr{D}\eta_1)(\boldsymbol{r},t) + [\kappa_2 - \kappa_1 - 2\kappa_4\eta_0(\boldsymbol{r},t)]\,\eta_1(\boldsymbol{r},t)\}\,dt + \sqrt{2}\left[\kappa_2\eta_0(\boldsymbol{r},t) - \kappa_4\eta_0^2(\boldsymbol{r},t\right]^{1/2} dW(\boldsymbol{r},t)\,, \tag{13.3.75}$$

$$d\eta_2(\boldsymbol{r},t) = \left\{(\mathscr{D}\eta_2)(\boldsymbol{r},t) + [\kappa_2 - \kappa_1 - 2\kappa_4\eta_0(\boldsymbol{r},t)]\,\eta_2(\boldsymbol{r},t) - \kappa_4\eta_1^2(\boldsymbol{r},t)\right\} dt + \frac{[\kappa_2 - 2\kappa_4\eta_0(\boldsymbol{r},t)]\eta_1(\boldsymbol{r},t)}{\sqrt{2}[\kappa_2\eta_0(\boldsymbol{r},t) - \kappa_4\eta_0^2(\boldsymbol{r},t)]^{1/2}}dW(\boldsymbol{r},t)\,. \tag{13.3.76}$$

Equation (13.3.74) has a homogeneous steady-state solution (13.3.68) so that

$$\kappa_2 - \kappa_1 - 2\kappa_4\eta_0 \equiv \kappa = [(\kappa_1 - \kappa_2)^2 + 4\kappa_4\kappa_3]^{1/2}\,. \tag{13.3.77}$$

Substituting this in (13.3.74–13.3.76) and taking the Fourier transforms, we get

$$\eta_1(\boldsymbol{q},t) = (2\kappa_1\eta_0)^{1/2} \int_0^t e^{-[\mathscr{D}(q^2)+\kappa](t-t')}\, d\tilde{W}(\boldsymbol{q},t')\,, \tag{13.3.78}$$

$$\tilde{\eta}_2(\boldsymbol{q},t) = -\kappa_4 \int d^3\boldsymbol{q}_1 \int_0^t dt' e^{-[\mathscr{D}(q^2)+\kappa](t-t')}\, \tilde{\eta}_1(\boldsymbol{q}-\boldsymbol{q}_1,t')\tilde{\eta}_1(\boldsymbol{q}_1,t') + \frac{\kappa_1+\kappa}{[2(\kappa_1\eta_0-\kappa_3)]^{1/2}} \int d^3\boldsymbol{q}_1 \int_0^t e^{-[\mathscr{D}(q^2)+\kappa](t-t')}\, \tilde{\eta}_1(\boldsymbol{q}-\boldsymbol{q}_1,t')\, d\tilde{W}(\boldsymbol{q}_1,t)\,, \tag{13.3.79}$$

where

$$d\tilde{W}(\boldsymbol{q},t) = \left(\frac{1}{2\pi}\right)^{3/2} \int e^{-i\boldsymbol{q}\cdot\boldsymbol{r}}\, dW(\boldsymbol{r},t)d^3\boldsymbol{r}\,, \tag{13.3.80}$$

and

$$\tilde{\eta}_1(\boldsymbol{q},t) = \left(\frac{1}{2\pi}\right)^{3/2} \int e^{-i\boldsymbol{q}\cdot\boldsymbol{r}}\, \eta_1(\boldsymbol{r},t)d^3\boldsymbol{r}\,, \tag{13.3.81}$$

etc., and $\mathscr{D}(q^2)$ is the Fourier transform $\mathscr{D}(|\boldsymbol{r}-\boldsymbol{r}'|)$. We have left out the trivial initial value terms in (13.3.78, 13.3.79).

To the lowest order, the mean concentration and the correlation function are given by (in the stationary state)

$$\langle\rho(\boldsymbol{r},t)\rangle = \eta_0\,, \tag{13.3.82}$$

$$\langle\rho(\boldsymbol{r},t)\rho(\boldsymbol{r}',t)\rangle - \langle\rho(\boldsymbol{r},t)\rangle\langle\rho(\boldsymbol{r}',t)\rangle = \eta_0\delta(\boldsymbol{r}-\boldsymbol{r}') + \langle\eta_1(\boldsymbol{r},t)\eta_1(\boldsymbol{r}',t)\rangle\,. \tag{13.3.83}$$

From (13.3.78) it follows that

$$\langle\tilde{\eta}_1(\boldsymbol{q},t)\tilde{\eta}_1(\boldsymbol{q}',t)\rangle = \frac{\kappa_1\eta_0\delta(\boldsymbol{q}+\boldsymbol{q}')}{\mathscr{D}(q^2)+\kappa}[1 - e^{-2[\mathscr{D}(q^2)+\kappa]t}]\,. \tag{13.3.84}$$

Hence, the lowest-order contributions to the correlation function in the steady state are given by

$$\langle \rho(\boldsymbol{r})\rho(\boldsymbol{r}') - \langle\rho(\boldsymbol{r})\rangle\langle\rho(\boldsymbol{r}')\rangle\rangle = \eta_0\delta(\boldsymbol{r}-\boldsymbol{r}') + \frac{\kappa_1\eta_0}{(2\pi)^3}\int d^3\boldsymbol{q}\frac{e^{i\boldsymbol{q}\cdot(\boldsymbol{r}-\boldsymbol{r}')}}{\mathscr{D}(q^2)+\kappa}\,. \tag{13.3.85}$$

If we assume that

$$(\mathscr{D}\eta)(\boldsymbol{r}) = D\nabla^2\eta(\boldsymbol{r})\,, \tag{13.3.86}$$

then

$$\mathscr{D}(q^2) = Dq^2\,. \tag{13.3.87}$$

Equation (13.3.85) then has exactly the same form as derived previously in (13.3.39), namely,

$$\langle\rho(\boldsymbol{r}),\rho(\boldsymbol{r}')\rangle_s = \langle\rho\rangle_0\delta(\boldsymbol{r}-\boldsymbol{r}') + \frac{\kappa_1\langle\rho_0\rangle}{4\pi D|\boldsymbol{r}-\boldsymbol{r}'|}\exp(-|\boldsymbol{r}-\boldsymbol{r}'|/l_c)\,, \tag{13.3.88}$$

where

$$l_c = \sqrt{D/\kappa}\,. \tag{13.3.89}$$

All the same conclusions will follow concerning local and global fluctuations, Poissonian and non-Poissonian fluctuations, since these are all consequences of the form (13.3.88) and not of the particular values of the parameters chosen.

Notice that if $\kappa_1 = \kappa_2$, as $\kappa_3 \to 0$ so does $\kappa \to 0$, and hence $l_c \to \infty$, and the same long-range correlation phenomena occur here as in that example.

a) Higher-Order Corrections—Divergence Problems: These are only lowest-order results, which can be obtained equally well by a system size expansion, Kramers-Moyal method, or by factorisation assumptions on correlation function equations.

The next order correction to the mean concentration is $\langle\eta_2(\boldsymbol{r},t)\rangle$. Now from (13.3.79)

$$\langle\tilde{\eta}_2(\boldsymbol{q},t)\rangle = -\int d^3\boldsymbol{q}_1\int_0^t dt' e^{-[\mathscr{D}(q^2)+\kappa](t-t')}\langle\eta_1(\boldsymbol{q}-\boldsymbol{q}_1,t')\eta_1(\boldsymbol{q}_1,t')\rangle\,. \tag{13.3.90}$$

In the steady state (13.3.90) gives

$$\langle\tilde{\eta}_2(\boldsymbol{q})\rangle = -\kappa_4\kappa_1\eta_0\delta(\boldsymbol{q})\int\frac{d^3\boldsymbol{q}_1}{\mathscr{D}(q^2)+\kappa}\,. \tag{13.3.91}$$

For the choice of $\mathscr{D}$ given by (13.3.87), equation (13.3.91) gives

$$\langle\eta_2\rangle = \frac{\kappa_4\kappa_1\eta_0}{(2\pi)^{3/2}}\int\frac{d^3\boldsymbol{q}_1}{Dq^2+\kappa}\,, \tag{13.3.92}$$

and this integral is divergent. The problem lies in the continuum form chosen, for if one preserves the discrete cells at all stages, one has two modifications:

i) $\int d^3\boldsymbol{q}$ is a sum over all allowed $\boldsymbol{q}$ which have a maximum value $|\boldsymbol{q}| \sim 1/l$.

ii) $\mathscr{D}(q^2)$ is some trigonometric function of q and l. If

$$d_{ij} = \begin{cases} 0, & i \text{ not adjacent to } j\,,\\ d, & \text{otherwise}\,, \end{cases} \tag{13.3.93}$$

then $\mathscr{D}(q^2)$ has the form

$$\mathscr{D}(q^2) = \frac{4D}{l^2}\left[\sin^2\left(\frac{lq_x}{2}\right) + \sin^2\left(\frac{lq_y}{2}\right) + \sin^2\left(\frac{lq_z}{2}\right)\right]. \tag{13.3.94}$$

iii) In this form, no divergence arises since the sum is finite, and for $l|\boldsymbol{q}| \ll 1$,

$$\mathscr{D}(q^2) \to Dq^2. \tag{13.3.95}$$

b) Nonlocal Reactions: The divergence, can also be avoided by preserving the continuum but modifying the $\eta^2(\boldsymbol{r})$ term in the equations. This term represents a collision picture of chemical reactions and will be nonlocal for a variety of reasons.

i) Molecules have a finite size, so a *strictly* local form ignores this.

ii) The time scale on which the system is Markovian is such that molecules will have travelled a certain distance in this time; the products would be the result of encounters at a variety of previous positions, and would be produced at yet another position.

iii) By making the replacement

$$\eta(\boldsymbol{r})^2 \to \int d^3\boldsymbol{r}' d^3\boldsymbol{r}'' m(\boldsymbol{r}-\boldsymbol{r},' - \boldsymbol{r}'')\eta(\boldsymbol{r}')\eta(\boldsymbol{r}''), \tag{13.3.96}$$

one finds that instead of (13.3.92) one has

$$\langle\eta_2\rangle = -\frac{\kappa_4\kappa_1\eta_0}{(2\pi)^{3/2}}\int\frac{d^3\boldsymbol{q}\,\tilde{m}(\boldsymbol{q},-\boldsymbol{q})}{Dq^2+\kappa}. \tag{13.3.97}$$

Here $\tilde{m}(\boldsymbol{q},\boldsymbol{q}')$ is the Fourier transform of $m(\boldsymbol{r},\boldsymbol{r}')$. If m is sufficiently nonlocal, at high $\boldsymbol{q}$ it follows that $\tilde{m}(\boldsymbol{q},-\boldsymbol{q})$ will decay and $\langle\eta_2\rangle$ will be finite.

13.4 Connection Between Local and Global Descriptions

We saw in Sect. 13.3.2 that the variance of the fluctuations in a volume element V is Poissonian for small V and for sufficiently large V, approaches that corresponding to the master equation without diffusion for the corresponding reaction. This arises because the reactions give rise to fluctuations which add to each other whereas diffusion, as a mere transfer of molecules from one place to another, has an effect on the fluctuations in a volume which is effective only on the surface of the volume.

There is a precise theorem which expresses the fact that if diffusion is fast enough, the molecules in a volume V will travel around rapidly and meet each other very frequently. Hence, any molecule will be equally likely to meet any other molecule. The results are summarised by *Arnold* [13.5] and have been proved quite rigorously.

Basically, the method of proof depends on adiabatic elimination techniques as developed in Chap. 8 and Chap. 9 and can be easily demonstrated by Poisson representation methods.

13.4.1 Explicit Adiabatic Elimination of Inhomogeneous Modes

We suppose we are dealing with a single chemical species which diffuses and reacts according to the cell model and thus has a Poisson representation Fokker-Planck equation:

$$\frac{\partial P}{\partial t} = \left\{ \sum_{i,j} \left[-\frac{\partial}{\partial x_i} D_{ij} x_j \right] + \sum_i \left[\frac{\partial}{\partial x_i} a(x_i) + \frac{\partial^2}{\partial x_i^2} b(x_i) \right] \right\} P \,. \tag{13.4.1}$$

We introduce the eigenvectors $f_i(\boldsymbol{q})$ of D_{ij}, which satisfy

$$\sum_j D_{ij} f_j(\boldsymbol{q}) = -D\lambda(\boldsymbol{q}) f_i(\boldsymbol{q}) \,. \tag{13.4.2}$$

The precise nature of the eigenfunctions is not very important and will depend on D_{ij}. For the simplest form, with transfer only between adjacent cells and with reflecting boundaries at the walls of the total system, assumed to be one dimensional and of length $L = Nl$ (with l the cell length), one has

$$f_j(q) \propto \cos(qlj) \,. \tag{13.4.3}$$

The reflecting boundary condition requires that q has the form

$$q = \frac{n\pi}{Nl} \quad (n = 0, 1, \ldots, N) \,, \tag{13.4.4}$$

and

$$\lambda(q) = \frac{4 \sin^2\left(\frac{1}{2} ql\right)}{l^2} \,. \tag{13.4.5}$$

Appropriate modifications must be made to take care of more dimensions, but the basic structure is the same, namely,

$$\lambda(\mathbf{0}) = 0 \,, \qquad \lambda(\boldsymbol{q}) > 0 \quad (\boldsymbol{q} \neq 0) \,, \tag{13.4.6}$$

and

$$f_i(\mathbf{0}) = \text{constant} = N^{-1/2} \,. \tag{13.4.7}$$

This last result represents the homogeneous stationary state of diffusion with the normalisation $N^{1/2}$ fixed by the completeness and orthogonality relations

$$\left. \begin{aligned} &\sum_j f_j(q) f_j(q') = \delta_{q,q'} \,, \\ &\sum_q f_j(q) f_j(q) = \delta_{i,j} \,. \end{aligned} \right\} \tag{13.4.8}$$

We now introduce the variables

$$x(q) = \sum_j f_j(q) x_j \,. \tag{13.4.9}$$

The variable of interest is proportional to $x(0)$, since

$$x(0) = N^{-1/2} \sum_j x_j \,, \tag{13.4.10}$$

is proportional to the total amount of substance in the system. The other variables are to be adiabatically eliminated. In anticipation of an appropriate choice of operators L_1, L_2, L_3, we define

$$y(q) = x(q)\sqrt{D}, \qquad (q \neq 0), \tag{13.4.11}$$

$$x = \frac{x(0)}{\sqrt{N}} = \frac{1}{N}\sum_j x_j. \tag{13.4.12}$$

The terms corresponding to spatial diffusion in the Fokker-Planck equation can now be written

$$\sum_{i,j}\frac{\partial}{\partial x_i}D_{ij}x_j = -D\sum_q \lambda(q)\frac{\partial}{\partial y(q)}y(q). \tag{13.4.13}$$

Define now

$$v_i = \sum_{q\neq 0} f_i(q)y(q), \tag{13.4.14}$$

then

$$\begin{aligned}\sum_i \frac{\partial}{\partial x_i}a(x_i) &= \frac{1}{N}\sum_i \frac{\partial}{\partial x}a\left(x+\frac{1}{\sqrt{D}}v_i\right)\\ &+\sqrt{D}\sum_{q\neq 0}\frac{\partial}{\partial y(q)}\sum_i f_i(q)a\left(x+\frac{1}{\sqrt{D}}v_i\right),\end{aligned} \tag{13.4.15}$$

$$\begin{aligned}\sum_i \frac{\partial^2}{\partial x_i^2}b(x_i) &= \frac{1}{N^2}\sum_i \frac{\partial^2}{\partial x^2}b\left(x+\frac{1}{\sqrt{D}}v_i\right)\\ &+\frac{1}{N}\sqrt{D}\frac{\partial}{\partial x}\sum_{\substack{i\\ q\neq 0}}\frac{\partial}{\partial y(q)}f_i(q)b\left(x+\frac{1}{\sqrt{D}}v_i\right)\\ &+D\sum_{\substack{i\\ q,q'\neq 0}}\frac{\partial^2}{\partial y(q)\partial y(q')}f_i(q)f_i(q')b\left(x+\frac{1}{\sqrt{D}}v_i\right).\end{aligned} \tag{13.4.16}$$

We now write this in decreasing powers of $\sqrt{D}$. Thus, define L_1 by

$$DL_1 = D\sum_{q\neq 0}\left[-\lambda(q)\frac{\partial}{\partial y(q)}y(q)+\frac{\partial^2}{\partial y(q)^2}b(x)\right], \tag{13.4.17}$$

which is the coefficient of D in an expansion of the terms above (lower-order terms are absorbed in L_2). We also set

$$L_3 = \left\langle \sum_i \left[\frac{1}{N}\frac{\partial}{\partial x}a\left(x+\frac{1}{\sqrt{D}}v_i\right)+\frac{1}{N^2}\frac{\partial^2}{\partial x^2}b\left(x+\frac{1}{\sqrt{D}}v_i\right)\right]\right\rangle, \tag{13.4.18}$$

where the average is over the stationary distribution of L_1. As D becomes large,

$$L_3 = \left[\frac{\partial}{\partial x}a(x)+\frac{1}{N}\frac{\partial^2}{\partial x^2}b(x)\right]+O\left(\frac{1}{\sqrt{D}}\right). \tag{13.4.19}$$

We define L_2 to order $\sqrt{D}$, which can be computed to involve only terms in which $\partial/\partial y(q)$ stands to the left. Thus, in carrying out an adiabatic elimination of the $y(q)$

variables as $D \to \infty$ to lowest order, there will be no $PL_2L_1^{-1}L_2$ contribution and we ultimately find the Fokker-Planck equation

$$\frac{\partial P}{\partial t} = \left[\frac{\partial}{\partial x}a(x) + \frac{1}{N}\frac{\partial^2}{\partial x^2}b(x)\right]P, \tag{13.4.20}$$

which corresponds to the global Master equation, since x is by (13.4.12) a concentration. Notice that the condition for validity of the global master equation will be

$$D\lambda(1) \gg K, \tag{13.4.21}$$

where K represents a typical rate of the chemical reaction. Alternatively, noticing that $\lambda(1) \approx (\pi/Nl^2)$, this can be written

$$D \gg \frac{KN^2l^2}{\pi^2}, \tag{13.4.22}$$

or

$$\sqrt{\frac{D}{K}} \gg \frac{L}{\pi}. \tag{13.4.23}$$

The left-hand side, by definition of diffusion, represents the root mean square distance travelled in the time scale of the chemical reaction, and the inequality says this must be very much bigger than the length of the system. Thus, diffusion must be able to homogenise the system. In [13.6] this result has been extended to include homogenisation on an arbitrary scale.

13.5 Phase-Space Master Equation

The remainder of this chapter deals with a stochastic version of kinetic theory and draws on the works of *van Kampen* [13.7] and *van den Broek* et al. [13.8].

We consider a gas in which molecules are described in terms of their positions $\boldsymbol{r}$ and velocities $\boldsymbol{v}$. The molecules move about and collide with each other. When a collision between two molecules occurs, their velocities are altered by this collision but their positions are not changed. However, between the collisions, the particles move freely with constant velocity.

Thus, there are two processes—collision and flow. Each of these can be handled quite easily in the absence of the other, but combining them is not straightforward.

13.5.1 Treatment of Flow

We suppose there is a large number of particles of mass m with coordinates $\boldsymbol{r}_n$ and velocities $\boldsymbol{v}_n$, which do not collide but move under the influence of an *external* force field $m\boldsymbol{A}(\boldsymbol{r})$. The equations of motion are

$$\dot{\boldsymbol{r}}_n = \boldsymbol{v}_n, \qquad \dot{\boldsymbol{v}}_n = \boldsymbol{A}(\boldsymbol{r}_n). \tag{13.5.1}$$

Then a phase-space density can be defined as

$$f(\boldsymbol{r},\boldsymbol{v},t) = \sum_n \delta(\boldsymbol{r}-\boldsymbol{r}_n(t))\delta(\boldsymbol{v}-\boldsymbol{v}_n(t)), \tag{13.5.2}$$

so that

$$\partial_t f(\boldsymbol{r},\boldsymbol{v},t) = \sum_n [\dot{\boldsymbol{r}}_n \cdot \nabla\delta(\mathbf{r}-\boldsymbol{r}_n(t))\delta(\boldsymbol{v}-\boldsymbol{v}_n(t)) + \delta(\boldsymbol{r}-\boldsymbol{r}_n(t))\,\boldsymbol{v}_n \cdot \nabla_v\delta(\boldsymbol{v}-\boldsymbol{v}_n(t))], \tag{13.5.3}$$

and using the properties of the delta function and the equations of motion (13.5.1), we get

$$\left[\partial_t + \boldsymbol{v}\cdot\nabla + \boldsymbol{A}(\boldsymbol{r})\cdot\nabla_v\right] f(\boldsymbol{r},\boldsymbol{v},t) = 0. \tag{13.5.4}$$

This is a deterministic flow equation for a phase-space density. If the particles are distributed according to some initial probability distribution in position and velocity, (13.5.4) is unaltered but $f(\boldsymbol{r},\boldsymbol{v},t)$ is to be interpreted as the average of $f(\boldsymbol{r},\boldsymbol{v},t)$ as defined in (13.5.2) over the initial positions and velocities.

Equation (13.5.4) is exact. The variable $f(\boldsymbol{r},\boldsymbol{v},t)$ can be regarded as a random variable whose time development is given by this equation, which can be regarded as a stochastic differential equation with zero noise. The number of particles in a phase cell, i.e., a six-dimensional volume element of volume Δ_i centred on the phase-space coordinate $(\boldsymbol{r}_i,\boldsymbol{v}_i)$ is, of course,

$$X(\boldsymbol{r}_i,\boldsymbol{v}_i) = \int_{\Delta_i} d^3\boldsymbol{r}\, d^3\boldsymbol{v}\, f(\boldsymbol{r},\boldsymbol{v}). \tag{13.5.5}$$

13.5.2 Flow as a Birth-Death Process

For the purpose of compatibility with collisions, represented by a birth-death Master equation, it would be very helpful to be able to represent flow as a birth-death process in the cells Δ_i. This cannot be done for arbitrary cell size but, in the limit of vanishingly small cell size, there is a birth-death representation of any flow process.

a) Formulation of Flow as a Limit of a Master Equation: Consider a density $\rho(r,t)$ in a one-dimensional system which obeys the flow equation

$$\partial_t\rho(r,t) = -\kappa\partial_r\rho(r,t). \tag{13.5.6}$$

This deterministic equation is the limit of a discrete equation for x_i, defined as

$$x_i(t) = \int_{\Delta_i} dr\, \rho(\boldsymbol{r},t) \simeq \lambda\rho(\boldsymbol{r}_i,t), \tag{13.5.7}$$

where Δ_i is an interval of length λ around a cell point at r_i. The flow equation is then the limit as $\lambda \to 0$ of

$$\partial_t x_i(t) = \frac{\kappa}{\lambda}[x_{i-1}(t) - x_i(t)], \tag{13.5.8}$$

i.e.,

$$\lambda\partial_t\rho(r_i,t) = \frac{\kappa}{\lambda}[\lambda\rho(r_i-\lambda,t) - \lambda\rho(r_i,t)], \tag{13.5.9}$$

whose limit as $\lambda \to 0$ is the flow equation (13.5.6). A stochastic version is found by considering particles jumping from cell i to cell $i+1$ with transition probabilities per unit time:

$$\left.\begin{aligned} t^{+}(\boldsymbol{x}) &= \kappa x_i/\lambda\,, \\ t^{-}(\boldsymbol{x}) &= 0\,. \end{aligned}\right\} \tag{13.5.10}$$

This is of the form studied in Sect. 11.5 , whose notation (11.5.4, 11.5.5) takes the form here of

$$\left.\begin{aligned} A &\to i\,, \\ N^i_j &\to \delta_{i,j}\,, \\ M^i_j &\to \delta_{i+1,j}\,, \\ r^i_j &\to \delta_{j,i+1} - \delta_{j,i}\,. \end{aligned}\right\} \tag{13.5.11}$$

We consider the Kramers-Moyal expansion and show that in the limit $\lambda \to 0$, all derivate moments except those of first order vanish. The drift coefficient is given by (11.5.33), i.e.,

$$A_a(\boldsymbol{x}) = \sum_i(\delta_{a,i+1} - \delta_{a,i})\frac{\kappa x_i}{\lambda} = \frac{\kappa}{\lambda}(x_{a-1} - x_a)\,, \tag{13.5.12}$$

and the diffusion matrix is given by (11.5.34)

$$B_{a,b}(\boldsymbol{x}) = \frac{\kappa}{\lambda}\Big[(\delta_{a,b} - \delta_{a-1,b})x_{a-1} + (\delta_{a,b} - \delta_{a,b-1})x_a)\Big]\,. \tag{13.5.13}$$

We now set $x_i = \lambda\rho(r_i)$ as in (13.5.7) and take a small λ limit; one finds

$$A_a(\boldsymbol{x}) \to \frac{\kappa}{\lambda}[\lambda\rho(r_a - \lambda, t) - \lambda\rho(r_a, t)]\,, \tag{13.5.14}$$

$$\to -\lambda\kappa\partial_r\rho(r,t)\,, \tag{13.5.15}$$

and the limiting value of $B_{a,b}(\boldsymbol{x})$ is found similarly, but also using

$$\delta_{a,b} \to \lambda\,\delta(r_a - r_b)\,, \tag{13.5.16}$$

to be

$$B_{a,b}(\boldsymbol{x}) \to \kappa\lambda^3\partial_r\partial_{r'}[\delta(r - r')\rho(r)]\,. \tag{13.5.17}$$

Thus, in this limit, $p(r,t)$ obeys a stochastic differential equation

$$d\rho(r,t) = -\kappa\partial_r\rho(r,t)\,dt + \lambda^{1/2}\,dW(r,t)\,, \tag{13.5.18}$$

where

$$dW(r,t)\,dW(r',t) = \kappa\,dt\,\partial_r\partial_{r'}[\delta(r - r')\rho(r,t)]\,. \tag{13.5.19}$$

We see that in the limit $\lambda \to 0$, the noise term in (13.5.18) vanishes, leading to the deterministic result as predicted.

b) Nature of the Deterministic Limit: It is interesting to ask why this deterministic limit occurs. It is not a system size expansion in the usual sense of the word, since neither the numbers of particles x_i nor the transition probabilities $t^{+}(\boldsymbol{x})$ become large in the small λ limit. However, the transition probability for a single particle at r_i to jump to the next cell does become infinite in the small λ limit, so in this sense, the motion becomes deterministic.

The reader can also check that this result is independent of dimensionality of space. Thus, we can find a representation of flow in phase space which, in the limit of small cell size, does become equivalent to the flow equation (13.5.4).

c) **Phase-Space Formulation:** Let us now consider a full phase-space formulation, including both terms of the flow equation. The cells in phase space are taken to be six-dimensional boxes with side length λ in position coordinates and ξ in velocity coordinates. We define the phase-space density in terms of the total number of molecules X in a phase cell by

$$f(\boldsymbol{r}^q, \boldsymbol{v}^q) = X(\boldsymbol{r}^q, \boldsymbol{v}^q)/\xi^3\lambda^3 . \tag{13.5.20}$$

We consider transitions of two types. For simplicity, consider these only in the x-direction and define

$$\boldsymbol{\lambda}_x = (\lambda, 0, 0) . \tag{13.5.21}$$

i) *Flow in position space* : This is described by the processes:

$$X(\boldsymbol{r}^i, \boldsymbol{v}^i) \to X(\boldsymbol{r}^i, \boldsymbol{v}^i) - 1 , \tag{13.5.22}$$

and either

$$X(\boldsymbol{r}^i + \boldsymbol{\lambda}_x \boldsymbol{v}^i) \to X(\boldsymbol{r}^i + \boldsymbol{\lambda}_x, \boldsymbol{v}^i) + 1 , \qquad (v_x > 0) \tag{13.5.23}$$

or

$$X(\boldsymbol{r}^i - \boldsymbol{\lambda}_x, \boldsymbol{v}^i) \to X(\boldsymbol{r}^i - \boldsymbol{\lambda}_x, \boldsymbol{v}^i) + 1 , \qquad (v_x < 0) . \tag{13.5.24}$$

Then we make the labelling $A \to (i, x, 1)$:

$$\left.\begin{aligned} N_a^{(i,x,1)} &= \delta(\boldsymbol{r}^i, \boldsymbol{r}^a)\,\delta(\boldsymbol{v}^i, \boldsymbol{v}^a) , \\ M_a^{(i,x,1)} &= \delta(\boldsymbol{r}^i - \boldsymbol{\lambda}_x, \boldsymbol{r}^a)\,\delta(\boldsymbol{v}^i, \boldsymbol{v}^a) , \\ r_a^{(i,x,1)} &= M_a^{(i,x,1)} - N_a^{(i,x,1)} , \end{aligned}\right\} \tag{13.5.25}$$

$$\left.\begin{aligned} t^+_{(i,x,1)}(\boldsymbol{X}) &= \lambda^2 |v_x| X(\boldsymbol{r}^i, \boldsymbol{v}^i)/\lambda^3 , \\ t^-_{(i,x,1)}(\boldsymbol{X}) &= 0 . \end{aligned}\right\} \tag{13.5.26}$$

The form (13.5.26) is written to indicate explicitly that the transition probability is the product of the end surface area of the cell (λ^2), the number of particles per unit volume, and the x component of velocity, which is an approximation to the rate of flow across this cell face.

Consequently, assuming $v_x > 0$,

$$A_a^{(x,1)} = v_x[X(\boldsymbol{r}^a + \boldsymbol{\lambda}_x, \boldsymbol{v}^a) - X(\boldsymbol{r}^a, \boldsymbol{v}^a)]/\lambda , \tag{13.5.27}$$

$$= \xi^3\lambda^3 v_x[f(\boldsymbol{r}^a + \boldsymbol{\lambda}_x, \boldsymbol{v}^a) - f(\boldsymbol{r}^a, \boldsymbol{v}^a)]/\lambda] , \tag{13.5.28}$$

$$\to \xi^3\lambda^3 v_x \frac{\partial}{\partial x} f(\boldsymbol{r}, \boldsymbol{v}) . \tag{13.5.29}$$

In a similar way to the previous example, the diffusion matrix $B_{a,b}$ can be shown to be proportional to λ and hence vanishes as $\lambda \to 0$.

ii) *Flow in velocity space* : Define

$$\boldsymbol{\xi}_x = (\xi, 0, 0). \tag{13.5.30}$$

Then we have

$$X(\boldsymbol{r}^i, \boldsymbol{v}^i) \to X(\boldsymbol{r}^i, \boldsymbol{v}^i) - 1, \tag{13.5.31}$$

and either

$$X(\boldsymbol{r}^i, \boldsymbol{v}^i + \boldsymbol{\xi}_x) \to X(\boldsymbol{r}^i, \boldsymbol{v}^i + \boldsymbol{\xi}_x) + 1, \qquad (A_x > 0) \tag{13.5.32}$$

or

$$X(\boldsymbol{r}^i, \boldsymbol{v}^i - \boldsymbol{\xi}_x) \to X(\boldsymbol{r}^i, \boldsymbol{v}^i - \boldsymbol{\xi}_x) + 1, \qquad (A_x < 0). \tag{13.5.33}$$

The labelling is

$$\left.\begin{aligned}
A \quad & \to (i, x, 2), \\
N_a^{(i,x,2)} &= \delta(\boldsymbol{v}^i, \boldsymbol{v}^a)\delta(\boldsymbol{r}^i, \boldsymbol{r}^a), \\
M_a^{(i,x,2)} &= \delta(\boldsymbol{v}^i - \boldsymbol{\xi}_x, \boldsymbol{v}^a)\delta(\boldsymbol{r}^i, \boldsymbol{r}^a), \\
r_a^{(i,x,2)} &= [\delta(\boldsymbol{v}^i - \boldsymbol{\xi}_x, \boldsymbol{v}^a) - \delta(\boldsymbol{v}^i, \boldsymbol{v}^a)]\delta(\boldsymbol{r}^i, \boldsymbol{r}^a), \\
t^+_{(i,x,2)}(\boldsymbol{X}) &= \xi^2 |A_x(\boldsymbol{r}^i)| X(\boldsymbol{r}^i, \boldsymbol{v}^i)/\xi^3, \\
t^-_{(i,x,2)}(\boldsymbol{X}) &= 0.
\end{aligned}\right\} \tag{13.5.34}$$

Consequently, assuming $A_x(\boldsymbol{r}^a) > 0$, the drift coefficient is

$$\begin{aligned}
A_a &= [X(\boldsymbol{r}^a, \boldsymbol{v}^a + \boldsymbol{\xi}_x)A_x(\boldsymbol{r}^a) - X(\boldsymbol{r}^a, \boldsymbol{v}^a)A_x(\boldsymbol{r}^a)]/\xi, & (13.5.35)\\
&= \xi^3\lambda^3 A_x(\boldsymbol{r}^a)[f(\mathbf{r}^a, \boldsymbol{v}^a + \boldsymbol{\xi}_x) - f(\boldsymbol{r}^a, \boldsymbol{v}^a)]/\xi, & (13.5.36)\\
&\to \xi^3\lambda^3 A_x(\boldsymbol{r})\frac{\partial}{\partial v_x} f(\boldsymbol{r}, \boldsymbol{v}). & (13.5.37)
\end{aligned}$$

Again, similar reasoning shows the diffusion coefficient vanishes as $\xi \to 0$. Putting together (13.5.29, 13.5.37), one obtains the appropriate flow equation (13.5.4) in the $\lambda, \xi \to 0$ limit.

13.5.3 Inclusion of Collisions—the Boltzmann Master Equation

We consider firstly particles in velocity space only and divide the velocity space into cells as before. Let $X(\boldsymbol{v})$ be the number of molecules with velocity $\boldsymbol{v}$ (where the velocity is, of course, discretised).

A collision is then, represented by a "chemical reaction",

$$X(\boldsymbol{v}_i) + X(\boldsymbol{v}_j) \to X(\boldsymbol{v}_k) + X(\boldsymbol{v}_l). \tag{13.5.38}$$

The collision is labelled by the index (i, j, k, l) and we have (using the notation of Sect. 11.5)

$$\left.\begin{aligned}
N_a^{ij,kl} &= \delta_{a,i} + \delta_{a,j}, \\
M_a^{ij,kl} &= \delta_{a,k} + \delta_{a,l}, \\
r_a^{ij,kl} &= -\delta_{a,i} - \delta_{a,j} + \delta_{a,k} + \delta_{a,l},
\end{aligned}\right\} \tag{13.5.39}$$

and the transition probabilities per unit time are taken in the form

$$t^+_{ij,kl}(X) = R(ij,kl)\,X(\boldsymbol{v}_i)\,X(\boldsymbol{v}_j)\,, \qquad t^-_{ij,kl}(X) = 0\,. \tag{13.5.40}$$

There are five collisional invariants, that is, quantities conserved during a collision, which arise from dynamics. These are:

i) The number of molecules—there are two on each side of (13.5.38);

ii) The three components of momentum: Since all molecules here are the same, this means in a collision

$$\boldsymbol{v}_i + \boldsymbol{v}_j = \boldsymbol{v}_k + \boldsymbol{v}_l\,; \tag{13.5.41}$$

iii) The total energy: this means that

$$\boldsymbol{v}_i^2 + \boldsymbol{v}_j^2 = \boldsymbol{v}_k^2 + \boldsymbol{v}_l^2\,. \tag{13.5.42}$$

The quantities $\boldsymbol{v}$, and $\boldsymbol{v}^2$ are known as additive invariants and it can be shown that any function which is similarly additively conserved is a linear function of them (with a possible constant term) [13.9]

In all molecular collisions we have time reversal symmetry, which in this case implies

$$R(ij,kl) = R(kl,ij)\,. \tag{13.5.43}$$

Finally, because of the identity of all particles, we have

$$R(ij,kl) = R(ji,kl) = R(ij,lk), \text{ etc.} \tag{13.5.44}$$

We now have a variety of possible approaches. It is impractical to work directly with the master equation, but both the system size expansion and the Poisson representation give realistic methods of attack.

a) System Size Expansion: Assume ξ^3, the volume in phase space of the phase cells is large, and we can write a Fokker-Planck equation using the Kramers-Moyal expansion. From (11.5.33), we can write a Fokker-Planck equation with a drift term

$$A_a(X) = \sum_{i,j,k,l} (-\delta_{a,i} - \delta_{a,j} + \delta_{a,k} + \delta_{a,l})X(\boldsymbol{v}_i)X(\boldsymbol{v}_j)R(ij,kl)\,, \tag{13.5.45}$$

and, utilising all available symmetries, we can write

$$A_a(X) = 2\sum_{j,k,l} R(aj,kl)[X(\boldsymbol{v}_k)X(\boldsymbol{v}_l) - X(\boldsymbol{v}_a)X(\boldsymbol{v}_j)]\,. \tag{13.5.46}$$

The diffusion coefficient can also be deduced using (11.5.34) as

$$B_{ab}(X) = \sum_{i,j,k,l} r_a^{ij,kl} r_b^{ij,kl} R(ij,kl)X(\boldsymbol{v}_i)X(\boldsymbol{v}_j)\,, \tag{13.5.47}$$

and again, utilising to the full all available symmetries,

$$\begin{aligned} B_{ab}(X) = {} & 2\delta_{a,b}\sum_{j,k,l} R(aj,kl)[X(\boldsymbol{v}_a)X(\boldsymbol{v}_j) + X(\boldsymbol{v}_k)X(\boldsymbol{v}_l)] \\ & + 2\sum_{i,j} R(ij,ab)[X(\boldsymbol{v}_i)X(\boldsymbol{v}_j) + X(\boldsymbol{v}_a)X(\boldsymbol{v}_b)] \\ & - 4\sum_{j,l} R(aj,bl)[X(\boldsymbol{v}_a)X(\boldsymbol{v}_j) + X(\boldsymbol{v}_b)X(\boldsymbol{v}_l)]\,. \end{aligned} \tag{13.5.48}$$

These imply a stochastic differential equation

$$dX(\boldsymbol{v}_a) = \left\{2 \sum_{j,k,l} R(aj,kl)[X(\boldsymbol{v}_k)X(\boldsymbol{v}_l) - X(\boldsymbol{v}_a)X(\boldsymbol{v}_j)]\right\} dt + dW(\boldsymbol{v}_a, t), \tag{13.5.49}$$

where

$$dW(\boldsymbol{v}_a, t)\, dW(\boldsymbol{v}_b, t) = B_{ab}(\boldsymbol{X})\, dt. \tag{13.5.50}$$

Neglecting the stochastic term, we recover the *Boltzmann Equation* for $X(\boldsymbol{v}_a)$ in a discretised form. As always, this Kramers-Moyal equation is only valid in a small noise limit which is equivalent to a system size expansion, the size being ξ^3, the volume of the momentum space cells.

b) Poisson Representation: The Boltzmann master equation is an ideal candidate for a Poisson representation treatment. Using the variable $\alpha(\boldsymbol{v}_a)$ as usual, we can follow through the results (12.1.6–12.1.9) to obtain a Poisson representation Fokker-Planck equation with a drift term

$$A_a(\boldsymbol{\alpha}) = 2 \sum_{j,k,l} R(aj,kl)[\alpha(\boldsymbol{v}_k)\alpha(\boldsymbol{v}_l) - \alpha(\boldsymbol{v}_a)\alpha(\boldsymbol{v}_j)], \tag{13.5.51}$$

and diffusion matrix

$$\begin{aligned} B_{ab}(\boldsymbol{\alpha}) = {} & 2\delta_{ab} \sum_{i,j,l} R(ij,kl)[\alpha(\boldsymbol{v}_i)\alpha(\boldsymbol{v}_j) - \alpha(\boldsymbol{v}_a)\alpha(\boldsymbol{v}_l)] \\ & + 2\sum_{i,j} R(ij,ab)[\alpha(\boldsymbol{v}_i)\alpha(\boldsymbol{v}_j) - \alpha(\boldsymbol{v}_a)\alpha(\boldsymbol{v}_b)] \\ & - \delta_{a,b} \sum_{j,l} R(aj,bl)[\alpha(\boldsymbol{v}_a)\alpha(\boldsymbol{v}_j) - \alpha(\boldsymbol{v}_b)\alpha(\boldsymbol{v}_l)]. \end{aligned} \tag{13.5.52}$$

The corresponding stochastic differential equation is

$$d\alpha(\boldsymbol{v}_a) = 2 \sum_{j,k,l} R(aj,kl)[\alpha(\boldsymbol{v}_k)\alpha(\boldsymbol{v}_l) - \alpha(\boldsymbol{v}_a)\alpha(\boldsymbol{v}_j)] + d\tilde{W}(\boldsymbol{v}_a, t), \tag{13.5.53}$$

where

$$d\tilde{W}(\boldsymbol{v}_a, t)\, d\tilde{W}(\boldsymbol{v}_b, t) = B_{ab}(\boldsymbol{\alpha})\, dt. \tag{13.5.54}$$

As emphasised previously, *this Poisson representation stochastic differential equation is exactly equivalent to the Boltzmann master equation assumed.* Unlike the Kramers-Moyal or system size expansions, it is valid for all sizes of velocity space cell ξ.

c) Stationary Solution of the Boltzmann Master Equation: We have chosen to write the Boltzmann master equation with $t^-_{ij,kl}(\boldsymbol{X})$ zero, but we can alternatively write

$$t^-_{ij,kl}(\boldsymbol{X}) = t^+_{ij,kl}(\boldsymbol{X}), \tag{13.5.55}$$

and appropriately divide all the R's by 2, since everything is now counted twice.

The condition for detailed balance (11.5.18) is trivially satisfied. Although we have set $t^-(ij,kl) = 0$, the reversed transition is actually given by $t^+(kl,ij)$. Hence,

$$k^+_{ij,kl} = k^-_{ij,kl} = R(ij,kl), \tag{13.5.56}$$

provided the time-reversal symmetry (13.5.43) is satisfied.

Under these conditions, the stationary state has a mean $\langle X \rangle = \alpha$ satisfying

$$\alpha(\boldsymbol{v}_i)\alpha(\boldsymbol{v}_j) = \alpha(\boldsymbol{v}_k)\alpha(\boldsymbol{v}_l) . \tag{13.5.57}$$

This means that $\log(\alpha(\boldsymbol{v}_i))$ is additively conserved and must be a function of the invariants (13.5.41, 13.5.42). Hence,

$$\alpha(\boldsymbol{v}) = \exp\left(\frac{\mu - (\boldsymbol{v} - \boldsymbol{U})^2/2m}{kT}\right) . \tag{13.5.58}$$

Here m is the mass of the molecules and μ, $\boldsymbol{U}$ and kT are parameters which are of course identified with the chemical potential, the mean velocity of the molecules, and the absolute temperature multiplied by Boltzmann's constant.

The stationary distribution is then a multivariate Poisson with mean values given by (13.5.58). The fluctuations in number are uncorrelated for different velocities.

13.5.4 Collisions and Flow Together

There is a fundamental difficulty in combining the treatment of flow and that of collisions. It arises because a stochastic treatment of flow requires infinitesimally small cells, whereas the Boltzmann master equation is better understood in terms of cells of finite size. This means that it is almost impossible to write down explicitly an exact stochastic equation for the system, except in the Poisson representation which we shall shortly come to.

To formally write a multivariate phase-space master equation is, however, straightforward when we assume we have phase-space cells of finite size $\lambda^3\xi^3$. We simply include all transitions available, i.e., those leading to flow in position space, flow in velocity space and collisions. The resultant Master equation thus includes the possible transitions specified in (13.5.25, 13.5.26, 13.5.34) and in a modified form (13.5.39, 13.5.40. Here, however, we have collisions within each cell defined by the transition probability per unit time

$$t^+_{ij,kl}(\boldsymbol{X}) = \delta(\boldsymbol{r}_i, \boldsymbol{r}_j)\delta(\boldsymbol{r}_k, \boldsymbol{r}_l)\delta(\boldsymbol{r}_i, \boldsymbol{r}_k)R(ij, kl)X_iX_j . \tag{13.5.59}$$

For finite $\lambda^3\xi^3$, there will be an extra stochastic effect arising from the finite cell size as pointed out in Sect. 13.5.2, which disappears in the limit of small λ and ξ when transfer from flow is purely deterministic.

The resulting master equation is rather cumbersome to write down and we shall not do this explicitly. Most work that has been done with it has involved a system size expansion or equivalently, a Kramers-Moyal approximation. The precise limit in which this is valid depends on the system size dependence of $R(ij, kl)$. The system size in this case is the six-dimensional phase-space volume $\lambda^3\xi^3$. In order to make the deterministic equation for the density, defined by

$$f(\boldsymbol{r}_i, \boldsymbol{v}_i) = X(\boldsymbol{r}_i, \boldsymbol{v}_i)/\lambda^3\xi^3 . \tag{13.5.60}$$

independent of cell size, $R(ij, kl)$ as defined in (13.5.59) must scale like $(\lambda^3\xi^3)^4$, i.e.

$$R(ij, kl) = \bar{R}(ij, kl)(\lambda^3\xi^3)^4 . \tag{13.5.61}$$

This is interpretable as meaning that $\bar{R}(ij, kl)$ is the mean collision rate per phase space volume element in each of the arguments.

Taking the conservation law (13.5.41, 13.5.42) into account, we can then write

$$\bar{R}(ij, kl) = 8\sigma[(\boldsymbol{v}_i - \boldsymbol{v}_j)^2, (\boldsymbol{v}_i - \boldsymbol{v}_j) \cdot (\boldsymbol{v}_k - \boldsymbol{v}_l)] \times \delta(v_i^2 + v_j^2 - v_k^2 - v_l^2)\delta(\boldsymbol{v}_l + \boldsymbol{v}_j - \boldsymbol{v}_k - \boldsymbol{v}_l)] , \tag{13.5.62}$$

and we have assumed that σ is a function only of scalars. [The fact that σ is only a function of $(\boldsymbol{v}_i - \boldsymbol{v}_j)^2$ and $(\boldsymbol{v}_i - \boldsymbol{v}_j) \cdot (\boldsymbol{v}_k - \boldsymbol{v}_l)$ is a result of scattering theory, and it follows from invariance with respect to the Galilean group of transformations, i.e., rotational invariance, and the fact that the laws of physics do not depend on the particular choice of unaccelerated frame of reference. We choose to keep the dependence in terms of scalar products for simplicity of expression in the fluctuation terms.]

a) Kramers-Moyal Expression: We now replace the summations by integrations according to

$$(\xi^3 \lambda^3) \sum_j \rightarrow \int d^3\boldsymbol{r}_j d^3\boldsymbol{v}_j , \tag{13.5.63}$$

and change the variables by

$$\left.\begin{aligned} \boldsymbol{v}_j &= \boldsymbol{v}_i , \\ \boldsymbol{v}_k &= \tfrac{1}{2}(\boldsymbol{p} + \boldsymbol{q}) , \\ \boldsymbol{v}_l &= \tfrac{1}{2}(\boldsymbol{p} - \boldsymbol{q}) , \end{aligned}\right\} \tag{13.5.64}$$

After a certain amount of manipulation in the collision term, the deterministic equation comes out in the form [from (13.5.4, 13.5.46)]

$$\begin{aligned} df(\boldsymbol{r}, \boldsymbol{v}) = \{ &-\boldsymbol{v} \cdot \nabla f(\boldsymbol{r}, \boldsymbol{v}) - A \cdot \nabla_{\boldsymbol{v}} f(\boldsymbol{r}, \boldsymbol{v}) \\ &+ \int d^3\boldsymbol{v}_1 \int \frac{d^3\boldsymbol{q}}{|\boldsymbol{q}|} \delta(|\boldsymbol{q}| - |\boldsymbol{v} - \boldsymbol{v}_1|) \sigma[q^2, \boldsymbol{q} \cdot (\boldsymbol{v} - \boldsymbol{v}_1)] \\ &\times \left[f\left[\boldsymbol{r}, \tfrac{1}{2}(\boldsymbol{v} + \boldsymbol{v}_1 - \boldsymbol{q})\right] f\left[\boldsymbol{r}, \tfrac{1}{2}(\boldsymbol{v} + \boldsymbol{v}_1 + \boldsymbol{q})\right] - f(\boldsymbol{r}, \boldsymbol{v}) f(\boldsymbol{r}, \boldsymbol{v}_1) \right] \} \, dt . \end{aligned} \tag{13.5.65}$$

The stochastic differential equation, arising from a Kramers-Moyal expansion, is obtained by adding a stochastic term $dW(\boldsymbol{r}, \boldsymbol{v}, t)$ satisfying

$$\begin{aligned} dW(\boldsymbol{r}, \boldsymbol{v}, t) \, dW(\boldsymbol{r}', \boldsymbol{v}', t) = \delta(\boldsymbol{r} - \boldsymbol{r}') \, dt \Big\{ &\delta(\boldsymbol{v} - \boldsymbol{v}') \int d^2\boldsymbol{v}_1 \frac{d^3\boldsymbol{q}}{|\boldsymbol{q}|} \delta(|\boldsymbol{q}| - |\boldsymbol{v} - \boldsymbol{v}_1|) \\ &\times \sigma(q^2, \boldsymbol{q} \cdot (\boldsymbol{v} - \boldsymbol{v}_1)] f\left[\boldsymbol{r}, \tfrac{1}{2}(\boldsymbol{v} + \boldsymbol{v}_1 - \boldsymbol{q})\right] f\left[\boldsymbol{r}, \tfrac{1}{2}(\boldsymbol{v} + \boldsymbol{v}_1 + \boldsymbol{q})\right] + f(\boldsymbol{r}, \boldsymbol{v}) f(\boldsymbol{r}, \boldsymbol{v}_1)] \\ &-2 \int d^3\boldsymbol{k} \, \delta[(\boldsymbol{v}_a - \boldsymbol{v}_b) \cdot \boldsymbol{k}] \sigma\left[(\boldsymbol{v}_a - \boldsymbol{v}_b)^2 + \frac{1}{4}\boldsymbol{k}^2, -(\boldsymbol{v}_a - \boldsymbol{v}_b)^2 + \frac{1}{4}\boldsymbol{k}^2 \right] \\ &\times [f(\boldsymbol{v}_a) f(\boldsymbol{v}_b - \boldsymbol{k}) + f(\boldsymbol{v}_b) f(\boldsymbol{v}_a - \boldsymbol{k})] \Big\} . \end{aligned} \tag{13.5.66}$$

Using such a Kramers-Moyal method, *van den Broek* et al. [13.8] have been able to apply a Chapman-Enskog expansion and have obtained fluctuating hydrodynamics from it.

The validity of this method, which depends on the largeness of the cells for the validity of the Kramers-Moyal expansion and the smallness of the cells for the validity of the modelling of flow as a birth-death process, is naturally open to question. However, since the Chapman-Enskog method is probably equivalent to the adiabatic elimination of the variables governed by the Boltzmann collision term, the result of adiabatic elimination is not likely to be very sensitive to the precise form of this operator. Thus, the Kramers-Moyal approximation may indeed be sufficiently accurate, even for very small cells.

b) Poisson Representation: The Poisson representation stochastic differential equation can be similarly obtained from (13.5.51, 13.5.52). We use the symbol $\phi(\boldsymbol{r},\boldsymbol{v})$ defined by

$$\phi(\boldsymbol{r},\boldsymbol{v}) = \alpha(\boldsymbol{r},\boldsymbol{v})/(\lambda^3\zeta^3), \tag{13.5.67}$$

We find

$$\begin{aligned} d\phi(\boldsymbol{r},\boldsymbol{v}) = dt\Bigg\{ &-\boldsymbol{v}\cdot\nabla\phi(\boldsymbol{r},\boldsymbol{v}) - \boldsymbol{A}\cdot\nabla_{\boldsymbol{v}}\phi(\boldsymbol{r},\boldsymbol{v}) + \int d^3v_1 \frac{d^3\boldsymbol{q}}{|\boldsymbol{q}|}\delta(|\boldsymbol{q}| - |\boldsymbol{v}-\boldsymbol{v}_1|) \\ &\times \sigma[q^2, \boldsymbol{q}\cdot(\boldsymbol{v}-\boldsymbol{v}_1)]\Big[f\Big[\boldsymbol{r}, \tfrac{1}{2}(\boldsymbol{v}+\boldsymbol{v}_1-\boldsymbol{q})\Big]f\Big[\boldsymbol{r}, \tfrac{1}{2}(\boldsymbol{v}+\boldsymbol{v}_1+\boldsymbol{q})\Big] \\ &- f(\boldsymbol{r},\boldsymbol{v})f(\boldsymbol{r},\boldsymbol{v}_1)\Big]\Bigg\} + dW(\boldsymbol{r},\boldsymbol{v},t), \end{aligned} \tag{13.5.68}$$

where

$$\begin{aligned} &dW(\boldsymbol{r},\boldsymbol{v},t)\,dW(\boldsymbol{r}',\boldsymbol{v}',t) \\ &\quad = dt\,\delta(\boldsymbol{r}-\boldsymbol{r}')\delta(\boldsymbol{v}-\boldsymbol{v}')\Bigg(\int d^3v_1 \frac{d^3\boldsymbol{q}}{|\boldsymbol{q}|}\delta(|\boldsymbol{q}| - |\boldsymbol{v}-\boldsymbol{r}_1|)\sigma[q^2, \boldsymbol{q}\cdot(\boldsymbol{v}-\boldsymbol{v}_1)] \\ &\qquad \times \Big\{\phi\Big[\boldsymbol{r}, \tfrac{1}{2}(\boldsymbol{v}+\boldsymbol{v}_1-\boldsymbol{q})\Big]\phi\Big[\boldsymbol{r}, \tfrac{1}{2}(\boldsymbol{v}+\boldsymbol{v}_1+\boldsymbol{q})\Big] - \phi(\boldsymbol{r},\boldsymbol{v})\phi(\boldsymbol{r},\boldsymbol{v}_1)\Big\} \\ &\qquad + \int \frac{d^3\boldsymbol{q}}{|\boldsymbol{q}|}\delta(|\boldsymbol{q}| - |\boldsymbol{v}-\boldsymbol{v}_1|)\sigma[q^2, \boldsymbol{q}\cdot(\boldsymbol{v}-\boldsymbol{v}_1)] \\ &\qquad \times \Big\{\phi\Big[\boldsymbol{r}, \tfrac{1}{2}(\boldsymbol{r}+\boldsymbol{r}'-\boldsymbol{q})\Big]\phi\Big[\boldsymbol{r}, \tfrac{1}{2}(\boldsymbol{v}+\boldsymbol{v}'+\boldsymbol{q})\Big] - \phi(\boldsymbol{r},\boldsymbol{v})\phi(\boldsymbol{r},\boldsymbol{v}')\Big\}\Bigg). \end{aligned} \tag{13.5.69}$$

The terms in (13.5.69) correspond to the first two terms in (13.5.52). The final term gives zero contribution in the limit that $\xi^3\lambda^3 \to 0$.

As always, we emphasise that this Poisson representation stochastic differential equation is exactly equivalent to the small cell size limit of the Boltzmann Master equation with flow terms added. Equations (13.5.68, 13.5.69) have not previously been written down explicitly, and so far, have not been applied. By employing Chapman-Enskog or similar techniques, one could probably deduce exact fluctuating hydrodynamic equations.

14. Bistability, Metastability, and Escape Problems

This chapter is devoted to the asymptotic study of systems which can exist in at least two stable states, and to some closely related problems. Such systems are of great practical importance, e.g., switching and storage devices in computers are systems which have this property. So do certain molecules which can isomerise, and more recently, a large number of electronic, chemical and physical systems which demonstrate related properties in rich variety have been investigated.

The problems of interest are:

i) How stable are the various states relative to each other?

ii) How long does it take for a system to switch spontaneously from one state to another?

iii) How is the transfer made, i.e., through what path in the relevant state space?

iv) How does a system relax from an unstable state?

These questions can all be answered relatively easily for one-dimensional diffusion processes—but the extension to several, but few dimensions is only recent. The extension to infinitely many variables brings us to the field of the liquid-gas transition and similar phase transitions where the system can be in one of two phases and arbitrarily distributed in space. This is a field which is not ready to be written down systematically from a stochastic point of view, and it is not treated here.

The chapter is divided basically into three parts: single variable bistable diffusion processes, one-step birth-death bistable systems and many-variable diffusion processes. The results are all qualitatively similar, but a great deal of effort must be invested for quantitative precision.

14.1 Diffusion in a Double-Well Potential (One Variable)

We consider once more the model of Sect. 5.5, where the probability density $p(x,t)$ of a particle obeys the Fokker-Planck equation

$$\partial_t p(x,t) = \partial_x[U'(x)p(x,t)] + D\partial_x^2 p(x,t)\,. \tag{14.1.1}$$

The shape of $U(x)$ is as shown in Fig. 14.1. There are two minima at a and c and in between, a local maximum. The stationary distribution is

$$p_s(x) = \mathcal{N}\exp[-U(x)/D]\,, \tag{14.1.2}$$

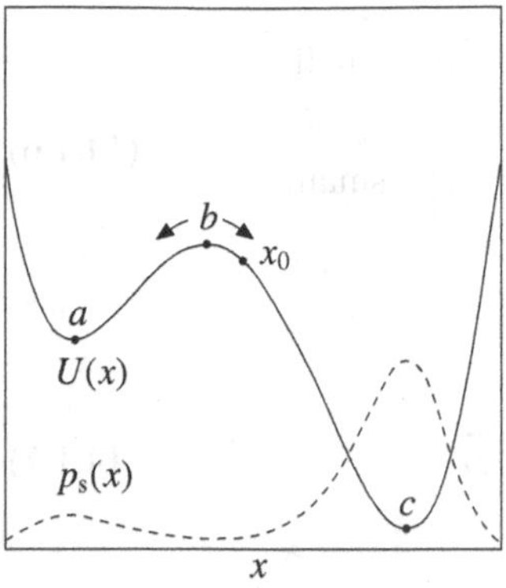

Fig. 14.1. Plot of $p_s(x)$ and $U(x)$ for a double well potential

and it is this that demonstrates the bistability. Corresponding to a, c and b are two maxima, and a central minimum as plotted in Fig. 14.1. The system is thus most likely to be found at a or c.

14.1.1 Behaviour for $D = 0$

In this case, $x(t)$ obeys the differential equation

$$\frac{dx}{dt} = -U'(x), \qquad x(0) = x_0 . \tag{14.1.3}$$

Since

$$\frac{dU(x)}{dt} = U'(x)\frac{dx}{dt} = -[U'(x)]^2 < 0 , \tag{14.1.4}$$

$x(t)$ always moves in such a way as to minimise $U(x)$, and stops only when $U'(x)$ is zero. Thus, depending on whether x_0 is greater than or less than b, the particle ends up at c or a, respectively. The motion follows the arrows on the figure.

Once the particle is at a or c it stays there. If it starts exactly at b, it also stays there, though the slightest perturbation drives it to a or c. Thus, b is an unstable stationary point and a and c are stable. There is no question of *relative* stability of a and c.

14.1.2 Behaviour if D is Very Small

With the addition of noise, the situation changes. The stationary state can be approximated asymptotically as follows. Assuming $U(x)$ is everywhere sufficiently smooth, we can write

$$U(x) \simeq \begin{cases} U(a) + \frac{1}{2}U''(a)(x-a)^2 , & |x-a| \ \text{small,} \\ U(c) + \frac{1}{2}U''(c)(x-c)^2 , & |x-c| \ \text{small.} \end{cases} \tag{14.1.5}$$

If D is very small, then we may approximate

$$
p_s(x) \simeq \begin{cases} \mathcal{N} \exp\left(-\dfrac{U(a) + \frac{1}{2}U''(a)(x-a)^2}{D}\right), & |x-a| \text{ small,} \\ \mathcal{N} \exp\left(-\dfrac{U(c) + \frac{1}{2}U''(c)(x-c)^2}{D}\right), & |x-c| \text{ small,} \\ 0, \quad \text{elsewhere,} & \end{cases} \tag{14.1.6}
$$

so that

$$
\mathcal{N}^{-1} \simeq \mathrm{e}^{-U(a)/D} \sqrt{2\pi D/U''(a)} + \mathrm{e}^{-U(c)/D} \sqrt{2\pi D/U''(c)} . \tag{14.1.7}
$$

Suppose, as drawn in Fig. 14.1,

$$
U(a) > U(c) . \tag{14.1.8}
$$

Then for small enough D, the second term is overwhelmingly larger than the first and $\mathcal{N}^{-1}$ can be approximated by it alone. Substituting into (14.1.6) we find

$$
\begin{aligned} p_s(x) &= \sqrt{\frac{U''(c)}{2\pi D}} \exp\left(-\frac{U''(c)(x-c)^2}{2D}\right), && |x-c| \sim \sqrt{D}, \\ &= 0, && \text{otherwise.} \end{aligned}
$$

This means that in the limit of very small D, the deterministic stationary state at which $U(x)$ has an absolute minimum is the more stable state in the sense that in the stochastic stationary state, $p_s(x)$, is very small everywhere except in its immediate vicinity.

Of course this result disagrees with the previous one, which stated that each state was equally stable. The distinction is one of time, and effectively we will show that the deterministic behaviour is reproduced stochastically if we start the system at x_0 and consider the limit $D \to 0$ of $p(x,t)$ for any finite t. The methods of Sect. 7.3 show this as long as the expansion about the deterministic equation is valid. Equations (7.3.9, 7.3.10) show that this will be the case provided $U'(x_0)$ is nonzero, or, in the case of any finite D, $U'(x_0)$ is of order D^0 (here, D replaces the ε^2 in Sect. 7.3.) This is true provided x_0 is not within a neighbourhood of width of order $D^{1/2}$ of a, c, or b. This means that in the case of a and c, fluctuations take over and the motion is given approximately by linearising the stochastic differential equation around a or c. Around b, the linearised stochastic differential equation is unstable. The particle, therefore, follows the Ornstein-Uhlenbeck Process until it leaves the immediate neighbourhood of $x = b$, at which stage the asymptotic expansion in $\sqrt{D}$ takes over.

However, for $t \to \infty$, the asymptotic expansion is no longer valid. Or, in other words, the $t \to \infty$ limit of the small noise perturbation theory does not reproduce the $D \to 0$ limit of the stationary state.

The process that can occur is that of escape over the central barrier. The noise $dW(t)$ can cause the particle to climb the barrier at b and reach the other side. This involves times of order $\exp(-\text{const}/D)$, which do not contribute to an asymptotic expansion in powers of D since they go to zero faster than any power of D as $D \to 0$.

14.1.3 Exit Time

This was investigated in Sect. 5.5.3. The time for the particle, initially near a, to reach the central point b is

$$T(a \to b) = \frac{\pi}{\sqrt{U''(a)\,|U''(b)|}} \exp\left\{\frac{U(b) - U(a)}{D}\right\}, \tag{14.1.9}$$

(half as long as the time for the particle to reach a point well to the right of b). For $D \to 0$, this becomes exponentially large. The time taken for the system to reach the stationary state will thus also become exponentially large, and on such a time scale development of the solutions of the corresponding SDE in powers of $D^{1/2}$ cannot be expected to be valid.

14.1.4 Splitting Probability

Suppose we put the particle at x_0: What is the probability that it reaches a before c, or c before a? This can be related to the problem of exit through a particular end of an interval, studied in Sect. 5.5.4. We put absorbing barriers at $x = a$ and $x = c$, and using the results of that section, find π_a and π_c, the *"splitting probabilities"* for reaching a or c first. These are (noting that the diffusion coefficient, D, is hence independent of x):

$$\pi_a(x_0) = \left[\int_{x_0}^{c} dx\, p_s(x)^{-1}\right] \Big/ \left[\int_{a}^{c} dx\, p_s(x)^{-1}\right], \tag{14.1.10}$$

$$\pi_c(x_0) = \left[\int_{a}^{x_0} dx\, p_s(x)^{-1}\right] \Big/ \left[\int_{a}^{c} dx\, p_s(x^{-1})\right]. \tag{14.1.11}$$

The splitting probabilities π_a and π_c can be viewed more generally as simply the probability that the particle, started at x_0, will fall into the left or right-hand well, since the particle, having reached a, will remain on that side of the well for a time of the same order as the mean exit time to b.

We now consider two possible asymptotic forms as $D \to 0$.

a) x_0 a Finite Distance from b: We first evaluate

$$\int_{a}^{c} dx\, p_s(x)^{-1} = \int_{a}^{c} dx\, \mathcal{N}^{-1} \exp[U(x)/D]. \tag{14.1.12}$$

This is dominated by the behaviour at $x \sim b$. An asymptotic evaluation is correctly obtained by setting

$$U(x) \simeq U(b) - \tfrac{1}{2}|U''(b)|/(b - x)^2. \tag{14.1.13}$$

As $D \to 0$, the limits at $x = a, c$ effectively recede to $\pm\infty$ and we find

$$\int_{a}^{c} dx\, p_s(x)^{-1} \sim \mathcal{N}^{-1} \sqrt{\frac{2\pi D}{|U''(b)|}} \exp[U(b)/D]. \tag{14.1.14}$$

Now suppose $x_0 < b$. Then $\int\limits_a^{x_0} dx\, p(x)^{-1}$ can be evaluated by the substitution

$$y = U(x), \tag{14.1.15}$$

with an inverse $x = W(y)$, and is asymptotically

$$\mathscr{N}^{-1} \int\limits_{-\infty}^{U(x_0)} e^{y/D} W'(y) dy \approx \mathscr{N}^{-1} D\, e^{U(x_0)/D}\, W'[U(x_0)] = \mathscr{N}^{-1} \frac{D\, e^{U(x_0)/D}}{U'(x_0)} . \tag{14.1.16}$$

Thus,

$$\pi_c \approx \frac{1}{U'(x_0)} \sqrt{\frac{|U''(b)|D}{2\pi}} \exp\left[\frac{U(x_0) - U(b)}{D}\right], \tag{14.1.17}$$

and

$$\pi_a = 1 - \pi_c . \tag{14.1.18}$$

We see here that the splitting probability depends only on x_0 and b. Thus, the probability of reaching c in this limit is governed entirely by the probability of jumping the barrier at b. The points at a and c are effectively infinitely distant.

b) x_0 Infinitesimally Distant from b: Suppose

$$x_0 = b - y_0 \sqrt{D} . \tag{14.1.19}$$

In this case, we can make the approximation (14.1.13) in both integrals. Defining [14.1]

$$\text{erf}(x) = \sqrt{\tfrac{\pi}{2}} \int\limits_0^x dt\, e^{-t^2} , \tag{14.1.20}$$

we find

$$\pi_c = 1 - \pi_a \approx \tfrac{1}{2}\left\{1 - \text{erf}\left[y_0 \sqrt{|U''(b)|}\right]\right\} , \tag{14.1.21}$$

$$= \frac{1}{2}\left\{1 - \text{erf}\left[(b - x_0) \sqrt{|U''(b)|\,/D}\right]\right\} . \tag{14.1.22}$$

Equation (14.1.22) is the result that would be obtained if we replaced $U(x)$ by its quadratic approximation (14.1.13) over the whole range.

c) Comparison of Two Regions: The two regions give different results, and we find that a simple linearisation of the stochastic differential equation [which is what replacing $U(x)$ by a quadratic approximation amounts to] gives the correct result only in the limit of large D and in a region of order of magnitude $\sqrt{D}$ around the maximum b.

14.1.5 Decay from an Unstable State

The mean time for a particle placed at a point on a potential to reach one well or the other is an object capable of being measured experimentally. If we use (14.1.1) for the process, then the mean time to reach a or c from b can be computed exactly using the formulae of Sect. 5.5.4. The mean time to reach a from b is the solution of

$$-U'(x)\partial_x[\pi_a(x)T(a,x)] + D\partial_x^2[\pi_a(x)T(a,x)] = -\pi_a(x)\,, \tag{14.1.23}$$

with the boundary conditions

$$\pi_a(a)T(a,a) = \pi_a(c)T(a,c) = 0\,, \tag{14.1.24}$$

and $\pi_a(x)$ given by (14.1.10).

The solution to (14.1.23) is quite straightforward to obtain by direct integration, but it is rather cumbersome. The solution technique is exactly the same as that used for (5.5.21) and the result is similar. Even the case covered by (5.5.21) where we do not distinguish between exit at the right or at the left, is very complicated.

For the record, however, we set down that

$$T(a,x) = \frac{\pi_c(x)\int_x^c dx'\, p_s(x')^{-1}\int_a^{x'}\pi_a(z)p_s(z)\,dz - \pi_a(x)\int_a^x dx'\, p_s(x')^{-1}\int_a^{x'}\pi_a(z)p_s(z)\,dz}{D\pi_a(x)} \tag{14.1.25}$$

where one considers that $\pi_a(x)$ is given by (14.1.10) and $p_s(z)$ by (14.1.2). It can be seen that even for the simplest possible situation, namely,

$$U(x) = -\tfrac{1}{2}kx^2\,, \tag{14.1.26}$$

the expression is almost impossible to comprehend. An asymptotic treatment is perhaps required. Fortunately, in the cases where $p_s(z)$ is sharply peaked at a and c with a sharp minimum at b, the problem reduces essentially to the problem of relaxation to a or to c with a reflecting barrier at b.

To see this note that

i) The explicit solution for $\pi_a(x)$ (14.1.10) means

$$\pi_a(x) \approx \begin{cases} 1\,, & (x<b)\,,\\ \tfrac{1}{2}\,, & (x=b)\,,\\ 0\,, & (x>b)\,, \end{cases} \tag{14.1.27}$$

and the transition from 1 to 0 takes place over a distance of order $\sqrt{D}$, the width of the peak in $p_s(x)^{-1}$.

ii) In the integrals with integrand $\pi_a(z)p_s(z)$, we let us first consider the case $x' > b$. In this case the estimates allow us to say

$$\int_a^{x'}\pi_a(z)p_s(z)\,dz \approx \tfrac{1}{2}n_a\,, \tag{14.1.28}$$

where $n_a = \int_{-\infty}^{b} p_s(z)\,dz$ and represents the probability of being in the left-hand well.

Now considering the case when $x' < a$, we may still approximate $\pi_a(z)$ by 1, so we get

$$\int_a^{x'} \pi_a(z) p_s(z)\,dz \simeq \int_a^{x'} p_s(z)\,dz = \frac{n_a}{2} - \int_{x'}^{b} p_s(z)\,dz\,. \qquad (14.1.29)$$

Substituting these estimates into (14.1.25) we obtain

$$T(a,b) \simeq D^{-1} \int_a^b dx'\, p_s(x)^{-1} \int_{x'}^{b} p_s(z)\,dz\,. \qquad (14.1.30)$$

which is the exact mean exit time from b to a in which there is a reflecting barrier at b. Similarly,

$$T(c,b) \simeq D^{-1} \int_b^c dx'\, p_s(x)^{-1} \int_b^{x'} p_s(z)\,dz\,. \qquad (14.1.31)$$

14.2 Equilibration of Populations in Each Well

Suppose we start the system with the particle initially in the left-hand well at some position x_i so that

$$p(x,0) = \delta(x - x_i)\,. \qquad (14.2.1)$$

If D is very small, the time for x to reach the centre is very long and for times small compared with the first exit time, there will be no effect arising from the existence of the well at c. We may effectively assume that there is a reflecting barrier at b.

The motion inside the left-hand well will be described simply by a small noise expansion, and thus the typical relaxation time will be the same as that of the deterministic motion. Let us approximate

$$U(x) \approx U(a) + \tfrac{1}{2} U''(a) x^2\,, \qquad (14.2.2)$$

then the system is now an Ornstein-Uhlenbeck process and the typical time scale is of the order of $[U''(a)]^{-1}$.

Thus, we expect a two time scale description. In the short term, the system relaxes to a quasistationary state in the well in which it started. Then on a longer time scale, it can jump over the maximum at b and the long-time bimodal stationary distribution is approached.

14.2.1 Kramers' Method

In 1940, *Kramers* [14.2] considered the escape problem from the point of view of molecular transformation. He introduced what is called the Kramers equation (Sect. 6.4.1) in which he considered motion under the influence of a potential $V(x)$ which was double welled. In the case of large damping, he showed that a corresponding Smoluchowski equation of the form found in (8.2.16) could be used, and hence, fundamentally, the escape problem is reduced to the one presently under consideration.

Kramers' method has been rediscovered and reformulated many times [14.3]. It will be presented here in a form which makes its precise range of validity reasonably clear.

Using the notation of Fig. 14.1, define

$$M(x,t) = \int_{-\infty}^{x} dx' p(x',t), \tag{14.2.3}$$

$$N_a(t) = 1 - N_c(t) = M(b,t), \tag{14.2.4}$$

and

$$N_0(t) = (c-a)p(x_0,t). \tag{14.2.5}$$

Further, define the corresponding stationary quantities by

$$\begin{aligned} n_a &= 1 - n_c = \int_{-\infty}^{c} p_s(x')\,dx', \\ n_0 &= (c-a)p_s(x_0). \end{aligned} \tag{14.2.6}$$

From the Fokker-Planck equation (14.1.1) and the form of $p_s(x)$ given in (14.1.2) we can write

$$\frac{dM(x,t)}{dt} = Dp_s(x)\frac{\partial}{\partial x}\left(\frac{p(x,t)}{p_s(x)}\right), \tag{14.2.7}$$

which can be integrated to give

$$\frac{d}{dt}\int_{a}^{x_0} dx\,\frac{M(x,t)}{p_s(x)} = D\left(\frac{p(x_0,t)}{p_s(x_0)} - \frac{p(a,t)}{p_s(a)}\right). \tag{14.2.8}$$

This equation contains no approximations. We want to introduce some kind of approximation which would be valid at long times.

We are forced to introduce a somewhat less rigorous argument than desirable in order to present the essence of the method. Since we believe relaxation within each well is rather rapid, we would expect the distribution in each well (in a time of order of magnitude which is finite as $D \to 0$) to approach the same shape as the stationary distribution, but the relative weights of the two peaks to be different. This can be formalised by writing

$$p(x,t) = \begin{cases} p_s(x)\dfrac{N_a(t)}{n_a}, & x < b, \\[2ex] p_s(x)\dfrac{N_c(t)}{n_c}, & x > b. \end{cases} \tag{14.2.9}$$

This would be accurate to lowest order in D, except in a region of magnitude $\sqrt{D}$ around b.

If we substitute these into (14.2.8), we obtain

$$\kappa(x_0)\dot{N}_a(t) = D\left(\frac{N_0(t)}{n_0} - \frac{N_a(t)}{n_a}\right), \tag{14.2.10}$$

$$\mu(x_0)\dot{N}_c(t) = D\left(\frac{N_0(t)}{n_0} - \frac{N_c(t)}{n_c}\right), \tag{14.2.11}$$

with

$$\kappa(x_0) = \int_a^{x_0} p_s(x)^{-1}[1-\psi(x)]\,dx\,, \tag{14.2.12}$$

$$\mu(x_0) = \int_{x_0}^{c} p_s(x)^{-1}[1-\psi(x)]\,dx\,, \tag{14.2.13}$$

and

$$\psi(x) = \begin{cases} n_a^{-1}\int_x^b p_s(z)\,dz\,, & x<b\,,\\ n_c^{-1}\int_b^x p_s(z)\,dz\,, & x>b\,. \end{cases} \tag{14.2.14}$$

Note that if x is finitely different from a or c, then $\psi(x)$ vanishes exponentially as $D \to 0$, as follows directly from the explicit form of $p_s(x)$. Hence, since x in both integrals (14.2.12, 14.2.13) satisfies this condition over the whole range of integration, we can set

$$\psi(x) = 0\,, \tag{14.2.15}$$

and use

$$\kappa(x_0) = \int_a^{x_0} p_s(x)^{-1}dx\,, \qquad \mu(x_0) = \int_{x_0}^{c} p_s(x)^{-1}\,dx\,. \tag{14.2.16}$$

a) Three State Interpretation: Equations (14.2.10, 14.2.11) correspond to a process able to be written as a chemical reaction of the kind

$$X_a \rightleftharpoons X_0 \rightleftharpoons X_c\,, \tag{14.2.17}$$

except that there is no equation for N_0, which would be interpreted as the proportion of X_0. By noting that $N_a + N_c = 1$, we find that (14.2.10, 14.2.11) require

$$N_0(t) = n_0\,\frac{\mu(x_0)N_a(t) + \kappa(x_0)N_c(t)}{\kappa(x_0)+\mu(x_0)}\,. \tag{14.2.18}$$

This is the same equation as would be obtained by adiabatically eliminating the variable $N_0(t)$ from (14.2.10, 14.2.11) and the further equation for $N_0(t)$

$$\dot{N}_0(t) = D\left\{\frac{N_a(t)}{n_a\kappa(x_0)} + \frac{N_c(t)}{n_c\mu(x_0)}\right\} - DN_0(t)\left\{\frac{1}{n_0\kappa(x_0)} + \frac{1}{n_0\mu(x_0)}\right\}. \tag{14.2.19}$$

Since

$$n_0 = p_s(x_0)(c-a) = \mathcal{N}(c-a)\exp\left(-U(x_0)/D\right), \tag{14.2.20}$$

we see that the limit $D \to 0$ corresponds to $n_0 \to 0$, and hence the rate constant in (14.2.19) multiplying $N_0(t)$ becomes exponentially large. Hence, adiabatic elimination is valid.

This three-state interpretation is essentially the transition state theory of chemical reactions proposed by *Eyring* [14.5].

b) Elimination of Intermediate States: Eliminating $N_0(t)$ from (14.2.10, 14.2.11) by adding the two equations, we get

$$\dot{N}_a(t) = -\dot{N}_c(t) = r_a\, N_a(t) + r_c\, N_c(t)\,, \tag{14.2.21}$$

with

$$r_a = D\left[n_a \int_a^c dx\, p_s(x)^{-1}\right]^{-1}, \qquad r_c = D\left[n_c \int_a^c dx\, p_s(x)^{-1}\right]^{-1}, \tag{14.2.22}$$

where r_a and r_c are independent of x_0. Thus, the precise choice of x_0 does not affect the interpeak relaxation.

Since $N_a + N_c = 1$, the relaxation time constant, τ_r, is given by

$$\tau_r^{-1} = r_a + r_c = \frac{D}{n_a\, n_c \int_a^c dx\, p_s(x)^{-1}}\,. \tag{14.2.23}$$

c) The Escape Probability Per Unit Time: For a particle initially near a to reach x_0, the escape probability per unit time is the decay rate of $N_a(t)$ under the condition that an absorbing barrier is at x_0. This means that in (14.2.10, 14.2.11) we set $N_0(t) = 0$ [but note that $p_s(x)$ is still defined by (14.1.2)]. Similar reasoning gives us

$$\dot{N}_a(t) = -DN_a(t)/n_a\kappa(x_0)\,, \tag{14.2.24}$$

so that the mean first passage time is given by

$$\tau_a = n_a D^{-1} \int_a^{x_0} dx\, p_s(x)^{-1}\,. \tag{14.2.25}$$

This result is essentially that obtained in (5.5.29) by more rigorous reasoning.

d) Dependence of Relaxation Time on Peak Populations: Equation (14.2.23) looks at first glance like a simple formula relating the relaxation time to n_a and $n_c = 1 - n_a$. One might think that all other factors were independent of n_a and $\tau_r \propto n_a(1 - n_a)$. However, a more careful evaluation shows this is not so. If we use the asymptotic evaluation (14.1.14) we find

$$\tau_r = \frac{n_a n_c}{\mathcal{N} D}\sqrt{\frac{2\pi D}{|U''(b)|}}\exp\left(\frac{U(b)}{D}\right), \tag{14.2.26}$$

and similarly, $\mathcal{N}$ can be evaluated asymptotically by taking the contribution from each peak. We obtain

$$\mathcal{N}^{-1} = \sqrt{2\pi D}\left\{\frac{1}{\sqrt{U''(a)}}\exp\left(-\frac{U(a)}{D}\right) + \frac{1}{\sqrt{U''(c)}}\exp\left(-\frac{U(c)}{D}\right)\right\}, \tag{14.2.27}$$

and similarly, by definition (14.2.6) of n_a and n_c,

$$\frac{n_a}{n_c} = \sqrt{\frac{U''(c)}{U''(a)}} \exp\left(\frac{U(c) - U(a)}{D}\right). \tag{14.2.28}$$

After a certain amount of algebra, one can rewrite (14.2.26) as

$$\tau_r = 2\pi H(b; a, c)[n_a n_c]^{1/2}, \tag{14.2.29}$$

with $H(b; a, c)$ a function which depends on the height of $U(b)$ compared to the average of $U(a)$ and $U(c)$: Explicitly,

$$H(b; a, c) = [|U''(b)|^{-1/2} U''(a)^{-1/4} U''(c)^{-1/4}] \exp\left(\frac{2U(b) - U(a) - U(c)}{2D}\right). \tag{14.2.30}$$

14.2.2 Example: Reversible Denaturation of Chymotrypsinogen

Chymotrypsinogen is a protein which can be transformed into a denatured form by applying elevated pressures of up to several thousand atmospheres, as demonstrated by *Hawley* [14.6]. The low pressure form is a globular protein formed from a chain folded back and forth on itself. This has a larger volume than the denatured form, in which the chain takes an extended form, and which is thus thermodynamically favoured at high pressures.

A somewhat unrealistic, but simple, model of this process is given by the equation.

$$dx = \frac{-U'(x)}{\gamma} dt + \sqrt{\frac{2kT}{\gamma}} dW(t), \tag{14.2.31}$$

where x is the volume of a molecule, $U(x)$ is the Gibbs Free energy per molecule and kT/γ takes the place of D. Here γ is a friction constant, which would arise by an adiabatic elimination procedure like that used to derive the Smoluchowski equation in Sect. 6.4. The stationary distribution is then

$$p_s(x) = \mathcal{N} \exp[-U(x)/kT], \tag{14.2.32}$$

as required by statistical mechanics.

The explicit effect of the variation of pressure is included by writing

$$U(x) = U_0(x) + x\delta p, \tag{14.2.33}$$

where δp is the pressure difference from the state in which n_a and n_c are equal. The term $x\delta p$ changes the relative stability of the two minima and is equivalent to the work done against the pressure δp. From (14.2.28) this requires $U_0(x)$ to satisfy

$$\sqrt{U_0''(a)} \exp[U_0(a)/kT] = \sqrt{U_0''(c)} \exp[U_0(c)/kT]. \tag{14.2.34}$$

The maxima and minima of $U(x)$ are slightly different from those of $U_0(x)$. If we assume that higher derivatives of $U_0(x)$ are negligible, then the maxima and minima of $U(x)$ are at points where $U'(x) = 0$ and are given by $a + \delta a$, $b + \delta b$, $c + \delta c$, where

$$\left.\begin{aligned} \delta a &= -\,\delta p/U_0''(a) \;= \beta_a \delta p\,, \\ \delta b &= \;\;\delta p/|U_0''(b)| = \beta_b \delta p\,, \\ \delta c &= -\,\delta p/U_0''(c) \;= \beta_c \delta p\,. \end{aligned}\right\} \tag{14.2.35}$$

We thus identify β_a and β_c as the *compressibilities* $\partial x/\partial p$ of the states a and c, which are negative, as required by stability. The quantity β_b is some kind of *incompressibility* of the transition state. Since this is unstable, β_b is positive.

The values of $U(a)$, $U(b)$, $U(c)$ of these minima are

$$\left.\begin{aligned} U(a+\delta a) &= U_0(a+\delta a) + (a+\delta a)\delta p\,, \\ &= U_0(a) + a\delta p + \frac{1}{2}\beta_a(\delta p)^2\,, \\ U(b+\delta b) &= U_0(b) + b\delta p + \frac{1}{2}\beta_b(\delta p)^2\,, \\ U(c+\delta c) &= U_0(c) + c\delta p + \frac{1}{2}\beta_c(\delta p)^2\,, \end{aligned}\right\} \tag{14.2.36}$$

and from (14.2.28) we obtain

$$\frac{n_a}{n_c} = \exp\left[-\frac{(a-c)}{kT}\delta p - \frac{(\beta_a-\beta_c)}{2kT}(\delta p)^2\right]. \tag{14.2.37}$$

This formula is exactly that obtained by thermodynamic reasoning and fits the data well.

The relaxation time τ_r to the stationary distribution has also been measured. We compute it using (14.2.29, 14.2.30). We find that

$$\tau_r(0) \;= \gamma^{-1}\pi {\beta_b}^{1/2} \exp\left[\frac{U(b)}{kT}\right], \tag{14.2.38}$$

$$\tau_r(\delta p) = (n_a n_c)^{1/2}\tau_r(0)\exp\left[-\frac{a+c-2b}{2kT}\,\delta_p + \left(\beta_b - \tfrac{1}{2}\beta_a - \tfrac{1}{2}\beta_c\right)\frac{\delta p^2}{2kT}\right]. \tag{14.2.39}$$

Notice that, in principle, a and c, the volumes of the two states and β_a and β_c, their compressibilities, are all measurable directly. The transition state data b, $U(b)$ and β_b are left as free parameters to be determined from lifetime measurements. Of course, the quadratic terms will only be valid for sufficiently small δp and for applications, it may be necessary to use a more sophisticated method.

In Fig. 14.2, $\tau_r(\delta p)$ and n_a/n_c are plotted for a set of possible values of parameters, as computed from (14.2.39).

Notice that the equilibration time reaches a maximum near the point at which natural and denatured forms are in equal concentration. Some skewing, however, can be induced by making the potential asymmetric. Hawley, in fact, notes that measurements in this region are limited by "instrumental stability and the patience of the investigator."

Finally, the curves with zero compressibility are given for comparison. The difference is so large at the wings that it is clear that the quadratic correction method is

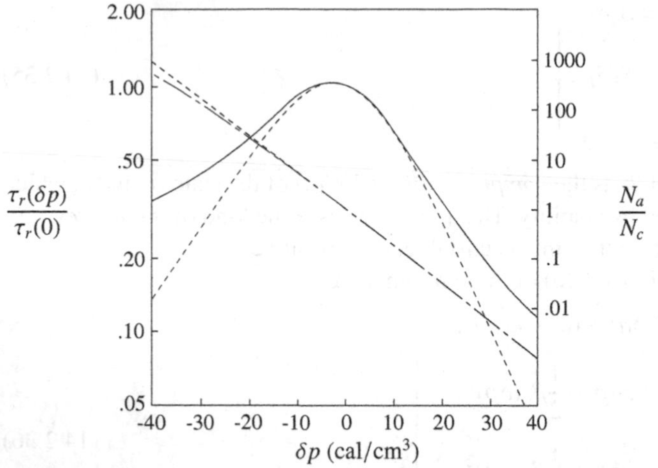

Fig. 14.2. Relaxation rate $\tau_r(\delta p)$ and ratio of concentrations of natural and denatured forms of chymotrypsinogen according to (14.2.39). Dashed line omits compressibility corrections

not valid for the $\tau_r(\delta p)$ curve. However, the corrections almost cancel for the n_a/n_c curve. A realistic treatment is given in in [14.7], where more variables are included. This preserves the qualitative nature of this simple description, but permits as well the possibility $\beta_b < 0$.

14.2.3 Bistability with Birth-Death Master Equations (One Variable)

The qualitative behaviour of bistable systems governed by one-step birth-death Master equations is almost the same as that for Fokker-Planck equations.

Consider a one step process with transition probabilities $t^+(x)$, $t^-(x)$ so that the Master equation can be written as

$$\frac{dP(x)}{dt} = J(x+1,t) - J(x,t)\,, \tag{14.2.40}$$

with

$$J(x,t) = t^-(x)P(X,t) - t^+(x-1)P(x-1,t)\,. \tag{14.2.41}$$

Suppose now that the stationary distribution has maxima at a, c, and a minimum at b, and in a similar way to that in Sect. 14.2.1, define

$$M(x,t) = \sum_{z=0}^{x-1} P(z,t)\,, \tag{14.2.42}$$

$$N_a(t) = 1 - N_c(t) = M(b,t)\,, \tag{14.2.43}$$

and if x_0 is a point near b,

$$N_a(t) = P(x_0,t)\,. \tag{14.2.44}$$

The corresponding stationary quantities are

$$n_a = 1 - n_c = \sum_{z=0}^{b-1} P_s(z)\,, \tag{14.2.45}$$

$$n_0 = P_s(x_0)\,. \tag{14.2.46}$$

We now sum (14.2.40) from 0 to $x-1$ to obtain

$$\frac{dM(x,t)}{dt} = J(x,t)\,, \tag{14.2.47}$$

[since $J(0,t) = 0$]. We now use the fact that the stationary solution $P_s(x)$ is in a one-step process obtained from the detailed balance equation

$$t^-(x)P_s(x) = t^+(x-1)P_s(x-1)\,, \tag{14.2.48}$$

to introduce an "integrating factor" for (14.2.47). Namely, define

$$\beta(x,t) = P(x,t)/P_s(x)\,. \tag{14.2.49}$$

Then (14.2.47) can be written

$$\frac{dM(x,t)}{dt} = P_s(x)t^-(x)[\beta(x,t) - \beta(x-1,t)]\,, \tag{14.2.50}$$

so that

$$\frac{d}{dt}\sum_{z=a+1}^{x_0}\left[\frac{M(z,t)}{P_s(z)t^-(z)}\right] = \beta(x_0,t) - \beta(a,t) = \frac{P(x_0,t)}{P_s(x_0)} - \frac{P(a,t)}{P_s(a)}\,. \tag{14.2.51}$$

Equation (14.2.51) is now in almost precisely the form of (14.2.8) for the corresponding Fokker-Planck process. It depends on the solution being obtained via a detailed balance method (14.2.48).

We make the same assumptions as Kramers; namely, we assume that only relaxation between peaks is now relevant and write

$$P(x,t) = \begin{cases} P_s(x)N_a(t)/n_a\,, & x < b\,, \\ P_s(x)N_c(t)/n_c\,, & x > b\,, \end{cases} \tag{14.2.52}$$

and obtain relaxation equations exactly like those in (14.2.10, 14.2.11)

$$\kappa(x_0)\dot{N}_a(t) = N_0(t)/n_0 - N_a(t)/n_a\,, \tag{14.2.53}$$

$$\mu(x_0)\dot{N}_c(t) = N_0(t)/n_0 - N_c(t)/n_c\,. \tag{14.2.54}$$

where

$$\kappa(x_0) = \sum_{z=a+1}^{x_0}[P_s(z)t^-(z)]^{-1}[1-\psi(z)]\,, \tag{14.2.55}$$

$$\mu(x_0) = \sum_{z=x_0+1}^{c}[P_s(z)t^{-1}(z)]^{-1}[1-\psi(z)]\,, \tag{14.2.56}$$

with

$$\psi(z) = \begin{cases} n_a^{-1} \sum\limits_{y=z}^{b} P_s(y) & z < b\,, \\ n_c^{-1} \sum\limits_{y=b+1}^{z+1} P_s(y) & z > b\,. \end{cases} \tag{14.2.57}$$

The only significant difference is that D appears on the right of (14.2.10, 14.2.11) but is here replaced by a factor $t^-(z)^{-1}$ in the definitions of $\kappa(x_0)$ and $\mu(x_0)$.

All the same approximations can be made, the only difficulty being a precise reformulation of the $D \to 0$ limit, which must here correspond to a large number limit, in which all functions change smoothly as x changes by ± 1. This is just the limit of the system size expansion, in which a Fokker-Planck description can be used anyway.

We shall not go into details, but merely mention that exact mean exit times are obtainable by the method of Sect. 11.4. By adapting the methods of Sect. 5.5.4 to this system one finds the splitting probabilities that the system initially at x_0, reaches points a, c are

$$\pi_a = \left\{ \sum_{z=x_0+1}^{c} [P_s(z)t^-(z)]^{-1} \right\} \Bigg/ \left\{ \sum_{z=a+1}^{c} [P_s(z)t^-(z)]^{-1} \right\}, \tag{14.2.58}$$

$$\pi_c = \left\{ \sum_{z=a+1}^{x_0} [P_s(z)t^{-1}(z)]^{-1} \right\} \Bigg/ \left\{ \sum_{z=a+1}^{c} [P_s(z)t^{-1}(z)]^{-1} \right\}. \tag{14.2.59}$$

Thus, for all practical considerations one might just as well model by means of a Fokker-Planck description. It is rare that one knows exactly what the underlying mechanisms are, so that any equation written down can be no more than an educated guess, for which purpose the simplest is the most appropriate.

14.3 Bistability in Multivariable Systems

There is a wide variety of possibilities when one deals with multivariable systems. If the system is described by a master equation, the possible variety of kinds of transition and state space is so bewilderingly rich, that one can hardly imagine where to start. However, since we saw in Sect. 14.2.3 that a master equation description is not very different from a Fokker-Planck description, it seems reasonable to restrict oneself to these, which, it turns out, are already quite sufficiently complicated.

The heuristic treatments of these problems, as developed mainly by the physicists *Langer, Landauer* and *Swanson* [14.8] have been made more rigorous by mathematical treatments by *Schuss* and *Matkowsky* [14.9] and others. The first rigorous treatment was by *Ventsel* and *Freidlin* [14.10] which, however, does not seem to have attracted much attention by applied workers since the rigour is used only to confirm estimates that have long been guessed, rather than to give precise asymptotic expansions, as do the more recent treatments.

We will consider here systems described in a space of l dimensions by a Fokker-Planck equation which is conveniently written in the form

$$\partial_t p = \nabla \cdot [-\boldsymbol{v}(\boldsymbol{x})p + \varepsilon \mathsf{D}(\boldsymbol{x}) \cdot \nabla p]\,, \tag{14.3.1}$$

whose stationary solution is called $p_s(\boldsymbol{x})$ and which is assumed to be known in much of what follows. It can, of course, be estimated asymptotically in the small ε limit by the method of Sect. 7.3.3.

14.3.1 Distribution of Exit Points

We will treat here only a simplified case of (14.3.1) in which $\mathsf{D}(\boldsymbol{x})$ is the identity:

$$\mathsf{D}(\boldsymbol{x}) = \mathbf{1}. \tag{14.3.2}$$

This does conceal features which can arise from strongly varying D but mostly, the results are not greatly changed.

We suppose that the system is confined to a region R with boundary S, and that the velocity field $\boldsymbol{v}(\boldsymbol{x})$ points inwards to a stationary point $\boldsymbol{a}$. The problem is the asymptotic estimate of the distribution of points $\boldsymbol{b}$ on S at which the point escapes from R. We use (6.6.49) for $\pi(\boldsymbol{b},\boldsymbol{x})$ (the distribution of escape points on S, starting from the point $\boldsymbol{x}$), which in this case takes the form

$$\boldsymbol{v}(\boldsymbol{x}) \cdot \nabla\pi(\boldsymbol{b},\boldsymbol{x}) + \varepsilon\nabla^2\pi(\boldsymbol{b},\boldsymbol{x}) = 0, \tag{14.3.3}$$

with boundary condition

$$\pi(\boldsymbol{b},\boldsymbol{u}) = \delta_s(\boldsymbol{b}-\boldsymbol{u}), \qquad (\boldsymbol{u} \in S). \tag{14.3.4}$$

An asymptotic solution, valid for $\varepsilon \to 0$, is constructed, following the method of *Matkowsky and Schuss* [14.9].

a) Solution Near $\boldsymbol{x} = \boldsymbol{u}$ and in the Interior of R: Firstly one constructs a solution valid inside R. For $\varepsilon = 0$ we have

$$\boldsymbol{v}(\boldsymbol{x}) \cdot \nabla\pi(\boldsymbol{b},\boldsymbol{x}) = 0, \tag{14.3.5}$$

which implies that $\pi(\boldsymbol{b},\boldsymbol{x})$ is constant along the flow lines of $\boldsymbol{v}(\boldsymbol{x})$, since it simply states that the derivative of $\pi(\boldsymbol{b},\boldsymbol{x})$ along these lines is zero. Since we assume all the flow lines pass through $\boldsymbol{a}$, we have

$$\pi(\boldsymbol{b},\boldsymbol{x}) = \pi(\boldsymbol{b},\boldsymbol{a}), \tag{14.3.6}$$

for any $\boldsymbol{x}$ inside. However, the argument is flawed by the fact that $\boldsymbol{v}(\boldsymbol{a}) = 0$ and hence (14.3.5) is no longer an appropriate approximation.

We consider, therefore, the solution of (14.3.3) within a distance of order $\sqrt{\varepsilon}$ of the origin. To assist in this, we introduce new coordinates (z, y_r), which are chosen so that z measures the distance from $\boldsymbol{a}$, while the y_r are a set of $l-1$ tangential variables measuring the orientation around $\boldsymbol{a}$.

More precisely, choose $z(\boldsymbol{x})$ and $y_r(\boldsymbol{x})$ so that

$$\left.\begin{aligned} \boldsymbol{v}(\boldsymbol{x}) \cdot \nabla z(\boldsymbol{x}) &= -\,z(\boldsymbol{x}), \\ \boldsymbol{v}(\boldsymbol{x}) \cdot \nabla y_r(\boldsymbol{x}) &= \ 0, \\ z(\boldsymbol{a}) &= \ 0. \end{aligned}\right\} \tag{14.3.7}$$

The negative sign in the first of these equations takes account of the fact that $\boldsymbol{a}$ is assumed stable, so that $\boldsymbol{v}(\boldsymbol{x})$ points towards $\boldsymbol{a}$. Thus, $z(\boldsymbol{x})$ *increases* as $\boldsymbol{x}$ travels further from $\boldsymbol{a}$.

Thus, we find, for any function f,

$$\nabla f = \nabla z(\boldsymbol{x})\frac{\partial f}{\partial z} + \sum_r \nabla y_r(\boldsymbol{x})\frac{\partial f}{\partial y_r}\,, \tag{14.3.8}$$

and hence,

$$\boldsymbol{v}(\boldsymbol{x})\cdot\nabla\pi = -z\frac{\partial \pi}{\partial z}\,, \tag{14.3.9}$$

and

$$\nabla^2\pi = \nabla z(\boldsymbol{x})\cdot\nabla z(\boldsymbol{x})\frac{\partial^2\pi}{\partial z^2} + 2\sum_r \nabla z(\boldsymbol{x})\cdot\nabla y_r(\boldsymbol{x})\frac{\partial^2\pi}{\partial z\partial y_r}$$
$$+\sum_{r,s}\nabla y_r(\boldsymbol{x})\cdot\nabla y_s(\boldsymbol{x})\frac{\partial^2\pi}{\partial y_r\partial y_s} + \nabla^2 z(\boldsymbol{x})\frac{\partial\pi}{\partial z} + \sum_r \nabla^2 y_r(\boldsymbol{x})\frac{\partial\pi}{\partial y_r}\,. \tag{14.3.10}$$

We now evaluate π asymptotically by changing to the scaled (or stretched) variable ξ defined by

$$z = \xi\sqrt{\varepsilon}\,. \tag{14.3.11}$$

Substituting (14.3.8–14.3.11) into (14.3.3) we find that, to lowest order in ε,

$$-\xi\frac{\partial\pi}{\partial\xi} + H\frac{\partial^2\pi}{\partial\xi^2} = 0\,, \tag{14.3.12}$$

where

$$H = \nabla z(\boldsymbol{a})\cdot\nabla z(\boldsymbol{a})\,. \tag{14.3.13}$$

We can now solve this equation, getting

$$\pi(\boldsymbol{b},\boldsymbol{x}) = C_1\int_0^{z/\sqrt{\varepsilon}} d\xi\exp(\xi^2/2H) + \pi(\boldsymbol{b},\boldsymbol{a})\,. \tag{14.3.14}$$

Because H is positive, we can only match this solution for π with the asymptotic constancy of $\pi(\boldsymbol{b},\boldsymbol{x})$ along flow lines for $\boldsymbol{x}\neq\boldsymbol{a}$ if $C_1 = 0$. Hence, for all $\boldsymbol{x}$ on the interior of R,

$$\pi(\boldsymbol{b},\boldsymbol{x}) = \pi(\boldsymbol{b},\boldsymbol{a})\,. \tag{14.3.15}$$

Notice, however, that if $\boldsymbol{v}(\boldsymbol{x})$ points the other way, i.e., *is unstable*, we omit the negative sign in (14.3.9) and find that $\pi(\boldsymbol{b},\boldsymbol{x})$ is given by (14.3.3) with $H\to -H$, and hence in a distance of order $\sqrt{\varepsilon}$ of $\boldsymbol{a}$, $\pi(\boldsymbol{b},\boldsymbol{x})$ changes its value significantly.

b) Solution Near the Boundary $\boldsymbol{S}$: We consider, with an eye to later applications, the solution of the slightly more general equation

$$\boldsymbol{v}(\boldsymbol{x})\cdot\nabla f(\boldsymbol{x}) + \varepsilon\nabla^2 f(\boldsymbol{x}) = 0\,, \tag{14.3.16}$$

with

$$f(\boldsymbol{u}) = g(\boldsymbol{u})\,,\quad \boldsymbol{u}\in\boldsymbol{S}\,, \tag{14.3.17}$$

of which the boundary value problem (14.3.4) is a particular case. If $\boldsymbol{u}$ is a point on $\boldsymbol{S}$, we introduce $\boldsymbol{\nu}(\boldsymbol{u})$, the normal (pointing out) at the point $\boldsymbol{u}$ to $\boldsymbol{S}$, so that the surface element can be written

$$dS(\boldsymbol{u}) = |dS(\boldsymbol{u})|\,\boldsymbol{\nu}(\boldsymbol{u}) \equiv dS\,\boldsymbol{\nu}(\boldsymbol{u})\,, \tag{14.3.18}$$

where the last notation will be used for brevity in what follows.

Two situations then arise:

i) $\boldsymbol{\nu}(\boldsymbol{x}) \cdot \boldsymbol{v}(\boldsymbol{x}) \neq 0$ *on* $\boldsymbol{S}$ *or anywhere in* R *except* $\boldsymbol{x} = \boldsymbol{a}$, *which is stable:* clearly, in any asymptotic method, the boundary condition (14.3.17) is not compatible with a constant solution. Hence, there must be rapid changes at the boundary.

Near a point $\boldsymbol{u}$ on $\boldsymbol{S}$ we can write

$$\boldsymbol{v}(\boldsymbol{u}) \cdot \nabla f(\boldsymbol{x}) + \varepsilon \nabla^2 f(\boldsymbol{x}) = 0\,. \tag{14.3.19}$$

We define a variable ρ by

$$\boldsymbol{x} = \boldsymbol{u} - \varepsilon \rho \boldsymbol{\nu}(\boldsymbol{u})\,, \tag{14.3.20}$$

and other variables y_r parallel to $\boldsymbol{S}$.

Then to lowest order in ε, (14.3.19) reduces to (at a point near $\boldsymbol{u}$ on $\boldsymbol{S}$)

$$[\boldsymbol{\nu} \cdot \boldsymbol{v}(\boldsymbol{u})]\frac{\partial f}{\partial \rho} + H(\boldsymbol{u})\frac{\partial^2 f}{\partial \rho^2} = 0\,, \tag{14.3.21}$$

with $H(\boldsymbol{u}) = \boldsymbol{\nu}(\boldsymbol{u})^2 = 1$.

The solution is then

$$f(\boldsymbol{x}) = g(\boldsymbol{u}) + C_1(\boldsymbol{u})\{1 - \exp[-\boldsymbol{\nu} \cdot \boldsymbol{v}(\boldsymbol{u})\rho]\}\,. \tag{14.3.22}$$

As $\rho \to \infty$, we approach a finite distance into the interior of R and thus

$$f(\boldsymbol{x}) \to g(\boldsymbol{u}) + C_1(\boldsymbol{u}) = C_0\,, \tag{14.3.23}$$

from the analysis in (a), so

$$C_1(\boldsymbol{u}) = C_0 - g(\boldsymbol{u})\,. \tag{14.3.24}$$

One must now fix C_0, which is the principal quantity actually sought.

This can be done by means of Green's theorem. For let $p_s(\boldsymbol{x})$ be the usual stationary solution of the forward Fokker-Planck equation. We know it can be written

$$p_s(\boldsymbol{x}) = \exp\left\{-\frac{1}{\varepsilon}[\phi(\boldsymbol{x}) + O(\varepsilon)]\right\}\,, \tag{14.3.25}$$

as has been shown in Sect. 7.3.3 . We take (14.3.19), multiply by $p_s(\boldsymbol{x})$ and integrate over R. Using the fact that $p_s(\boldsymbol{x})$ satisfies the forward Fokker-Planck equation, this can be reduced to a surface integral

$$0 = \int_R d\boldsymbol{x}\, p_s(\boldsymbol{x})[\boldsymbol{v}(\boldsymbol{x}) \cdot \nabla f(\boldsymbol{x}) + \varepsilon \nabla^2 f(\boldsymbol{x})]\,, \tag{14.3.26}$$

$$= \int dS\left\{p_s(\boldsymbol{x})\boldsymbol{\nu} \cdot \boldsymbol{v}(\boldsymbol{x}) f(\boldsymbol{x}) + \varepsilon[p_s(\boldsymbol{x})\boldsymbol{\nu}\nabla f(\boldsymbol{x}) - f(\boldsymbol{x})\boldsymbol{\nu} \cdot \nabla p_s(\boldsymbol{x})]\right\}. \tag{14.3.27}$$

Noting that, to lowest order in ε.

$$\boldsymbol{\nu} \cdot \nabla f(\boldsymbol{x}) = -\frac{1}{\varepsilon}\frac{\partial f}{\partial \rho} = -\boldsymbol{\nu} \cdot \boldsymbol{v}(x)[C_0 - g(\boldsymbol{x})]\,, \tag{14.3.28}$$

and

$$\boldsymbol{\nu} \cdot \nabla p_s(\boldsymbol{x}) = -\frac{1}{\varepsilon} \boldsymbol{\nu} \cdot \nabla \phi(\boldsymbol{x}) \exp[-\phi(\boldsymbol{x})/\varepsilon] , \tag{14.3.29}$$

we deduce that

$$C_0 = \frac{\oint_S d\boldsymbol{S}(\boldsymbol{u}) \cdot [2\boldsymbol{v}(\boldsymbol{u}) + \nabla\phi(\boldsymbol{u})]\, \mathrm{e}^{-\phi(\boldsymbol{u})/\varepsilon} g(\boldsymbol{u})}{\oint_S d\boldsymbol{S}(\boldsymbol{u}) \cdot \boldsymbol{v}(\boldsymbol{u})\, \mathrm{e}^{-\phi(\boldsymbol{u})/\varepsilon}} . \tag{14.3.30}$$

Recalling that for this problem, $g(\boldsymbol{u}) = \delta_s(\boldsymbol{u} - \boldsymbol{b})$, we find that, if $\boldsymbol{x}$ is well in the interior of R,

$$\pi(\boldsymbol{x}, \boldsymbol{b}) = C_0 = \frac{\boldsymbol{\nu}(\boldsymbol{b}) \cdot [2\boldsymbol{v}(\boldsymbol{b}) + \nabla\phi(\boldsymbol{b})]\, \mathrm{e}^{-\phi(\boldsymbol{b})/\varepsilon}}{\oint_S d\boldsymbol{S}(\boldsymbol{u}) \cdot \boldsymbol{v}(\boldsymbol{u})\, \mathrm{e}^{-\phi(\boldsymbol{u})/\varepsilon}} . \tag{14.3.31}$$

We see here that the exit distribution is essentially $[\exp -\phi(\boldsymbol{b})/\varepsilon]$, i.e., approximately the stationary distribution. If the Fokker-Planck equation has a potential solution, then

$$\boldsymbol{v}(\boldsymbol{b}) = -\nabla\phi(\boldsymbol{b}) , \tag{14.3.32}$$

and

$$\pi(\boldsymbol{x}, \boldsymbol{b}) = \frac{\boldsymbol{\nu}(\boldsymbol{b}) \cdot \boldsymbol{v}(\boldsymbol{b})\, \mathrm{e}^{-\phi(\boldsymbol{b})/\varepsilon}}{\oint_S d\boldsymbol{S}(\boldsymbol{u}) \cdot \boldsymbol{v}(\boldsymbol{u})\, \mathrm{e}^{-\phi(\boldsymbol{u})/\varepsilon}} , \tag{14.3.33}$$

and we simply have a kind of average flow result.

ii) *The case of a Separatrix—$\boldsymbol{\nu}(\boldsymbol{u}) \cdot \boldsymbol{v}(\boldsymbol{u}) = 0$ for all $\boldsymbol{u}$ on $\boldsymbol{S}$* : This problem is more directly related to bistability, since midway between two stable points $\boldsymbol{a}$ and $\boldsymbol{c}$, a curve $\boldsymbol{\nu} \cdot \boldsymbol{v}(\boldsymbol{x}) = 0$ which separates the two regions and is known as *a separatrix* is expected.

The method is much the same except that near $\boldsymbol{u}$ on $\boldsymbol{S}$, we expect

$$\boldsymbol{v} \cdot \boldsymbol{\nu}(\boldsymbol{x}) \sim \boldsymbol{\nu} \cdot (\boldsymbol{x} - \boldsymbol{u})\kappa(\boldsymbol{u}) , \tag{14.3.34}$$

where $\kappa(\boldsymbol{u})$ is a coefficient which depends on $\boldsymbol{v}(\boldsymbol{x})$ and is assumed to be nonzero.

The situation is now like that at $\boldsymbol{x} = \boldsymbol{a}$ and it is appropriate to substitute

$$\boldsymbol{x} = \boldsymbol{u} - \sqrt{\varepsilon}\, \rho \boldsymbol{\nu}(\boldsymbol{u}) , \tag{14.3.35}$$

and to lowest order in ε (14.3.19) reduces to (at a point near $\boldsymbol{u}$ on $\boldsymbol{S}$)

$$\kappa(\boldsymbol{u})\rho \frac{\partial f}{\partial \rho} + \frac{\partial^2 f}{\partial \rho^2} = 0 , \tag{14.3.36}$$

so that

$$f(\boldsymbol{x}) = g(\boldsymbol{u}) + C_1 \int \rho_0 d\rho \exp\left[-\tfrac{1}{2}\kappa(\boldsymbol{u})\rho^2\right] , \tag{14.3.37}$$

and letting $\rho \to \infty$, we find

$$C_1 = [C_0 - g(\boldsymbol{u})] \sqrt{\frac{2\kappa(\boldsymbol{u})}{\pi}} . \tag{14.3.38}$$

The result for $\pi(\boldsymbol{x}, \boldsymbol{b})$ is now completed as before: one gets

$$\pi(\boldsymbol{x}, \boldsymbol{b}) = \frac{\boldsymbol{\nu}(\boldsymbol{b}) \cdot \left[\left(1 + \sqrt{\frac{2\kappa(\boldsymbol{b})}{\pi}}\right)\boldsymbol{v}(\boldsymbol{b}) + \nabla\phi(\boldsymbol{b})\right] \mathrm{e}^{-\phi(\boldsymbol{b})/\varepsilon}}{\int d\boldsymbol{S}(\boldsymbol{u}) \cdot \boldsymbol{v}(\boldsymbol{u}) \sqrt{\frac{2\kappa(\boldsymbol{u})}{\pi}} \mathrm{e}^{-\phi(\boldsymbol{u})/\varepsilon}} . \tag{14.3.39}$$

14.3.2 Asymptotic Analysis of Mean Exit Time

From our experience in one-dimensional systems, we expect the mean exit time from a point within R to be of order $\exp(K/\varepsilon)$; for some $K > 0$, as $\varepsilon \to 0$. We therefore define

$$\tau(\boldsymbol{x}) = \exp(-K/\varepsilon) T(\boldsymbol{x}) , \tag{14.3.40}$$

where $T(x)$ is the mean escape time from R starting at $\boldsymbol{x}$, and $\tau(\boldsymbol{x})$ satisfies, from Sect. 6.6,

$$\boldsymbol{v}(\boldsymbol{x}) \cdot \nabla\tau(\boldsymbol{x}) + \varepsilon\nabla^2\tau(\boldsymbol{x}) = -\mathrm{e}^{-K/\varepsilon} , \tag{14.3.41}$$

$$\tau(\boldsymbol{u}) = 0 , \qquad \text{for } \boldsymbol{u} \in \boldsymbol{S} . \tag{14.3.42}$$

If this scaling is correct, then any expansion of $\tau(\boldsymbol{x})$ in powers of ε will not see the exponential, so the equation to lowest order in ε will be essentially (14.3.19).

As in that case, we show that $\tau(\boldsymbol{x})$ is essentially constant in the interior of R and, in the case $\boldsymbol{\nu} \cdot \boldsymbol{v}(\boldsymbol{x}) \neq 0$ on $\boldsymbol{S}$, can be written as

$$\tau(\boldsymbol{x}) \sim C_0\{1 - \exp[-\boldsymbol{\nu} \cdot \boldsymbol{v}(\boldsymbol{u})\rho]\} , \tag{14.3.43}$$

near $\boldsymbol{S}$.

We multiply (14.3.41) by $p_s(\boldsymbol{x}) = \exp[-\phi(\boldsymbol{x})/\varepsilon]$ and use Green's theorem to obtain [in much the same way as (14.3.27) but with $\tau(\boldsymbol{x}) = 0$ on $\boldsymbol{S}$]

$$-\mathrm{e}^{-K/\varepsilon} \int d\boldsymbol{x}\, \mathrm{e}^{-\phi(\boldsymbol{x})/\varepsilon} = -\int d\boldsymbol{S}\, \mathrm{e}^{-\phi(\boldsymbol{x})/\varepsilon}[C_0 \boldsymbol{\nu}(\boldsymbol{x}) \cdot \boldsymbol{v}(\boldsymbol{x})] , \tag{14.3.44}$$

i.e.,

$$C_0 = \frac{\int_R d\boldsymbol{x}\, \mathrm{e}^{-[K+\phi(\boldsymbol{x})]/\varepsilon}}{\int_R d\boldsymbol{S}(\boldsymbol{x}) \cdot \boldsymbol{v}(\boldsymbol{x})\, \mathrm{e}^{-\phi(\boldsymbol{x})/\varepsilon}} . \tag{14.3.45}$$

By hypothesis, C_0 does not change exponentially like exp (A/ε). In the numerator of (14.3.45) the main contribution comes from the minimum of $\phi(x)$ which occurs at the point $\boldsymbol{a}$, whereas in the denominator, it occurs at the point on $\boldsymbol{S}$ where $\phi(\boldsymbol{x})$ is a minimum, which we shall call $\boldsymbol{x}_0$. Thus, the ratio behaves like

$$\exp\{[\phi(\boldsymbol{a}) - \phi(\boldsymbol{x}_0) - K]/\varepsilon\} . \tag{14.3.46}$$

and hence for C_0 to be asymptotically constant,

$$K = \phi(\boldsymbol{a}) - \phi(\boldsymbol{x}_0) , \tag{14.3.47}$$

and, for x well into the interior of R, we have

$$T(x) = \frac{\int_R dx\, e^{\phi(x)/\varepsilon}}{\int_R dS\, [e^{-\phi(x)/\varepsilon} \nu \cdot v(x)]} . \tag{14.3.48}$$

In the case of a sepratrix, that is, where $\nu \cdot v(x) = 0$ on all of S, we correspondingly write

$$T(x) \sim C_0 \int_0^\rho d\rho \exp\left[-\tfrac{1}{2}\kappa(u)\rho^2\right] , \tag{14.3.49}$$

and hence in the interior,

$$T(x) \sim C_0 \sqrt{\frac{\pi}{2\kappa(u)}} . \tag{14.3.50}$$

The analysis proceeds similarly and we find, for x well in the interior of R,

$$T(x) \sim \sqrt{\frac{\pi}{2\kappa(b)}} \frac{\int_R dx\, e^{\phi(x)/\varepsilon}}{\int_R dS\, e^{-\phi(x)/\varepsilon}} . \tag{14.3.51}$$

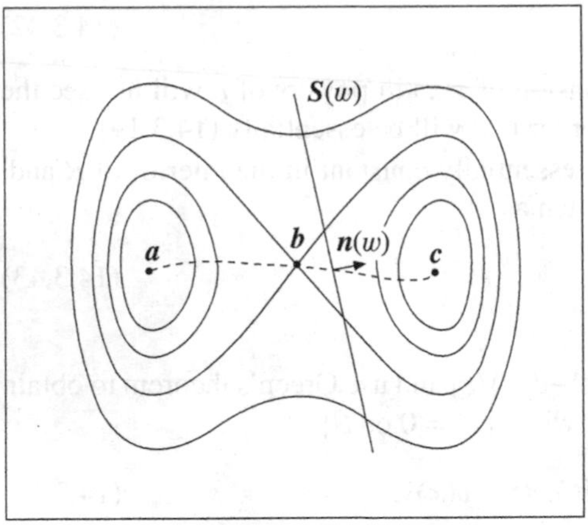

Fig. 14.3. Contours of the stationary distribution function $P_s(x)$. The plane $S(w)$ is oriented so that $P_s(x)$ has a unique maximum there, and the curve $x = u(w)$ (dashed line) is the locus of these maxima

14.3.3 Kramers' Method in Several Dimensions

The generalisation of Kramers' method is relatively straightforward. We consider a completely general Fokker-Planck equation in l dimensions [we use $P(x)$ for the probability density for notational ease]

$$\partial_t P = \nabla \cdot [-v(x)P + \varepsilon \mathsf{D}(x) \cdot \nabla P] , \tag{14.3.52}$$

whose stationary solution is to be called $P_s(x)$ and can only be exhibited explicitly if (14.3.52) satisfies potential conditions. We assume that $P_s(x)$ has two well-defined maxima at a and c and well-defined saddlepoint at b (Fig. 14.3). We assume that the value at the saddlepoint is very much smaller than the values at a and c. We introduce a family of $(l-1)$ dimensional planes $\boldsymbol{S}(w)$, where w is a parameter which labels the planes. We choose $\boldsymbol{S}(a)$ to pass through $\boldsymbol{a}$, $\boldsymbol{S}(b)$ through $\boldsymbol{b}$ and $\boldsymbol{S}(c)$ through $\boldsymbol{c}$. The planes $\boldsymbol{S}(w)$ are assumed to be oriented in such a way that $P_s(\boldsymbol{x})$ has a unique maximum when restricted to any one of them. We define, similarly to Sect. 14.2.1

$$M[\boldsymbol{S}(w)] = \int_{L(w)} d\boldsymbol{x}\, P(\boldsymbol{x}), \tag{14.3.53}$$

where $L(w)$ is the region of space to the left of the plane $\boldsymbol{S}(w)$; then

$$\dot{M}[\boldsymbol{S}(w)] = \int_{\boldsymbol{S}(w)} d\boldsymbol{S} \cdot [-\boldsymbol{v}(\boldsymbol{x})P + \varepsilon \mathsf{D}(\boldsymbol{x}) \cdot \nabla P]. \tag{14.3.54}$$

The current in stationary state is defined by

$$\boldsymbol{J}_s = -\boldsymbol{v}(\boldsymbol{x})P_s + \varepsilon \mathsf{D}(\boldsymbol{x}) \cdot \nabla P_s. \tag{14.3.55}$$

Assumption I: we exclude cases in which finite currents $\boldsymbol{J}_s$ occur where P_s is very small. Because of $\nabla \cdot \boldsymbol{J}_s = 0$, we can write

$$\boldsymbol{J}_s = -\varepsilon \nabla \cdot (\mathsf{A} P_s), \tag{14.3.56}$$

where A is an antisymmetric tensor. *We require that* A *be of the some order of magnitude as* $\mathsf{D}(\boldsymbol{x})$*, or smaller.*

Relaxation equations are derived in two stages. Define a quantity $\beta(\boldsymbol{x})$ by

$$\beta(\boldsymbol{x}) = \frac{P(\boldsymbol{x},t)}{P_s(\boldsymbol{x})} \simeq \begin{cases} N_a(t)/n_a, & \boldsymbol{x} \text{ near } \boldsymbol{a}, \\ N_c(t)/n_c, & \boldsymbol{x} \text{ near } \boldsymbol{c}. \end{cases} \tag{14.3.57}$$

This is the assumption that all relaxation within peaks has ceased. Substitute now in (14.3.54), integrate by parts discarding terms at infinity and obtain

$$\dot{M}[\boldsymbol{S}(w)] = \varepsilon \int_{\boldsymbol{S}(w)} d\boldsymbol{S} \cdot [\mathscr{D}(\boldsymbol{x}) \cdot \nabla \beta] P_s(\boldsymbol{x}), \tag{14.3.58}$$

with

$$\mathscr{D}(\boldsymbol{x}) = \mathsf{D}(\boldsymbol{x}) + \mathsf{A}(\boldsymbol{x}). \tag{14.3.59}$$

Assumption II: $P_s(\boldsymbol{x})$ is sharply singly peaked on $\boldsymbol{S}(w)$ so we may make the approximate evaluation

$$\dot{M}[\boldsymbol{S}(w)] = \left\{\varepsilon[\boldsymbol{n}(w) \cdot \mathscr{D}(\boldsymbol{x}) \cdot \nabla\beta]_{\boldsymbol{u}(w)} + \delta(w)\right\} \left| \int_{\boldsymbol{S}(w)} dS\, P_s(\boldsymbol{x}) \right|, \tag{14.3.60}$$

where $\delta(w)$ is expected to be very much smaller than the term in square brackets. Here $\boldsymbol{u}(w)$ is the position at which $P_s(\boldsymbol{x})$ has its maximum value when restricted to $\boldsymbol{S}(w)$, and $\boldsymbol{n}(w)$ is the normal to $\boldsymbol{S}(w)$.

Assumption III: The direction of $\boldsymbol{n}(w)$ can be chosen so that $\mathscr{D}^{\mathrm{T}}(\boldsymbol{x}) \cdot \boldsymbol{n}(w)$ is parallel to the tangent at w to the curve $\boldsymbol{x} = \boldsymbol{u}(w)$—without violating the other assumptions. Hence,

$$\mathscr{D}^{\mathrm{T}}[\boldsymbol{u}(w)] \cdot \boldsymbol{n}(w) = d(w)\partial_w \boldsymbol{u}(w). \tag{14.3.61}$$

Defining now

$$p(w) = \left| \int_{S(w)} dS\, P_s(\boldsymbol{x}) \right| , \tag{14.3.62}$$

which is (up to a slowly varying factor) the probability density for the particle to be on the plane $\boldsymbol{S}(w)$ and is expected to have a two-peaked shape with maxima at $w = a$ and $w = c$ and a minimum at $w = b$.

Assumption IV: These are assumed to be sharp maxima and minima. Neglecting $\delta(w)$, making the choice (14.3.61), and noting

$$\partial_w \boldsymbol{u}(w) \cdot \nabla\beta[u(\boldsymbol{w})] = \partial_w \beta[\boldsymbol{u}(\boldsymbol{w})] , \tag{14.3.63}$$

we find

$$\frac{1}{\varepsilon} \int_a^{w_0} dw \{\dot{M}[\boldsymbol{S}(w)]/[p(w)d(w)]\} = \beta(w_0) - \beta(a) . \tag{14.3.64}$$

Using the sharp peaked nature of $p(w)^{-1}$, (14.3.64) can now be approximated by taking the value at the peak, using (14.3.57) and

$$N(a, t) = M[\boldsymbol{S}(b), t] , \tag{14.3.65}$$

as well as defining

$$\kappa(w_0) = \int_s^{w_0} [p(w)]^{-1} dw , \tag{14.3.66}$$

$$\mu(w_0) = \int_{w_0}^{b} [p(w)]^{-1} dw , \tag{14.3.67}$$

to obtain the relaxation equations

$$\kappa(w_0)\dot{N}_a(t) = \varepsilon d(w_0)[N_0(t)/n_0 - N_a(t)/n_a] , \tag{14.3.68}$$

$$\mu(w_0)\dot{N}_c(t) = \varepsilon d(w_0)[N_0(t)/n_0 - N_c(t)/n_c] . \tag{14.3.69}$$

These are of exactly the same form as those in the one-variable case and all the same interpretations can be made.

14.3.4 Example: Brownian Motion in a Double Potential

We consider Brownian motion in velocity and position as outlined in Sect. 6.4. Thus, we consider the Fokker-Planck equation

$$\frac{\partial P(x, p, t)}{\partial t} = -p\frac{\partial P}{\partial x} + U'(x)\frac{\partial P}{\partial p} + \gamma\left[\frac{\partial P}{\partial p}pP + \frac{\partial^2 P}{\partial p^2}\right] . \tag{14.3.70}$$

In the notation of the previous section we have

$$\left.\begin{aligned}
\boldsymbol{x} &= (x, p)\,,\\
\boldsymbol{v}(\boldsymbol{x}) &= (p, -U'(x) - \gamma p)\,,\\
\varepsilon &= 1\,,\\
\mathsf{D}(\boldsymbol{x}) &= \begin{bmatrix} 0 & 0 \\ 0 & \gamma \end{bmatrix},\\
P_\mathrm{s}(\boldsymbol{x}) &= \mathcal{N}_2 \exp\left[-\frac{1}{2}p^2 - U(x)\right],\\
\mathcal{N}_2 &= (2\pi)^{-1/2}\mathcal{N}_1\,,\\
\mathcal{N}_1 &= \left\{\int_{-\infty}^{\infty} dx \exp[-U(x)]\right\}^{-1}.
\end{aligned}\right\} \tag{14.3.71}$$

Hence, we can write

$$\boldsymbol{v}(\boldsymbol{x}) = \begin{bmatrix} 0 & -1 \\ 1 & \gamma \end{bmatrix} \cdot \nabla(\log P_\mathrm{s})\,, \tag{14.3.72}$$

and the current in the stationary state is

$$\boldsymbol{J}_\mathrm{s} = -\boldsymbol{v}P_\mathrm{s} + \mathsf{D}\cdot\nabla P_\mathrm{s} = -\nabla\cdot\left\{\begin{bmatrix} 0 & -1 \\ 1 & 0 \end{bmatrix} P_\mathrm{s}\right\}, \tag{14.3.73}$$

so that A exists, and

$$\mathsf{A} = \begin{bmatrix} 0 & -1 \\ 1 & 0 \end{bmatrix}. \tag{14.3.74}$$

Thus, assumption I is satisfied.

The plane $\boldsymbol{S}(w)$ can be written in the form

$$\lambda x + p = w\,. \tag{14.3.75}$$

Assumption II requires us to maximise $P_\mathrm{s}(x)$ on this plane, which is equivalent to maximising $-\frac{1}{2}p^2 - U(\boldsymbol{x})$ on this plane. Using standard methods, we find that maxima must lie along the curve $\boldsymbol{u}(w)$ given by

$$\boldsymbol{u}(w) = \begin{bmatrix} x(w) \\ p(w) \end{bmatrix} = \begin{bmatrix} x(w) \\ w - \lambda x(w) \end{bmatrix}, \tag{14.3.76}$$

where $x(w)$ satisfies

$$U'[x(w)] + \lambda^2 x(w) - \lambda w = 0\,. \tag{14.3.77}$$

Whether $P_\mathrm{s}(\boldsymbol{x})$ is sharply peaked depends on the nature of $U(x)$.

We now implement Assumption III.

The parameter λ is a function of w on the particular set of planes which satisfy (14.3.61). The tangent to $\boldsymbol{u}(w)$ is parallel to

$$\left[\frac{dx}{dw}, 1 - \lambda\frac{dx}{dw} - x\frac{d\lambda}{dw}\right], \tag{14.3.78}$$

and differentiating (14.3.77) we have

$$\frac{dx}{dw} = (U'' + \lambda^2)^{-1}\left[\lambda - \frac{d\lambda}{dw}(2\lambda x - w)\right]. \tag{14.3.79}$$

The normal to (14.3.75) is parallel to $(\lambda, 1)$. Hence,

$$\mathscr{D}^{\mathrm{T}}\boldsymbol{n} = (1+\lambda^2)^{-1/2}\begin{bmatrix} 0 & 1 \\ -1 & \gamma \end{bmatrix}\begin{bmatrix} \lambda \\ 1 \end{bmatrix} = (1+\lambda^2)^{-1/2}\begin{bmatrix} 1 \\ \gamma - \lambda \end{bmatrix}, \tag{14.3.80}$$

and this is parallel to (14.3.78) if

$$\frac{dx}{dw}\Big/1 = \left[1 - \lambda\frac{dx}{dw} - x\frac{d\lambda}{dw}\right]\Big/(\gamma - \lambda). \tag{14.3.81}$$

We can now solve (14.3.79, 14.3.81) simultaneously, to get

$$\frac{dx}{dw} = \frac{1}{\gamma} - \frac{x}{\gamma^2}\left[\frac{U'' - \gamma\lambda + \lambda^2}{x(U'' + \lambda^2) - (2\lambda x - w)}\right], \tag{14.3.82}$$

$$\frac{d\lambda}{dw} = \frac{1}{\gamma}\left[\frac{U'' - \lambda\gamma + \gamma^2}{x(U'' + \lambda^2) - (2\lambda x - w)}\right], \tag{14.3.83}$$

The saddle point is at $(x, p) = (0, 0)$ and thus $w = 0 \Longleftrightarrow x = 0$. Using this in (14.3.82) we see that we must have

$$x \approx w/\gamma \quad \text{as} \quad w \to 0. \tag{14.3.84}$$

Near $x = 0$, we write approximately

$$U[x] \simeq -\tfrac{1}{2}U_2 x^2, \tag{14.3.85}$$

and substituting (14.3.84, 14.3.85) in (14.3.77), we see that

$$\lambda^2 - \gamma\lambda + U''(0) = 0, \tag{14.3.86}$$

which determines

$$\lambda(0) = \frac{\gamma}{2} \pm \sqrt{\frac{\gamma^2}{4} + U_2}. \tag{14.3.87}$$

We now see that (14.3.83) tells us that $d\lambda/dw = 0$ when $w = 0$. Thus, λ will not change significantly from (14.3.87) around the saddle point, and we shall from now on approximate λ by (14.3.87).

Only one of the roots is acceptable and physically, this should be $\lambda \to \infty$ in the high friction limit which would give Kramers' result and requires the positive sign. The other root corresponds to taking a plane such that we get a minimum of $\boldsymbol{P}_s(\boldsymbol{x})$ on it.

We now integrate (14.3.61) and determine $d(w)$. Notice that $d(w)$ must be defined with $\boldsymbol{n}(w)$ a unit vector. Direct substitution in (14.3.80) and using (14.3.84) yields

$$(1+\lambda^2)^{-1/2} = \frac{dx}{dw}(w=0)d(0) = \frac{1}{\gamma}d(0), \tag{14.3.88}$$

so that

$$d(0) = \gamma(1+\lambda^2)^{-1/2}. \tag{14.3.89}$$

Further,

$$p(w) = \int_{S(w)} |d\mathbf{S}\, P_s(x)| = \int S(w)\sqrt{dx^2 + dp^2}\, P_s(x,p)\,,$$

$$= \mathcal{N}_2 \frac{[1+\lambda^2]^{1/2}}{\lambda} \int dp \exp\left[-\frac{p^2}{2} - U\left(\frac{w-p}{\lambda}\right)\right]. \tag{14.3.90}$$

An exact evaluation depends on the choice of $U(x)$. Approximately, we use

$$U(x) \approx U_0 - \tfrac{1}{2}U_2 x^2\,, \tag{14.3.91}$$

and evaluate the result as a Gaussian. We get

$$p(w) = \frac{(1+\lambda^2)^{1/2}}{\lambda}\mathcal{N}_2\, \mathrm{e}^{-U_0} \exp\left[\frac{U_2 w^2}{2(\lambda^2 - U_2)}\right], \tag{14.3.92}$$

and thus, using the definitions (14.3.66, 14.3.67),

$$\kappa(0) \equiv \int_{-\infty}^{0} p(w)^{-1} dw = \tfrac{1}{2}\mathcal{N}_2^{-1} \frac{\lambda\gamma}{(1+\lambda^2)^{1/2}} \frac{\mathrm{e}^{U_0}}{\sqrt{U_2}} = \mu(0)\,. \tag{14.3.93}$$

Thus, from (14.2.24) adapted to the many dimensional theory, we have for the mean first passage time from one well to the point $x = 0$,

$$\tau_0 = \kappa(0) d(0)^{-1} = \frac{\lambda}{2}\mathrm{e}^{U_0}\mathcal{N}_1^{-1}\sqrt{\frac{2\pi}{U_2}}\,, \tag{14.3.94}$$

i.e.,

$$\tau_0 = \frac{1}{2}\left(\frac{\gamma}{2} + \sqrt{\frac{\gamma^2}{4} + U_2}\right)\mathrm{e}^{U_0}\mathcal{N}_1^{-1}\sqrt{\frac{2\pi}{U_2}}\,. \tag{14.3.95}$$

Comparisons with Other Results

a) Exact One-Dimensional Mean First Passage Time Using the Smoluchowski Equation: Using the methods of Chap. 8, one reduces Kramers' equation in the large friction limit to the Smoluchowski equation for

$$\hat{P}(x,t) = \int dv\, P(x,v,t)\,, \tag{14.3.96}$$

that is, to

$$\frac{\partial \hat{P}(x,t)}{\partial t} = \frac{1}{\gamma}\frac{\partial}{\partial x}\left[U'(x)\hat{P} + \frac{\partial \hat{P}}{\partial x}\right], \tag{14.3.97}$$

and the *exact* result for the mean first passage time from $x = a$ to $x = 0$ for this approximate equation is

$$\tau_1 = \gamma \int_a^0 dx \exp[U(x)] \int_{-\infty}^{x} dz \exp[-U(z)]\,. \tag{14.3.98}$$

This result can be evaluated numerically.

b) Kramers' Result: This is obtained by applying our method to the one dimensional Smoluchowski equation (14.3.97) and making Gaussian approximations to all integrals. The result is

$$\tau_2 = \frac{1}{2}\gamma e^{U_0}\mathcal{N}_1^{-1}\sqrt{\frac{2\pi}{U_2}}, \tag{14.3.99}$$

which differs from (14.3.95) for τ_0 by the replacement $\lambda \to \gamma$, which is clearly valid in a large γ limit. In this limit,

$$\tau_0 \simeq (1 + U_2\gamma^{-2})\tau_2 . \tag{14.3.100}$$

c) Corrected Smoluchowski Equation: A more accurate equation than the Smoluchowski equation (14.3.1) is the *corrected Smoluchowski equation* (8.2.75);

$$\frac{\partial \hat{P}}{\partial t} = \frac{1}{\gamma}\frac{\partial}{\partial x}\left\{[1+\gamma^{-2}U''(x)]\left[U'(x)\hat{P} + \frac{\partial \hat{P}}{\partial x}\right]\right\} . \tag{14.3.101}$$

One now calculates the exact mean first passage time for this equation using standard theory; it is

$$\tau_3 = \gamma \int_a^0 dx[1+\gamma^{-2}U''(x)]\exp[U(x)] \int_{-\infty}^x dz \exp[-U(z)] . \tag{14.3.102}$$

Note however, that the principal contribution to the x integral comes from near $x = 0$ so that the small correction term $\gamma^{-2}U''(x)$ should be sufficiently accurately evaluated by setting

$$U''(x) \simeq U''(0) = -U_2 , \tag{14.3.103}$$

in (14.3.102). We then find the *corrected Smoluchowski result*,

$$\tau_3 = (1-\gamma^{-2}U_2)^{-1}\tau_1 \simeq (1+\gamma^{-2}U_2)\tau_1 . \tag{14.3.104}$$

Notice that in this limit,

$$\frac{\tau_3}{\tau_1} = \frac{\tau_0}{\tau_2}, \tag{14.3.105}$$

which means that in the limit that all integrals may be evaluated as sharply peaked Gaussians, our result is in agreement with the corrected Smoluchowski.

d) Numerical Simulations: By computer simulation of the equivalent stochastic differential equations

$$dx = p\,dt , \tag{14.3.106}$$
$$dp = -[\gamma p + U'(x)]dt + \sqrt{2\gamma}\,dW(t) , \tag{14.3.107}$$

we can estimate the mean first passage time *to the plane* $\mathbf{S}_0$, i.e., to the line

$$p = -\lambda x . \tag{14.3.108}$$

The results have to be computed for a given set of potentials. In order to assess the effect of the sharpness of peaking, we consider different temperatures T, i.e., we consider

$$dx = p\,dt\,, \tag{14.3.109}$$
$$dp = -[\gamma p + U'(x)]dt + \sqrt{2\gamma T}\,dW(t)\,. \tag{14.3.110}$$

By the substitutions

$$p \to pT^{1/2}\,, \qquad x \to xT^{1/2}\,, \tag{14.3.111}$$

we obtain

$$\left.\begin{aligned} dx &= p\,dt\,,\\ dp &= -[\gamma p + V'(x,T)]\,dt + \sqrt{2\gamma}\,dW(t)\,, \end{aligned}\right\} \tag{14.3.112}$$

where

$$V(x,T) = U(xT^{1/2})\,. \tag{14.3.113}$$

The simulations were performed with

$$U(x) = \frac{1}{4}(x^2-1)^2\,, \tag{14.3.114}$$

and the results are shown in Fig. 14.4. They separate naturally into two sets: curved, or straight lines. The best answer is the corrected Smoluchowski which agrees with the simulations at all temperatures, and at low temperatures, agrees with our method. Thus, we confirm the validity of the method in the region of validity expected, since low temperature corresponds to sharply peaked distributions.

Notice also that the choice of the plane $\boldsymbol{S}_0$ as the separatrix is appropriate on another ground. For, near to $x = 0$, $p = 0$, we can write

$$dx = p\,dt\,, \tag{14.3.115}$$
$$dp = (-\gamma p + U_2 x)\,dt + \sqrt{2\gamma T}\,dW(t)\,. \tag{14.3.116}$$

The condition that the deterministic part of (dx, dp), namely, $(p, -\gamma p + U_2 x)$ is in the direction connecting the point (x, p) to the origin is

$$\frac{p}{-\gamma p + U_2 x} = \frac{x}{p}\,. \tag{14.3.117}$$

Putting $p = -\lambda x$, we find

$$\lambda^2 - \lambda\gamma - U_2 = 0\,, \tag{14.3.118}$$

which is the same as (14.3.86) near $x = 0$. The two solutions correspond to the deterministic motion pointing towards the origin (positive root) or pointing away from the origin (negative root).

Thus, when the particle is on the separatrix, in the next time interval dt, only the random term $dW(t)$ will move it off this separatrix and it will move it right or left with equal probability, i.e., this means that the splitting probability, to left or right, should be 1:1 on this plane.

This separatrix definition also agrees with that of Sects. 14.1, 14.2 where $\boldsymbol{v}(x)$ should be perpendicular to the normal to $\boldsymbol{S}$.

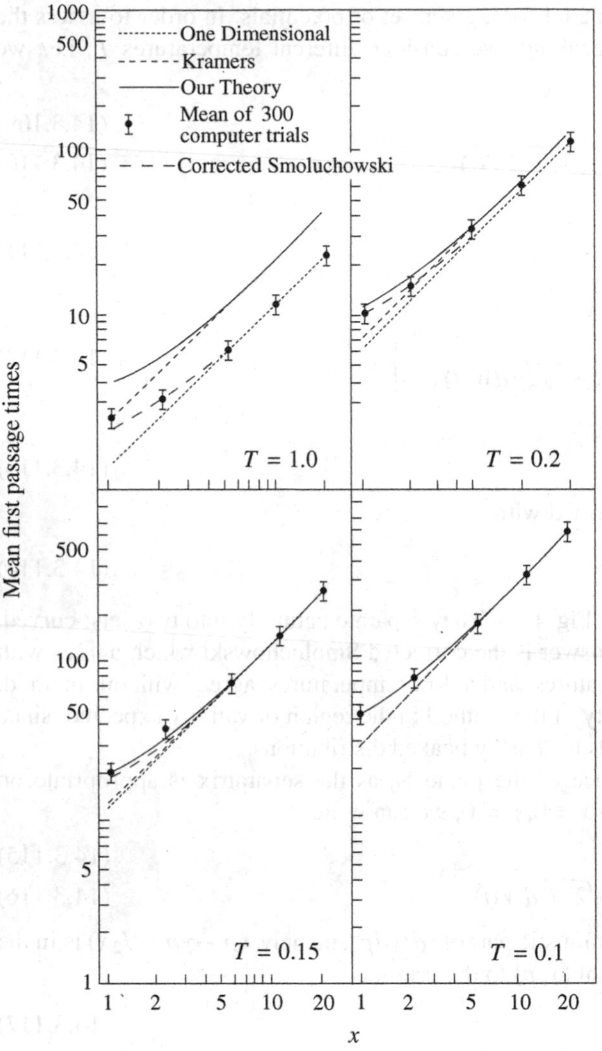

Fig. 14.4. Comparison of various estimates of the mean exit time from the double well potential of Sect. 14.3.4

15. Simulation of Stochastic Differential Equations

Simulating stochastic differential equations is something that can now be realistically attempted, in contrast to the situation when this book was first written. While the dramatic increase in the power and availability of computers is the most obvious reason, another is the development of a better understanding of the theoretical basis and efficiency of algorithms for solving stochastic differential equations. Unfortunately there is no easy adaptation of all but the simplest algorithms used for ordinary differential equations to the solution of stochastic differential equations, and it is also not possible to obtain the same order of convergence. However the same basic theoretical approach to algorithms can be used, and the differences can be made clear by quite straightforward arguments. As in the case of ordinary differential equations, the three main issues are *speed*, *accuracy* and *stability*. The first two of these are obvious requirements, and are of course connected with each other—a highly accurate algorithm can be used with a larger time step than a less accurate one, and thus the process of solution should proceed more rapidly. The aspect of stability is less obvious, but soon becomes apparent in practice if an unstable algorithm is used, in which errors in computation can become, often quite suddenly, so large that the solution found bears no relation to the real one.

The major differences in the simulation of stochastic differential equations arise from the *non differentiability* of the noise term, rather than its stochastic nature, and in this regard one has to keep a sense of realism, since the white noise which we use is an idealisation of a physical noise which may have a finite correlation time and may be differentiable. There is no point in trying to simulate accurately aspects of a model system which are valid on a time scale much shorter than the correlation time of the actual noise. The physical basis for the white noise limit, given in Chap. 6, can be used to assess whether some implementation of a non-white noise source should be used in a given case—very often the best way to implement the non-white noise is to use a supplementary stochastic differential equation. We will find, however, that there is one set of very important cases in which it is possible to use algorithms appropriate to ordinary differential equations, and that is the case when the noise terms are independent of the variable for which one is solving. In fact the class is somewhat wider than that, and the noise form in this class is known as *commutative noise*. As shown in Sect. 15.5.2c, there are non-trivial examples of equations of this kind.

There are no extensive libraries of stochastic differential equation algorithms available for use, and unfortunately uninformed intuition in this field can lead to considerable difficulty, extensive waste of time or even complete failure to solve the problem under study. This chapter is intended to address the main issues and give some advice on appropriate algorithms and strategies. It is not a comprehensive study of either all possible algorithms or of the theory underlying them. The theoretical basis is extensively canvassed in the books by *Kloeden and Platen* [15.1] and by *Kloeden, Platen and Schurz* [15.2]. The reference section to this chapter gives some sources which I hope will be helpful, and in the last section I have given an assessment of possible sources of software.

15.1 The One Variable Taylor Expansion

Intuition is a bad guide in the development of simulation algorithms for stochastic differential equations, so we develop the formal theoretical basis for them immediately, with no further preliminaries.

For simplicity, we consider first a one-variable, time-homogeneous stochastic differential equation, which can be written as the integral equation

$$x(t) = x(t_0) + \int_{t_0}^{t} a(x(s))\,ds + \int_{t_0}^{t} b(x(s))\,dW(s)\,. \tag{15.1.1}$$

Ito's formula (4.3.14) for an arbitrary function $f(x(t))$ can be written in integral form as

$$\begin{aligned} f(x(s)) = f(x(t_a)) &+ \int_{t_a}^{s} \left[a(x(s'))f'(x(s')) + \tfrac{1}{2}b(x(s'))^2 f''(x(s'))\right] ds' \\ &+ \int_{t_a}^{s} b(x(s'))f'(x(s'))\,dW(s')\,. \end{aligned} \tag{15.1.2}$$

This formula now gives a procedure for an iterative solution by substituting for $a(x(s))$ and $b(x(s))$ in the first equation in terms of $a(x(0))$ and a "remainder." This proceeds in various orders.

15.1.1 Euler Methods

The lowest order approximation is to set $t_a \to t_0$, neglect the integrals in (15.1.2), leading to the *explicit Euler algorithm*

$$x(t) \approx x(t_0) + a(x(t_0))(t - t_0) + b(x(t_0))[W(t) - W(t_0)]\,. \tag{15.1.3}$$

15.1.2 Higher Orders

The next order is quite easy in principle, but it is best to use an abbreviated notation, which we take in the form given by [15.2]. Introduce two operators

$$L_0 f = af' + \tfrac{1}{2}b^2 f'', \tag{15.1.4}$$

$$L_1 f = bf'\,. \tag{15.1.5}$$

In terms of these operators we find that (15.1.2) takes the form

$$f(x(s)) = f(x(t_a)) + \int_{t_a}^{s} L_0\, f(x(s'))\, ds' + \int_{t_a}^{s} L_1\, f(x(s'))\, dW(s') . \tag{15.1.6}$$

The strategy to be followed in developing an iterative expansion is governed by the fact that $dW(t)$ is of order of magnitude $\sqrt{dt}$. This means, roughly speaking, that we should evaluate the stochastic integrals to twice the order of the ordinary integrals. Therefore, one can use (15.1.6) to get the next higher order approximation to the stochastic integral in (15.1.1), getting

$$\begin{aligned} x(t) &= x(t_0) + a(x(t_0)) \int_{t_0}^{t} ds + b(x(t_0)) \int_{t_0}^{t} dW(s) \\ &\quad + L_1 b(x(t_0)) \int_{t_0}^{t} \int_{t_0}^{s} dW(s)\, dW(s') + \mathscr{R} . \end{aligned} \tag{15.1.7}$$

Here $\mathscr{R}$ is a remainder, and is given by

$$\begin{aligned} \mathscr{R} &= \int_{t_0}^{t} ds \int_{t_0}^{s} dW(s')\, L_1 a(x(s')) + \int_{t_0}^{t} dW(s) \int_{t_0}^{s} ds'\, L_0 b(x(s')) \\ &\quad + \int_{t_0}^{t} dW(s) \int_{t_0}^{s} dW(s') \int_{t_0}^{s'} dW(s'')\, (L_1)^2 b(x(s'')) \\ &\quad + \int_{t_0}^{t} ds \int_{t_0}^{s} ds'\, L_0 a(x(s')) \\ &\quad + \int_{t_0}^{t} dW(s) \int_{t_0}^{s} dW(s') \int_{t_0}^{s'} ds''\, L_0 L_1 b(x(s'')) . \end{aligned} \tag{15.1.8}$$

15.1.3 Multiple Stochastic Integrals

The double stochastic integral in (15.1.7) can be evaluated explicitly as in Sect. 4.2.2

$$\begin{aligned} \int_{t_0}^{t} \int_{t_0}^{s} dW(s)\, dW(s') &= \int_{t_0}^{t} [W(s) - W(t_0)]\, dW(s) \\ &= \tfrac{1}{2} \left\{ [W(t) - W(t_0)]^2 - (t - t_0) \right\} . \end{aligned} \tag{15.1.9}$$

It is clear that further iterates of the process will lead to a variety of stochastic integrals, such as

$$\int_{t_0}^{t} ds \int_{t_0}^{s} dW(s') , \tag{15.1.10}$$

$$\int_{t_0}^{t} ds \int_{t_0}^{s} dW(s') , \tag{15.1.11}$$

$$\int_{t_0}^{t} dW(s) \int_{t_0}^{s} dW(s') \int_{t_0}^{s'} dW(s''), \text{ etc.} \tag{15.1.12}$$

It may surprise the reader to be told that *none of these can be expressed directly in terms of Wiener process* $W(t)$. In practice this means that higher order simulations must include an algorithm for computing at least some of these integrals. Fortunately such algorithms have been developed by Kloeden and Platen [15.1, 15.2].

15.1.4 The Euler Algorithm

The rigorous estimation of the accuracy of an algorithm is quite a complex and detailed process, which is available in the book of Kloeden and Platen [15.1]—here we give a non-rigorous estimation of the error. To implement the algorithm we consider the time interval $(0, T)$ divided into N subintervals of size $\tau \equiv T/N$ at points $\tau_n = n\tau$, so that the function $x(t)$ is to be evaluated at the points

$$\tau_0, \tau_1, \tau_2, \tau_3, \ldots, \tau_{N-1}, \tau_N . \tag{15.1.13}$$

The corresponding Wiener increments are

$$\Delta W_n \equiv W(\tau_{n+1}) - W(\tau_n) . \tag{15.1.14}$$

Let us use the notation y_n for the solutions of the algorithm, which can be written as

$$y_{n+1} = y_n + a_n \tau + b_n \Delta W_n, \tag{15.1.15}$$
$$a_n = a(y_n), \tag{15.1.16}$$
$$b_n = b(y_n) . \tag{15.1.17}$$

The exact solution at the same points is written as $x_n \equiv x(\tau_n)$, which satisfies the exact equation on the interval (τ_n, τ_{n+1}), which we can write approximately using (15.1.7)

$$x_{n+1} \approx x_n + a(x_n)\tau + b(x_n)\Delta W_n + \tfrac{1}{2}L_0 b(x_n)(\Delta W_n{}^2 - \tau) , \tag{15.1.18}$$

where terms in the remainder $\mathscr{R}$ are of higher order in τ.

We can now consider at least two measures of the accuracy of the algorithm, which are related to the different measures of convergence, as discussed in Sect. 2.9, where four different definitions of convergence were presented.

a) Strong Order of Convergence: The most natural measure of the error is the root mean square of the difference between the exact solution and the solution of the Euler algorithm after a finite time interval T.

At the nth time step, the difference between the exact solution and the solution of the Euler algorithm is

$$e_n = x_n - y_n . \tag{15.1.19}$$

We can now subtract (15.1.15) from (15.1.18), to get

$$e_{n+1} = e_n + [a(y_n + e_n) - a_n]\tau + [b(y_n + e_n) - b_n]\Delta W_n \\ + \tfrac{1}{2}L_0 b(y_n + e_n)(\Delta W_n{}^2 - \tau) . \tag{15.1.20}$$

We now approximate using Taylor's theorem to lowest necessary order in each term to get

$$e_{n+1} = e_n(1 + a'_n \tau + b'_n \Delta W_n) + \tfrac{1}{2} L_0 b_n (\Delta W_n{}^2 - \tau). \tag{15.1.21}$$

Notice that

i) $\langle \Delta W_n^2 - \tau \rangle = 0$,

ii) $\langle \left(\Delta W_n^2 - \tau\right)^2 \rangle = 2\tau^2$,

iii) $\langle \Delta W_n \left(\Delta W_n^2 - \tau\right) \rangle = 0$,

iv) The solution of this equation makes e_n statistically independent of ΔW_n.

Using these properties, we see that

$$\langle e_{n+1}^2 \rangle = \langle e_n^2 \rangle \left[(1 + a'_n \tau)^2 + b'_n{}^2 \tau\right] + \tfrac{1}{2} (L_0 b_n)^2 \tau^2, \tag{15.1.22}$$

$$\leq \langle e_n^2 \rangle \left[(1 + A\tau)^2 + B\tau\right] + C\tau^2, \tag{15.1.23}$$

where A, B, C are the maxima of $a'_n, b'_n{}^2, \frac{1}{2}(L_0 b_n)^2$ over the range inside which the solution takes place.

Using the initial condition $e_0 = 0$, we find (using $N = T/\tau$) that $\langle e_N^2 \rangle$ has the bound

$$\langle e_N^2 \rangle \leq \frac{\left[(1 + A\tau)^2 + B\tau\right]^{T/\tau} - 1}{[(1 + A\tau)^2 + B\tau] - 1} C\tau^2, \tag{15.1.24}$$

$$\approx \frac{\exp[(2A + B)T] - 1}{2A + B} C\tau, \quad \text{for sufficiently large } N. \tag{15.1.25}$$

We now take as a measure of the error in the solution at time T the root mean square value

$$E(T) \equiv \sqrt{\langle e_N^2 \rangle} \sim \tau^{1/2}. \tag{15.1.26}$$

Thus, we say the strong order of convergence of the Euler algorithm is $\tau^{1/2}$.

b) Weak Order of Convergence: The measure of weak convergence is the rate at which the average of a smooth function of the variable approaches its exact value, and it is normally more rapid than the order of strong convergence.

First we consider the mean error; from (15.1.21), we note that the last term has zero mean, so that we add in as well the contribution from $\mathscr{R}$ in (15.1.8) to get

$$\langle e_{n+1} \rangle \approx \langle e_n \rangle (1 + a'_n \tau) + P_n \tau^2. \tag{15.1.27}$$

The final term derives from $\langle \mathscr{R} \rangle$—it is easy to see from (15.1.8) that the only non-zero contribution to the mean comes from the last two terms, both of which are of order τ^2. Thus we can deduce that there is a constant D such that

$$|\langle e_{n+1} \rangle| \leq |\langle e_n \rangle| |1 + D\tau| + |P|\tau^2. \tag{15.1.28}$$

Carrying out reasoning similar to that used above for strong convergence, this shows that

$$|\langle e_N \rangle| \sim \tau. \tag{15.1.29}$$

Now take a smooth function $f(x)$ and consider the difference between its mean as evaluated exactly, and as evaluated from the Euler algorithm

$$\langle f(x_N) - f(y_N)\rangle \approx \langle f'(y_N)\rangle\langle e_N\rangle + \tfrac{1}{2}\langle f''(y_N)\rangle\langle e_N^2\rangle + \dots \tag{15.1.30}$$

However, from (15.1.25, 15.1.29) the terms on the right hand side are both of order τ^1, so we deduce that the order of weak convergence of this algorithm is also τ^1.

15.1.5 Milstein Algorithm

Neglecting the remainder $\mathscr{R}$ gives us the *Milstein algorithm*

$$\begin{aligned} x(t) \approx x(t_0) &+ \left[a(x(t_0)) - \tfrac{1}{2}b(x(t_0))b'(x(t_0))\right](t-t_0) \\ &+ b(x(t_0))\,[W(t)-W(t_0)] + \tfrac{1}{2}b(x(t_0))b'(x(t_0))\,[W(t)-W(t_0)]^2 . \end{aligned} \tag{15.1.31}$$

The accuracy of the implementation of the Milstein algorithm can be evaluated in much the same way as the Euler algorithm. The algorithm takes the form

$$y_{n+1} = y_n + a_n\tau + b_n\Delta W_n + c_n(\Delta W_n^2 - \tau)\,, \tag{15.1.32}$$

$$c(x) = \tfrac{1}{2}L_1 b(x) = \tfrac{1}{2}b(x)b'(x)\,, \tag{15.1.33}$$

$$c_n = c(y_n)\,. \tag{15.1.34}$$

For the exact solution we write

$$x_{n+1} \approx x_n + a(x_n)\,\tau + b(x_n)\,\Delta W_n + c(x_n)[\Delta W_n{}^2 - \tau] + R_n\,. \tag{15.1.35}$$

The remainder R_n is a stochastic quantity defined using (15.1.8) over the interval (τ_n, τ_{n+1}), and it is easy to see that

$$\langle R_n\rangle \sim P_n\tau^2, \tag{15.1.36}$$

$$\langle R_n{}^2\rangle \sim Q_n\tau^3, \tag{15.1.37}$$

$$\langle R_n\Delta W_n\rangle \sim U_n\tau^2, \tag{15.1.38}$$

$$\langle R_n(\Delta W_n{}^2 - \tau)\rangle \sim V_n\tau^3\,, \tag{15.1.39}$$

in the sense that there are such quantities P_n, Q_n, U_n, V_n such that this gives the typical leading behaviour for small τ.

As for the Euler algorithm, we get the recursion relation

$$e_{n+1} \approx \left\{1 + a_n'\tau + b_n'\Delta W_n + c_n'[\Delta W_n{}^2 - \tau]\right\} e_n + R_n\,. \tag{15.1.40}$$

The error term e_n is statistically independent of ΔW_n, so we deduce

$$\langle e_{n+1}\rangle = (1 + a_n'\tau)\langle e_n\rangle + P_n\tau^2\,, \tag{15.1.41}$$

$$\begin{aligned} \langle e_{n+1}{}^2\rangle &= \left[(1+a_n'\tau)^2 + b_n'{}^2\tau + 2c_n'{}^2\tau^2\right]\langle e_n{}^2\rangle \\ &\quad + 2\left[b_n'U_n\tau^2 + c_n'V_n\tau^3\right]\langle e_n\rangle + Q_n\tau^3\,. \end{aligned} \tag{15.1.42}$$

From the first recursion relation we can deduce that $\langle e_n\rangle \sim \tau$, so that the contribution of the term in $\langle e_n\rangle$ in the second recursion relation is of order τ^3. Using the same methods as for the Euler algorithm we can deduce that the error measure for the Milstein algorithm is

$$E(T) \equiv \sqrt{\langle {e_N}^2 \rangle} \sim \tau . \tag{15.1.43}$$

This shows that the order of strong convergence is τ^1, and similar arguments to those used for the Euler algorithm show that the order of weak convergence is also τ^1.

15.2 The Meaning of Weak and Strong Convergence

The fact that the Euler algorithm (and indeed other algorithms as well) has different rates of convergence depending on the kind of convergence chosen is mathematically not surprising, since we have already seen in Sect. 2.9 that there are several different and inequivalent definitions of convergence for random variables. Clearly, what is called "the order of strong convergence" in this chapter corresponds to the *mean square limit* of Sect. 2.9, while "the order of weak convergence" corresponds to the *limit in distribution.*

What does this mean intuitively? Consider a number of exact solution trajectories $x^r(t)$ corresponding to different sample values $W^r(t)$ of a Wiener process, and the corresponding solutions $y^r(t)$ for a given time step value τ, produced using the Euler algorithm. The strong convergence criterion means essentially that the average value of $[x^r(t)-y^r(t)]^2 \sim \tau$; and this will mean that for most trajectories $x^r(t)-y^r(t) \sim \pm\tau^{1/2}$.

When we take averages of functions, our proof in Sect. 15.1.4b shows that the positive and negative terms of order $\pm\tau^{1/2}$ cancel, leaving only a residual error of order τ^1. Using the Milstein algorithm gives a better estimate of the actual paths, but does not improve the estimate of the averages.

The correspondence between reality and a stochastic differential equation is most logically given in terms of averages, since these are all we can measure. The precise trajectory of a given Wiener process used to generate a particular realisation of a stochastic differential equation is completely unknown and unknowable. If one looks carefully at the demonstration of the white noise limit of a non-white noise stochastic differential equation in Sect. 8.1, it will be clear that the derivation uses the *limit in distribution*. Thus, the trajectories of the physical non-white noise process are, in general, not in correspondence with those of the white noise process; as shown in Sect. 8.1.1, the same white noise stochastic differential equation can arise from quite different underlying non-white noise processes.

15.3 Stability

In practical situations the concept of *stability* of an algorithm is very important. By this is meant some measure of the extent to which an error in a solution will propagate as the algorithm is iterated. In order to discuss this more precisely one needs a definition of stability. To find a criterion which can be applied practically to all situations is essentially impossible, and in practice one often uses a generalisation of

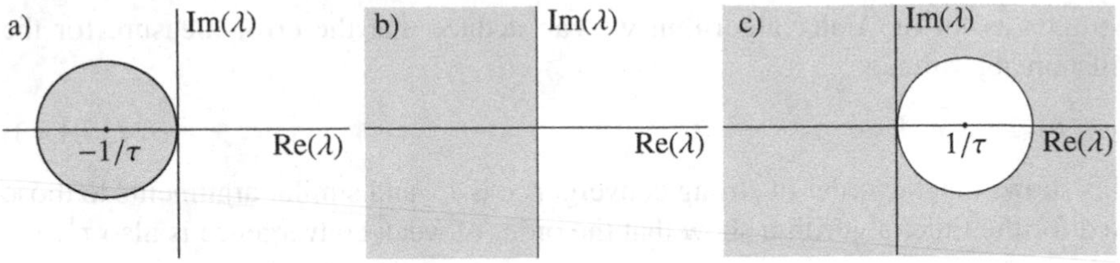

Fig. 15.1. Regions of stability (shaded grey) of: a) The explicit Euler algorithm; b) The semi-implicit algorithm; c) The fully implicit Euler algorithm

the concept of an *A-stable* algorithm, as adapted for stochastic differential equations [15.1, 15.2]. We consider the stability of a given algorithm when it is applied to the complex-valued linear stochastic differential equation

$$dx(t) = \lambda x(t) + dW(t) . \qquad (15.3.1)$$

a) Stability of the Explicit Euler Algorithm: For the Euler algorithm we would get from (15.1.15)

$$y_{n+1} = (1 + \lambda\tau)y_n + \Delta W_n . \qquad (15.3.2)$$

If $\bar{y}_n$ is a solution which starts at $\bar{y}_0$, then we can write

$$|y_{n+1} - \bar{y}_{n+1}| = |1 + \lambda\tau|\, |y_n - \bar{y}_n| , \qquad (15.3.3)$$

so that

$$|y_n - \bar{y}_n| = \left|1 + \lambda\tau\right|^n |y_0 - \bar{y}_0| . \qquad (15.3.4)$$

The difference $|y_0 - \bar{y}_0|$ can be seen as an error in the initial condition, and from this equation we deduce that the error will increase exponentially unless

$$|1 + \lambda\tau| < 1 . \qquad (15.3.5)$$

This is called the region of stability, and in the complex λ-plane it is the *interior* of a disc centred on $\lambda = -1/\tau$ and passing through the origin. As shown in Fig. 15.1a, this is a relatively small region of the complex λ-plane, but it is one which increases as the time step τ is reduced.

b) Stability of a Fully Implicit Euler Algorithm: In order to get a larger region of stability, one can introduce implicit algorithms, in which the formula required to compute the next value of x depends on that next value. The simplest of these is the fully implicit Euler algorithm, which comes from writing the approximation

$$x(t) \approx x(t_0) + a(x(t))(t - t_0) + b(x(t))[W(t) - W(t_0)] . \qquad (15.3.6)$$

To implement this algorithm for an arbitrary equation requires one to solve the implicit equation for $x(t)$ at every step, which in practice can only be done approximately. The advantage of an implicit algorithm is its much greater stability—we can show that the range of situations in which errors will start to grow uncontrollably is rather limited.

For the implementation of the implicit scheme (15.3.6), we would write instead of (15.1.15–17)

$$y_{n+1} = y_n + a_{n+1}\,\tau + b_{n+1}\,\Delta W_n, \tag{15.3.7}$$

$$a_n = a(y_n), \tag{15.3.8}$$

$$b_n = b(y_n)\,. \tag{15.3.9}$$

Applying this now to the test equation (15.3.1) we get in this case

$$y_{n+1} = y_n + \lambda\tau y_{n+1} + \Delta W_n\,, \tag{15.3.10}$$

leading to

$$|y_n - \bar{y}_n| = \left|1 - \lambda\tau\right|^{-n} |y_0 - \bar{y}_0|\,. \tag{15.3.11}$$

The region of stability is now

$$\left|1 - \lambda\tau\right| > 1\,, \tag{15.3.12}$$

which is the *exterior* of a disk centred on $\lambda = 1/\tau$, and passing through the origin, as shown in Fig. 15.1c.

c) Stability of a Semi-implicit Algorithm: Finally, one can introduce a compromise *semi-implicit* algorithm corresponding to the average of the implicit and explicit algorithms

$$y_{n+1} = y_n + \frac{a_n + a_{n+1}}{2}\tau + \frac{b_{n+1} + b_n}{2}\Delta W_n\,, \tag{15.3.13}$$

which gives

$$|y_n - \bar{y}_n| = \left|\frac{1 + \lambda\tau/2}{1 - \lambda\tau/2}\right|^n |y_0 - \bar{y}_0|\,. \tag{15.3.14}$$

This is stable when

$$|1 + \lambda\tau/2| < |1 - \lambda\tau/2|\,, \tag{15.3.15}$$

which corresponds to the left hand half plane, $\mathrm{Re}(\lambda) < 0$, as shown in Fig. 15.1b.

d) Definition of *A*-stability: An algorithm is said to be *A-stable* if the region of stability of the test equation (15.3.1) includes the whole of the left half of the complex λ-plane. For this equation, this means that wherever the exact solution does not grow, nor does the error, and we can see that both the fully implicit and the semi-implicit algorithms are A-stable. In fact the fully implicit method is stable even in regions where the solution itself can grow exponentially.

The implementation of any kind of implicit algorithm requires the introduction of a method of solving the implicit equation—that is, the equations (15.3.7, 15.3.13)—for the update of $x_n \to x_{n+1}$ which occurs at each time step, and this itself can be a nontrivial numerical problem. In practice, the semi-implicit concept appears to be the best compromise [15.3–15.5], in that it is A-stable, and also it is in practice more rapidly convergent than a fully implicit method.

15.3.1 Consistency

Kloeden and Platen [15.1] have introduced the term *consistency* to mean that the algorithm should converge in the limit of zero time step to the correct solution. While this seems an obvious requirement, it is in fact not satisfied by the implicit or the semi-implicit algorithms as introduced above, unless $b(x)$ is independent of x. However, these algorithms can be corrected, and the correction required is essentially of the same kind as used to relate the Ito to the Stratonovich version of the stochastic differential equation.

15.4 Implicit and Semi-implicit Algorithms

We consider an algorithm for the solution of the stochastic differential equation (15.1.1) given by the rule

$$\bar{y}_n = \epsilon y_{n+1} + (1-\epsilon) y_n, \tag{15.4.1}$$

$$\bar{A}_n = A(\bar{y}_n), \tag{15.4.2}$$

$$\bar{b}_n = b(\bar{y}_n), \tag{15.4.3}$$

$$y_{n+1} = y_n + \bar{A}_n \tau + \bar{b}_n \Delta W_n\,, \tag{15.4.4}$$

and here $A(x)$ is a function which will be determined so as to give the correct solution for the stochastic differential equation. The value of ϵ will characterise the degree of implicitness of the algorithm

To implement (15.4.4) requires us to solve the set of equations (15.4.1–15.4.4) for y_{n+1} at each time step by some algorithm which we do not specify at this time.

The convergence of the algorithm can be investigated by solving the system approximately to sufficient accuracy. Thus, if we write

$$z_n = y_{n+1} - y_n\,, \tag{15.4.5}$$

so that

$$\bar{y}_n = y_n + \epsilon z_n\,. \tag{15.4.6}$$

We then expand in powers of z_n

$$A(\bar{y}_n) = A(y_n) + \epsilon z_n A'(y_n) + \frac{\epsilon^2 z_n^2}{2} A''(y_n)\,, \tag{15.4.7}$$

$$b(\bar{y}_n) = b(y_n) + \epsilon z_n b'(y_n) + \frac{\epsilon^2 z_n^2}{2} b''(y_n)\,. \tag{15.4.8}$$

The set (15.4.1–15.4.8) yields a quadratic equation for z_n, and this can be solved up to second order to give

$$y_{n+1} = y_n + (A_n + \epsilon b_n b'_n)\tau + \epsilon b_n b'_n \left(\Delta W_n^2 - \tau\right) + b_n \Delta W_n\,. \tag{15.4.9}$$

The choice of $A(x)$ is now determined as

$$A(x) = a(x) - \epsilon b(x)b'(x). \tag{15.4.10}$$

a) Semi-implicit Algorithm: For $\epsilon = 1/2$, this choice makes the algorithm exactly equivalent to the Milstein algorithm (15.1.33); this is the correct form of the *semi-implicit algorithm.* It clearly has the same weak and strong orders of convergence, τ^1, as the Milstein algorithm.

b) Stratonovich Interpretation: If we take the Stratonovich stochastic differential equation

$$S\,dx = A(x)\,dt + b(x)\,dW(t), \tag{15.4.11}$$

then the semi-implicit algorithm corresponds to an algorithm of the same kind which one would use for an ordinary differential equation, namely

$$\bar{y}_n = \tfrac{1}{2}(y_{n+1} + y_n), \tag{15.4.12}$$

$$\bar{A}_n = A(\bar{y}_n), \tag{15.4.13}$$

$$\bar{b}_n = b(\bar{y}_n), \tag{15.4.14}$$

$$y_{n+1} = y_n + \bar{A}_n\tau + \bar{b}_n\Delta W_n. \tag{15.4.15}$$

It is clear too that this algorithm corresponds directly to the definitions of the Stratonovich stochastic integral and the Stratonovich stochastic differential equation.

For convenience, we also rewrite the algorithm in a form which is more explicitly as it would be executed on a computer

$$\bar{y}_n = y_n + \tfrac{1}{2}A(\bar{y}_n)\tau + \tfrac{1}{2}b(\bar{y}_n)\Delta W_n, \tag{15.4.16}$$

$$y_{n+1} = 2\bar{y}_n - y_n. \tag{15.4.17}$$

In this form, it is clear that the first equation is an implicit equation for $\bar{y}_n$, which must be solved by some root finding mechanism. Often simple iteration is used—about three iterates are usually sufficient.

15.5 Vector Stochastic Differential Equations

The treatment of stochastic differential equations with several variables, and more importantly, several noise terms, is not at all straightforward, and the development of algorithms for them is still an active subject. In this section we shall not do much more than sketch out the issues, and present the simpler algorithms. The methodology for proving the orders of convergence of the simple algorithms presented here is much the same as that for the one variable autonomous equations presented in the earlier sections of this chapter—it is simply somewhat more tedious.

15.5.1 Formulae and Notation

We want to consider the simulation of systems of stochastic differential equations, such as given in (4.3.21), but written in integral equation form

$$x(t) = x(t_0) + \int_{t_0}^{t} A(x(s), s)\, ds + \int_{t_0}^{t} \mathsf{B}(x(s), s)\, dW(s)\,. \tag{15.5.1}$$

We introduce the multidimensional operators, analogues of those defined in (15.1.4–15.1.5),

$$L_0 = \frac{\partial}{\partial t} + \sum_{k=1}^{d} A_k \frac{\partial}{\partial x_k} + \frac{1}{2} \sum_{k,l=1}^{d} \sum_{j=1}^{m} B_{k,j} B_{l,j} \frac{\partial^2}{\partial x_k \partial x_l}, \tag{15.5.2}$$

$$L_j = \sum_{k=1}^{d} B_{k,j} \frac{\partial}{\partial x_k}\,. \tag{15.5.3}$$

Here m is the dimension of the noise vector $dW(t)$, while d is the dimension of the variable $x(t)$.

In the case of an arbitrary function $f(x(t), t)$, Ito's formula (4.3.17) can be written in integral form as

$$f(x(s), s) = f(x(t_a), t_a) + \int_{t_a}^{s} L_0 f(x(s'), s')\, ds' + \sum_{j=1}^{m} \int_{t_a}^{s} L_j f(x(s'), s')\, dW_j(s')\,. \tag{15.5.4}$$

The procedure now is the same as in Sect. 15.1.2; we obtain an expansion analogous to (15.1.7)

$$x_i(t) = x_i(t_0) + A_i(x(t_0), t_0) \int_{t_0}^{t} ds + \sum_{j=1}^{m} B_{i,j}(x(t_0), t_0) \int_{t_0}^{t} dW_j(s) + \sum_{j,j'=1}^{m} L_{j'} B_{i,j}(x(t_0), t_0) \int_{t_0}^{t} dW_j(s) \int_{t_0}^{s} dW_{j'}(s') + \mathscr{R}\,. \tag{15.5.5}$$

The remainder term is analogous to that in (15.1.7), but more difficult to write down.

15.5.2 Multiple Stochastic Integrals

There is a major qualitative difference in the many variable case, occasioned by the appearance of the double stochastic integral

$$I_{i,j}(t, t_0) \equiv \int_{t_0}^{t} dW_i(s) \int_{t_0}^{s} dW_j(s')\,. \tag{15.5.6}$$

Only if $j = j'$ can this be explicitly evaluated as a function of the increments

$$\Delta W_j(t, t_0) \equiv \int_{t_0}^{t} dW_j(s)\,, \tag{15.5.7}$$

as in (15.1.9).

The symmetrised integral can be evaluated, but the evaluation requires some care, since we must use Ito integration. An adaptation of the proof in Sect. 4.2.2 gives the result

$$I_{i,j}(t,t_0) + I_{j,i}(t,t_0) = \int_{t_0}^{t} dW_i(s) \int_{t_0}^{s} dW_j(s') + \int_{t_0}^{t} dW_j(s) \int_{t_0}^{s} dW_i(s'), \tag{15.5.8}$$

$$= \int_{t_0}^{t} dW_i(s) \int_{t_0}^{t} dW_j(s') - \delta_{i,j}(t-t_0), \tag{15.5.9}$$

$$= \Delta W_i(t,t_0)\,\Delta W_j(t,t_0) - \delta_{i,j}(t-t_0). \tag{15.5.10}$$

a) General Stochastic Multiple Integrals: This is only one of a whole class of stochastic multiple integrals which arise in a general expansion to higher orders of (15.5.5). Following the formulation of Kloeden and Platen [15.1, 15.2], these can be concisely written by defining

$$W_0(t,t_0) \equiv t - t_0, \tag{15.5.11}$$

and the general integral to be evaluated can be written

$$I_{(i_1,i_2,i_3,\dots,i_n)}(t,t_0) \equiv \int_{t_0}^{t} dW_{i_1}(s_1) \int_{t_0}^{s_1} dW_{i_2}(s_2) \int_{t_0}^{s_2} dW_{i_3}(s_3) \dots \int_{t_0}^{s_{n-1}} dW_{i_n}(s_n). \tag{15.5.12}$$

There are also similarly defined Stratonovich multiple integrals, which are more convenient for the development of higher order algorithms. Kloeden and Platen give methods for their approximate evaluation in their books.

b) The Antisymmetric Second Order Integral: When account is taken of the relationships (15.5.8–15.5.10) it is convenient in lower order algorithms to use the definition

$$E_{i,j}(t,t_0) \equiv \int_{t_0}^{t} \int_{t_0}^{s} \left[dW_i(s) dW_j(s') - dW_j(s) dW_i(s') \right], \tag{15.5.13}$$

so that

$$I_{i,j}(t,t_0) = \tfrac{1}{2}\{\Delta W_i(t,t_0)\,\Delta W_j(t,t_0) + E_{i,j}(t,t_0)\}. \tag{15.5.14}$$

c) Commutative and Non-commutative Noise: The term involving the double stochastic integral in (15.5.5) *can* be expressed in terms of Wiener increments if the coefficient is symmetric

$$L_{j'} B_{i,j} = L_j B_{i,j'}. \tag{15.5.15}$$

Using the definition (15.5.3), this requirement can be written

$$\sum_{k=1}^{d} B_{k,j'} \frac{\partial B_{i,j}}{\partial x_k} = \sum_{k=1}^{d} B_{k,j} \frac{\partial B_{i,j'}}{\partial x_k}, \tag{15.5.16}$$

and if this condition is satisfied, the set of equations is said to have *commutative noise*. While not a particularly transparent condition, it is not uncommonly satisfied. Most notable are the cases

i) When the coefficients $B_{i,j}$ are independent of the variable $\boldsymbol{x}(t)$, but may have an explicit time dependence. This is known as *additive noise*.

ii) When there is only one noise term, so that the indices j, j' are given by $j = j' = 1$.

iii) If the noise coefficient can be written in the form

$$B_{i,j}(\boldsymbol{x}) = x_i G_{i,j}\,, \tag{15.5.17}$$

where $G_{i,j}$ is an arbitrary constant matrix, then

$$\sum_k B_{k,j'} \frac{\partial B_{i,j}}{\partial x_k} = \sum_k x_k G_{k,j'} \delta_{i,k} G_{i,j} = x_i G_{i,j} G_{i,j'} = \sum_k B_{k,j} \frac{\partial B_{i,j'}}{\partial x_k}. \tag{15.5.18}$$

iv) There is the special case of the previous condition when $d = m$ and $G_{i,j} \propto \delta_{i,j}$, which is called *diagonal noise* by Kloeden and Platen.

15.5.3 The Vector Euler Algorithm

With time steps at $\tau_n = n\tau$, and corresponding Wiener increments

$$\Delta \boldsymbol{W}^n \equiv \boldsymbol{W}(\tau_{n+1}) - \boldsymbol{W}(\tau_n)\,, \tag{15.5.19}$$

we use the notation $\boldsymbol{y}_n$ for the solutions of the algorithm. The vector Euler algorithm is then defined by

$$\boldsymbol{y}^{n+1} = \boldsymbol{y}^n + \boldsymbol{A}^n \tau + \mathsf{B}^n \Delta \boldsymbol{W}^n, \tag{15.5.20}$$

$$\boldsymbol{A}^n = \boldsymbol{A}(\boldsymbol{y}^n, \tau_n)\,, \tag{15.5.21}$$

$$\mathsf{B}^n = \mathsf{B}(\boldsymbol{y}^n, \tau_n)\,. \tag{15.5.22}$$

This algorithm has strong order of convergence $\tau^{1/2}$, and weak order of convergence τ^1. A proof along the lines of that given in Sect. 15.1.4 can be straightforwardly developed.

15.5.4 The Vector Milstein Algorithm

The vector Milstein algorithm uses the stochastic integrals of Sect. 15.5.2

$$y_i^{n+1} = y_i^n + A_i^n \tau + \sum_k B_{i,k}^n \,\Delta W_k^n + \sum_{j,j'} C_{i,j,j'}\, \Delta W_{j,j'}^n\,. \tag{15.5.23}$$

In the last term we need the quantity

$$C_{i,j,j'} \equiv \sum_{k=1}^{d} B_{k,j'}(\boldsymbol{y}^n, \tau_n) \frac{\partial B_{i,j}(\boldsymbol{y}^n, \tau_n)}{\partial x_k}\,, \tag{15.5.24}$$

and the stochastic integral, which is defined as

$$\Delta W_{j,j'}^n \equiv \int_{\tau_n}^{\tau_{n+1}} dW_j(s) \int_{\tau_n}^{s} dW_{j'}(s')\,. \tag{15.5.25}$$

It is more convenient to write the algorithm in terms of the symmetric and antisymmetric parts of this quantity; thus we write

$$\Delta W^n_{j,j'} = \tfrac{1}{2}\left\{\Delta W^n_j\,\Delta W^n_{j'} - \tau\,\delta_{j,j'} + E^n_{j,j'}\right\}, \tag{15.5.26}$$

in which we have used (15.5.10) for the symmetric part, and we define the antisymmetric part by

$$E^n_{j,j'} \equiv \int_{\tau_n}^{\tau_{n+1}} dW_j(s)\int_{\tau_n}^{s} dW_{j'}(s') - \int_{\tau_n}^{\tau_{n+1}} dW_{j'}(s)\int_{\tau_n}^{s} dW_j(s'). \tag{15.5.27}$$

Provided one can simulate the antisymmetric stochastic integral $E^n_{j,j'}$ the Milstein algorithm has both weak and strong order of convergence τ^1.

15.5.5 The Strong Vector Semi-implicit Algorithm

There are two kinds of vector semi-implicit algorithm, known as weak and strong. These semi-implicit algorithms have greater stability than the Euler and Milstein algorithms, and thus are greatly preferred for practical implementation.

The strong vector semi-implicit algorithm can be put in the form

$$\begin{aligned}
\bar{y}^n &= \tfrac{1}{2}\left(y^{n+1} + y^n\right), && (15.5.28)\\
\bar{\tau}_n &= \tfrac{1}{2}\left(\tau_{n+1} + \tau_n\right), && (15.5.29)\\
\bar{A}^n &= A(\bar{y}^n, \bar{\tau}_n), && (15.5.30)\\
\bar{\bar{A}}^n_i &\equiv \bar{A}^n_i - \tfrac{1}{2}\sum_j \bar{C}_{i,j,j}, && (15.5.31)\\
\mathsf{B}^n &= \mathsf{B}(\bar{y}^n, \bar{\tau}_n), && (15.5.32)\\
\bar{C}_{i,j,j'} &= C_{i,j,j'}(\bar{y}^n, \bar{\tau}_n), && (15.5.33)\\
y^{n+1}_i &= y^n_i + \bar{\bar{A}}^n_i\,\tau + \sum_k \bar{B}^n_{i,k}\,\Delta W^n_k + \sum_{j,j'} \bar{C}_{i,j,j'}\,E^n_{j,j'}\,. && (15.5.34)
\end{aligned}$$

It can be shown to have both strong and weak order of convergence τ^1, but requires one to simulate the antisymmetric stochastic integral $E^n_{j,j'}$ unless the condition of commutative noise, (15.5.16), is satisfied.

15.5.6 The Weak Vector Semi-implicit Algorithm

The weak vector semi-implicit algorithm arises by noting that the mean of $E^n_{j,j'}$ vanishes, so omitting this term gives an algorithm which has weak order of convergence τ^1; explicitly, using the same form as (15.4.16, 15.4.17), it can be written

$$\bar{\tau}_n = \tfrac{1}{2}\left(\tau_{n+1} + \tau_n\right), \tag{15.5.35}$$

$$\bar{\boldsymbol{A}}^n = \boldsymbol{A}(\bar{\boldsymbol{y}}^n, \bar{\tau}_n), \tag{15.5.36}$$

$$\bar{\bar{A}}_i^n \equiv \bar{A}_i^n - \tfrac{1}{2}\sum_j \bar{C}_{i,j,j}, \tag{15.5.37}$$

$$\mathsf{B}^n = \mathsf{B}(\bar{\boldsymbol{y}}^n, \bar{\tau}_n), \tag{15.5.38}$$

$$\bar{C}_{i,j,j'} = C_{i,j,j'}(\bar{\boldsymbol{y}}^n, \bar{\tau}_n), \tag{15.5.39}$$

$$y_i^{n+1} = y_i^n + \bar{\bar{A}}_i^n\,\tau + \sum_k \bar{B}_{i,k}^n\,\Delta W_k^n. \tag{15.5.40}$$

$$\boldsymbol{y}^{n+1} = 2\bar{\boldsymbol{y}}^n - \boldsymbol{y}^n. \tag{15.5.41}$$

This algorithm has the significant advantage that the antisymmetric stochastic integral does not need to be simulated.

15.6 Higher Order Algorithms

a) Special Structural Features: There are quite a number of higher order algorithms available, but the increased accuracy is obtained only at the cost of increasing complexity and the requirement to compute stochastic multiple integrals—and the number of different kinds of integrals which can turn up increases exponentially with the order of convergence. The advice given by Kloeden [15.6] puts the situation in perspective

> Software for general SDEs is likely to be inefficient unless it takes into account special structural features of specific SDEs which can *avoid* the need to generate *all* multiple stochastic integrals in the scheme.

The principal "special structural features" which assist in this way arise from the presence of commutative noise.

b) Simplified Noise for Weak Algorithms: There are weakly convergent algorithms which replace the full Gaussian noise with a simplified noise, such as a dichotomic random noise which takes on only the values ± 1, thereby achieving a speed advantage.

c) Computation of Derivatives: Computation of derivatives analytically is something which it is often felt worthwhile avoiding, since they can produce expressions of significant complexity. Thus, all but the Euler schemes discussed above require the computation of $C_{i,j,j'}$, as defined by (15.5.23). However, provided the noise is commutative, and provided we are using a semi-implicit scheme, this only appears in the coefficient of τ, and is equivalent to the Ito correction required when converting to a Stratonovich stochastic differential equation. In many situations in the natural sciences the equation is in fact most naturally defined as a Stratonovich stochastic differential equation, so in fact *no derivatives need be computed*. However, such is not necessarily the case in the kinds of models used in finance.

d) Algorithms of the Runge-Kutta Type: *Milstein* and *Tretyakov* [15.7] have considered various more sophisticated algorithms for propagating stochastic differential equations with dynamic random processes. In particular, they have shown that for situations where the influence of the dynamic noise on the system is very much less than the deterministic evolution (the small noise limit), one can accurately describe the total evolution using a relatively simple modification to the fourth-order Runge-Kutta algorithm. Essentially, in this method one calculates the deterministic evolution using the fourth-order Runge-Kutta algorithm, while the noise is calculated using an Euler type derivative calculation based on the state of the system at the start of each time step.

e) Use of Higher-order Algorithms: One can probably conclude that higher order algorithms are best left to experts in writing software. The collection of software available in conjunction with [15.2] provides all the basic tools needed if it is felt desirable to use such algorithms. On the other hand the *XMDS package* [15.8], which can be used to simulate a wide (but not exhaustive) range of problems employs the weak semi-implicit algorithm (15.5.35–15.5.41) for all stochastic differential equations since:

i) This algorithm is strongly convergent for commutative noise;

ii) If the equation is posed in Stratonovich form, there is no need to compute derivatives;

iii) The algorithm has good stability properties;

iv) The algorithm handles high dimensional situations very well, since it does not need to compute stochastic integrals.

When one takes into account the rapidly increasing speed of modern computers, and the time taken to write bug-free software, one comes to the conclusion that the semi-implicit algorithm is by far the best general purpose choice.

15.7 Stochastic Partial Differential Equations

Spatially distributed systems, such as those described in Chap. 8, naturally produce stochastic partial differential equations of the kind

$$da(x,t) = \left\{ D\frac{\partial^2 a(x,t)}{\partial x^2} + f(a(x,t),x,t) \right\} dt + g(a(x,t),x,t)\, dW(x,t)\,, \qquad (15.7.1)$$

and here the noise function satisfies the equation

$$dW(x,t)\, dW(x',t) = H(x,x')\, dt\,. \qquad (15.7.2)$$

In Chap. 8 $H(x,x')$ was typically a delta function, but we want to consider a more general possibility here. In fact, as noted in Sect. 13.3.3, delta-function correlated noise can give rise to divergences in some nonlinear cases, and therefore has to be smoothed in some way. Thus (15.7.2) is much more realistic.

Similar kinds of equations appear in quantum optics and the physics of Bose-Einstein condensation, though there these tend to be nonlinear versions of the Schrödinger equation with noise included, in contrast to the equations of Chap. 13, which were nonlinear relatives of the diffusion equation. However, their basic structure is the same, and the same kinds of algorithms are in general applicable. Even so, this example should not be considered to be the most general—generalisations to include higher dimensions and different orders of spatial derivatives are also possible.

In any numerical implementation of this sort of problem a *spatial lattice* of points x_i must be introduced, in which case the equation reduces to a vector stochastic differential equation of the kind considered in Sect. 15.5. However, there are two special features which must be taken into account, which make the treatment of this kind of stochastic differential equation worthy of special attention.

i) The only operator which connects different points on the spatial lattice is the spatial derivative, which is considerably simpler than a general function. The most obvious benefit is the possibility of using the fast Fourier transform to implement the derivative,

ii) The number of lattice points is normally quite large—in a two dimensional system a lattice of $512 \times 512 \approx 250,000$ points is quite common. This means that an algorithm which requires the evaluation of the antisymmetric noise $E^n_{j,j'}$ defined in (15.5.27) needs to compute not only 250,000 noise terms, but also about 30,000,000,000 antisymmetric noises.

   The extra speed gained by such an algorithm may well be destroyed by the added labour of the evaluation of these antisymmetric noises. Higher order algorithms which employ higher order stochastic integrals will behave even worse.

In this section we describe the procedure used by Drummond and co-workers [15.9], which has been very successful in practice even in three dimensional situations, and which has been implemented very conveniently in the package *XMDS* [15.8],in a form which automatically produces fast code for a reasonably general class of equations.

15.7.1 Fourier Transform Methods

In any practical problem the range of x is finite, and we take it to be the interval $(-L/2, L/2)$. Correspondingly, the Fourier transform notation used is

$$\mathscr{F}[a](k) \equiv \frac{1}{\sqrt{2\pi L}} \int_{-L/2}^{L/2} a(x)\mathrm{e}^{-ikx}\,dx, \quad k = \frac{2\pi r}{L}, \quad r \text{ integral.} \tag{15.7.3}$$

$$\equiv \tilde{a}(k)\,. \tag{15.7.4}$$

The corresponding inverse is written

$$\mathscr{F}^{-1}[\tilde{a}] \equiv \frac{1}{\sqrt{2\pi L}} \sum_k \tilde{a}(k)\mathrm{e}^{ikx}\,. \tag{15.7.5}$$

We can write derivative operators using the Fourier transform operator as, for example

$$\frac{\partial^2}{\partial x^2}a(x) = \mathscr{F}^{-1}\left[-k^2\mathscr{F}[a]\right](x). \qquad (15.7.6)$$

Such a Fourier transform can be carried out very quickly using a fast Fourier transform method, and makes spatial derivative operators into relatively easily computed linear operators. Of course in practical implementations the range of k must be restricted, and this is enforced by the requirement that the lattice points x_i must form a discrete grid.

Although the time overhead is not much different from using a finite difference method of sufficiently high order to be accurate—possibly up to order 10 or more—in practice one finds that algorithms which use Fourier transform methods are more stable than those that use finite difference methods.

15.7.2 The Interaction Picture Method

The name of this method comes from quantum mechanics, in which a similar technique is used to simplify the description of a system described by the sum of two Hamiltonians. Its application to the efficient numerical solution of partial differential equations was made by *Ballagh* [15.10]. It is most convenient to take the equation in the Stratonovich form, since this gives the most natural way of writing the semi-implicit algorithm. Thus, we consider an equation of the form

$$(S)\, da(x,t) = \{\mathscr{L}a(x,t) + f(a(x,t),x,t)\}\, dt + g(a(x,t),x,t)\, dW(x,t), \qquad (15.7.7)$$

in which $\mathscr{L}$ is a general partial differential operator with respect to x. The idea now is to carry out the evolution induced by $\mathscr{L}$ exactly, in alternation with the remainder of the evolution.

Thus, the solution of

$$dp(x,t) = \mathscr{L}p(x,t)\, dt \qquad (15.7.8)$$

can be written as

$$p(x,t) = \exp\{(t-t_0)\mathscr{L}\}\, p(x,t_0) \qquad (15.7.9)$$

$$= \mathscr{F}^{-1}\left[\exp\left\{(t-t_0)\tilde{\mathscr{L}}(k)\right\}\tilde{p}(k,t_0)\right], \qquad (15.7.10)$$

where $\tilde{\mathscr{L}}(k)$ is the Fourier transform expression of $\mathscr{L}$. We use this to define the "interaction picture" function $b(x,t)$ offset by an initial time T; thus

$$b(x,t) \equiv \exp\{-(t-T)\mathscr{L}\}\, a(x,t). \qquad (15.7.11)$$

This satisfies the stochastic differential equation

$$(S)\, db(x,t) = \exp\{-(t-T)\mathscr{L}\}\left[f(a(x,t),x,t)\, dt + g(a(x,t),x,t)\, dW(x,t)\right]. \qquad (15.7.12)$$

To this equation we now apply the weak vector semi-implicit algorithm (15.5.35–15.5.41). We choose the time T to be given by the mid-point time of the step

$$T = \bar{t}_n \equiv t_n + \frac{\tau}{2}, \tag{15.7.13}$$

since when we evaluate the functions f and g this will be at the mid point time, and with this choice we have there

$$a(x, \bar{t}_n) = b(x, \bar{t}_n), \tag{15.7.14}$$

since the exponential in (15.7.11) becomes 1.

The algorithm then takes the form

$$b(x, t_n) = \exp\{(t_n - \bar{t}_n)\mathscr{L}\}\, a(x, t_n), \tag{15.7.15}$$

$$b(x, \bar{t}_n) = b(x, t_n) + \tfrac{1}{2} f(b(x, \bar{t}_n), x, T)\,\tau + \tfrac{1}{2} g(a(x, T), x, \bar{t}_n)\,\Delta W_n(x), \tag{15.7.16}$$

$$b(x, t_{n+1}) = 2b(x, \bar{t}_n) - b(x, t_n), \tag{15.7.17}$$

$$a(x, t_{n+1}) = \exp\{(t_{n+1} - \bar{t}_n)\mathscr{L}\}\, b(x, t_{n+1}). \tag{15.7.18}$$

We remind the reader that this algorithm requires the original equation (15.7.7) to be in the *Stratonovich* form. It is of strong order $\tau^{1/2}$ and of weak order τ^1.

If the equation *in the interaction picture* satisfies the condition (15.5.16) for commutative noise, we can expect the algorithm to be of strong order τ^1—but it may not be easy to verify whether or not this condition is satisfied. However, it seems likely that if the noise is commutative in the original equation(15.7.1), then the degree to which the noise is not commutative in the interaction picture would be small.

15.8 Software Resources

There are no extensive libraries available for stochastic differential equations, so one cannot expect to find a ready made package to solve a particular problem. The following appear to be the major sources at the time of writing.

The major development of higher-order algorithms has been done by Kloeden, Platen and co-workers, and is available via Kloeden's home page [15.11]. In particular, one can find there the software (in Borland Pascal) for the book by Kloeden, Platen and Schurz [15.2], as well as software for Maple [15.12]. Some rather elementary software for Matlab is presented in [15.13] and somewhat more advanced version for Matlab as well as Maple in [15.14], with software for both of these downloadable from [15.15].

The XMDS package is a package designed for solving the kinds of ordinary and partial differential equations, both stochastic and non-stochastic which turn up in theoretical physics. The aim to is to provide high quality compiled C-code programs using a markup language based on XML. The package and documentation are available are downloadable from [15.8] and run on a variety of computer operating systems.

References

Chapter 1

1.1 R. Brown, *A brief account of microscopical observations... on the particles contained in the pollen of plants...*, Phil. Mag. **4**, 121 (1828).

1.2 A. Einstein, *Über die von der molekular-kinetischen Theorie der Wärme geforderte Bewegung von in ruhenden Flüssigkeiten suspendierten Teilchen*, Ann. Phys. (Leipzig) **17**, 549 (1905).

1.3 M. von Smoluchowski, *Zur kinetischen Theorie der Brownsche Bewegung*, Ann. Phys. (Leipzig) **21**, 756 (1906).

1.4 Lord Rayleigh, *Scientific Papers III* (Cambridge University Press, Cambridge, 1899–1920), p. 473 (Phil. Mag. 1891).

1.5 P. Langevin, *Sur la théorie du mouvement brownien*, Comptes Rendues **146**, 530 (1906).

1.6 M. von Smoluchowski, *Drei Vorträge über Diffusion, Brownsche Molekularbewegung und Koagulation kolloider Lösungen*, Phys. Z. **16**, 321 (1916); **17**, 557 (1916).

1.7 B. J. Berné and R. Pecora, *Dynamic Light Scattering* (Wiley, New York, 1976).

1.8 M. L. Bachelier, *Théorie de la Spéculation* (1995 reprint by Éditions Jacques Gabay, Paris, 1900).

1.9 P. A. Samuelson, *Rational Theory of Warrant Pricing*, Industrial Management Review **6**, 13 (1965).

1.10 R. C. Merton, *Continuous Time Finance* (Blackwell, Oxford, 1992).

1.11 F. Black and M. Scholes, *The Pricing of Options and Corporate Liabilities*, Journal of Political Economy **81**, 637 (1973).

1.12 R. C. Merton, *Theory of Rational Option Pricing*, Bell Journal of Economics and Management Science **4**, 141 (1973).

1.13 A. J. Lotka, *Elements of Physical Biology* (Williams and Wilkins, Baltimore, 1925) [Reissued as *Elements of Mathematical Biology*, Dover, 1956].

1.14 V. Volterra, *Variazioni e fluttuazioni del numero d'individui in specie animali conviventi*, Mem. Acad. Lincei **2**, 31 (1926).

1.15 C. B. Huffaker, *Experimental studies on predation: dispersion factors and predator-prey oscillations*, Hilgardia **27**, 343 (1958).

1.16 M. P. Hassell, *The Dynamics of Arthropod Predator-Prey Systems* (Monographs in Population Biology No.13) (Princeton University Press, Princeton, 1978).

1.17 S. O. Rice, *Mathematical analysis of random noise*, Bell Syst. Tech. J. **23**, 282 (1944); **24**, 46 (1945).

1.18 W. Schottky, *Über spontane Stromschwankungen in verschiedenen Elektrizitätsleitern*, Ann. Phys. (Leipzig) **57**, 541 (1918).

1.19 G. I. Taylor, *Diffusion by Continuous Movements*, Proc. London. Math. Soc. **s2-20**, 196 (1922).

1.20 N. Wiener, *Generalized harmonic analysis*, Acta. Math. **55**, 117 (1930).

1.21 A. Khinchin, *Korrelationstheorie der stationären stochastischen Prozesse*, Math. Annalen **109**, 604 (1934).

1.22 J. B. Johnson, *Thermal Agitation of Electricity in Conductors*, Phys. Rev. **32**, 97 (1928).

1.23 H. Nyquist, *Thermal Agitation of Electric Charge in Conductors*, Phys. Rev. **32**, 110 (1928).

Chapter 2

2.1 W. Feller, *An Introduction to Probability Theory and its Applications*, 2nd ed. (Wiley, New York, 1974).
2.2 A. Papoulis, *Probability, Random Variables, and Stochastic Processes* (McGraw-Hill, New York, 1965, 1984, 1991).
2.3 A. N. Kolmogorov, *Foundations of the Theory of Probability* (Chelsea, New York, 1950) [The German original appeared in 1933].
2.4 B. V. Gnedenko, *The Theory of Probability* (Chelsea, New York, 1963).
2.5 R. L. Stratonovich, *Introduction to the Theory of Random Noise* (Gordon and Breach, New York, 1963).
2.6 E. Meeron, *Series Expansion of Distribution Functions in Multicomponent Fluid Systems*, J. Chem. Phys **27**, 1238 (1957).
2.7 N. G. van Kampen, *A cumulant expansion for stochastic linear differential equations*, Physica **74**, 215 (1973); **74**, 239 (1973).
2.8 J. Marcinkiewicz, *Sur une propriété de la loi de Gauss*, Math. Zeits. **44**, 612 (1939).
2.9 A. K. Rajagopal and E. C. G. Sudarshan, *Some generalizations of the Marcinkiewicz theorem and its implications to certain approximation schemes in many-particle physics*, Phys. Rev. A **10**, 1852 (1974).

Chapter 3

3.1 P. E. Protter, *Stochastic Integration and Differential Equations*, 2nd ed. (Springer, New York, 2004).
3.2 F. Haake, *Springer Tracts in Modern Physics, Vol 66* (Springer, Berlin, Heidelberg, New York, 1973).
3.3 H. Spohn, *Kinetic equations from Hamiltonian dynamics: Markovian limits*, Rev. Mod. Phys. **52**, 569 (1980).
3.4 I. F. Gihman and A. V. Skorokhod, *The Theory of Stochastic Processes* Vols. I, II, III. (Springer, Berlin, Heidelberg, New York, 1975).
3.5 See [3.4], Vol II.
3.6 G. E. Uhlenbeck and L. S. Ornstein, *On the Theory of the Brownian Motion*, Phys. Rev. **36**, 823 (1930).
3.7 A. Papoulis, *Probability, Random Variables, and Stochastic Processes* (McGraw-Hill, New York, 1965).

Chapter 4

4.1 I. F. Gihman and A. V. Skorokhod, *Stochastic Differential Equations* (Springer, Berlin, Heidelberg, New York, 1972).
4.2 R. L. Stratonovich, *Introduction to the Theory of Random Noise* (Gordon and Breach, New York, London, 1963).
4.3 L. Arnold, *Stochastic Differential Equations* (Wiley-Interscience, New York, 1974, reprinted by Krieger, Malabar Florida, 1991).
4.4 R. Kubo, in *Stochastic Processes in Chemical Physics*, edited by K. E. Shuler (Wiley-Interscience, New York, 1969).
4.5 N. G. van Kampen, *Stochastic differential equations*, Phys. Rept **24C**, 171–228 (1976).

Chapter 5

5.1 A. D. Fokker, *Die mittlere Energie rotierender elektrischer Dipole im Strahlungsfeld*, Ann. Phys. (Leipzig) **42**, 310 (1914).

5.2 M. Planck, *Über einen Satz der statischen Dynamik und seine Erweiterung in der Quantentheorie*, Sitzungsber. Preuss. Akad. Wiss. Phys. Math. Kl. 324 (1917).

5.3 A. N. Kolmogorov, *Über die analytischen Methoden in der Wahrscheinlichkeitsrechnung*, Math. Ann. **104**, 415–418 (1931).

5.4 W. Feller, *The Parabolic Differential Equations and the Associated Semi-Groups of Transformations*, No. 3 in *Second Series*, The Annals of Mathematics **55**, 468 (1952).

5.5 A. Erdelyi, *Higher Transcendental Functions,* Vols. I–III (McGraw-Hill, New York, 1953).

5.6 C. W. Gardiner, *Diffusive Traversal Time in a One Dimensional Medium*, J. Stat. Phys. **49**, 827 (1987).

Chapter 6

6.1 N. G. van Kampen, *Derivation of the phenomenological equations from the master equation*, Physica **23**, 707 (1957); **23**, 816 (1957).

6.2 R. Graham and H. Haken, *Generalized thermodynamic potential for Markoff systems in detailed balance and far from thermal equilibrium*, Zeits. Phys. **243**, 289–302 (1971).

6.3 H. Kramers, *Brownian motion in a field of force and the diffusion model of chemical reactions*, Physica **7**, 284 (1940).

6.4 L. Onsager, *Reciprocal Relations in Irreversible Processes. I.*, Phys. Rev **37**, 405 (1931).

6.5 H. B. G. Casimir, *On Onsager's Principle of Microscopic Reversibility*, Rev. Mod. Phys. **17**, 343 (1945).

Chapter 7

7.1 R. Graham and T. Tel, *On the weak-noise limit of Fokker-Planck models*, J. Stat. Phys **35**, 729 (1984).

7.2 R. Graham and T. Tel, *Weak-noise limit of Fokker-Planck models and nondifferentiable potentials for dissipative dynamical systems*, Phys. Rev. A **31**, 1109 (1985).

Chapter 8

8.1 G. C. Papanicolaou and W. Kohler, *Asymptotic theory of mixing stochastic ordinary differential equations*, Comm. Pure Appl. Math. **27**, 641–668 (1974).

8.2 R. L. Stratonovich, *Introduction to the Theory of Random Noise* (Gordon and Breach, New York, London, 1963).

8.3 H. Haken, *Synergetics* An Introduction (Springer, Heidelberg, New York, 1978, 1981, 1983).

8.4 H. C. Brinkman, *Brownian motion in a field of force and the diffusion theory of chemical reactions*, Physica **22**, 29 (1956).
8.5 G. Wilemski, *On the derivation of Smoluchowski equations with corrections in the classical theory of Brownian motion*, J. Stat. Phys. **14**, 153–170 (1976).
8.6 U. M. Titulaer, *A systematic solution procedure for the Fokker-Planck equation of a Brownian particle in the high-friction case*, Physica **91 A**, 321–344 (1978).
8.7 T. W. Marshall and E. J. Watson, *A drop of ink comes from my pen... It comes to earth I know not when*, J. Phys. A **18**, 3531 (1985), A **20**, 1345 (1987).
8.8 J. V. Selinger and U. M. Titulaer, *The kinetic boundary layer for the Klein-Kramers equation; A new numerical approach*, J. Stat. Phys **36**, 293 (1984).
8.9 U. M. Titulaer, *Nonuniversal nonanalytic density profiles for Brownian particles near a partially absorbing wall*, Phys. Lett. A **108**, 19 (1985).
8.10 U. M. Titulaer, *The density profile for the Klein-Kramers equation near an absorbing wall*, J. Stat. Phys. **37**, 589 (1985)
See also [8.9] for a treatment of partial absorption and [8.7] for an exact solution.

Chapter 9

9.1 C. Bloch, *Sur la théorie des perturbations des états liés*, Nuclear Physics **6**, 329 (1958).
9.2 U. Titulaer, *The Chapman-Enskog procedure as a form of degenerate perturbation theory*, Physica A **100**, 234 (1980).
9.3 U. Titulaer, *Corrections to the Smoluchowski equation in the presence of hydrodynamic interactions*, Physica A **100**, 251 (1980).

Chapter 10

10.1 L. Bachelier, *Théorie de la Spéculation* (Gauthier-Villars, Paris, 1900).
10.2 P. A. Samuelson, *Rational Theory of Warrant Pricing*, IMR; Industrial Management Review **6**, 13 (1965).
10.3 M. F. M. Osborne, *Brownian Motion in the Stock Market*, Operations Research **7**, 145 (1959).
10.4 M. F. M. Osborne, *Reply to* Comments on 'Brownian Motion in the Stock Market', Operations Research **7**, 807 (1959).
10.5 D. Bernoulli, *Specimen Theoriae Novae de Mensura Sortis*, Commentarii Academiae Scientarum Imperialis Petropolitanae **V**, 175 (1738).
10.6 D. Bernoulli, *Exposition of a New Theory on the Measurement of Risk*, Econometrica **22**, 23 (1954).
10.7 B. Mandelbrot, *The Variation of Certain Speculative Prices*, The Journal of Business **36**, 394 (1963).
10.8 F. Black and M. Scholes, *The Pricing of Options and Corporate Liabilities*, The Journal of Political Economy **81**, 637 (1973).
10.9 R. C. Merton, *Theory of Rational Option Pricing*, The Bell Journal of Economics and Management Science **4**, 141 (1973).
10.10 D. Applebaum, *Lévy Processes and Stochastic Calculus* (Cambridge University Press, Cambridge, 2004).
10.11 R. Cont and P. Tankov, *Financial Modelling with Jump Processes* (Chapman and Hall, London, 2004).

10.12 S. Raible, *Lévy processes in finance: Theory, numerics, and empirical facts* (Ph. D thesis, University of Freiburg, Freiburg, 2000).
10.13 P. Lévy, *Théorie de l'Addition des Variables Aléatoires*, 2nd. ed. (Gauthier-Villars, Paris, 1954).
10.14 E. Eberlein and U. Keller, *Hyperbolic Distributions in Finance*, Bernoulli **1**, 281 (1995).
10.15 O. E. Barndorff-Nielsen and N. Shephard, *Non-Gaussian Ornstein-Uhlenbeck-based models and some of their uses in financial economics*, Journal of the Royal Statistical Society B **63**, 167 (2001).
10.16 O. E. Barndorff-Nielsen, *Infinite Divisibility of the Hyperbolic and Inverse Gaussian Distributions*, Zeits. Wahrscheinlichkeitstheorie verw. Gebiete **38**, 309 (1977).
10.17 R. Cont, *Empirical properties of asset returns: stylized facts and statistical issues*, Quantitative Finance **1**, 223 (2001).
10.18 E. F. Fama, *Mandelbrot and the Stable Paretian Hypothesis*, The Journal of Business **36**, 420 (1963).
10.19 E. Eberlein, U. Keller, and K. Prause, *New Insights into Smile, Mispricing, and Value at Risk: The Hyperbolic Model*, The Journal of Business **71**, 371 (1998).

Chapter 11

11.1 I. S. Matheson, D. F. Walls, and C. W. Gardiner, *Stochastic Models of First Order Nonequilibrium Phase Transitions in Chemical Reactions*, J. Stat. Phys. **12**, 21 (1975), H. K. Janssen, Z. Phys. **260**, 67 (1974).
11.2 N. G. van Kampen, *A Power Series Expansion of the Master Equation*, Can. J. Phys. **39**, 551–567 (1961).
11.3 H. Kramers, *Brownian motion in a field of force and the diffusion model of chemical reactions*, Physica **7**, 284 (1940).
11.4 J. E. Moyal, *Stochastic Processes and Statistical Physics*, J. R. Stat. Soc. B **11**, 151–210 (1949).
11.5 A. Einstein, *Über die von der molekular-kinetischen Theorie der Wärme geforderte Bewegung von in ruhenden Flüssigkeiten suspendierten Teilchen*, Ann. Phys. (Leipzig) **17**, 549 (1905).
11.6 T. G. Kurtz, *The Relationship between Stochastic and Deterministic Models for Chemical Reactions*, The Journal of Chemical Physics **57**, 2976 (1972).
11.7 N. G. van Kampen, *The expansion of the master equation*, Adv. Chem. Phys. **34**, 245–309 (1976).
11.8 S. Karlin and H. M. Taylor, *A First Course in Stochastic Processes* (Academic Press, New York, 1975).
11.9 H. Haken, *Synergetics* An Introduction (Springer, Heidelberg, New York, 1978).
11.10 J. Schnakenberg, *Network theory of microscopic and macroscopic behavior of master equation systems*, Rev. Mod. Phys **48**, 571–586 (1976).

Chapter 12

12.1 C. W. Gardiner and S. Chaturvedi, *The Poisson representation I. A New Technique for Chemical Master Equations*, J. Stat. Phys **17**, 429–468 (1977).

12.2 P. D. Drummond, *Gauge Poisson representations for birth/death master equations*, Eur. Phys. J. B **38**, 614 (2004).

12.3 A. J. Drummond and P. D. Drummond, Extinction in a self-regulating population with demographic and environmental noise, arXiv:0807.4772, 2008.

12.4 P. D. Drummond and C. W. Gardiner, *Generalised P-Representations in Quantum Optics*, J. Phys. A **13**, 2353 (1980).

12.5 C. W. Gardiner and P. Zoller, *Quantum Noise*, 3rd ed. (Springer, Heidelberg, Berlin, 2004).

12.6 W. Bernard and H. B. Callen, *Irreversible Thermodynamics of Nonlinear Processes and Noise in Driven Systems*, Rev. Mod. Phys. **31**, 1017 (1959), W. Bernard and H. B. Callen, Phys. Rev. **118**, 1466 (1960).

12.7 K. Hochberg, *A Signed Measure on Path Space Related to Wiener Measure*, Ann. Prob. **3**, 433–458 (1978).

Chapter 13

13.1 R. F. Curtain, "Stochastic Partial Differential Equations", in *Stochastic Nonlinear Systems*, edited by L. Arnold and R. Lefever (Springer, Berlin, Heidelberg, New York, 1981).

13.2 J. Keizer, *Dissipation and fluctuations in nonequilibrium thermodynamics*, J. Chem Phys. **64**, 1679–1687 (1976).

13.3 G. Nicolis, *Fluctuations around nonequilibrium states in open nonlinear systems*, J. Stat. Phys. **6**, 195 (1972).

13.4 F. Schlögl, *Chemical reaction models for non-equilibrium phase transitions*, Z. Phys. A **253**, 147 (1972).

13.5 L. Arnold, "Consistency of Models of Chemical reactions", in *Dynamics of Synergetic Systems*, edited by H. Haken (Springer, Berlin, Heidelberg, New York, 1980).

13.6 C. W. Gardiner and M. L. Steyn-Ross, *Adiabatic Elimination in Stochastic Systems. II: Application to Reaction-Diffusion and Hydrodynamic-Like Systems*, Phys. Rev. A **29**, 2823 (1984).

13.7 N. G. van Kampen, "Fluctutations in Continuous Systems", in *Topics in Statistical Mechanics and Biophysics*, edited by R. A. Piccirelli (Am. Inst. of Physics, New York, 1976).

13.8 C. van den Broek, W. Horsthemke, and M. Malek-Mansour, *On the diffusion operator of the multivariate master equation*, Physica **89 A**, 339–352 (1977), L. Brenig and C van den Broek, Phys. Rev. A **21**, 1039 (1980).

13.9 H. Grad, "Principles of the Kinetic Theory of Gases", in *Handbuch der Physik*, edited by S. Flügge (Springer, Berlin, Göttingen, New York, 1958), Vol. 12.

Chapter 14

14.1 M. Abramowitz and I. Stegun, *Handbook of Mathematical Functions* (Dover, New York, 1964).

14.2 H. Kramers, *Brownian motion in a field of force and the diffusion model of chemical reactions*, Physica **7**, 284 (1940).

14.3 A summary of the situation is given in the appendix to [14.4].

14.4 M. Büttiker and R. Landauer, in *Nonlinear Phenomena at Phase Transitions and Instabilities*, edited by T. Riste (Plenum, New York, London, 1982).
14.5 H. Eyring, *The Activated Complex in Chemical Reactions*, J. Chem. Phys. **3**, 107 (1935).
14.6 S. A. Hawley, *Reversible pressure-temperature denaturation of chymotrypsinogen*, Biochemistry **10**, 2436 (1971),
S. A. Hawley and R. M. Mitchell, Biochem. **14**, 3257 (1975).
14.7 C. W. Gardiner, *A Stochastic Model of Reversible Pressure Denaturation of Chymotrypsinogen*, J Chem Phys **78**, 2549 (1983).
14.8 J. S. Langer, *Statistical theory of the decay of metastable states*, Ann. Phys. N. Y. **54**, 258 (1969),
R. Landauer and J.A. Swanson, Phys. Rev. **121**, 1668 (1961).
14.9 Z. Schuss, *Singular Perturbation Methods in Stochastic Differential Equations of Mathematical Physics*, S.I.A.M. Rev. **22**, 119 (1980),
B. J. Matkowsky and Z, Schuss, S.I.A.M. J. Appl. Math **33**, 365 (1977).
14.10 A. D. Ventsel and M. I. Freidlin, *On Small Random perturbations of Dynamical Systems*, Russian Math. Surveys **25**, 1 (1970).

Chapter 15

15.1 P. E. Kloeden and E. Platen, *Numerical Solution of Stochastic Differential Equations* (Springer, Berlin, Heidelberg, NewYork, 1992).
15.2 P. E. Kloeden, E. Platen, and H. Schurz, *Numerical Solution of SDE through Computer Experiments* (Springer, Berlin, Heidelberg, NewYork, 1994).
15.3 P. D. Drummond and I. K. Mortimer, *Computer simulations of multiplicative stochastic differential equations*, J. Comp. Phys. **93**, 144 (1991).
15.4 K. J. McNeil and I. J. D. Craig, *Positive P-representation for second harmonic generation: Analytic and computational results*, Phys. Rev. A **41**, 4009 (1990).
15.5 A. M. Smith and C. W. Gardiner, *Simulations of nonlinear quantum damping using the positive P-representation*, Phys. Rev. A **39**, 3511 (1989).
15.6 P. E. Kloeden, "Stochastic differential equations and their numerical solution", Durham Summer School 2000 Transparencies, available from 11, (2000).
15.7 G. N. Milstein and M. V. Tretyakov, *Stochastic Numerics for Mathematical Physics* (Springer-Verlag, Berlin, Heidelberg, NewYork, 2004).
15.8 XmdS eXtensible multi-dimensional Simulator, Available at
`www.xmds.org` and
`sourceforge.net/projects/xmds`, (2008).
15.9 M. J. Werner and P. D. Drummond, *Robust algorithms for solving stochastic partial differential equations*, J. Comp. Phys. **131**, 312 (1997).
15.10 R. J. Ballagh, "Computational Methods for Nonlinear Partial Differential Equations"
`www.physics.otago.ac.nz/research/uca/resources`
`/comp_lectures_ballagh.html`, (2008).
15.11 Peter Kloeden's home page
`www.math.uni-frankfurt.de/~numerik/kloeden`, (2008).
15.12 S. O. Cyganowski, L. Grüne, and P. E. Kloeden, in *Proceedings of the IX-th summer school in Numerical Analysis*, edited by J. F. Blowey, J. P. Coleman, and A. W. Craig (Springer, Berlin, Heidelberg, NewYork, 2001), pp. 127–178.

15.13 D. J. Higham, *An algorithmic introduction to numerical simulation of stochastic differential equations*, SIAM Review **43**, 525 (2001).
15.14 D. J. Higham and P. E. Kloeden, in *Programming Languages and Systems in Computational Economics and Finance*, edited by S. S. Neilsen (Kluwer, Amsterdam, 2002), pp. 233–270.
15.15 Available at
`www.maths.strath.ac.uk/~aas96106/algfiles.html`, (2008).

Bibliography

Aoki, Masanao, *New approaches to macroeconomic modeling : evolutionary stochastic dynamics, multiple equilibria, and externalities as field effects* (Cambridge University Press, Cambridge New York 1996)

Methods such as those used in statistical mechanics and as presented in *Stochastic Methods* are developed in the context of macroeconomics.

Arnold, L., *Stochastic Differential Equations—Theory and Applications* (Wiley, New York 1974)

The most readable mathematical text on stochastic differential equations. It is recommended for readers interested in a rigorous, but understandable account of stochastic differential equations.

Arnold, L. and Lefever, R. (eds.), *Stochastic Nonlinear Systems* (Springer, Berlin, Heidelberg, New York 1981)

A coverage of the field of applied stochastic methods, with references to most relevant fields of application.

Arnold, L., *Random Dynamical Systems* (Springer, Berlin, Heidelberg 1998)

Contains, extends and unites various developments in the intersection of probability theory and dynamical systems. Mainly rigorous mathematics, but pays considerable attention to numerical and qualitative aspects, including aspects of bifurcation theory.

Beran, Jan, *Statistics for long-memory processes* (Chapman & Hall, New York 1994)

A specialised book covering a class of processes not covered in *Stochastic Methods*, or in any standard book.

Bharucha-Reid, A. T., *Elements of the Theory of Markov Processes and their Applications* (McGraw-Hill, New York 1960)

A useful book for applications, with a coverage of population biology, nuclear processes, astronomy, chemistry, and queues. Many exact solutions given to models which approximate real systems.

Cox, D. R. and Miller, H. D., *The Theory of Stochastic Processes* (Methuen, London 1965)

A book in the English tradition, more oriented to applications in statistics rather than in science. Very easy to read.

Feller, W., *An Introduction to Probability Theory and its Applications, Vol. II, 2nd ed.* (Wiley, New York 1971)

A very readable and even entertaining book, mainly on the foundations of stochastic processes, by one of those who created the subject. Not particularly useful for applications but rather as a reminder to the applied worker of the beauty of the subject itself

Erdi, P. and Toth J., *Mathematical models of chemical reactions: theory and applications of deterministic and stochastic models* (Princeton University Press, Princeton, N.J. 1989)

A book which treats in much more depth the material on chemical reactions such as is presented in Chap. 7 of *Stochastic Methods* .

Gammaitoni, L., Hänggi, P., Jung, P. and Marchesoni F., *Stochastic Resonance* (Reviews of Modern Physics **70** 223–287, 1998)

A major development in the field of dynamical systems over the last two decades has been the field of *stochastic resonance*, in which the existence of noise can actually improve the performance of a system. The archetypical model is that of simple symmetric bistable process driven by both an additive random noise, for simplicity, white and Gaussian, and an external sinusoidal bias. This is major review of the subject, with extensive references to the original literature

Gihman, I. I. and Skorohod, A. V., *Stochastic Differential Equations* (Springer, Berlin, Heidelberg, New York 1972)

A thorough Russian work. Not easy to read but worth the effort. Develops almost all Markov process theory as stochastic differential equations, as well as some non-Markov processes.

Gihman, I. I. and Skorohod, A. V., *The Theory of Stochastic Processes, Vols. I–III* (Springer, Berlin, Heidelberg, New York 1974)

A massive work in the very thorough and rigorous Russian tradition. Not really useful in applications, but can be regarded as the most modern rigorous work. With effort, it can be understood by a non-mathematician.

Goel, N. S. and Richter-Dyn, N., *Stochastic Models in Biology* (Academic, New York 1974)

An applied book, mainly on birth and death processes of populations. Does use some approximate methods.

Graham, R., *Statistical Theory of Instabilities in Stationary Nonequilibrium Systems with Applications to Lasers and Non Linear Optics (Springer Tracts in Modem Physics, Vol. 66)* (Springer, Berlin, Heidelberg, New York 1973)

Applications of stochastic methods in quantum optics and other nonequilibrium systems.

Haken, H., *Synergetics, An Introduction, 2nd ed.* (Springer, Berlin, Heidelberg, New York 1978)

A major reference on applications of stochastic processes to a wide variety of cooperative phenomena. Many elementary examples, as well as advanced work.

Hänggi, P. and Thomas, H., *Stochastic Processes: Time Evolution, Symmetries and Linear Response* (Physics Reports **88**, 207-319, 1982)

A review of the application of Stochastic processes to physics. Contains some parts which treat non-Markov processes.

Honerkamp, J., *Stochastic dynamical systems: concepts, numerical methods, data analysis* (Wiley, New York 1994)

An interesting book which includes material on physical applications and algorithmic methods.

Karlin, S., Taylor, H. M., *A First Course in Stochastic Processes, 2nd ed.* (Academic, New York 1975)

Quite a readable mathematical book with many worked examples. Emphasis on Markov chains and jump processes.

Kloeden, Peter E. and Platen, Eckhard, *Numerical solution of stochastic differential equations* (Springer, Berlin New York 1992)

An exhaustive canvassing of numerical and simulation methods for stochastic differential equations. This is a topic which is rarely covered in texts on stochastic processes, and one whose relevance is now much more significant than formerly because of the availability of sufficiently powerful computers at a reasonable price.

Lamberton, Damien, *Introduction to stochastic calculus applied to finance* (Chapman and Hall, London 1995)

An advanced text which provides excellent introduction to the modern theory of financial mathematics and in particular to the pricing of options, including the Black-Scholes formula and its generalisations.

Lax, M., *Classical Noise I—Rev. Mod. Phys. 32, 25* (1960); *II—Phys. Chem. Solids 14, 248* (1960) *(with P. Mengert); III—Rev. Mod. Phys. 38, 359* (1966); *IV—Rev. Mod. Phys. 38, 541* (1966)

A pioneering, but rather discursive treatment, of the application of stochastic methods to physical systems, with an emphasis on solid state physics.

Mandelbrot, Benoit B., *Fractals and Scaling in Finance—Discontinuity, Concentration, Risk* (Springer, Berlin New York 1997)

This is a wonderful book, not because it relates in any direct way to the subject of *Stochastic Methods*, but for what it reveals about the mind of the man who broadened our concepts of curves and surfaces to include the ideas of fractional dimensionality. This book is a compendium of his extensive writing on financial and other markets, in which we can see how he struggles with data which always hints at an underlying model, but never reveals exactly what it is. Mandelbrot's famous paper introducing the Paretian process description of "certain speculative prices" is included here, with the entertaining story of how it came to published. His ideas have not met with mainstream acceptance in the financial world, neither has mainstream thought on the subject met with any acceptance from him. Mandelbrot is a man who uses pure mathematics as a creative tool to study the world; one for whom rigour is a servant, not a master.

Mikosch, Thomas, *Elementary stochastic calculus with finance in view* (World Scientific, Singapore 1998)

An book written for commerce students who needed to know more about the application of Stochastic methods to financial theory, and in particular, to the Black-Scholes theory of option pricing. The treatment is on the whole at about the same level as *Stochastic Methods.*

Nicolis, G. and Prigogine, I., *Self Organisation in Nonequilibrium Systems* (Wiley, New York 1977)

Mainly about self-organisation; this book does have a readable and useful account of the stochastic methods of the Brussels school. Very recommendable as a source of actual current work on applications of stochastic methods to reaction-diffusion systems and nonequilibrium statistical mechanics.

Oppenheim, I., Shuler, K. E. and Weiss, G. H., *Stochastic Processes in Chemical Physics—the Master Equation* (M.I.T. Press, Cambridge 1977)

A brief introduction to stochastic methods, followed by a very interesting and useful compendium of basic papers on the subject, including the basic papers in which projection techniques were first introduced by Zwanzig.

Papoulis, A., *Probability, Random Variables, and Stochastic Processes* (McGraw-Hill, New York 1965)

An excellent reference for learning practical but reasonably rigorous probability methods. Oriented mainly towards electrical engineering in terms of applications.

Paul, Wolfgang and Baschnagel, Jörg, *Stochastic processes: from physics to finance* (Springer, Berlin New York 1999)

A book by physicists which extends the material of *Stochastic Methods* into the realm of mathematical finance. It also includes material on Levy flights and their applications.

Risken, H., *The Fokker Planck Equation* (Springer, Berlin, Heidelberg, New York 1984)

A very thorough treatment of methods of solving and applying the Fokker-Planck equation. Contains a treatment of continued fraction methods and motion in periodic potentials, and a great deal of material not included in less specialised books.

Stratonovich, R. L., *Topics in the Theory of Random Noise, Vols. I and II* (Gordon and Breach, New York 1963)

A compendium of a radio engineer's view of stochastic methods. Contains many beautiful results in a characteristic style which is often very illuminating. Stochastic differential equations are introduced as limits of real processes. One of the fundamental books on applied stochastic differential equations.

Tapiero, Charles S., *Applied stochastic models and control for finance and insurance* (Kluwer, Boston 1998)

A book on a similar level to *Stochastic Methods*, but with an emphasis on control theory, optimisation and finance.

van Kampen, N. G., *Stochastic Processes in Physics and Chemistry* (North Holland, Amsterdam, New York, Oxford 1981, 2001)

A book on the application of stochastic processes to physical and chemical phenomena. Characterised by clear reasoning, and a precise style, with particular attention to evaluation of the conditions of validity of approximations used commonly in physical and chemical applications.

Wax N., *Selected Papers on Noise and Stochastic Processes* (Dover, New York 1954)

A fundamental historical reference, containing the original work of Uhlenbeck and Ornstein on Brownian Motion, and Chandrasekhar's classic paper *Stochastic Problems in Physics and Astronomy*, as well as other classic papers.

Author Index

Symbol Index

The pages referred to are normally the first mention of the symbol in text. The order is alphabetical, lower case first, and ignoring typeface. There are three parts to the index, Latin, Greek, and Mathematical.

Latin Symbols

Greek Symbols

Mathematical Symbols

Subject Index